AF838331

# Learning Management System Technologies and Software Solutions for Online Teaching:
## Tools and Applications

Yefim Kats
*Ellis University, USA & Rivier College, USA*

Information Science REFERENCE
**INFORMATION SCIENCE REFERENCE**
Hershey · New York

| | |
|---|---|
| Director of Editorial Content: | Kristin Klinger |
| Director of Book Publications: | Julia Mosemann |
| Acquisitions Editor: | Lindsay Johnston |
| Development Editor: | Elizabeth Ardner |
| Typesetter: | Gregory Snader |
| Production Editor: | Jamie Snavely |
| Cover Design: | Lisa Tosheff |
| Printed at: | Yurchak Printing Inc. |

Published in the United States of America by
Information Science Reference (an imprint of IGI Global)
701 E. Chocolate Avenue
Hershey PA 17033
Tel: 717-533-8845
Fax: 717-533-8661
E-mail: cust@igi-global.com
Web site: http://www.igi-global.com/reference

Copyright © 2010 by IGI Global. All rights reserved. No part of this publication may be reproduced, stored or distributed in any form or by any means, electronic or mechanical, including photocopying, without written permission from the publisher.

Product or company names used in this set are for identification purposes only. Inclusion of the names of the products or companies does not indicate a claim of ownership by IGI Global of the trademark or registered trademark.

Library of Congress Cataloging-in-Publication Data

Learning management system technologies and software solutions for online teaching : tools and applications / Yefim Kats, editor.
p. cm.
Includes bibliographical references and index.
Summary: "This book gives a general coverage of learning management systems followed by a comparative analysis of the particular LMS products, review of technologies supporting different aspect of educational process, and, the best practices and methodologies for LMS-supported course delivery"--Provided by publisher.
ISBN 978-1-61520-853-1 (hardcover) -- ISBN 978-1-61520-854-8 (ebook) 1. Internet in higher education. 2. Education, Higher--Computer-assisted instruction. 3. Organizational effectiveness. 4. System analysis. I. Kats, Yefim.
LB2395.7.L43 2010
371.33'44678071--dc22
2009040623

British Cataloguing in Publication Data
A Cataloguing in Publication record for this book is available from the British Library.

All work contributed to this book is new, previously-unpublished material. The views expressed in this book are those of the authors, but not necessarily of the publisher.

*In memory of my parents, Pinhas and Mura*

*To Baoqing*

## Editorial Advisory Board

Daniel Ferens, *Data and Analysis Center for Software, ITT IS, USA*
James Geller, *New Jersey Institute of Technology, USA*
Wolfgang Hommel, *Leibniz Supercomputing Centre, Germany*
Antony Pina, *The Sullivan University System, USA*
Vladimir Riabov, *Rivier College, USA*
Kate Kouprianova, *Strayer University, USA*
Judy Van Biljon, *University of South Africa, South Africa*
Tobias Zimmerman, *Zentrum für Hochschuldidaktik und Erwachsenenbildung, Switzerland*

## List of Reviewers

Larbi Esmahi, *Athabasca University, Canada*
Linzi Kemp, *American University of Sharjah, UAE*
Patricia McGee, *The University of Texas at San Antonio, USA*
Geraldine Torrisi-Steele, *Logan Campus, Griffith University, Australia*
Adam Westerski, *Ciudad Universitaria, Spain*

# Table of Contents

# Detailed Table of Contents



**Section 1**
**Learning Management Systems: An Overview**



A broadly-based overview focused on the LMSs typically used in academia. Among the LMSs reviewed are Blackboard, Desire2Learn, Angel, eCollege, Sakai, and Moodle. This chapter introduces the reader to key features found in most LMSs and presents a comparative analysis of the most commonly used products.



The chapter is an impressive attempt at a new practical definition of an e-Learning platform, aimed at overcoming the current disparity in evaluation criteria for existing and emerging e-Learning software solutions. The begins with the discussion of the early PLATO systems and proceeds to the currently used technological solutions such as Moodle and OLAT, thus paving the way to the comprehensive definition of an e-Learning platform.



The chapter is intended to provide guidance to developers as well as system and network administrators on integrating a newly adopted LMS into existing infrastructure. Examples of successful security measures are discussed in the context of the popular open-source LMSs OLAT and Moodle. Dr. Hommel presents an interesting analysis of how Federated Identity Management could support secure communication between e-Learning entities.

## Section 2
## Selecting a Suitable Learning Management System: Challenges and Solutions



This chapter is focused on a variety of issues related to the process of selecting a LMS platform that satisfies a set of chosen criteria such as student and faculty profiles, affordability, adaptability, and robustness, among others. The author brings into focus his experience at the University of Applied Sciences Northwestern Switzerland. In this context, he compares two open-source LMSs, Moodle and OLAT, and shows how OLAT better suits the requirements of his home institution.



The authors analyze and discuss in detail the motivation behind the adoption of a popular open-source LMS Sakai at the University of Limerick, Ireland. They further proceed showing how the integration of Adobe Connect Pro software enhances students' and faculty's experience by providing a rich synchronous learning environment. The chapter includes a thorough examination of the pedagogical, technological, and social issues arising from the adoption of online tutorial pedagogy.



The chapter presents a case study with the focus on preparing faculty for the transition from the commercial LMS WebCT to another upgraded version of this popular e-Learning platform. The research is empirically based and includes surveys of faculty at the authors' home institution. The authors generalize their findings by identifying the factors crucial for the successful transition to a new learning platform and provide recommendations on training and support of faculty.

The author describes tracking tools used in LMSs, using Blackboard as a typical example. He presents and analyzes ongoing research efforts focused on improving current tracking practices and integrating them into existing e-Learning platforms. In this context, Dr. Douglas examines attempts to use data mining and visualization techniques to enhance tracking methodologies as well as overall course management. Among the technologies discussed are a visualization tool CoureVis with its recent implementation GISMO and a discussion analysis tool DAT.



The chapter offers a discussion and analysis of tools used for synchronous course delivery and support. The chapter covers software solutions supporting instant messaging, conferencing, and collaboration. Among the popular tools discussed are a variety of instant messaging and Internet chat tools, and conferencing software such as FM Live Communication and DimDim. The use of synchronous communication software is examined from the perspectives of instructors, students, and technical support staff.



The chapter investigates the problem of file management and sharing in an increasingly complex e-Learning environment. The authors link this issue to the growing popularity of Web 2.0 tools allowing speedy exchange of text, audio, and video files by both faculty and students. The different aspects of file storage, sharing, and management are considered from the perspective of systemic integration of a variety of tools, including Web 2.0 technologies, LMS solutions, and standard word processing and research software. The study is based on the University of Illinois' experience and focused on using Xythos system in conjunction with the Blackboard LMS platform.



Paloma Moreno-Clari and Esteban Sanchis-Kilders offer a comprehensive analysis of plug-in online assessment tools for the popular open-source LMS dotLearn based on OpenACS architecture. The authors emphasize that the tools described are enabled to utilize a variety of third-party interfaces (API) and a wide range of Web resources such as Google, Flickr, and YouTube. The chapter is a valuable source of information for LMS administrators and software developers.

The chapter presents a dialogue oriented teaching methodology for hybrid courses delivered utilizing an open-source LMS, OLAT. The authors argue that this approach allows instructors to successfully overcome the challenge of handling large classes by enhancing social learning, creative thinking, and problem-solving skills. They emphasize the role of a powerful LMS such as OLAT for the implementation of dialogue learning in large classes with up to several hundred students. The bulk of the chapter offers a detailed implementation of the dialogue didactic methodology.



The chapter examines a variety of free software tools used by the authors for online synchronous and asynchronous course delivery in computer science. They place a special emphasis on using the tools for virtual labs and student course projects. In particular, the authors review tools for the implementation of Unified Modeling Language diagrams, computer programs written in C/C++/JAVA programming languages; they also cover popular free tools for classes in data communication, database systems, networking, and Web development. The chapter concludes with examples of students' research papers and projects.



The authors explore case-study libraries in the context of implementation of case-based learning methodology. The authors begin by outlining the advantages and limitations of case-based learning in educational process and identify the key elements necessary for successful implementation of case studies. They discuss an example of the case-study library supporting engineering programs and further illustrate this approach by presenting the interactive usability case library developed by the authors and implemented in a server-client architecture framework.



The chapter presents experiences and opinions of online students and identifies competencies important for a successful e-Learning environment. The study is based on the survey administered online to American, Israeli, Mexican, and Japanese students. The author discusses the complex relationship between technologies used in online education and successful pedagogical methodologies. The appendix to the chapter lists fifty-eight questions used in the survey.

# Preface

Over the last two decades, the landscape of higher education changed considerably, with more and more resources spent on technology supporting both face-to-face and distance learning educational models. This trend is influenced, to a large extent, by the tremendous growth of online learning, due to the unique role played by computer-related technologies for online course delivery and associated activities including technical support.

Nowadays, online courses are offered not only by 'pure' online educational institutions but also by traditional schools. According to the research by Allen and Seaman (2008) prepared for the recent Sloan Report, in the United States more than "twenty percent of all U.S. higher education students … [or] over 3.9 million students were taking at least one online course during the fall 2007 term; a 12-percent increase over the number reported the previous year" (p. 1). Moreover, distance learning courses are currently being offered not only to incoming college students and adult learners but also to K-12 students, increasingly covering practically all population groups from children to mature adults. We can expect that the recent economic downturn in the United States (and across the globe) will make a further positive impact on online enrollment due to rising fuel costs and rapid occupational shifts in the job market.

It is important to emphasize that the term distance learning can be applied either to pure online course offerings, where 100 percent of course content is delivered online, or to courses with only some of the content delivered online. The latter include blended or hybrid courses as well as traditional courses supported by Web-based content. Thus, there is no universal consensus on the use of terminology related to online forms of course delivery. For example, the Sloan Report quoted above considers online courses as "those in which at least 80 percent of the course content is delivered online" (p. 4).

While courses supported by online content can rely on commonly used multiple Web-enabled technologies, such as e-mail and instant messaging for communication between instructors and students or static Web content for displaying syllabi and course announcements, all Web-based and Web-supported forms of course delivery increasingly use specialized educational technologies called learning management systems (LMS) or course management systems (CMS). These technologies are widely used not only for 'online' course delivery but also for computerized support of traditional courses, as more and more students expect at least some access to all their classes through the World Wide Web.

## LEARNING MANAGEMENT SYSTEMS: AN OVERVIEW

The use of computer technologies in educational practice goes back a few decades to the dawn of the 'information revolution.' While in the 1980s the importance of computers for education was still a matter of debate (Cuban, 1986), in the 1990s the explosive proliferation of personal computers, educational software, and the World Wide Web placed computer-assisted pedagogy at the center of the educational process.

This trend gave rise to rapid development of learning management systems – full-scale learning platforms supporting multiple facets of an educational process, from administrative functions to course delivery and assessment. Some researchers (Mallon et al., 2009) trace the emergence of LMSs to rudimentary "training management systems," which later evolved into the full-scaled "e-learning platforms;" others (Bailey, 1993) emphasize the importance of integrated learning systems as precursors of contemporary learning management systems.

However, as with the term online course, there is no universal consensus on the semantic scope of the term learning management system, often used interchangeably with the term course management system. In the context of this publication, we do not find it productive to try to find one all embracing 'essential' definition of what a LMS is or to differentiate between a LMS and a CMS. Consequently, the two terms are often used interchangeably. In fact, the book can be considered as an attempt at the 'ostensive' definition of a LMS, covering a wide range of commercial and open-source LMSs and related technologies used in the institutions of higher education. At the same time, it is not our intention to cover in this publication LMSs used in the business world for corporate training and related purposes.

Currently, multiple LMS solutions for higher education can be purchased from the commercial providers such as the Blackboard Corporation, the dominant player in the United States with approximately 57 percent of the educational market, or obtained from the open-source community without any licensing cost. The most popular open-source products, Moodle, Sakai, and OLAT, are addressed in this book from multiple perspectives. Some institutions, especially 'for profit' schools, develop their own in-house proprietary e-Learning platforms not covered in this book. With a relative maturity of the LMS market, some participants are looking to move from their previously acquired LMS to a new product. The switch to a new product is usually motivated either by an attempt to reduce the cost of licensing and maintaining an expensive technology or by dissatisfaction with the existing LMS. The book should help both the novice in the area of educational software and the seasoned user to either install their first or to switch to a new, more suitable, LMS.

All LMS products, commercial and open-source, share virtually the same capabilities, providing administrative functions such as student registration and assessment as well as different forms of content management. Typically LMSs utilize advanced relational database software such as Oracle, Microsoft SQL Server, or (especially for open-source systems) MySQL. The use of relational databases with their emphasis on data independence greatly enhances security of LMSs, incorporating a variety of login 'roles' such as an instructor, a student, or a guest, among others. An instructor can either privately interact with the course participants or create discussion groups and teams with different profiles.

With the advance of so-called Web 2.0, modern LMSs decisively moved towards allowing integration with multiple Web 2.0 enhanced technologies such as Facebook, Twitter, visual and audio tools, and ePortfolio supporting software. Moreover, some schools conduct classes in synchronous mode, making extensive use of conferencing software such as Adobe Connect or open-source DimDim.

As a related development, the proliferation of LMSs and supporting technologies made a definite impact on teaching methodologies. As students increasingly take courses in either online or hybrid format, they typically expect to have access to at least some course material online in all their classes. Consequently, even an instructor of a traditional face-to-face class must be able to use a LMS to one extent or another. At the same time, while faculty needs to master new computerized technologies, software developers also should be able to accommodate best educational practices and methodologies. In particular, Web 2.0 applications may be used to enhance student interaction and collaboration (Beldarrain, 2006). In this respect, among the important challenges to LMSs mentioned by researchers is a necessity to incorporate constructivist pedagogy, active and collaborative learning, and personalized attention to all course participants. That is why this book, along with the coverage of LMSs and related technologies, also includes a thorough discussion of the interaction between technology and educational practice. In the next section, we provide a detailed guide to the book's content.

## BOOK CHAPTERS OVERVIEW

The book "Learning Management Systems for Online Teaching: Tools and Applications" is organized into four sections, beginning with the general coverage of LMSs and, then, consistently moving to the comparative analysis of the particular LMS products, review of technologies supporting different aspect of educational process, and, finally, to the best practices and methodologies for LMS-supported course delivery.

Section One, "Learning Management Systems: An Overview," begins with a broadly-based overview by Anthony Pina focused on the LMSs typically used in academia. The chapter places a special emphasis on the commercial products. Among the LMSs reviewed are Blackboard, Desire2Learn, Angel, eCollege, Sakai, and Moodle. In this chapter, Dr. Pina introduces the reader to key features found in most LMSs and presents a comparative analysis of the most commonly used products. The chapter also serves as an introduction to Section Two which focuses on issues associated with selecting an appropriate LMS platform.

In the second chapter, Michael Piotrowski analyzes the concept of the e-Learning platform as a central technological component of online learning infrastructure. He first presents e-Learning platforms from the historical perspective tracing their development to the 1960s. He begins with the discussion of the early PLATO systems and proceeds to the currently used technological solutions such as Moodle and OLAT, thus paving the way to the comprehensive definition of an e-Learning platform. The chapter is an impressive attempt at a new practical definition of an e-Learning platform, aimed at overcoming the current disparity in evaluation criteria for existing and emerging e-Learning software solutions.

Section One concludes with the chapter by Wolfgang Hommel, where he addresses security and privacy of the e-Learning environment. He considers security and privacy management from the systemic perspective of sophisticated distributed environment prone to a variety of security risks. Examples of successful security measures are discussed in the context of the popular open-source LMSs OLAT and Moodle. Dr. Hommel presents an interesting analysis of how Federated Identity Management could support secure communication between e-Learning entities. The chapter is intended to provide guidance to developers as well as system and network administrators on integrating a newly adopted LMS into existing infrastructure.

Section Two, "Selecting a Suitable Learning Management System: Challenges and Solutions," is focused on a variety of issues related to the process of selecting a LMS platform that satisfies a set of chosen criteria such as student and faculty profiles, affordability, adaptability, and robustness, among others. This section begins with a chapter by Cerstin Mahlow on choosing an appropriate e-Learning environment for a university. The author brings into focus her experience at the University of Applied

Sciences Northwestern Switzerland. In this context, she compares two open-source LMSs, Moodle and OLAT, and shows how OLAT better suits the requirements of her home institution.

In the next chapter, Juley McGourty and Angelica Risquez analyze the adoption of a popular open-source LMS Sakai at the University of Limerick, Ireland. The authors discuss in detail the motivation behind adopting a particular version of Sakai e-Learning platform at their home institution. They further proceed showing how the integration of Adobe Connect Pro software enhances students' and faculty's experience by providing a rich synchronous learning environment. The chapter includes a thorough examination of the pedagogical, technological, and social issues arising from the adoption of online tutorial pedagogy.

In the last chapter of Section Two, Danilo Baylen and his coauthors present a case study with the focus on preparing faculty for the transition from the commercial LMS WebCT to another upgraded version of this popular e-Learning platform. As the authors note, the change was further complicated after the purchase of WebCT by the Blackboard Corporation. The research is empirically based and includes surveys of faculty at the authors' home institution. The authors generalize their findings by identifying the factors crucial for the successful transition to a new learning platform and provide recommendations on training and support of faculty.

Section Three of the book, "Supporting Technologies for Student Tracking, Evaluation, and Synchronous Course Delivery," offers a detailed discussion of software tools supporting a wide variety of LMS functionalities. In the first chapter, Sergey Butakov and Vladislav Scherbinin review plagiarism detecting tools and methodologies, including cross language plagiarism detection. The authors' research covers popular software solutions, such as Turnitin and SafeAssign, which are often integrated into commercial and open-source LMSs. Furthermore, the chapter describes a new architecture for plagiarism detection infrastructure and its implementation as a plug-in tool for the LMS Moodle. The comprehensive analysis of plagiarism detection is an excellent source of information for faculty and college administrators alike.

The next chapter by Steven Tello and Luvai Motiwalla emphasizes the importance and examines the role of LMS functionalities in the implementation of assessment processes in higher education. The authors briefly review a number of commercial tools available for a variety of assessment activities such as planning, tracking, analysis, and reporting. The bulk of the chapter presents a case study focused on the design and implementation of a LMS-based assessment system, eOutcomes, at the University of Massachusetts Lowell, USA. The LMS used in conjunction with this project is Blackboard Vista 3.x.

The chapter by Daniel Tan, Adrian Lu, and Sheryl Wong from Nanyang Technological University in Singapore describes a new campus-wide e-Learning platform eUreka, designed and implemented at their home institution. They emphasize that the motivation behind eUreka arises from the project-based learning pedagogical paradigm adopted by their faculty. In this context, the authors address a variety of project-focused activities supported by eUreka and pay special attention to its ability to integrate Web 2.0 applications and features, such as an active ownership of the content created by students and enhanced student participation and collaboration.

In the next chapter, Ian Douglas considers the issue of tracking students' participation in online courses. He describes tracking tools used in LMSs, using Blackboard as a typical example. He further presents and analyzes ongoing research efforts focused on improving current tracking practices and integrating them into existing e-Learning platforms. In this context, Dr. Douglas examines attempts to use data mining and visualization techniques to enhance tracking methodologies as well as overall course management. Among the technologies discussed are a visualization tool CourseVis with its recent implementation GISMO and a discussion analysis tool DAT.

Clark Shah-Nelson in his chapter offers a discussion and analysis of tools used for synchronous course delivery and support. The popularity of such tools has been growing with the increasing availability and decreasing cost of broadband Internet access. The chapter covers software solutions supporting instant messaging, conferencing, and collaboration. Among the popular tools discussed are a variety of instant messaging and Internet chat tools, and conferencing software such as FM Live Communication and DimDim. The author emphasizes that the latter can be easily integrated into the Moodle LMS platform. The use of synchronous communication software is examined from the perspectives of instructors, students, and technical support staff.

The chapter by Vickie Cook and Kara McElwrath investigates the problem of file management and sharing in an increasingly complex e-Learning environment. The authors link this issue to the growing popularity of Web 2.0 tools allowing speedy exchange of text, audio, and video files by both faculty and students. The different aspects of file storage, sharing, and management are considered from the perspective of systemic integration of a variety of tools, including Web 2.0 technologies, LMS solutions, and standard word processing and research software. The study is based on the University of Illinois' experience and focused on using Xythos system in conjunction with the Blackboard LMS platform.

Paloma Moreno-Clari and Esteban Sanchis-Kilders offer a comprehensive analysis of plug-in online assessment tools for the popular open-source LMS dotLearn based on OpenACS architecture. The authors emphasize that the tools described are enabled to utilize a variety of third-party interfaces (API) and a wide range of Web resources such as Google, Flickr, and YouTube. The project, conceived and implemented at one of the largest European universities - the University of Valencia, is a valuable source of information for LMS administrators and software developers.

Section Four of the book, "Learning Management Systems and Best Practices in Online Education" begins with a broadly-based study by Nory Jones and Christian Graham and focused on improving hybrid and online course delivery with emerging technologies. The chapter examines how distance learning teaching methodologies could be enhanced by emerging software technologies, in particular Web 2.0 tools and applications. Among the tools and applications reviewed are Weblogs, a variety of social networking tools, Second Life, videoconferencing, and wireless technologies. The use of emerging technologies is linked to the active learning paradigm.

In his chapter, Kam Vat offers a comprehensive study of electronic portfolios as assessment and teaching tools. He addresses the issue from a variety of perspectives and links the use of electronic portfolios to the constructivist teaching methodology. The author further examines the implementation of electronic portfolios with readily available free tools and such LMSs as Desire2Learn and Sakai. The chapter concludes with an interesting and well-presented case study of electronic portfolio design for the broadly-based programming course in computer science.

The chapter by Ricardo Rademacher Mena offers a thorough examination of successful practices for teaching and design of online science classes. Though the author's research is focused primarily on physics and mathematics courses, his findings can be easily generalized to any science classroom. He reviews the best practices for designing non-graded as well as graded assignments such as research papers, online discussions, and lab simulations. The study is supported by multiple examples presented using a popular open- source LMS Moodle.

The chapter by Tobias Zimmermann, Karen-Lynn Bucher, and Daniel Hurtado presents a dialogue oriented teaching methodology for hybrid courses delivered utilizing an open-source LMS, OLAT. The authors argue that this approach allows instructors to successfully overcome the challenge of handling large classes by enhancing social learning, creative thinking, and problem-solving skills. They emphasize the role of a powerful LMS such as OLAT for the implementation of dialogue learning in large classes with up to several hundred students. The bulk of the chapter offers a detailed implementation of the dialogue didactic methodology.

The chapter by Vladimir Riabov and Brian Higgs examines a variety of free software tools used by the authors for online synchronous and asynchronous course delivery in computer science. They place a special emphasis on using the tools for virtual labs and student course projects. In particular, the authors review tools for the implementation of Unified Modeling Language diagrams, computer programs written in C/C++/JAVA programming languages; they also cover popular free tools for classes in data communication, database systems, networking, and Web development. The chapter concludes with examples of students' research papers and projects.

In their chapter, Hao Jiang, Craig Ganoe, and John Carroll explore case-study libraries in the context of implementation of case-based learning methodology. The authors begin by outlining the advantages and limitations of case-based learning in educational process and identify the key elements necessary for successful implementation of case studies. They discuss an example of the case-study library supporting engineering programs and further illustrate this approach by presenting the interactive usability case library developed by the authors and implemented in a server-client architecture framework.

Michael Beaudoin, as the title of his chapter suggests, presents experiences and opinions of online students and identifies competencies important for a successful e-Learning environment. The study is based on the survey administered online to American, Israeli, Mexican, and Japanese students. The author discusses the complex relationship between technologies used in online education and successful pedagogical methodologies. He notes that, as the study suggests, technological medium is often less important to online learners than instructional support. The appendix to the chapter lists fifty-eight questions used in the survey.

Finally, Melanie Shaw, Kelley Walters, and David Long investigate activities and assessment techniques typical for online and hybrid courses. Their study is not only based on cutting edge academic research but also utilizes results of the survey conducted by the authors. Among the survey parameters are LMS usage, course length, programs of study, and user preferences. The LMSs used by the survey participants include Blackboard, WebCT, Angel, Moodle, and eCollege among others.

## CONCLUSION

With the rapid proliferation of distance learning, schools confront the difficult problem of choosing and managing an appropriate technological environment that fits their budget, technical resources, curriculum, pedagogy, and profile of the student body. In this context, the book is intended to fill the gap in the current literature on the interaction between LMSs, supporting technologies, and relevant teaching methodologies. This book is intended for administrators, faculty, subject specialists, and all those looking to launch a new or to expand an existing distance learning program. In particular, it covers commercial and open-source LMSs as well as technologies used for synchronous and asynchronous course delivery, and it offers a comprehensive discussion of factors influencing the transition from one LMS to another. The reader will also find coverage of virtual labs, electronic portfolios, and technological solutions related to the problems of plagiarism, student tracking, assessment, and security of e-learning environment. The thorough scholarly research is complemented by interesting case studies. We hope that this book will prove to be a comprehensive guide on the available technological solutions in the area of online education.

## REFERENCES

Allen, E., & Seaman, J. (2008). Staying the Course: Online Education in the United States, 2008. BABSON Survey Research Group, The Sloan Consortium.

Bailey, G. D. (Ed.). (1993). Computer Based Integrated Learning Systems. Englewood Cliffs, NJ: Educational Technology Publications.

Beldarrain, Y. (2006). Distance Education Trends: Integrating new technologies to foster student interaction and collaboration. Distance Education, 27(3), 139 – 153.

Cuban, L. (1986). Teachers and Machines: The classroom use of technology since 1920. New York: Teachers College Press.

Mallon, D., Bersin, J., Howard C., & O'Leonard, K. (2009). Learning Management Systems 2009. Executive Summary. Bersin and Associates Research Report.

# Acknowledgment

The author wishes to express gratitude to contributors, reviewers, and all those who helped with this research. Special thanks to the Editorial Advisory Board members for their support, encouragement, and insightful comments on the reviewed chapters. Last but not least I would like to thank everyone at IGI Global who worked hard to ensure the timely production of this book. I am grateful to Editorial assistants Elizabeth Ardner and Dave DeRicco. Elizabeth provided valuable editorial suggestions and patient responses to numerous emails.

*Yefim Kats*

# Section 1
# Learning Management Systems:
## An Overview

# Chapter 1
# An Overview of Learning Management Systems

**Anthony A. Piña**
*Sullivan University System, USA*

## ABSTRACT

*In this chapter, the reader is taken through a macro level view of learning management systems, with a particular emphasis on systems offered by commercial vendors. Included is a consideration of the growth of learning management systems during the past decade, the common features and tools contained within these systems, and a look at the advantages and disadvantages that learning management systems provide to institutions. In addition, the reader is presented with specific resources and options for evaluating, selecting and deploying learning management systems. A section highlighting the possible advantages and disadvantages of selecting a commercial versus an open source system is followed by a series of brief profiles of the leading vendors of commercial and open source learning management systems.*

## INTRODUCTION

Learning Management Systems, such as Blackboard, Desire2Learn, Angel, eCollege, Sakai and Moodle, have become nearly ubiquitous at colleges and universities (Dabbagh & Bannan-Ritland, 2005). Both faculty and administrators point to the popularity of these systems as evidence that e-learning has become institutionalized within higher education (Piña, 2008a). Harrington and his colleagues have observed that no other innovation in higher education has resulted in such rapid and widespread adoption as the learning management system (Harrington, Gordon, & Schibik, 2004). Although the media comparison literature attests to a near century long history of technology-delivered instruction, the current generation of learning management systems is just entering its second decade (Piña, 2008b).

From a technical standpoint, a learning management system (hereinafter referred to as an LMS) is a server-based software program that interfaces with a database containing information about users, courses and content. In that sense, it resembles

DOI: 10.4018/978-1-61520-853-1.ch001

Copyright © 2010, IGI Global. Copying or distributing in print or electronic forms without written permission of IGI Global is prohibited.

other systems designed for e-commerce, human resources, payroll, and student records. What makes an LMS unique is its instructional nature. An LMS provides a place for learning and teaching activities to occur within a seamless environment, one that is not dependent upon time and space boundaries (Ullman & Rabinowitz, 2004). These systems allow educational institutions to manage a large number of fully online or blended (part online and part face-to-face) courses using a common interface and set of resources. Face-to-face courses that use an LMS for required or supplemental activities are often referred to as web-enhanced courses (Schmidt, 2002).

Learning management systems are known in the literature by several different names. These include course management systems, virtual learning environments and e-learning courseware (Gibbons, 2005). Some authors recognize distinctions between course management systems and learning management systems (e.g. Ceraulo, 2005, Watson & Watson, 2007), while others argue that the term "course management system" should be abandoned, since the acronym CMS is also used for content management systems and may cause confusion (Piña, Green & Eggers, 2008). Notwithstanding these minor controversies, the vast majority of U.S. based journals and other printed and digital media tend to use the terms "learning management system" (LMS) and "course management system" (CMS) interchangeably, while the designation "virtual learning environment" (VLE) is most popular in Europe and Asia.

## LMS GROWTH

The Campus Computing Project reported in 2002 that approximately three-quarters of all colleges and universities in the U.S. had adopted an LMS and that nearly one-fifth of all college courses used an LMS (Campus Computing Project, 2002). By 2006, LMS adoption had increased to 90% (Bassett & Burdt, 2006).

Bersin & Associates, researching industry trends in North American LMS usage, note that between the years 2004 and 2006, the LMS market enjoyed a growth of 26% and generated an estimated 480 million dollars in annual revenues (O'Leonard & Bersin, 2006). Data provided by Eduventures suggests that higher education institutions may have accounted for up to one-half of LMS revenues (Bassett & Burdt, 2006). Between 2006 and 2008, growth was a more modest 10.6%, likely due to market saturation and a slowing economy; however, 2009 revenues were projected to be at least 715 million (Bersin, Howard, O'Leonard & Mallon 2009).

During its first five years (2001-2006), Desire-2Learn experienced a 2,117% growth in revenue (Deloitte & Touche, 2007; Kempfert, 2003). Angel Learning became the fastest growing LMS among community colleges, with its market share increasing over 10% from 2007 to 2008 (Lokken, 2009). Blackboard, Inc's acquisition of WebCT in 2006 created an entity with a client base representing an 80% share of the educational LMS market (Mangan, 2008) and its purchase of Angel Learning in 2009 further cemented its dominance (Carter, 2009).

## FEATURES OF AN LMS

Dabbagh & Bannan-Ritland (2005) identified the most common features of an LMS by categorizing them as pedagogical tools for:

- content creation
- communication
- assessment
- administration

**Content creation** and display tools allow instructors to generate course content within an embedded text/HTML editor or to upload documents, spreadsheets, presentations, images, animations, audio or video into the LMS. Hyperlinks can point to websites or documents residing outside LMS. Assignments or drop boxes provide a place for students to submit assigned materials to their instructors for grading and feedback. Instructors can organize content into folders and subfolders and can use the content release feature to display or hide folders and individual content items; thereby giving the instructor control over when content is viewable by students.

Communication tools found in an LMS allow instructors to incorporate student-instructor and student-student interaction into the course. Asynchronous (non real-time) tools include course announcements, student web pages, e-mail to instructors and class members, threaded discussion boards, wikis, blogs, and file sharing. Synchronous (real-time) tools found in a typical LMS include text chat, whiteboard, and a sharable web browser. Groups of students can be placed into virtual teams or groups, which may include text chat, threaded discussion and file sharing ability that can be seen only by the members of the group and by the instructor.

Assessmenttools provide instructors with a number of ways to test, survey and track student achievement and activity in the course. Common tools include a test/assessment manager for creating and deploying exams, a generator for creating different types of questions (multiple choice, true/false, essay, short answer, matching, etc.) and question pools or test banks to store questions that can be used for multiple exams. Questions in an exam (and choices in a multiple-choice question) can be randomized and can be displayed one-at-a-time or all at once. Instructors can give a time limit for exams and can specify the type and amount of feedback that students receive for correct and incorrect answers. Exams can be graded, ungraded or delivered as anonymous surveys with aggregated results. An electronic grade book for managing student assignments and displaying student grades is a feature of virtually every LMS and is the most highly valued LMS feature by students (Kvavnik & Caruso, 2005). Less valued by students, but highly valued by instructors, is the ability to track student activity within an LMS, including logins, time spent and specific areas visited.

Administrative tools for instructors include control panels with the ability to manage the settings for the content creation, communication and assessment tools, customize the look of the course, make tools, content and resources available or unavailable to users, manage files and move or copy content. Administrative tools for LMS system administrators allow them to manage the creation of user accounts and courses, enrollment of instructors and students into courses, enabling and disabling of accounts and courses, and tracking activity in the system.

## Add-On Components

In addition to the standard tools provided within the LMS, most vendors offer additional products that can be integrated with the system. A common tool offered by many LMS vendors is an e-portfolio, a personal storage area allotted to individual students that resides apart from the online courses. The e-portfolio allows students to archive the assignments and projects created in their courses (often referred to as "artifacts") for use in later courses, for meeting graduation requirements, or for perusal by potential employers. The e-portfolio has become popular with teacher education programs and is gaining popularity in other disciplines (Burnett & Williams, 2009).

A learning object repository also allows for content to be stored outside of the individual online courses. The purpose of a learning object repository differs from that of an e-portfolio in that the e-portfolio facilitates the viewing of content by different *users*, while a learning object repository provides for the sharing of content among different *courses*. A personal repository permits an instructor to save content apart from a particular course and then import the content into one or more other courses. Many learning object repositories also allow instructors the option to link to items inside the repository, rather than having to copy the item each time into their courses. The advantage to this is that if the instructor wishes to edit or modify the item residing in the repository, the linked items within the courses will also be changed.

As accreditation and administrative demands for measurable outcomes data on systems and students become a higher priority for institutions, many LMS vendors are responding by offering institutional outcomes assessment systems that work in tandem with their learning management system. These systems pull learning outcomes data across all courses and track student improvement according to institutional objectives and standards.

## LMS ADVANTAGES

Before learning management systems became widely available, the delivery of online instruction required faculty or instructional designers to master Hypertext Markup Language (HTML) or web page authoring programs (Hill, Wiley, Nelson & Han, 2004). The ability of students to navigate the site intuitively, the quality of the visual design, and the overall instructional effectiveness of the course, were often dependent upon the coding and formatting expertise of the faculty member. An instructor wishing to provide a more interactive learning experience would be required to search for standalone software programs for discussion forums, chat or whiteboards and would need to cobble them together and discover which would (or would not) work with the other. Turning in assignments generally involved e-mailing the professors, while assessment was done manually, with the results often entered into spreadsheets. Delivery of content online was limited, as publicly available websites were not protected from copyright violation by fair use guidelines (Piña & Eggers, 2006).

The major advantage of the LMS is that it brought content delivery, communication, assessment and administration of online instruction into a single secure platform that could be accessed by anyone on the Internet (Dabbagh & Bannan-Ritland, 2005, Ullman & Rabinowitz, 2004). The standardized interface of an LMS made it easy for students to navigate through different online courses and for faculty to put their content online. The secure password-protected nature of an LMS limits access of instructional resources to users enrolled in the course, which permits instructors to take greater advantage of Fair Use and Teach Act protections for instructional use of materials than would be possible using the institution's public web site (Gibbons, 2005). The ability of several LMS products to deliver instruction to mobile devices takes advantage of research indicating that many users are increasingly relying upon mobile devices, rather than computers, to access the Internet and perform other technology tasks (Ranie & Keeter, 2006).

In 2005, EDUCAUSE conducted a study of the LMS use of over 12,000 college students. Researchers found that student attitudes toward their CMS were overwhelmingly positive (Kvavnik & Caruso, 2005). The features of the CMS that students used most often and rated the most valuable were the ability to keep track of grades, access sample quizzes, class readings and syllabi, taking tests online, turning in assignments online and getting assignments back with feedback and grades. Interestingly, students did not rate online discussion board assignments nearly as high (Kvavnik & Caruso, 2005). These findings support the notion that students place the highest value in those features that make their lives easier and their learning more convenient (Piña, Green & Eggers, 2008).

## LMS LIMITATIONS

As with most new and evolving technologies, the LMS is not without its limitations and disadvantages. In a review of several studies of LMS use in higher education, Ioannu and Hannafin (2008) reported that many users found that the systems were often slow, confusing, and focused more upon administrative needs than student needs. Another common complaint was that the LMS interface was dull and rigid, compared to more engaging online social environments, such as MySpace, Facebook and YouTube (Ioannu & Hannafin, 2008; Piña, 2007). Siemens (2004) noted that the LMS interface is not friendly to many users and should be simplified and made more intuitive. Lane (2008) found that current systems were designed to function primarily as a repository of materials and did not support sound pedagogical practice, while Piña, Green and Eggers (2008) lamented the lack of instructional design guidance and the absences of tools for the development of rich multimedia-based instruction. Although a number of teaching tools exist within a LMS (Dabbagh & Bannan-Ritland, 2005), these do not include tools to guide the instructor in the design of online instruction and in sound pedagogical practice (Ioannu & Hannafin, 2008).

## EVALUATING AND SELECTING AN LMS

At many colleges and universities, the suggestion to consider adoption of a new LMS is met by apocalyptic visions of crashing systems, lost data, cancelled courses and faculty and student rebellion. In actuality, changing an institution's LMS has become a rather commonplace occurrence, with surveys showing a sizeable number of colleges and universities that are considering such a change (Ioannu & Hannafin, 2008). A Google search combining "LMS," "learning management system" or "course management system" with "evaluation" or "comparison," will result in several college and university websites documenting the processes that they have undertaken in their quest to find the right LMS (e.g. Smart & Meyer, 2003). While there is no "cookie cutter" approach to evaluating and selecting an LMS, a number of strategies have been found to be successful for different institutions.

### Determine Needs

The selection process of the best LMS for a given institution is determined primarily by institutional needs, preferences and how the system will be used. A college of university that offers no fully online courses but makes the LMS available to faculty who wish to web-enhance their face-to-face courses will likely need a feature set that differs from the institution that offers several advanced degrees online. In short, a needs analysis determines the gap that exists between the "optimal" condition (i.e. what the institution and its people should be able to do with a fully functional LMS) and the "actual" condition (i.e. current state of the learning management at the institution). The difference between the optimal and actual conditions is the institution's needs (Rossett, 1991).

### Include Relevant Constituencies

Adoption of an LMS affects a number of individuals and groups across an institution. The committee or task force commissioned to evaluate and recommend an LMS platform should include representation from faculty and student users and from the various departments that will be administering, supporting or interfacing with the LMS. These could include academic affairs, centers for teaching and learning, educational technology or distance learning departments, network services, database management or academic affairs.

### Seek Outside Expertise

Most commercial and non-commercial LMS products have an official or unofficial network of user groups or communities of practice. Many of these feature active listservs and some meet regularly and hold conferences. These groups can become

*Table 1.*

| | |
|---|---|
| Communication Tools<br>• Discussion Forum<br>• Discussion Management<br>• File Exchange<br>• Internal Email<br>• Online Journal/Notes<br>• Real-time Chat<br>• Whiteboard<br>Productivity Tools<br>• Bookmarks<br>• Calendar/Progress Review<br>• Searching Within Course<br>• Work Offline/Synchronize<br>• Orientation/Help<br>Student Involvement Tools<br>• Group work<br>• Community Networking<br>• Student Portfolios<br>Course Delivery Tools<br>• Test Types<br>• Automated Testing Management<br>• Automated Testing Support<br>• Online Marking Tools<br>• Online Grade book<br>• Course Management<br>• Student Tracking | Content Development Tools<br>• Accessibility Compliance<br>• Content Sharing/Reuse<br>• Course Templates<br>• Customized Look and Feel<br>• Instructional Design Tools<br>• Instructional Standards Compliance<br>Hardware/Software<br>• Client Browser Required<br>• Database Requirements<br>• UNIX Server<br>• Windows Server<br>Administration Tools<br>• Authentication<br>• Course Authorization<br>• Registration Integration<br>• Hosted Services<br>Company Details/Licensing<br>• Company Profile<br>• Costs / Licensing<br>• Open Source<br>• Optional Extras |

an invaluable source of real information "from the trenches" that is unfiltered by the manufacturers. An example of such a group is SLATE (Supporting Learning and Technology in Education) which is a community of practice of professionals involved in LMS issues at over 60 schools, colleges and universities (Piña, Sadowski, Scheidenhelm & Heydenburg, 2008). Although listed as the Midwest Blackboard Users Group, SLATE includes member who use various different systems.

## Compare Feature Sets

While searching vendor websites or seeing demonstrations at conferences or on-campus can be beneficial, it is often difficult to compare systems side-by-side in this manner. A highly convenient resource is the EduTools Home Page, operated by the Western Cooperative for Educational Telecommunications (WCET) at www.edutools.info (EduTools, 2009). EduTools is a community-driven site that includes comparisons and reviews several e-Learning products. The "Course Management Systems Comparison" features over three dozen systems (a few are different releases of the same system) allowing users to make side-by-side comparisons of the LMS features that they find important. The EduTools site includes descriptions of the following features in Table 1.

## Field Testing

Trying out the LMS in an actual field test is where the "rubber meets the road". This allows users to experience the LMS interface, navigation, logic and performance of the instructional and administrative tools of the system. Most institutions will choose a limited number of courses to try out the systems over a term and collect evaluation data from the various constituencies.

## Infrastructure

Research involving both distance learning faculty and program administrators, found that higher education institutions do a good job of making the LMS available to students and faculty (Piña, 2008a). However, for distance learning to be successfully institutionalized, the most important factor was that the technical infrastructure was sufficient to support the distance learning activity (Piña, 2008b). A successful LMS implementation is dependent upon a reliable physical plant, adequate bandwidth and staffing to manage and support the system. How the system is deployed will influence greatly how well it operates. This is discussed below.

## DEPLOYING AN LMS

The feature set of an LMS, though important, is not the only consideration in the selection and deployment of system for learning management. In addition to determining *which* system to use, decisions must be made regarding *how* the system will be used.

### Standalone vs. Integrated

Three important administrative functions of an LMS are 1) the creation and management of user accounts; 2) the creation and management of course shells; and 3) the enrollment of students and instructors into their courses. These functions can be performed using a standalone or integrated LMS strategy.

A standalone LMS is one that does not have a direct interface with other campus administrative or academic systems. In a standalone system, the creation and management of user accounts (including usernames and passwords), the creation of new courses, and the enrollment of instructors and students into courses occur manually. This can be done by having an administrator enter the user and course data into the system to create a single user account, course, or enrollment. A more efficient way to run a standalone system is by using a batch file feed for the creation of user accounts, new courses and enrollment. In this scenario, a text file with user and/or course information gathered from the campus student information system (SIS) or enterprise resource planning (ERP) system--such as Banner, PeopleSoft, Datatel or Jenzabar—is uploaded manually into the system. This allows for the creation of multiple accounts, courses or enrollments, rather than having to do them one at a time.

Integrating the LMS with the SIS or ERP systems is the preferred method at many colleges and universities, due to its increased efficiency. In an integrated system, the files for user accounts, course creation, student enrollment, and instructor assignment are fed directly from the SIS or ERP system into the LMS. This can occur in one of two ways:

- A "snapshot" approach, where files are fed into the LMS at regular intervals (e.g. every day at 9:00 am and 4:00 pm). The file contains a "snapshot" of the user accounts, courses and enrollments at that point in time. The LMS is updated each time a new snapshot file is fed into the system.
- An "event-driven" or real-time approach, where the LMS is updated any time a change is made in the SIS or ERP system. For example, if a student enrolls in a course using an online form tied to the SIS or ERP, the student is placed automatically in the corresponding online course in the LMS.

Integrating the LMS with campus administrative systems eliminates time-consuming data entry or manual uploading of multiple text files. Another advantage is that instructors do not have to manually enroll or drop students into their online courses. Students who drop or who have an academic hold placed on their accounts are made invisible to the instructor (i.e. they disappear from the class roster). When the hold is lifted, the students' accounts are automatically enabled--without having lost their previous assignments or grades. The main disadvantages of integrated systems are the extra costs involved in programming the systems to work together and licensing the event-driven or snapshot hardware or software "tool" from the LMS vendor.

### Self-Hosted vs. Vendor-Hosted

The institution also has a choice regarding the physical location and configuration of the LMS hardware and software. An LMS can be hosted and maintained on the institution's own servers or the choice can be made to contract with an outside vendor to provide hosting and maintenance through an application service provider (ASP) agreement.

Self-hosting allows institutions to retain a greater level of control over the operation, timeline, upgrades and maintenance of the system and will usually result in significantly lower fees paid to the vendor. This option is most desirable for organizations with sufficient in-house application and server expertise and staff and where the I.T. culture of the institution places a high priority on supporting instructional technologies. However, when trouble occurs or it is time to upgrade the LMS version, self-hosted institutions often find themselves paying premium fees to the vendor for upgrading help and technical assistance (Piña & Eggers, 2006).

An ASP-hosted arrangement may be a more feasible solution for many organizations, since the annual fee paid to the vendor is usually far less than the cost of periodic servicing of hardware and hiring technical personnel to manage and maintain the LMS servers. Other reasons to consider ASP hosting is that the LMS would not have to compete with other campus entities for limited technology resources, personnel, support, and bandwidth. Most ASP vendors provide 24/7 technical support--something that educational organizations typically cannot afford to do. A disadvantage of not hosting the LMS in-house is the lack of control of the "back end" of the system, since most vendor-hosted clients do not have "root level" administrative rights to the LMS server. ASP hosting and support can be provided by an LMS vendor, such as Blackboard or Desire2Learn or by an outside technology support company like Presidium or Embanet.

## COMMERCIAL ACADEMIC LMS VENDORS

Users and institutions who wish to utilize an LMS for online, blended or web-enhanced courses have a plethora of options. These vary widely in price, features and installed user base. Brandon-Hall LMS KnowledgeBase lists over 90 different learning management systems, exclusive of those profiled below (Brandon-Hall, 2009). Although so many systems are in existence, the majority of these are vying currently for a relatively small percentage of the academic LMS market.

Below are brief profiles of the vendors whose systems are dominant in academia. Blackboard's purchase of WebCT in 2006 and Angel in 2009 has narrowed the field of widely-used commercial learning management systems. WebCT is no longer being developed as a separate product; however Blackboard has committed to the development of and support of a separate Angel LMS at least through the year 2012 (Chasen, 2009), which is why Angel is included in the profiles below.

### Blackboard

By the end of the first decade of the new millennium, Blackboard, Inc. had established itself as the clear market leader among commercial LMS vendors in the K-12 and higher education arenas. Established in 1997, Blackboard's growth was fueled by its merger with CourseInfo in 1999 and by the strategic acquisition of several LMS and non LMS companies, including MadDuck Technologies (makers of Web Course in a Box) in 2000, CampusWide by ATT in 2000, Prometheus in 2002, SA Cash in 2003, WebCT in 2006, NTI Group in 2008, and Angel Learning in 2009 (Blackboard, 2009). Blackboard's acquisition strategy has resulted in a wide-reaching product line that includes learning management systems, LMS add-on systems, a system for conducting various campus business transactions, and a system for inter-campus communications and emergency alerts.

Students navigate through a Blackboard course by a left side menu of content and tool pages. Instructors can add to and modify the number of these pages. Instructors upload and manage most content via a control panel, although content pages can be edited at the individual page level for convenience. One of Blackboard's advantages is the seamless integration of graded assignments and discussions with the Grade Center. When instructors create new graded assignments on content pages or discussion boards, these are generated automatically in the Grade Center.

Since WebCT's purchase by Blackboard has influenced both Blackboard's development and the LMS market in general, a brief synopsis of WebCT will be offered here:

First developed in 1996 at the University of Calgary, Canada, WebCT was Blackboard's largest competitor for the academic LMS market during its decade of existence. WebCT's LMS was known for its ability to be customized by instructors and its advanced tools for the delivery of instruction. It was also a complex program, with multiple control panels for creating and importing course content, which made it challenging for many instructors to learn. The decision as to which LMS a university would adopt was often characterized as a choice between ease of use (Blackboard) or sophistication (WebCT). When Blackboard acquired WebCT in 2006, it was estimated that WebCT's market share was 35-40%, while Blackboard's was 45-50% (O'Hara, 2005). For the next few years, Blackboard sold WebCT's two LMS products, Campus Edition (CE) and Vista, alongside its own Blackboard Enterprise System. In 2009 Blackboard Learn Version 9 was introduced, which incorporated functionality from both Blackboard and WebCT.

Blackboard's business and product development strategy of building through acquisition (also used by Microsoft and many other companies) has influenced the architecture of its LMS. Rather than attempting to build all available capabilities and features into is base system, Blackboard has adopted a modular approach. The LMS is "extended" by the incorporation of "building blocks," which are applications built by third parties--often programmers at universities who are Blackboard customers. Other building blocks are created by vendors who wish to integrate their programs seamlessly into Blackboard. While a few building blocks are available free of charge, most are purchased or licensed annually. Occasionally, the latest version of Blackboard will incorporate new features that had previously been third-party building blocks.

Going beyond the individual capabilities of building blocks are the larger Blackboard Content System, Blackboard Community System and Blackboard Outcomes System. These are separate add-on systems to the core LMS (Blackboard Learn). While the tools included within the Blackboard Learn LMS are course-centric, the Content and Community Systems add institutional-based content and learning object management, e-library reserves, an e-portfolio, and group and user role management. The Outcomes System is Blackboard's institutional assessment tool to track and report learning outcomes throughout the organization. Due to the significant annual investment required for the Content, Community and Outcomes Systems (which can triple or quadruple the initial licensing fee for the Learn system), many customers opt to forego the institutional capabilities and use only the core Blackboard Learn LMS.

Blackboard's NG (Next Generation) LMS strategy involves building upon its history of acquisition and modularization. Two of its latest acquisitions, Angel Learning and SafeAssign, promise to add new capabilities to the LMS. SafeAssign is a plagiarism detection tool that is now incorporated into the Blackboard Learn system. By 2012, a single Blackboard LMS product incorporating technologies from Blackboard, WebCT and Angel is expected to be available (Chasen, 2009). Interfacing with other technologies and even competing products is another strategy of Blackboard NG. Two initial applications are Blackboard Sync and linking to open source courses. Blackboard Sync permits a student, who is logged into Facebook, to see whether new assignments or discussions have been posted within any of their Blackboard courses. Courses that have been created in either Moodle or Sakai can now be launched from within the students' Blackboard home page, alongside their Blackboard courses.

## Angel

The Angel Learning Platform was developed at Indiana University-Purdue University, Indianapolis in 1996 and incorporated in 2000, first as CyberLearning Labs, and later as Angel Learning (Ceraulo, 2005). During its first decade, Angel's annual growth was a healthy 100%, but did not match the rapid initial growth of Blackboard, WebCT or Desire2Learn (Angel Learning, 2009). Following Blackboard's 2006 acquisition of WebCT and copyright infringement lawsuit against Desire2Learn, Angel's market share began to rise at an accelerated rate, particularly in the K-12, career college and community college arenas. Some of Angel notable clients include Penn State University, Michigan State University, the 98,000 student Jefferson County Public School System in Kentucky. A recent study of LMS use in the community colleges found that Angel's market share among the institutions surveyed rose from 9 to nearly 21 percent in a single year (Lokken, 2009). Michael Chasen, the CEO of Blackboard, cited Angel Learning's advanced technology, its superior customer service culture, and its affordability as reasons for Blackboard's interest in merging with Angel (Chasen, 2009). In May, 2009, Blackboard acquired Angel Learning for $95 million.

Since its inception, Angel's focus has been on how to enhance teaching and learning in academia, with over 90% of its client base in schools, colleges and universities (Angel Learning, 2009). Angel was one of the first systems to incorporate Web 2.0 technologies--such as wikis, blogs, RSS feeds and iTunes integration--and to include the ability to display on mobile devices and assess whether courses meet Section 508 accessibility guidelines. Angel's user environment features a portal-type interface with channels that can be added, arranged and customized. Students may access course materials and content items in either a sequential order, by clicking a "next" link, or non-sequentially by clicking directly on the item.

Angel's LMS provides a robust set of tools for instructors, including the ability to utilize many types of content and convenient editing menus that do not require the instructor to go to a control panel or task pane. Particularly useful for instructors is the ability to view the course in different roles (e.g. instructor, student, teacher's assistance and guest) and a rubric generator that facilitates grading of essays, reports and projects. Interactive "agent" technology allows instructors to set various parameters for releasing content to individuals or groups of students, allowing for instruction to be either automated or highly personalized. Both instructors and administrators benefit greatly from Angel's advanced reporting tools, which can query the database provide in-depth tracking and reporting of student, instructor, course and system activities, without the user having to know SQL programming.

In contrast to Blackboard's modular approach to LMS architecture, Angel has striven to build both course and enterprise functionality into its core LMS product. Its institutional directory allows the administration of the system to be built according to the organizational structure of the institution, including segmentation by multiple schools, campuses, sites, offices, departments or other groups. The learning object repository--typically an add-on product for learning management systems--is also built into the Angel LMS.

## Desire2Learn

Desire2Learn (D2L) holds a unique place in the short history of learning management systems. While D2L's rapid growth during its first half decade has been the topic of a number of favorable articles (e.g. Deloitte & Touche, 2007), it has been D2L's legal battles with Blackboard that have dominated the press and the blogosphere since 2006 (e.g. Mangan, 2008). The following is a brief summary of the major events of the litigation:

- In 1999 and 2000, Blackboard applied to the U.S. Patent Office for patents related to LMS technology, including the ability for users to be assigned multiple roles in different courses (e.g. instructor, student, teacher assistant).
- The U.S. Patent Office granted the Blackboard's patent in 2006. The day that the patent was issued, Blackboard sued Desire2Learn, Inc. for patent infringement. Blackboard later announced that it would not enforce its copyright nor pursue litigation against open source developers.
- In February, 2008 a jury in Texas awarded Blackboard a 3.1 million dollar settlement and issued an injunction against D2L.
- In March, 2008 the U.S. Patent Office issued a preliminary finding that rejected all 44 claims of Blackboard patent.
- D2L issued a new "workaround" version of its LMS claiming that its product no longer included the disputed Blackboard technology. Blackboard disagreed.
- In July, 2009, the U.S. Court of Appeals issued a ruling in favor of D2L, invalidating all of Blackboard's claims against D2L.
- By late 2009, the issue remains unresolved.

As with many learning management systems, Desire2Learn's roots are in academia. Developed in 1999 by a student at the University of Waterloo, Canada, D2L achieve a revenue growth of 2,117% between 2000 and 2005, making it one of Canada's fastest growing companies (Deloitte & Touche, 2007; Kempfert, 2003).

During this time, the dominant systems were Blackboard and WebCT, the former perceived to be easy but limited and the latter considered robust but complicated. Desire2 Learn successfully staked out a middle ground between the two, featuring more customization and sophisticated delivery than Blackboard and an intuitive interface that was more user-friendly than WebCT (Piña & Eggers, 2006). The strategy proved very effective, with several institutions choosing D2L after performing side-by-side comparisons and evaluations with other LMS products (Kempfert, 2003).

D2L's biggest successes have included becoming the LMS of choice for several large and well-known universities and for state- and system-wide consortia. These include the University of Arizona, University of Iowa and the Ohio State University and the Minnesota State Colleges and Universities, Oklahoma State Regents for Higher Education, South Dakota Board of Regents, Tennessee Board of Regents, University of Wisconsin System (Angel Learning, 2009) and most of the Seventh-Day Adventist affiliated colleges and universities in the United States and Canada (Piña & Eggers, 2006). Although D2L's percentage of the academic LMS market was less than 10% at the time of the Blackboard/WebCT merger, it was widely considered to be Blackboard's biggest competitor after WebCT (Johnson, 2007; Kempfer, 2003).

Content delivery in D2L is structured like a table of contents, which permits the learner to progress in a logical manner through the course. Course development is facilitated by a file manager, allowing for different content items, such as documents, web pages and graphics, to be loaded in the system one time and deployed multiple times within the course, increasing performance and saving storage space. This is particularly effective for the instructor who creates web page content outside the LMS, as links to images are not broken or lost when the HTML and image files are uploaded into the file manager and deployed across the course (Piña & Heydenburg, 2008). Course editing tools for instructors are available on the same page as the content, either on left-side task pane or icons to right of the content, eliminating the need to go "out" to a separate control panel page. Administrators have a great deal of control over the look and feel of a course and can grant create different "levels" of instructors, each with increasing amounts of access to customize their courses and access course features.

## Pearson eCollege

Pearson eCollege, a division of publishing giant Pearson Education, operates under a business model that differs from that of other LMS vendors. While most companies offer the LMS software as a product, with the options for self hosting or vendor hosting and support, Pearson views itself as the provider of a total institutional solution for distance learning. An eCollege package is hosted by Pearson and includes the learning management system and a large number of value-added services, such as instructional design and development, marketing and promotion, student and faculty technical support, system and trend reporting, program evaluation and staff development. In essence, eCollege can become an institution's distance learning department (Simba Information, 2007).

Founded in 1996 as Real Education and later renamed eCollege, the company was acquired for $477 million by Pearson Education in 2007 (Carnevale, 2007). The "total package" strategy of Pearson eCollege has made it an attractive choice for colleges without an internal e-learning office or department and for large distributed institutions, such as DeVry University, National University and Kaplan University that combine a large online program with multiple physical campuses.

The eCollege LMS is structured similarly to Blackboard's system, with a left-side navigation pane that is navigated sequentially by the student and easily modified by the instructor. Instructor content tools are accessed by a series of buttons on the top of the window and allow the instructor to upload documents, images and presentations into the LMS. The eCollege LMS is part of Pearson's larger e-Learning strategy and is used as the engine for several Pearson products, including online laboratory programs, student placement and assessment exams, standalone e-Learning courses and training products. Pearson also partners with other LMS vendors, so its customers will often be using its eCollege-based products alongside Blackboard, Angel Desire2Learn, Sakai or Moodle.

## COMMERCIAL NON-ACADEMIC LMS VENDORS

Although this report focuses primarily on learning management systems used in academia, Josh Bersin and his Associates issue an annual report on 30 LMS vendors who products cater to industry, military and other non-academia clients (Bersin, Howard, O'Leonard & Mallon, 2009). This group includes vendors that use their systems to deliver their own proprietary content, such as SkillSoft and ElementK, and full-featured ERP systems for managing an organization's finances, human resources and training management, such as Oracle and SAP. Many of the industry-centric learning management systems include enrollment management (fro tracking fees), talent management and other capabilities not generally included in an academic-based LMS. The marketplace leaders in this arena are Saba and SumTotal Systems, each of which reports a worldwide base of 17 million users (Bersin, Howard, O'Leonard & Mallon, 2009).

## OPEN SOURCE LMS

As commercial LMS companies grow larger and as their products become more complex and expensive, many schools, colleges and universities are questioning whether their needs can be better met by open source products, rather than by a commercial system (Piña, 2007). Open source has financial and programmatic appeal (Stewart, 2007). For those who subscribe to a social constructivist point of view where reality is constructed from the collective experiences of groups (Driscoll, 2007), open source also has a philosophical appeal. In an open source environment, the source code of the product is made available to the user without charge. Software licensing fees, which can be substantial, are eliminated. Open source software frees the user from a contractual agreement with a specific vendor. A program or system based on open source software may be customized and branded according to a user or institution's needs and desires--rather than to a vendor's current priorities. In the case of learning management systems, there exists a vibrant and active community of developers for Moodle and Sakai, the two most popular systems, and for several others (see Other Open Source Systems below).

While it is true that the source code of an open source LMS is free, the implementation of an open source LMS may involve a substantial investment in infrastructure. This would include server hardware and software, server administration, database administration, programming and technical support that would otherwise be supplied by the vendor of a commercial system. In order to leverage the advantages of being able to customize the LMS (a primary "selling" point for open source software), an institution running Moodle would require in-house expertise in MySQL and PHP programming, while Sakai would require Java programmers (Piña, 2007).

Operating a customized and institution-specific LMS has its own potential pitfalls. The "closed" nature of commercial learning management systems limits internal modifications and provides a cadre of users whose systems operate more or less the same. While the 2006 shipment of a bug-laden WebCT Vista product proves that commercial releases sometime occur hastily (Chasen, 2009), commercial learning management systems are typically field tested prior to public release and tend to be backed by a warranty and contractual obligations of customer support and down-time limitations. An open source LMS that has been heavily customized by local programmers and developers (for whom the LMS may be one of may competing duties), might be quite unlike those at other institutions any and users may find themselves alone if a customization goes awry. Open source code carries no guarantee or warranty.

Many of the above issues could be mitigated by contracting with an emerging cottage industry of third-party open source support vendors. These provide hosting, custom programming and support services for open source learning management systems. Two of the most popular are Moodlerooms for Moodle and rSmart for Sakai. This type of arrangement mirrors the ASP relationships that institutions have with commercial LMS vendors and include many of the same advantages. However, it is also true that an ASP relationship means that the institution may have merely switch one commercial company for another and could be locked into using that company's version of the open source LMS.

### Moodle

Available to the public since 2002, Moodle (Modular Object-Oriented Dynamic Learning Environment) was developed in Australia as a free alternative to commercial learning management systems. Unlike an LMS from a commercial vendor, the system is developed and supported by an active community of developers, users and administrators that keeps the software evolving at a steady pace. Moodle is based upon the open source MySQL database and PHP scripting language (Valade, 2002) and is available for download--free of charge--at http://moodle.org. Users may install Moodle on a single computer or on a department or institutional server. The ease of installation, testing and adoption has contributed to making Moodle the fastest growing LMS (Cole & Foster, 2007).

Moodle.org reports the number of registered Moodle sites at over 35,000, with over 2 million courses and over 25 million users--more than any other single LMS. These numbers can be misleading, however, when compared to those of commercial LMS vendors. A single institution (e.g. a large university) will typically maintain a single campus-wide license of a commercial LMS and thus be considered equivalent to a single registered Moodle site. At the same institution, it is possible (and common) for multiple departments or individual faculty to be running their own unique instances of Moodle, each of which would be considered a registered site. Thus a university may have one licensed instance of Blackboard, but dozens of instances of Moodle. Recently, a number of notable institutions, such as the UK Open University in England, Athabasca University on Canada, the University of California at Los Angeles (UCLA) and San Francisco State University in the United States, have switched from their previous LMS to Moodle.

The Moodle interface contains a feature set similar to those of a commercial LMS (e.g. announcements, course outline, navigation and administration links). However, Moodle's interface is indicative of its constructivist roots. Rather than focusing on document management and delivery, Moodle is geared more toward facilitating communication and social interaction, with channels such as "people," "events" and "latest news" featured prominently on the course home page. Moodle's structure resembles that of many social networking sites, such as Facebook.

## Sakai

Sakai, arguably the second most popular open source LMS, has very different roots than Moodle. Sakai was established in 2004 by a consortium of major research universities as an alternative to home-grown or commercial learning management systems. These institutions included the Massachusetts Institute of Technology (M.I.T.), Stanford University, Indiana University, Michigan State University and University of California at Berkeley.

Rather than using a MySQL/PHP architecture like Moodle, Sakai is built upon the Java language platform, similar to Blackboard. Sakai's interface and feature set resemble its commercial LMS counterparts and contains tools for content creation, communication, assessment and administration (Dabbagh & Bannan-Ritland, 2005). In addition to being an LMS for online, blended or web-enhanced courses, the developers of Sakai envision its engine to be utilized as a collaboration platform for research and other projects outside the scope of a virtual class. One of Sakai's most popular applications is an e-portfolio product that can be used in addition to its LMS or by an institution that uses a non-Sakai LMS. Sakai also features a module that links library resources into its LMS.

The Sakai Project website (www.sakaiproject.org) gives a list of over 150 institutions--primarily in the United States and Eastern Europe—that are using Sakai as the primary campus LMS. One of the most successful installations is Etudes (http://etudes.org), a non-profit organization that provides and supports its version of Sakai to over two dozen California Community Colleges.

## Other Open Source Systems

Outside the United States and Canada, there are several open source learning management systems in use at schools, colleges and universities and in business and industry. These include Claroline (Belgium), eFront (Greece), Docebo (Italy), Dokeos (Belgium), ILIAS (Germany) and OLAT (Switzerland). An exhaustive examination of each of these systems is beyond the scope of this chapter; however the reader is invited to learn more by following the links found in the LMS Companies and Organizations section below.

## CONCLUSION

Research conducted by the Sloan Consortium and the National Center for Education Statistics indicates that online student enrollments is the fastest growing sector of higher education, increasing at a rate ten times higher than total college and university enrollments (Allen & Seaman, 2007). The vast majority of these online courses are being delivered via learning management systems, one of the true success stories in the long history of technology use in education (Piña, 2008b). The LMS's first decade has been one of fierce competition, with companies appearing, disappearing or bought by other companies. A few companies have emerged currently as dominant players; however, the history of technology is populated with companies, such as WordStar and WordPerfect, which once dominated their respective domains and later disappeared (Bergin, 2006).

New research is showing that the emergent wave of minority users are favoring mobile technologies over desktop and notebook computers and are adept at using the advanced features of these technologies for both entertainment and communication (Ranie & Madden, 2005; Ranie & Keeter, 2006). Angel and Desire2Learn have made their systems capable of display on iPods and PDAs and the other LMS manufacturers are following suit. It is likely that as both LMS and mobile device technologies advance, current challenges, such as screen and input device size, will become a thing of the past.

The LMS market is just beginning to respond to the explosion of Web 2.0 activity that is becoming a routine part of students' lives. Most commercial learning management systems now include blogs and wikis among their instructional tools; however, the web is becoming ever more participatory and users are now active content authors, collaborators and contributors. The LMS was originally designed as a document repository and a system of delivering content authored by the instructor to the user. This paradigm is now being challenged. Even as questions arise as to whether open source, social networking software or Google will signal the demise of the commercial learning management system (Young, 2009), it is not hard to imagine that somewhere right now, some young genius is tinkering away at his or her computer, just about to discover the next generation of (or perhaps replacement to) the LMS.

## LMS COMPANIES AND ORGANIZATIONS

**ANGEL Learning**
6510 Telecom Drive
Suite 400
Indianapolis, IN 46278
www.angellearning.com
**Blackboard**
650 Massachusetts Avenue N.W.
6th Floor
Washington, DC 20001-3796
www.blackboard.com

**Claroline**
www.claroline.net
**Desire2Learn**
305 King Street West
Suite 200
Kitchener, Ontario, Canada
N2G 1B9
www.desire2learn.com
**Docebo**
Via A. Moro 1 - 20050
Macherio (MI) – Milano
ITALY
www.docebo.com/doceboCms/
**Dokeos e-learning**
Rue Victor Hugo, 201
B-1030 Brussels
BELGIUM
www.dokeos.com
**eFront**
46 Salaminos Str,
Aspropyrgos, 19300, Athens
GREECE
www.efrontlearning.net
**Element K**
500 Canal View Boulevard
Rochester, NY 14623
www.elementk.com
**Embanet**
1000 East Woodfield Road
Suite 240, Schaumburg, IL 60173
www.embanet.com
**ILIAS**
www.ilias.de
**Moodle**
www.moodle.org
**Moodlerooms**
1101 East 33rd Street
Suite A306
Baltimore, MD 21218
www.moodlerooms.com
**OLAT**
www.olat.org/website/en/html/index.html
**Oracle Learning Management**
500 Oracle Parkway
Redwood Shores, CA 94065
http://www.oracle.com
**Presidium**
1810 Samuel Morse Drive
Reston, Virginia 20190
www.presidiuminc.com
**Pearson eCollege**
4900 S. Monaco Street
Suite 200
Denver, CO 80237
www.ecollege.com
**Rsmart**
4343 East Camelback Rd.
Suite 210
Phoenix, AZ 85018
www.rsmart.com
**Saba Software**
2400 Bridge Parkway
Redwood Shores, CA 94065
www.saba.com
**Sakai Foundation**
P. O. Box 130256
Ann Arbor, MI 48113-0256
www.sakaiproject.org
**SAP Enterprise Learning**
3999 West Chester Pike
Newtown Square, PA 19073
http://www.sap.com
**SkillSoft**
107 Northeastern Blvd.
Nashua, NH 03062
www.skillsoft.com
**SumTotal Systems**
1808 North Shoreline Boulevard
Mountain View, CA 94043
www.sumtotalsystems.com

## REFERENCES


Allen, I. E., & Seaman, J. (2007). *Online nation: Five years of growth in online learning*. Needham, MA: The Sloan Consortium.

Angel Learning. (2009). *About angel learning*. Retrieved June 1, 2009, from http://www.angel-learning.com

Bassett, E., & Burndt, C. (2006). Beyond course management systems. *EDTECH: Focus on Higher Education October/November 2006*. Retrieved May 9, 2009, from http://www.edtechmag.com/higher/october-november-2006/tech-outlook-2.html

Bergin, T. (2006). The proliferation and consolidation of word processing software: 1985-1995. *IEEE Annals of the History of Computing, 28*(4).

Bersin, J., Howard, C., O'Leonard, K., & Mallon, D. (2009). *Learning management systems 2009: Facts, practical analysis, trends and provider profiles*. Oakland, CA: Bersin and Associates.

Blackboard, Inc. (2009). *Blackboard media center press releases*. Retrieved June 10, 2009, from http://www.blackboard.com/Company/Media-Center/Press-Releases.aspx

Brandon-Hall. (2009). *Brandon-Hall Research LMS KnowedgeBase 2009*. Retrieved June 10, 2009, from http://www.brandon-hall.co/publications/lmskb/lmskb_firms.shtml

Burnett, M. N., & Williams, J. M. (2009). Institutional uses of rubrics and e-portfolios: Spelman College and Rose-Hulman institute. *Peer Review, 11*(1).

Campus Computing Project. (2002). *The campus computing report*. Encino, CA: Campus Computing.

Carnevale, D. (2007). Pearson's purchase of eCollege signals publishers' interest in online education. *The Chronicle of Higher Education, 53*(38).

Carter, D. (2009, May 7). Blackboard-angel deal gives pause to some. *eSchool News*. Retrieved June 1, 2009, from http://www.eschoolnews.com/news/top-news/?I=58675.

Ceraulo, S. (2005). Benefits of upgrading to LMS. *Distance Education Report, 9*(9).

Chasen, M. (2009). *Future of blackboard and angel*. Keynote presentation at the Angel User Conference, Chicago, IL.

Cole, J., & Foster, H. (2007). *Using moodle* (2nd ed.). Sebastapol, CA: O'Reilly Media.

Dabbagh, N., & Bannan-Ritland, B. (2005). *Online learning: Concepts, strategies, and applications*. Upper Saddle River, NJ: Pearson Prentice Hall.

Deloitte & Touche. LLP (2007). *2006 Canadian technology fast 50 winners*. Retrieved June 11, 2009, from http://www.deloitte.com.

Driscoll, M. P. (2007). Psychological theories of instructional design. In Reiser, R. A., & Dempsey, J. V. (Eds.), *Trends and issues in instructional design and technology* (2nd ed.). Upper Saddle River, NJ: Pearson Merrill/Prentice-Hall.

EduTools. (2009). *CMS: CMS home*. Retrieved June 10, 2009 from http://www.edutools.info/static.jsp?pj=4&page=HOME

Eduventures. (2007). *Postsecondary academic computing: Identifying opportunities beyond LMS*. Retrieved from http://www.eduventures.com

Gibbons, S. (2005). Library course-management systems: an overview. *ALA Library Technology Reports, 41*(3).

Harrington, C. F., Gordon, S. A., & Schibik, T. J. (2004). Course management system utilization and implications for practice: A national survey of department chairpersons. *Online Journal of Distance Learning Administration, 7*(4).

Hill, J. R., Wiley, D., Nelson, L. M., & Han, S. (2004). Exploring research on internet-based learning: From infrastructure to interactions. In Jonassen, D. H. (Ed.), *Handbook of research on educational communications and technology* (2nd ed.). New York: Simon and Schuster/Macmillan.

Ioannou, A., & Hannafin, R. (2008). Deficiencies of course management systems: Do students care? *Quarterly Review of Distance Education, 9*(4).

Johnson, J. (2007). Technology at the technical college. *University Business, 10*(2).

Kempfert, T. (2003). UW selects Desire2Learn as its new e-learning system. *Teaching With Technology Today, 9*(7).

Kvavnik, R., & Caruso, J. (2005). *ECAR study of students and technology 2005: Convenience, connection, control and learning*. Boulder, CO: EDUCAUSE Center for Applied Research.

Lane, L. M. (2008). Toolbox or trap? Course management systems and pedagogy. *EDUCAUSE Quarterly, 31*(2).

Lokken, F. (2009). *2008 distance education survey results: Tracking the impact of eLearning at community colleges*. Washington, DC: Instructional Technology Council.

Mangan, K. (2008, February 13). Courtroom showdown begins in blackboard's patent dispute with desire2learn. *Chronicle of Higher Education*.

O'Hara, T. (2005, November 26). Blackboard's webct deal spurs antitrust questioning. *The Washington Post.*

O'Leonard, K., & Bersin, J. (2006). *Learning management systems 2006: Facts, practical analysis, trends and vendor profiles*. Oakland, CA: Bersin & Associates, Pike, G. H. (2009). The fences of a patent. *Information Today, 26*(2).

Piña, A. A. (2005). Course management systems: Introduction and implications for libraries. *Library Hi Tech News, 24*(5).

Piña, A. A. (2008a). How Institutionalized is Distance Learning? A Study of Institutional Role, Locale and Academic Level. *Online Journal of Distance Learning Administration, 11*(1).

Piña, A. A. (2008b). *Distance Learning and the Institution: Foundations, Importance and Implementation.* Saarbrücken, Germany: VDM Publishing.

Piña, A. A., & Eggers, M. R. (2006). *Teaching, administering and supporting blackboard, webct and desire2learn.* Paper presented at the annual Technology in Education (TechEd) Conference, Pasadena, CA.

Piña, A. A., Green, S., & Eggers, M. R. (2008). *Learning management systems: Lessons from the front lines.* Paper presented at the annual Technology in Education (TechEd) Conference, Ontario, CA.

Piña, A. A., & Heydenburg, P. R. (2008). *Learning management systems: Platform-independent course design.* Paper presented at the annual Supporting Learning and Technology in Education (SLATE) Conference, Chicago, IL.

Piña, A. A., Sadowski, K. P., Scheidenhelm, C. L., & Heydenburg, P. R. (2008). SLATE: A community of practice for supporting learning and technology in education. *International Journal of Instructional Technology and Distance Learning, 5*(7).

Rainie, L., & Madden, M. (2005). *Podcasting catches on.* Washington, DC: Pew Research Center.

Ranie, L., & Keeter, S. (2006). *How Americans use their cell phones.* Washington, DC: The Pew Research Center.

Rossett, A. (1991). Needs assessment. In Anglin, G. J. (Ed.), *Instructional technology*. Englewood, CO: Libraries Unlimited.

Schmidt, K. (2002). The web-enhanced classroom. *Journal of Information Technology*, *18*(2).

Siemens, G. (2004). *Learning management systems: The wrong place to start learning*. Retrieved May 4, 2009, from http://www.elearnspace.org/Articles/lms.htm

Simba Information. (2007). Pearson expands with eCollege. *Electronic Education Report, 14*(10).

Smart, K. A., & Meyer, K. A. (2003). Changing course management systems: Lessons learned. *EDUCAUSE Quarterly*, *28*(2).

Stewart, B. (2007). Why Athabasca chose moodle. *Distance Education Report, 11*(3).Changing Course Management Systems: Lessons Learned Changing Course Management Systems: Lessons LearnedChanging

Ullman, C., & Rabinowitz, M. (2004). Course management systems and the reinvention of instruction. *Technological Horizons in Education Journal*. Retrieved May 11, 2009, from http://www.thejournal.com/articles/17014

Valade, J. (2002). *PHP & MySQL for dummies*. New York: Wiley.

Watson, W. R., & Watson, S. L. (2007). An argument for clarity: What are learning management systems, what are they not, and what should they become? *TechTrends*, *51*(2).

Young, J. (2009, June 1). Colleges consider using blogs instead of blackboard. *Chronicle of Higher Education*.

# Chapter 2
# What is an E-Learning Platform?

**Michael Piotrowski**
*ZHAW Zurich University of Applied Sciences, Switzerland*

## ABSTRACT

*In this chapter, the author looks at an indispensable component of e-learning, namely e-learning platforms, i.e., the software that provides the technical infrastructure on which e-learning activities can take place. However, there is no universally accepted definition of what an e-learning platform actually is. The author analyzes existing definitions and note several problems with them. He then proposes a new definition, which abstracts from implementation details and instead relies on the idea that there are six activities that e-learning platforms must support. In addition, the author proposes a visualization technique based on this definition, which may be applied for comparing and evaluating e-learning platforms.*

## SOFTWARE FOR E-LEARNING

E-learning relies on technology: It requires hardware, software, and network infrastructure. Most e-learning environments today are *Web-based*, i.e., they are accessed via Web browsers (using HTTP) over a TCP/IP network such as the Internet or an intranet (e.g., a university campus network).

Thus, in general, e-learning today does not have any special hardware or networking requirements: In theory, only Internet access and a computer capable of running a Web browser is necessary to access Web-based e-learning applications. In practice, however, many applications make use of client-side scripting (using JavaScript, Adobe Flash, or Java Applets) or contain media or documents requiring proprietary software (such as Apple QuickTime or Microsoft Windows Media players for movies, or Microsoft PowerPoint for presentations). This means that a certain amount of computing power must be available and the choice of operating systems may be restricted.

An institution offering e-learning also only needs standard server hardware and Internet connectivity, both of which must, of course, be sized

DOI: 10.4018/978-1-61520-853-1.ch002

Copyright © 2010, IGI Global. Copying or distributing in print or electronic forms without written permission of IGI Global is prohibited.

according to demand, i.e., it depends on factors such as the number of students simultaneously using the system and the type and amount of media being served.

In principle, many types of software and network services can be used for e-learning; examples include e-mail, Usenet, chats, discussion forums, wikis, blogs, collaboration (CSCW) tools, simulation software, testing and assessment software, e-portfolios, vocabulary trainers, and games.

These applications can be used individually or in various combinations for e-learning; for example, Graziadei (1996) describes a "Virtual Instructional Classroom Environment in Science" from the early 1990s based on a variety of programs and services, including e-mail, VAX Notes[1], Questionmark Perception[2], and HyperCard[3].

However, the drawbacks of such a setup are obvious: The most serious are the lack of common user management and authentication, varying user interfaces, and limited interoperability between the tools. With the advent of the Web and the institutionalization of e-learning, Web-based e-learning platforms were created to provide a single, consistent user interface for all aspects of a course.

The functionality of e-learning platforms typically includes access to learning content and tests, communication and collaboration tools for students, and course management and assessment facilities for instructors. E-learning platforms may also include administrative functionality or interfaces to administrative systems (often called "campus management systems") for managing student admissions and enrollment (sometimes termed "student life cycle management"), for resource planning, accounting, etc.

Numerous e-learning platforms are available today. Some of today's most popular platforms are the commercial systems Blackboard, Clix, and Desire2Learn, and the open-source platforms ILIAS, Moodle, OLAT, and Sakai.[4]

The vast array of available e-learning platforms makes it difficult for institutions to select the platform that best suits their needs. Furthermore, there is no single, universally agreed upon definition of what an e-learning platform actually is. This poses a problem for both research and practice: Without a solid definition, there is no objective reference framework for describing, comparing, and evaluating systems. It is standard practice in science to first define the terms and their meaning before proceeding.

Unfortunately, as e-learning is still a relatively young field of research and, as Conole and Martin (2007) point out, "not as yet a rigorously defined area" (p. 12), it has not yet developed a conventional terminology. What we currently have are numerous terms, many of which are almost—but not quite—synonymous, and which are often only vaguely defined.

Very often, we see examples in place of definitions. While examples can be helpful for illustrating a concept, they are no substitutes for definitions, as they fail to abstract from the concrete. In the case of e-learning platforms, saying, for instance, that an e-learning platform is "something like Blackboard," implicitly compares every other platform to Blackboard, making this system the reference. In fact, Blackboard is merely an example, and it is dubious whether its features and functionality are actually those that define e-learning platforms. Working without an independent definition is especially dangerous when evaluating systems, e.g., as a basis for a decision on what system to deploy at an institution. When an evaluation is not based on sound criteria, there is a high risk that decisions are eventually made on the basis of unreasonable criteria, such as the absolute number of features or personal preferences. Before considering systems in detail, it is thus important to have, on the one hand, a clear idea of the actual requirements, and, on the other hand, a clear reference framework, so that requirements and systems can be compared in a meaningful way. When selecting an e-learning platform, the main questions are which kinds of activities are to be supported by the platform and how well different platforms support these activities.

In this chapter we will develop a definition for the term e-learning platform based on a review of previous attempts at definitions and of actual e-learning platforms. We will then describe how this definition can serve as a reference system for classifying and evaluating e-learning platforms.

## WHAT IS AN E-LEARNING PLATFORM?

Unfortunately, there is as much terminological disagreement about how to name software systems facilitating or supporting e-learning as there is about the term *e-learning* itself.

Consequently, there are many names for such systems, such as *learning management system* (LMS), *learning content management system* (LCMS), *course management system* (CMS), *virtual learning environment* (VLE), *managed learning environment* (MLE), *technology-enhanced learning environment* (TELE), or *learning support system* (LSS). Usage is not consistent: Depending on the author, the vendor, or local custom, some of these terms may describe different types of systems, or they may be used more or less interchangeably. For example, Holyfield (2003) notes that there are "some significantly different versions of what an MLE means, ranging from an all encompassing vision, to a more limited view that sees it as a sort of enhanced Virtual Learning Environment (VLE)" (p. 1). Stansfield and Conolly (2009), on the other hand, treat the terms as equivalent and state that a virtual learning environment "is also called: learning management system (LMS); course management system (CMS); learning content management system (LCMS); managed learning environment (MLE); learning support system (LSS); and learning platform (LP)" (p. 31).

We prefer the term *e-learning platform*; first, it shows the relationship to the activity it is supposed to facilitate or support, i.e., e-learning. Second, the term *platform* is generic enough not to suggest a certain implementation or structure, and, unlike *management system*, it does not put the focus on managing.

Before we look at today's e-learning platforms, we should first take a step back to get a larger perspective and briefly consider the history of e-learning platforms.

### Some Historical Notes

With the recent surge in e-learning-related activities at universities worldwide and the increasing media coverage, one may assume that e-learning platforms are a new phenomenon that appeared with the spread of the Web. In fact, today there is an implicit understanding that an e-learning platform has a Web user interface; this is consistent with the general assumption that e-learning *per se* is Web-based.

However, the ancestry of today's e-learning platforms can be traced back to earlier computer-assisted instruction (CAI) systems, which started to be developed in the 1960s. Early such systems, e.g., PLATO I (Bitzer et al., 1961), are certainly not e-learning platforms in the modern sense, but later mainframe-based systems pioneered a number of key concepts. Especially noteworthy are the later versions of PLATO, namely, PLATO III and PLATO IV.[5]

The PLATO system was originally developed at the University of Illinois; from 1976 until the late 1980s it was marketed commercially by Control Data Corporation. The PLATO project received funding from the U.S. National Science Foundation (NSF) and was formally evaluated by the Educational Testing Service (ETS) in 1978 (Murphy and Appeal, 1978).

PLATO ran on Control Data Cyber 70 series mainframes, capable of serving about 500 concurrent users (Hart, 1995, p. 18), and used highly innovative, microprocessor-based terminals with bitmapped plasma graphics displays, touch screens, and local processing capabilities; Figure 1 shows one such terminal, called the PLATO V terminal. While most terminals at the time were *dumb*, i.e., hard-wired to interpret only a limited number of simple control codes, without any programmability and local computing capabilities, the PLATO V terminal incorporated an Intel 8080 microprocessor (Stifle, 1978) and was thus able to execute programs downloaded from the host computer. This is essentially the same approach as taken by Web applications today: The bulk of the processing takes place on the server, but for some functions, the Web browser (which corresponds to a terminal in the mainframe architecture) downloads JavaScript code from the server and executes it locally.

The plasma display was also an invention of the group developing the PLATO system, the Computer-Based Education Research Laboratory (CERL) at the University of Illinois (Bitzer et al., 1971). For audio output, a synthesizer could be connected to the terminal. The PLATO synthesizer developed by Sherwin Gooch was one of the first computer-controlled music synthesizers (Gooch, 1980) and an innovation in itself. The synthesizer could be used both as a musical instrument and for speech synthesis. It could thus be employed for a wide variety of educational applications, such as melodic dictation in music education (Pembrook, 1986) or text to speech (TTS) applications in language teaching (Sherwood, 1981).

On the software side, PLATO introduced concepts such as online forums and message boards, online testing, e-mail, chat rooms, instant messaging, remote screen sharing, and multi-player online games.

While the origins of the PLATO project were in programmed instruction, PLATO III and PLATO IV were not dedicated to a single paradigm of instruction. The availability of the TUTOR programming language (Avner and Tenczar, 1969) certainly played an important role, as it allowed instructors to design their own lessons with relative ease. At the same time, it was effectively a general-purpose programming language, so it could be used to implement any approach considered adequate, or for writing software that was not directly learning-related, including (non-educational) games.

## E-Learning Platforms Today

Today, hundreds of e-learning platforms are available. In contrast to PLATO, today's e-learning platforms are based on standard technology: Practically all e-learning platforms today are Web applications, i.e., students and instructors interact with it through a Web browser. Most e-learning platforms rely on a standard relational database management system (RDBMS) for storing information; typical requirements are MySQL or PostgreSQL for open-source systems and Oracle or Microsoft SQL Server for commercial systems. For user authentication, most e-learning platforms can connect to directory services such as LDAP (Zeilenga, 2006) or Shibboleth (Morgan et al., 2004).

Current e-learning platforms are implemented in a wide variety of programming languages and use different system architectures. Many systems are written in PHP (e.g., ILIAS and Moodle), some in Java (e.g., OLAT and Sakai), others are based on Microsoft.NET (e.g., IntraLearn). Blackboard is written in a mix of Perl and Java.

With respect to licensing, e-learning platforms can be separated into open-source and closed-source systems; the former (including ILIAS, Moodle, OLAT, and Sakai) are usually developed by communities, released under the GNU Public License (GPL) or a similar license, and are available for free, whereas the latter are typically developed and distributed by commercial vendors. Vendors usually do not make the source code of their products available and use various license agreements; cost is often dependent on the number of users.

Universities are increasingly migrating toward open-source platforms. One important factor for

*Figure 1. PLATO V terminals featured plasma touch screens and local processing capabilities. (Photography contributed to Wikipedia by user Mtnman79; licensed under the creative commons attribution 3.0 license)*

this tendency are the license costs associated with commercial systems. After the acquisition of its competitors WebCT and ANGEL Learning, Blackboard, Inc. is reported to have a market share of over 80% (Carter, 2009). The acquisition has also further reduced the number of commercial choices for institutions and raised concerns with many institutions, which fear a monopoly position of Blackboard and vendor lock-in (Jaschik, 2009). Open-source software mitigates the risk of vendor lock-in and allows universities to adapt the software to their special needs without having to develop a complete system from scratch. Typical requirements are interfaces to campus management systems or alternative authentication solutions.

We will now briefly discuss and compare two current open-source e-learning platforms, Moodle and OLAT, which can be considered representative for the current state of the art.[6]

## Moodle

Moodle development was started in 1999 by Martin Dougiamas, who was working as a WebCT administrator at Curtin University of Technology, Perth, Australia, out of dissatisfaction with WebCT.[7] Version 1.0 was released in 2002. Moodle is licensed under the GNU Public License (GPL). Moodle uses a traditional architecture based on the PHP programming language, the Apache Web server, and the MySQL database management system.[8] Even though it is possible to program in an object-oriented style with PHP, the design of Moodle is purposely *not* object-oriented (despite the name); Dougiamas writes:

A large developer community is now working on Moodle. Moodle has a significant user base; as of July 2009, the Moodle Web site[9] speaks of over 39,800 "registered validated" sites. Even though it is clear that not all of them are actually in productive use, the number at least indicates an enormous interest in Moodle; equally impressive is the growth from about 15,000 registered sites in 2006.

From a user's perspective, most of the activities in Moodle take place inside courses, i.e., few functions can be used outside the context of a specific course. Courses (see Figure 2) themselves have a flat (i.e., non-hierarchical) structure. Course authors can only choose from three predefined course "formats," namely "Social" (oriented around one main forum), "Topics" (organized into topic sections, with each topic section consisting of activities), and "Weekly" (organized week by week, with start and end dates, and each week consisting of activities). *Activities* is the Moodle term for functions such as assignments, chats, forums, quizzes, wikis, etc.

## OLAT

Development of OLAT[10] also started in 1999, at the University of Zurich (Switzerland). The first version, developed by three student tutors[11] to help tutors cope with increasing numbers of students in computer science, was put into production in 2000.

Like Moodle, the system was written in PHP and based on a similar architecture. This version was awarded the Medida-Prix[12] in 2000. Other departments at the University of Zurich quickly picked up OLAT and the IT services started to offer central hosting.

With increasing numbers of users, the developers realized that the existing architecture did not offer the modularity needed to allow the clean implementation of requested features and that it would not scale as required. A complete redesign, based on the experience made with the first version, and a complete reimplementation in Java was thus undertaken, leading to version 3.0 in 2004, which was released as open source software under the Apache Open-Source License. Also in 2004, the University of Zurich decided to use OLAT as their "strategic learning management system."

The redesign separated the infrastructure from the e-learning "business logic": OLAT 3.0 and subsequent releases (currently version 6.x) are based on a component-based architecture, i.e., OLAT is actually an application consisting of several components on top of a general-purpose framework. Even though it is not advertised as such, the framework can be used to build other Web applications.[13]

Due to the experience with large numbers of concurrent users in the PHP-based version, the reimplementation also focused on scalability. From the end users' (i.e., the instructors' and students') point of view, OLAT differs from Moodle in many respects (see Figure 3 for a screenshot). OLAT provides instructors and students with numerous tools that can be used outside of courses, adding groupware capabilities (such as synchronous and asynchronous communication, file sharing, and instant messaging) to the e-learning platform.

Courses in OLAT can have arbitrary hierarchical structures and there are no prescribed structures; instead, instructors can define sequencing rules, if desired.

According to the Web site, there are about 150 OLAT installations running worldwide as of July 2009. The largest OLAT installation is located at the University of Zurich (maintained by the Multimedia & E-Learning Services group); this installation is also used by other Swiss universities, such as the University of Basel, the University of Bern, the University of Lucerne, and the Swiss Federal Institutes of Technology in Zurich and Lausanne. It currently (as of March 2009) hosts over 2,800 course and serves over 51,000 users, with about 400 to 1200 concurrent users on average.

OLAT development is centered at the University of Zurich; contributions from outside developers have to be approved and integrated by the development team in Zurich.

## Comparison

Moodle and OLAT are two e-learning platforms that represent the current state of the art. Both systems have many basic features in common: Instructors can create courses and fill them with content; there are facilities for user management, for handling assignments, and for assessment; there are functions for creating surveys and glossaries; and there are forums and chats available.

Much of this functionality was already available in PLATO—of course, with a quite different user interface. Obviously, PLATO differs from today's systems in a number of ways as it was developed in a quite different environment, both with respect to computing and to teaching.

On closer inspection, it becomes apparent that while Moodle and OLAT offer similar features, they implement them in very different ways, and they are based on different philosophies. For example, most activities in Moodle take place inside courses. OLAT, on the other hand, provides groupware-like functionality that can also be used outside of courses. Moodle requires instructors to choose from three predefined course structures. OLAT, on the other hand, allows instructors to define their own structures and sequencing rules (which can be saved as "course templates"). Moodle advertises that it is "designed using sound pedagogical principles,"[14] whereas OLAT emphasizes that it is pedagogically neutral.

To summarize: In this section we have briefly described a historical system, PLATO, and two current e-learning platforms, Moodle and OLAT. We have found that—despite the differences between the systems—all three systems essentially provide the same functionality. This is not to say that the systems have identical features, but rather that they all provide support for certain activities, such as delivery of learning content, discussions, assessment, or user management, even though the concrete realization of this support differs. Furthermore, although it is rarely acknowledged, we can, in many respects, consider PLATO a direct precursor of today's e-learning platforms.

## E-LEARNING PLATFORMS: TOWARDS A PRACTICAL DEFINITION

The goal of the brief historical excursus and the description of Moodle and OLAT in the preceding section was to put the concept of an e-learning platform into a broader scope and to help to abstract from the concrete systems with which we are familiar.

We will now try to find a practical definition for the term *e-learning platform*. To do so, we will first critically review a number of interesting definitions from the literature and discuss their positive and negative aspects. On this basis, we will then develop our own definition, which should be practically usable as a reference framework, e.g., for comparing and evaluating systems.

### Review of Previous Definitions

The first definition we would like to consider is that of Collis and Moonen (2001). These authors refer to Robson (1999) and define a *WWW-based course-management systems* as a

> *comprehensive software package that supports some or all aspects of course preparation, delivery and interaction, and allows these aspects to be accessible via a network. (Collis and Moonen, 2001, p. 78)*

The vagueness of this definition ("supports some or all aspects") already gives an indication that it is hard to give a precise definition of what an e-learning platform exactly is. Many similar definitions can be found in the literature, but while they are better than examples, we do not think that this type of definition is very helpful for *practical* use, as the features it lists are too generic. We thus have to look for other approaches.

What type of definition are we looking for? Despite the vagueness of the commonly given definitions, we note that, nevertheless, most researchers and practitioners are likely to agree on a set of *typical* e-learning platforms. Thus, there clearly is a class of software systems identifiable as "e-learning platforms," but it is obviously not defined by a "checklist" of necessary and sufficient features that a system must possess in order to be an instance of the class.

It is therefore necessary to rather take a "prototype view," i.e., e-learning platforms—as a concept—are characterized by a number of properties—not all of which are equally important—which actual systems fulfill to various degrees: Membership in the class of e-learning platforms is thus not binary, but gradual.

Robson (1999), to which Collis and Moonen refer, in fact lists five common features of e-learning platforms (Robson uses the term course-support system): Computer-mediated communication, navigational tools, course management, assess-

*Figure 2. Screenshot of the Moodle "features" demo course at http://moodle.org/ Early on I made the decision to avoid using a class-oriented design: Again, to keep it simple to understand for novices. Code reuse is instead achieved by libraries of clearly-named functions and consistent layout of script files. (http://docs.moodle.org/en/Moodle_architecture (accessed 2008-09-10))*

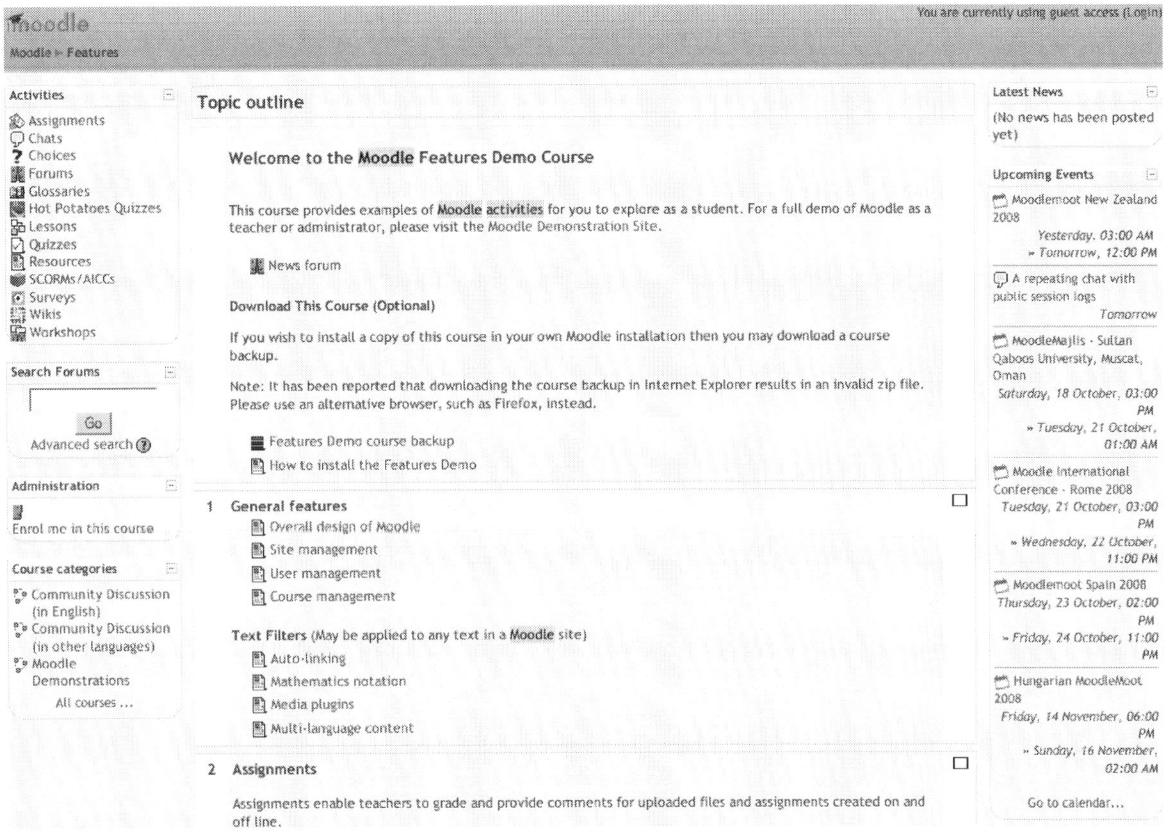


ment, and authoring tools. Robson also gives the following descriptions of these features:

- Computer-mediated communication. Examples are threaded discussion forums, email messaging systems, and chat rooms. Computer-mediated communication allows learners and instructors to communicate with each other publicly, privately, and (sometimes) in pre-set groups.
- Navigational tools. Navigation tools are used to organize a WWW course-support site into the equivalent of modules and lessons. The organizational structures tell students what to do next. Navigational tools also provide buttons that grant access to the various features of the course-support system and are generally part of a container that lends a unified look and feel to the HTML pages supporting a given course.
- Course management. In the classroom setting this means keeping track of students and their records. In the electronic environment it is also necessary to manage security and support a variety of user types such as instructors, students, and guests.
- Assessment. The most common form of WWW-based assessment is the on-line quiz which returns immediate, pre-determined feedback to the student. Feedback

*Figure 3. OLAT home page at the University of Zurich http://www.olat.uzh.ch/ (accessed 2008-10-17)*

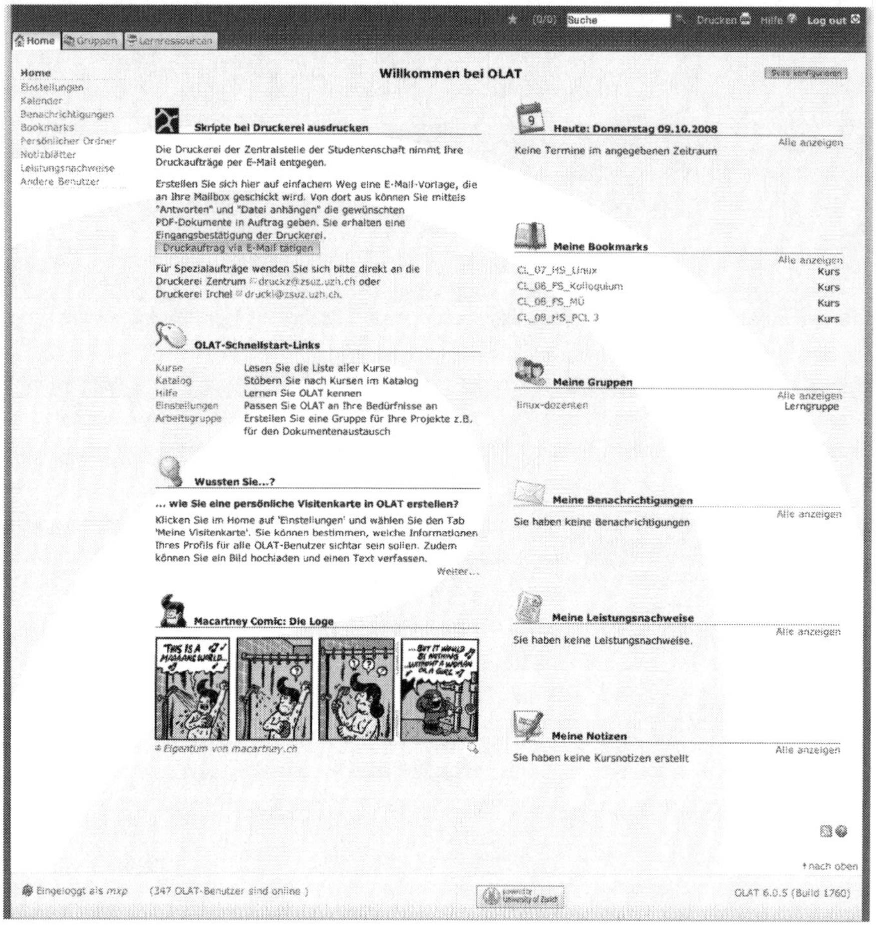


usually includes a score and written comments. Questions may be randomly selected and may take on a number of different forms: true/false, multiple choice, fill-in-the-blank, matching questions, essay questions (not usually machine-graded), and perhaps others. Tools for alternative forms of assessment, such as portfolio assessment, are rare.

- Authoring tools. Almost no course-support system offers a full authoring environment, but many allow instructors to upload and organize material, create discussions, create and edit on-line quizzes, and otherwise control the features offered by the environment. Increasingly, commercial products are attempting to integrate commercial authoring tools to bridge this gap.

While Robson primarily *describes* existing systems, his list goes beyond generic descriptions and abstracts from specific features. Another interesting definition is given by de Boer (2004), who describes an e-learning platform (de Boer uses the term *course-management system*) as an

> *integrated combination of Web-based tools specifically focused on the educational support of distributing content and enabling communication and organization and pedagogical support within courses. (de Boer, 2004, p. 23)*

To describe the differences between various systems, de Boer considers the design and development of e-learning platforms to be affected by four "lines of influence," namely: (1) Communication systems (e.g., e-mail and chats), (2) knowledge management systems, (3) computer-based training, and (4) computer-supported collaborative work (CSCW).

From these four sets of influences, de Boer derives four "dimensions"[15] and visualizes them as quadrants of a system of coordinates, where each "dimension" represents a type of activity: Communicating, organizing, delivering, and creating. Different e-learning platforms can then be thought of as being located at different points in this system and covering different areas.

This analysis already goes significantly beyond most definitions found in the literature, and provides a basic reference system that allows classifying e-learning platforms with respect to abstract criteria instead of by comparison to other implementations.

However, de Boer's diagrams are only rough sketches intended as an explanation of the general idea. A more significant shortcoming is that by concentrating on influences from non-e-learning applications, de Boer fails to account for the activities *specific* to e-learning; in particular, two important e-learning activities are missing, namely *collaboration* (which is also missing from Robson's list) and *assessment* (which Robson does list). While de Boer names CSCW systems as an "influence" on e-learning platforms, he does not consider collaboration as an activity on its own. However, we consider collaboration a very specific activity, and it is not covered by any of de Boer's four dimensions.

De Boer's dimensions for the most part match the features given by Robson (1999) (see above), and in fact he maps Robson's features to his four dimensions (see de Boer, 2004, p. 27). Here Robson's assessment feature is assigned to the dimensions "creation" and "content delivery." While assessment does involve creation and delivery of tests, it is much more than that and it is certainly an important educational activity. Therefore, de Boer's definition can be considered a refined version of Robson's, but it still has some significant shortcomings, as it does not cover collaboration and assessment properly.

We will therefore have a look at yet another definition. For research into the use of e-learning platforms, Malikowski et al. (2007) propose a model for describing and evaluating features. Their model consists of five categories:

1. Transmitting course content, i.e., delivery of content and course-related information, including grades,
2. evaluating students, i.e., assessment,
3. evaluating courses and instructors, i.e., surveys and questionnaires,
4. creating class discussions, i.e., communication functionality such as forums, and
5. creating computer based instruction, i.e., facilities for sequencing learning content.

Malikowski et al. do not attempt to give an abstract definition of what an e-learning platform is (they use the term *course management system*), but rather describe the features found in current systems (Blackboard, Desire2Learn, and WebCT are explicitly mentioned). The five categories of their model can be partially matched to Robson's features and de Boer's activities; Malikowski et al. specifically list the evaluation of courses and instructors, and category 5, "creating computer-based instruction," combines a number of functions from different items on Robson's list (primarily from navigational tools, assessment, and authoring tools). On the other hand, Malikowski et al. do not have a category covering student collaboration—we would argue that category 4, "creating class discussions," describes only a small portion of collaboration adequately.

Thus Malikowski et al.'s model can be considered as a further refinement, but, as we have seen, it also fails to account for some essential aspects of e-learning, in particular student collaboration.

We will now discuss one last definition, namely that given by the U.K. Department for Children, Schools and Families. The term learning platform is defined as follows:

*It is an umbrella term that describes a broad range of ICT systems used to deliver and support learning. As a minimum, we expect it to combine communication and collaboration tools, secure individual online working space, tools to enable teachers to manage and tailor content to user needs, pupil progress tracking and anytime/anywhere access. You might hear the term learning platform being applied to a virtual learning environment (VLE) or to the components of a managed learning environment (MLE). (United Kingdom. Department for Education and Skills, 2005, p. 18)*

This definition acknowledges that there are many different types of systems and it poses certain minimal requirements. Like the model of Malikowski et al., it lists relatively concrete functionality, including collaboration, which is missing in Malikowski et al.'s model. However, the requirements seem to be somewhat *ad hoc*; for example, the definition lists "pupil progress tracking," which may be thought to include assessment, but assessment is not specifically mentioned. Similarly, it is not clear what it means to tailor content to user needs—in particular, does it include the *creation* of content?

To summarize: In this section, we have briefly outlined and reviewed several definitions of e-learning platforms. We have shown that they all miss some important aspects that characterize e-learning. In the next section, we will thus build on the definitions presented in this section to define a comprehensive definition for e-learning platforms.

## A New Definition

We have noted in the previous section that each of the definitions we have reviewed misses some important aspect of e-learning. To arrive at a comprehensive definition, we suggest to abstract from concrete functionality and to instead start from the activities that characterize e-learning and that should be supported by e-learning platforms.

After analyzing numerous definitions (some of which we have discussed in the preceding section) and actual systems, we propose that there are *six activities*: *Creation*, *organization*, *delivery*, *communication*, *collaboration*, and *assessment*. Some of these activities correspond to activities mentioned in the definitions reviewed in the previous section, sometimes under different names.

One important difference is that we consider collaboration and assessment to be defining aspects of e-learning, which cannot be subsumed under other activities. For example, assessment involves the *creation* of questions, the *organization* into tests, the *delivery* of tests to students, and *communication* between instructors and students; however, scoring, grading, and the interpretation and further use of test results are distinctive aspects of assessment, which make it a unique activity.

We may thus describe the six activities as follows:

- **Creation** refers to the production of learning and teaching materials by instructors.
- **Organization** refers to the arrangement of the materials for educational purposes (e.g., combining them into modules or courses).
- **Delivery** refers to the publication and presentation of the materials, so that they can be accessed by students.
- **Communication** refers to the computer-mediated communication between students and instructors and among students.
- **Collaboration** refers to students jointly working on files or projects; it also includes collaboration between instructors.
- **Assessment** refers to the formative and summative evaluation of learning progress and outcomes, including feedback.

It is important to note that these are *activities*, not functions, and that e-learning platforms are characterized by supporting these activities. This support may be implemented in different ways and by different functions. Furthermore, as we have noted above, actual e-learning platforms support these activities to varying degrees. However, to be classifiable as an e-learning platform, a system will need to support all of these activities to some degree, while other types of systems are likely to support only few of these activities.

We have now gathered all necessary elements, and building on our preceding analysis, we thus arrive at the following definition:

- **Definition:** An e-learning platform is a system which provides integrated support for the six activities—creation, organization, delivery, communication, collaboration, and assessment—in an educational context.

The term *integrated* here means that, regardless of the actual implementation, an e-learning platform must *function* as one system, and it must appear as a single system to the users. Thus, if it consists of separate subsystems, end users should not be aware of the individual subsystems; in any case, data exchange between the constituent elements of the e-learning platform must be fully automatic. Furthermore, an e-learning platform must offer more than one service and provide most of the facilities required for e-learning.

For example, an application offering *only* multiple-choice tests (such as Respondus or Hot Potatoes) is, according to our definition, not an e-learning platform. Neither is a loose collection of tools an e-learning platform in the sense of our definition. For example, the ELBA E-Learning Toolbox[16] is a set of e-learning tools offered to the teaching staff at ETH Zurich and the University of Zurich. The Toolbox includes, among others, the above-mentioned Hot Potatoes tool for quizzes, Phorum as a discussion board, and VTSurvey for Web-based surveys. Instructors using tools from the ELBA E-Learning Toolbox have a single administrative point of contact (see Sengstag and Miller, 2005), but the constituent tools are in no way technically integrated: Every tool has its own user management and no data is shared.[17]

The six activities and systems' support for each of them can be visualized by a *radar plot*, as shown in Figure 4: The six axes represent the six activities that e-learning platforms should support. The radial coordinates should be read as "no support" at the center and "extensive support" at the outermost circle. When graphing actual systems, each system is characterized by a distinctive shape that depends on the level of support for each activity.

For illustration, Figure 4 shows approximate graphs for OLAT (an e-learning platform, described above), BSCW (a system for computer-supported collaborative work not specifically oriented towards e-learning) (Appelt, 1999), and Respondus (an assessment tool). As can be seen from the graphs, the e-learning platform offers medium to high support for all of the six e-learning activities, while the other systems only provide support for some of them.

In this section we have presented a new definition for the term *e-learning platform*. This definition is based on an analysis of other definitions on the one hand, and, on the other hand, on the systematic abstraction from concrete functionality to six activities which characterize e-learning and which an e-learning platform should support. We have also presented a visualization based on this definition, which interprets the six activities as dimensions of a coordinate system; this visualization allows graphically rendering the level of support a system offers for each of the six activities.

Thus, our definition differs from others in that it is (1) based on abstract activities rather than concrete functions, and (2) offers a tool for graphically representing, comparing, and evaluating e-learning platforms in a common

*Figure 4. Radar plots can visualize the level of support systems provide for different activities*

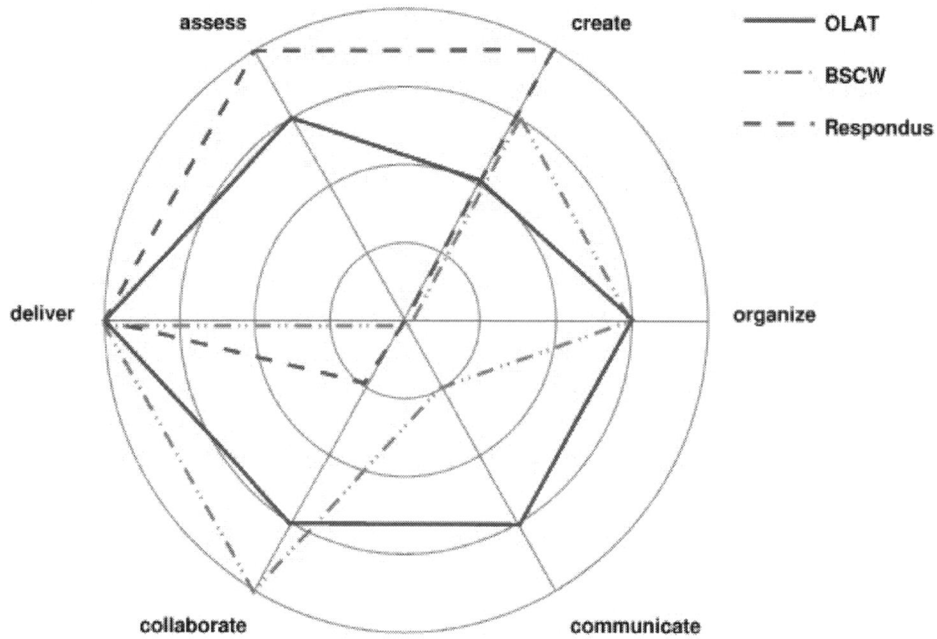


reference system independent from some concrete implementation.

The next section is an outlook to applications—in particular in the evaluation of e-learning platforms—and future research.

## OUTLOOK AND FUTURE RESEARCH DIRECTIONS

There are hundreds of e-learning platforms today, both closed-source commercial products and open-source systems. The report of Baumgartner et al. (2002) reflects the situation in 2002 but still gives a good overview of the number and diversity of systems available. Several academic institutions have also published their evaluation reports (e.g., Monnard and Brugger, 2003; Catalyst IT Ltd., 2004; Di Domenico et al., 2005).

When reviewing e-learning platforms, there are, as for any software, three main views to consider: (1) The view of the end users (i.e., the students and instructors), (2) the administrator's view, and (3) the implementation view.

End users of e-learning platforms require support for the six activities listed in the definition in the preceding section: Creation, organization, delivery, communication, collaboration, and assessment. While details differ, all of the major e-learning platforms support each of these activities to some degree. E-learning platforms thus can no longer be evaluated for their appropriateness in a certain environment by simply comparing feature lists. As Monnard and Brugger note, "every product does have its specific strengths, where it outdoes most of its competitors, whereas it is lacking in other aspects," so that direct comparison is "nearly impossible." (Monnard and Brugger, 2003, p. 2)

Our definition presented in the previous section offers an alternative approach to the evaluation of e-learning platforms. The six activities on which the definition is based could also help evaluators to discover and to organize their requirements. Once the specific requirements for each area of activity have been fixed, e-learning platforms can systematically be analyzed to determine the level of support they offer for each activity. The results can be visualized in a radar plot (see Figure 4), which helps to discuss and communicate the results and to motivate the final decision.

The graphs in Figure 4 are approximations for illustration purposes based on a general assessment of the systems. Obviously, when using this approach for an actual evaluation, the coordinates must be based on objective criteria. These criteria, however, can only be determined in the context of that evaluation, since they depend on the concrete requirements of the future users, i.e., the institution and the individual instructors and students. The evaluation criteria thus have to differ from one evaluation to another. For example, the possibility to include mathematical formula in discussions may be an essential feature for some institutions and thus influence the rating for the Communication activity. For another institution, such a feature may be merely "nice to have," or it may be considered completely irrelevant—and support for threaded discussions may be seen as a more important feature. Previous evaluations should of course be taken into account when setting up evaluation criteria, but there is no set of universal, objective evaluation criteria for e-learning platforms, which one could simply apply.

Thus, as we have noted above, the six activities we have proposed in this chapter can be used as a framework that can help to structure and systematize requirements and criteria—but this framework will have to be filled with concrete criteria for concrete evaluations.

## CONCLUSION

E-learning research does not yet have an established, universally accepted terminology. Thus, for many key concepts there are numerous different terms, which are also used with different meanings and connotations by different authors. In addition, many of these terms are only vaguely defined. However, objective, reliable research requires solid definitions for its basic concepts and must clearly state the assumptions on which it relies. One problematic concept is the software that facilitates e-learning. We call this the *e-learning platform*. In this chapter, we have reviewed a number of previous insufficient definitions and we have devised a new definition, which relies on the idea that there are *six key activities* in e-learning (creation, organization, delivery, communication, collaboration, and assessment) that an e-learning platform must support. We hope that our definition helps to improve communication about and evaluation of e-learning platforms, and that it thus helps to advance the state of the art in e-learning research.

## REFERENCES

Anklam, P. (2001). The Camelot of collaboration: The case of VAX Notes. *Knowledge Management Magazine, 5*(2).

Appelt, W. (1999). WWW based collaboration with the BSCW system. In *SOFSEM '99: Proceedings of the 26th Conference on Current Trends in Theory and Practice of Informatics* (LNCS 1725, pp. 66–78). Berlin: Springer.

Avner, R. A., & Tenczar, P. (1969). *The TUTOR manual*. CERL Report X-4, University of Illinois, Urbana, IL.

Baumgartner, P., Häfele, H., & Maier-Häfele, K. (2002). *E-Learning Praxishandbuch: Auswahl von Lernplattformen. Marktübersicht – Funktionen – Fachbegriffe*. Innsbruck, Austria: StudienVerlag.

Bitzer, D. L., Braunfeld, P. G., & Lichtenberger, W. W. (1961). PLATO: An automatic teaching device. *IRE Transactions on Education, 4*, 157–161. doi:10.1109/TE.1961.4322215

Bitzer, D. L., Slottow, H. G., & Willson, R. H. (1971). *A gaseous discharge display and memory mechanism*. United States Patent 3,559,190. Filing Date: December 22, 1966.

Bizoňová, Z., & Ranc, D. (2007). Model driven LMS platform integration. In *The Third Advanced International Conference on Telecommunications, 2007 (AICT'07)* (pp. 25–29). Washington, DC: IEEE Computer Society.

Carter, D. (2009, May 7). Blackboard–Angel deal gives pause to some. *eSchool News*. Retrieved July 26, 2009, from http://www.eschoolnews.com/news/top-news/?I=58675

Catalyst, I. T. Ltd. (2004). Technical evaluation of selected learning management systems. Technical report, The Open Polytechnic of New Zealand, Wellington, New Zealand.

Collis, B., & Moonen, J. (2001). *Flexible Learning in a Digital World: Experiences and Expectations. Open & Distance Learning Series*. London: Kogan Page.

Conole, G., & Oliver, M. (2007). Introduction. In Conole, G., & Oliver, M. (Eds.), *Contemporary Perspectives in E-Learning Research: Themes, Methods and Impact on Practice* (pp. 3–20). London: Routledge.

De Boer, W. (2004). *Flexibility support for a changing university*. PhD thesis, University of Twente, Enschede, The Netherlands.

Di Domenico, F., Panizzi, E., Sterbini, A., & Temperini, M. (2005). *Analysis of commercial and experimental e-learning systems. Technical report*. Trondheim, Norway: TISIP Research Foundation.

Gnägi, F. (2006). *OLAT–Moodle Comparison*. Retrieved July 31, 2009, from http://www.frentix.com/de/olat/docu/OLAT-MOODLE-en.pdf

Gooch, S. J. (1980). *Cybernetic music system*. United States Patent 4,206,675. Filing Date: February 28, 1977.

Graziadei, W. D. (1996). VICE in REST. In Harrison, T. M., & Stephen, T. D. (Eds.), *Computer Networking and Scholarly Communication in Twenty-First-Century University* (pp. 257–276). Albany, NY: SUNY Press.

Hart, R. S. (1995). The Illinois PLATO Foreign Languages Project. *CALICO Journal*, *12*(4), 15–37.

Holyfield, S. (2003). *Developing a Shared Understanding of the Managed Learning Environment (MLE) – the role of diagramming and requirements gathering*. Commissioned JISC report. Retrieved July 28, 2009, from http://www.jisc.ac.uk/uploaded_documents/ACF450.pdf

Jaschik, S. (2009, May 7). Blackboard buys Angel. *Inside Higher Ed*. Retrieved July 26, 2009, from http://www.insidehighered.com/news/2009/05/07/bb

Malikowski, S. R., Thompson, M. E., & Theis, J. G. (2007). A model for research into course management systems: Bridging technology and learning theory. *Journal of Educational Computing Research*, *36*(2), 149–173. doi:10.2190/1002-1T50-27G2-H3V7

Monnard, J., & Brugger, R. (2003). Web based course platforms: Evaluation report. Technical report, Edutech mandate of the Swiss Virtual Campus program, Freiburg, Switzerland.

Morgan, R. L., Cantor, S., Carmody, S., Hoehn, W., & Klingenstein, K. (2004). Federated security: The Shibboleth approach. *EDUCAUSE Quarterly*, *27*(4), 12–17.

Murphy, R. T., & Appeal, L. R. (1978). Evaluation of the PLATO IV computer-based education system in the community college. *SIGCUE Outlook*, *12*(1), 12–28. doi:10.1145/963847.963849

Pembrook, R. G. (1986). Some implications of students' attitudes toward a computer-based melodic dictation program. *Journal of Research in Music Education, 34*(2), 121–133. doi:10.2307/3344740

Robson, R. (1999). WWW-based course-support systems: The first generation. *International Journal of Educational Telecommunications, 5*(4), 271–282.

Sengstag, C., & Miller, D. (2005). Von der klassischen Vorlesung zur Bologna-kompatiblen Lehrveranstaltung: Redesign einer Lehrveranstaltung. *Zeitschrift für Hochschuldidaktik, 2005*(4), 63–74.

Sherwood, B. (1981). Speech synthesis applied to language teaching. *Studies in Language Learning, 3*(1), 171–181.

Stansfield, M., & Connolly, T. (2009). *Institutional Transformation through Best Practices in Virtual Campus Development: Advancing E-Learning Policies*. Hershey, PA: IGI.

Stifle, J. E. (1978). *The PLATO V terminal*. CERL Report X-50, University of Illinois, Urbana, IL.

United Kingdom Department for Education and Skills. (2005). *Learning platforms (Secondary)*. Ref. 2102-2005DBW-E.

Zeilenga, K. D. (2006). *Lightweight Directory Access Protocol (LDAP). Technical Specification Road Map. RFC 4510*. Internet Engineering Task Force.

## ENDNOTES

1 VAX Notes, later called DECnotes, was an early computer conferencing system—inspired by the Notes system of PLATO (see below)—developed in the 1980s at Digital Equipment Corp. (see Anklam, 2001).

2 http://questionmark.com/ (accessed 2009-07-31)

3 HyperCard, first released in 1987, was a hypermedia programming environment by Apple Computer, Inc. It was one of the first widely used hypertext systems before the Web.

4 Web sites of the mentioned systems: Blackboard: http://blackboard.com/, Clix: http://www.im-c.com/, Desire2Learn: http://desire2learn.com/, ILIAS: http://ilias.de/, Moodle: http://moodle.org/, OLAT: http://olat.org/, Sakai: http://sakaiproject.org/.

5 PLATO stands for "Programmed Logic for Automatic Teaching Operations."

6 The information used in the following discussion of Moodle and OLAT comes from a number of online sources, primarily the project Web sites http://moodle.org/ and http://www.olat.org/ (accessed 2009-07-31), and the documentation of the systems. Further sources are Gnägi (2006), Bizoňová and Ranc (2007), and the author's personal experience with OLAT at the University of Zurich and Moodle at the Zurich University of Applied Sciences.

7 *Moodle* stands for "Modular Object-Oriented Dynamic Learning Environment."

8 Other Web servers and DBMSs are also supported, but the combination of PHP, Apache, and MySQL—typically on a Linux platform—is the most common setup (often referred to as "the LAMP architecture").

9 See http://moodle.org/stats (accessed 2009-07-31).

10 OLAT stands for "Online Learning And Training."

11 Florian Gnägi, Sabina Jeger, and Franziska Schneider.

12 The Medida-Prix is a trinational (Germany, Austria, Switzerland) competitive award (endowed with €100,000 of prize money total) for e-learning projects organized by the Gesellschaft für Medien in der Wissenschaft (GMW).

[13] The framework is actually used for other applications.

[14] http://docs.moodle.org/en/Philosophy (accessed 2009-07-31)

[15] Unfortunately, de Boer's terminology is somewhat inconsistent: He variously refers to the "dimensions" as "structure," "characteristics," or "lines of influence."

[16] http://www.elba.ethz.ch/index_EN (accessed 2009-07-31)

[17] Web sites of the systems mentioned in this paragraph: Respondus: http://respondus.com/, Hot Potatoes: http://hotpot.uvic.ca/, Phorum: http://phorum.org/, VTSurvey: http://vtsurvey.sourceforge.net/.

# Chapter 3
# Security and Privacy Management for Learning Management Systems

**Wolfgang Hommel**
*Leibniz Supercomputing Centre, Germany*

## ABSTRACT

*Once a prototype for a Learning Management System (LMS) has successfully been set up and tested, the demand for putting it into production use rises. However, seamlessly integrating an LMS into existing data center infrastructures is a challenging task whose complexity is often underestimated. In this chapter we take a risk-driven approach ("what could go wrong?") to discuss the real-world operation of a fully-featured LMS from the perspective of security and privacy management. First, the authors analyze their LMS-specific security goals and the related threats to LMS components. They then investigate how an LMS security policy should be established and which technical controls can be used for implementation, enforcement, and auditing by the LMS administrators as well as by the system and network administrators. Finally, the authors discuss the benefits of inter-organizational LMS usage when it is based on identity federation technologies, and the new security and privacy challenges it brings.*

## MOTIVATION

Projects to establish E-Learning typically start with a strong focus on content and suitable didactical methods for the delivery of the carefully crafted learning material. This obviously is a commendable approach because it prioritizes both the learners' and the instructors' requirements, and thus results in a system that addresses the users' needs. However, it also often leads to an evolutionary growth of the Learning Management System (LMS) infrastructure. For example, additional web servers are often not considered before a downtime caused by hardware failure leads to a first wave of user complaints. Once we have achieved hardware redundancy, we might figure out that the new stand-by machine is not used during usage peaks, and so we rather need a load balancing solution – which, of course, must not only span the web servers, but also the backend, e.g. the streaming and the database servers. In a

DOI: 10.4018/978-1-61520-853-1.ch003

Copyright © 2010, IGI Global. Copying or distributing in print or electronic forms without written permission of IGI Global is prohibited.

nutshell, building an LMS prototype, which lets us concentrate on the software and our E-Learning project goals, and operating a fully-featured LMS in a production environment are completely different tasks.

This chapter deals with two specific aspects that we need to keep in mind from the very beginning when we plan to put our LMS into production use and subsequently enter the operation phase: security and privacy. Security is at stake because an enhanced and feature-rich LMS is a complex distributed system and – as we will see – a lucrative target for various types of attackers. Privacy protection must not be neglected because various LMS functionalities, e.g. personalization and authorization based on information about the learner's study course and study progress, depend on sensitive personally identifiable information (PII).

Security and privacy have many facets, and the regulatory requirements – such as country- and state-specific data protection laws – as well as the measures that are necessary to achieve high user acceptance are, to a large degree, specific to each individual scenario. Thus, we cannot present an one-fits-all solution to LMS security and privacy here. Instead, we discuss successful best practices that provide guidance for the security and privacy aware implementation, deployment, and operation of real-world LMS infrastructures.

We will initially investigate which technical components an LMS is typically composed of. This is a prerequisite for the following discussion of LMS security risks, based on the attack surfaces of these components and the various LMS-specific threats, including a broad range of potential attackers, their motivation and sophistication, and their means. We will then see that security and privacy are twofold: On the one hand, there are measures required on the management level, such as specifying an LMS security policy. On the other hand, technical controls must be applied to prevent, detect, and react to attempted or successful attacks. We will discuss both types of measures under the assumption that the LMS is not a green field experiment, but shall rather be seamlessly integrated into an existing information and communication technology (ICT) infrastructure, e.g. the one provided by a university's computing centre or a company's IT department.

Furthermore, we discuss the upcoming inter-organizational usage of LMS infrastructures. Modern technologies, such as Federated Identity Management, allow us to selectively grant LMS access to external users, e.g. other universities' students, without requiring a manual or self-registration-based creation of local LMS accounts. We discuss the benefits of this technology for higher education institutions as well as for 3rd party LMS hosting services and learning content providers. After an overview of the technical aspects, we will discuss the new security and privacy challenges such scenarios introduce and show the tools that can be used to deal with them successfully. Finally, we will give an outlook to security and privacy topics LMS researchers and practitioners will have to address in the next few years.

## SECURITY AND PRIVACY GOALS IN LMS INFRASTRUCTURES

If there was only one thing we should keep in mind about security and privacy, then it would be that neither of them is a feature that can easily be added to an already existing system at runtime. Instead, security and privacy must be integrated into the overall design from the very beginning in order to work properly and be efficient: Neither implementation nor configuration tricks can make up for a bad architectural design. Getting the design right requires that we know what we have to protect, and which threats we actually want to protect the LMS from. In other words, we need to know our assets and the risks to our LMS infrastructure. These risks can then be dealt with by either eliminating, reducing, delegating, or simply living with them. While eliminating all risks may be the most obvious solution, we have to adjust to two constraints in practice: There is no such thing as 100% security, and our budget is limited.

In this section, we will first identify our assets, i.e. the technical components that our LMS infrastructure is typically composed of, and the valuable information that is stored therein, e.g. learning objects and digital user identities. Then, we will discuss the goals of computer security and of information security management, and analyze how they specifically need to be applied to LMS deployments. A discussion of LMS specific attackers, threats, and the impact of successful attacks is the foundation for the security implementation strategy laid out in the next section.

## LMS Assets: Infrastructure Components and Valuable Information

There are various commercial and open source LMS products available to choose from, and still quite a few E-Learning projects prefer to build their very own LMS from open source components and own programming efforts to glue them together. Thus, there is not *the one* software product whose security properties we could discuss. However, we generally can assume that our LMS is not a large single, monolithic software application, but rather consists of several components. At the very practical minimum, the LMS architecture will follow the classical 3-tier concept and thus include the following components, which are also shown in figure 1:

- A graphical user interface as *frontend*. This can either be dedicated client software that needs to be installed on each user's machine ("fat client" model), or – more widely adopted – a web (HTTP/HTTPS) server product like Apache that delivers the content to the user's local web browser ("thin client" model). The frontend provides access to rich functionality at least for learners as well as instructors, and often is also the primary administration interface. For example, it will provide an overview of the courses that are available to the learners with an option to sign up for participation; it is also the primary user interface to conduct the courses and let the learners retrieve the learning material. It may also be used for assessments, such as electronic tests, and provide community features such as a dedicated discussion bulletin for the students enrolled in each of the courses. For authors, instructors, and tutors it is the primary tool to publish their content, schedule their courses, and to support their course participants. For LMS and system administrators, it will offer management functionality, e.g. to assign roles and permissions to users, and to fine-tune the LMS functionality and design.
- The *business logic layer*, which actually performs the operations that are triggered by the LMS users in the frontend. This piece of software is typically run by an application server or a Java servlet container such as Tomcat or Glassfish. From a technology perspective, this is the heart of the LMS and unique to each LMS product. Security-wise it is the central component that must provide application-level security functionality, such as user, role, and privilege management.
- The *persistency layer*, in which all the LMS-related data is stored. It is often based on a relational database management system, such as MySQL. Along with the information assets we discuss below, it typically also stores all the LMS metadata, e.g. the LMS configuration.

For very small installations, these three components can be deployed and run on the same physical machine, but due to the demand for storage

*Figure 1. Components of a typical learning management system infrastructure*

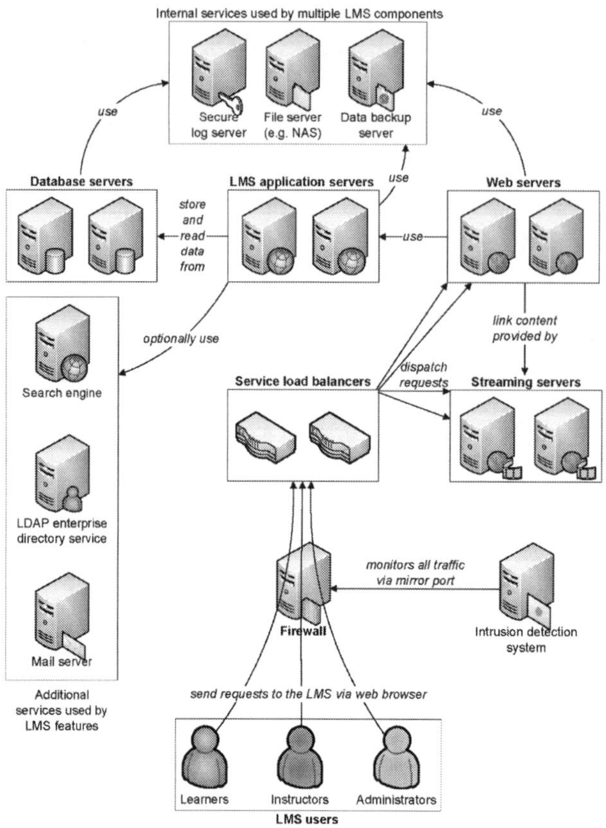


capacity and processing power, usually separate machines are used. Even for this minimum setup we must be aware that also the operating systems on which these components are installed need to be considered security-wise, e.g. regarding software updates for the OS itself.

Larger installations will typically make use of several additional components, some of which may already be available when an existing data center infrastructure is used:

- A *streaming server* is used to deliver multimedia content to the client. Unlike the download of full movies, streams can be viewed while the download is in progress, and the user can skip content she is not interested in without wasting the network bandwidth for it. Various file formats and streaming server applications are available, and picking a common file format for all learning material is one of the major design decisions of the E-Learning project that must be made quite early.
- A *mail server* is required to notify the LMS users about events by sending them an email, e.g. about the availability of new learning material in a certain course the learner has taken. Instant messaging, such as short text messages to the user's mobile phone, or groupware features, such as pop-up messages on the client PCs, are sometimes also used for this purpose, but are not yet in such wide-spread use.
- A dedicated *search engine* can be used when larger sets of learning objects are available, or when advanced semantic

search functionality is required that is not offered by the core LMS product. For example, courses in the field of medicine may be associated with a larger numbers of hospital patient photos that need to be searched using algorithms based on image recognition.

- An external user database, often referred to as identity repository or *LDAP enterprise directory service*, may be available. It can be used for authentication, so LMS users do not need separate passwords for their LMS account, and allows the import of basic user data, such as the learner's name and email address. However, it usually is neither intended nor usable as a replacement for the LMS user database.

Additional components can also be required to improve the availability and the service quality of the LMS infrastructure. For example, multiple web, application, and database servers can be used to spread the load evenly across multiple physical machines and to keep the LMS online when a single hardware component, e.g. one web server, is failing. The load distribution is performed by *service load balancers*, which – for obvious reasons – need to be highly available themselves. Central *file servers*, such as network attached storage (NAS) filers, are used to make the same files available to multiple machines, and a common *backup* solution, such as a tape archive, can further be used to prevent data loss. A central *log server*, which aggregates event messages from all the involved machines, not only simplifies the debugging of the infrastructure, e.g. when a user needs assistance regarding a technical problem, but is also essential for security-specific auditing as we discuss below. Given this variety, the scope and size of fully-featured real-world LMS infrastructures can quickly exceed the handful of server machines someone can manage as a part-time job, and we did not even discuss all the important security specifics yet.

Besides our hardware and the software that it is running, information is our primary asset that we must protect. We basically have to distinguish between three types of data:

- *User data*: As we already discussed above, data records about our learners contain a lot of highly sensitive PII; similar to a company's customer database, our LMS user database has a high counter value, and thus leaks, such as the disclosure of a list of our users' email addresses, must be avoided. Furthermore, for example course instructors and LMS administrators possess so-called *privileged* accounts, since they may use functionality that the majority of users may not, and so they should not become the weakest link in our security chain. User data also includes roles and entitlement assignments, and so we must be aware of mutually exclusive roles; for example, an instructor should not be allowed to participate in his own courses as a learner to avoid that he gets credit points from the electronic tests created by him.
- *Learning objects*: The creation of dedicated learning material is hard work, and intellectual property rights apply. This is especially important when 3rd party material is licensed from external content providers. While this is obvious in enterprise scenarios where learning material is purchased for on-the-job training, it has also been a major issue at universities, e.g. in their medicine faculties; access control must be established to ensure that only students enrolled in appropriate study courses may retrieve this data.
- *Metadata*: For simplicity, we assume that everything else can be called LMS metadata. For example, the LMS configuration itself, which may include customization efforts to match the organization's corporate design, falls into this category and must

obviously be protected from unauthorized modifications.

Since all these assets are of value to our organization, we must develop a strategy to avoid their loss or damage to them. This is the primary motivation for security on both the technical and the organizational level, which we discuss in the following section.

## Computer Security and Information Security Management Goals

Information security as a whole is too big a subject to be discussed in depth here; we thus only recapitulate the basics that everyone involved in an LMS infrastructure should be aware of, and substantiate the covered topics in the next sections.

On the technical level, computer security is often characterized by the acronym CIA, which stands for confidentiality, integrity, and availability. These are the very basic goals we address by applying security controls, i.e. tools and methodologies like network firewalls and data encryption. *Confidentiality* means that the information we protect must only be accessed by authorized parties; for example, leaking a user's complete record to the internet would typically be considered a breach of confidentiality. *Integrity* is provided when protected information cannot be modified unnoticed by unauthorized parties; for example, if learners can manipulate their own examination results, the LMS integrity would be violated. Service and data *availability* is a security goal due to the consequences their unavailability would have; for example, there could be legal consequences if an online exam cannot be performed because an attacker is sabotaging the LMS.

To be able to achieve these basic goals, a couple of security services must be established. For example, reliable user *authentication* is required as a basis for effective *authorization*, which in turn is a prerequisite to ensure the data's confidentiality and integrity. The security property of *non-repudiation* ensures that we can prove that someone has performed a certain action; for example, a learner must not be able to successfully deny his participation after failing a test. The scope of *access control* is the specification of authentication mechanisms, such as passwords or smartcards, and authorization rules, as well as the run-time monitoring of the LMS usage, e.g. by evaluating log files and checking them for suspicious activities.

*Privacy* is closely related to the confidentiality of the users' PII. At the very minimum, users must give their informed consent to the use of their personal data, and this data must only be used for purposes the users have agreed to. The amount of PII and the purposes it is used for will typically be specified in a privacy policy or online privacy statement, in which also the data retention time should be defined. On the one hand, a user record could be deleted from the LMS database once the user has finished all of her courses and the results have been sent to the university's examination office; on the other hand, new paradigms such as life-long online learning and dedicated courses targeted at alums are a motivation for longer retention times. From a privacy-enhancing perspective, the user should be put in control and have the option to decide about the removal of her PII by herself.

Privacy is also an issue for LMS that can be used anonymously, in which case the tracking of users, e.g. based on their IP address or by using HTTP cookies, may violate regional laws. We exclude such LMS from our further discussion and refer interested readers to related work, such as (Borcea, Donker, Franz, Pfitzmann, & Wahrig, 2006). However, certain features of LMS infrastructures may benefit from anonymity, such as when questions to the instructors can be asked anonymously, since some users might be too shy to ask otherwise; more detailed concepts are discussed in (Anwar, Greer, & Books, 2006).

Not each of these technical security goals will be equally important in each environment. Thus, requirements need to be elicited early in the project in order to decide about priorities; this especially affects how conflicting goals are addressed, such as non-repudiation vs. anonymity. However, information security needs also to be dealt with on the management level. Standards, such as ISO/IEC 27001:2005 (ISO/IEC, 2005), describe the organizational measures to plan, build, operate, and continuously improve information security within an organization. Among various others, it addresses the following issues:

- A *security policy* needs to be established, which has to provide management directives and must reflect the organization's requirements and the laws as well as regulations that apply. It also demonstrates the upper management's commitment to information security and privacy.
- There needs to be an *internal organization* of information security, including clearly assigned responsibilities, confidentiality agreements, and security-specific infrastructure reviews. Furthermore, measures to improve the staff's information security awareness, such as regular trainings, should be specified.
- An information security *incident management* process needs to be set up. For example, a workflow for countermeasures during ongoing denial-of-service attacks should be defined, and the steps to be taken after e.g. the user database has been compromised should be prepared in advance, i.e. before the catastrophic event has occurred. While it may be hard to specify the reactions to all the possible events in a sound and complete manner, this approach provides additional insight about the impact of attacks, which in turn helps us to prioritize the technical controls that are required to prevent them.

Furthermore, information security management also includes compliance management, i.e. it must specify measures to ensure that, for example, privacy regulations and intellectual property rights are addressed. While in sum there is a large number of information security aspects that need to be considered, LMS security management usually does not start from scratch; instead, the LMS management should encourage the adoption of locally applied best practices and strive for the integration into the existing security management frameworks.

## LMS Risk Analysis: Attackers, Threats, and Impacts

Knowing about potential attackers is the first step towards selecting security controls on the technical level, and an absolutely essential requirement for this goal. For example, how should we know whether we need a firewall and how we should configure it, if we do not know which attacks it shall protect us from? An *attacker model* discusses the known types of attackers, their motivation, means, skills, and the position from which they could launch an attack. For an LMS consisting of the components discussed above, basic attacker models include:

- *LMS administrators*: Since administrators have full access to all the data stored in the LMS, they can easily manipulate the LMS configuration and invade the users' privacy by snooping around the user database or intercepting passwords. While this type of misuse should be prohibited and penalized by means of the security policy, there is the risk of a compromise of an administrator's account, or of a disgruntled administrator wrecking havoc. These dangers must be addressed by a *privileged account management* strategy, which typically includes strong authentication, extended auditing and the enforced application of the four-eyes-principle for certain management operations.

- *LMS users*: Rogue users will try to circumvent the access control mechanisms that are in place. Their motivation is usually vandalism, the need for recognition, some sort of playing instinct, or the prospect of personal advantage. As discussed by (Graf, 2002), the readiness of a regular learner to cheat is much higher in an LMS than in most other ICT systems because the results may directly influence his future career. Instructors and tutors have privileged accounts for selected parts of the LMS; they might want to bypass these restrictions to get access to other courses' and users' data.
- *External attackers*: We consider any attacker who is neither an LMS administrator nor a regular LMS user an external attacker. However, we must be aware that further differentiation may be required; for example, someone external to the LMS might still be an insider of the organization and thus have knowledge about the infrastructure and its weaknesses, which a remote internet attacker does not have, at least not from the beginning. Similarly to the knowledge, also the sophistication with which an attack is performed may greatly vary.

Since, for example, a web-based LMS has an architecture and security properties that are, to a certain degree, similar to e-commerce web sites, the LMS may face attacks that often have been discussed in the context of web applications. We refer the interested reader to (OWASP-Project, 2009) and (Martin, 2009) for a thorough analysis of the related vulnerabilities, threats, and remedies, and focus on LMS specific security risks in the following discussion. Based on the risk analysis done by (von Solms, 2005) and current best practices, we need to estimate the likelihood and impact of the following potential security incidents in our security concept:

- *Unauthorized access to course material*: The lack of suitable access restrictions or the failure to properly enforce access control policies on the technical level may lead to unauthorized modifications of the course content or to an unwanted or premature disclosure of course material to LMS users; in turn, this may violate intellectual property rights. The modifications performed by the attacker may be hard to spot if they are not deletions or bogus content, but subtle changes, e.g. to complex formulas. Thus, we must facilitate the preservation of the course material's integrity.
- *Unauthorized access to user contributed material*, such as submitted homework assignments: We must ensure that an attacker cannot copy or delete the material submitted by other users unless explicitly required otherwise.
- *Unauthorized access to tests and test results*: If the LMS is used to perform electronic tests, e.g. with all learners gathered in a local lab's computer pool, we must make sure that the test cannot be accessed before it is officially started, and that the test content cannot be manipulated or deleted by an attacker. Similarly, the manipulation of test results must be prevented; an easy opportunity to improve one's own grades is a temptation probably not too many learners will overcome.
- *Unauthorized access to user data*: From the users' perspective, privacy-related incidents have the largest impact because once their PII has been disclosed, fixing the security leak cannot undo what happened. Even in an LMS that highly encourages community-driven learning experiences, each individual should be put in control of which PII is visible to the other users and the instructors.

- *Denial of service attacks* against LMS functionality or the LMS infrastructure: Due to the number of involved components, there are several ways to attack the availability of an LMS, such as overwhelming the LMS frontend with millions of bogus requests or exploiting a vulnerability that crashes the underlying relational database management system. Usually such attacks have only temporary consequences, but if an electronic test cannot be performed at the scheduled time, or a homework assignment cannot be submitted a whole weekend long because the LMS is unavailable, users and instructors will start to be uncontent.

The impact often exceeds the damage caused directly by a successful attack: As (Eibl, von Solms, & Schubert, 2006) discuss, especially the institution's credibility and reputation suffer from LMS security incidents. Regaining the users' trust and their acceptance of the LMS infrastructure after an incident can take a very long time, even when the technical problems have been solved. However, there also can be legal consequences, for example when learners miss deadlines for exam registration due to LMS unavailability, access to licensed 3rd party material has not been controlled properly, or exam results have to be revoked because of cheating.

Reliable protection against any kind of attacker and any type of attack is hard to achieve and involves time-consuming as well as cost-intensive efforts; thus, there must be management directives regarding the priorities and the necessary resources must be made available. It also must not be neglected that information security is a dynamic process and not a one-time concept. For example, when new functionalities, such as online exams, or new user groups, such as students from partner universities, are added to the LMS, the security requirements and threats need to be re-evaluated.

### Summary: Getting Started with Security and Privacy Management

We have seen in this section that LMS infrastructures are complex distributed systems, which consist of so many different components that their security and privacy specific management is a major and by no means trivial challenge. Besides the server machines and ICT services involved in the operation of an LMS, we especially need to address the security properties of the data stored therein, such as the learning objects and the user data. After an overview of the basic computer security and information security management goals, we discussed how the attackers and threats need to be analyzed with a focus on the impact of successful attacks in order to derive priorities for the implementation of security controls.

This basic methodology for security requirements engineering should be applied as soon as possible in the LMS design phase, because – as we discussed – security will not work as an afterthought. One of the deliverables in this project phase should be the LMS security policy: It states the selected security and privacy goals, explains their priorities, assigns responsibilities to the involved parties, and defines the internal processes and workflows, e.g. in the case of security incidents. Documenting these design decisions and their rationale as well as tying the upper management's commitment and necessary resources to the LMS security and privacy aspects is the first step towards the successful selection and implementation of concrete security controls, which we discuss in the following section.

## IMPLEMENTING AND OPERATING LMS SECURITY AND PRIVACY

Now that we have reviewed our security and privacy goals, discussed LMS specific threats, and investigated the basic organizational measures to set up an LMS security management framework, we need to analyze how specific security controls can be chosen from those available on the security market and how their long-term operation should be handled.

We will first address the selection of security controls based on a simple vulnerability and attack lifecycle; our goal is not the recommendation of certain products, but to give an outline how the LMS-specific security tool selection can be based on current best practices. Afterwards, we discuss the security management workflows required to make full use of these tools without neglecting the preparation of reactions to eventually successful attacks.

## Selecting From the Pool of LMS Security Controls

From the computer security perspective, any type of attack targets an assumed or indeed existing vulnerability of the LMS infrastructure. Due to the infrastructure's complexity, vulnerabilities may affect the network infrastructure, the server operating systems, implementation errors in the various server software products, or service misconfiguration. Such vulnerabilities may exist for a long time before they are detected, and threats are most dangerous in the time between the disclosure of the vulnerability, especially when instructions how to exploit it become publically known, and the time a suitable security fix has been released and applied.

Obviously, proper LMS configuration by means of the LMS administration frontend cannot cover the whole LMS infrastructure in a fine-grained way, but only the core LMS application itself. Thus, additional security controls are often chosen with a focus on how security breaches can be avoided. However, the assumption that any types of attacks can be stopped with the limited amount of security controls that we can implement and operate with our budget would be naive. Instead of just addressing the prevention of attacks, we must also ensure that ongoing attacks can be detected, and we must specify how we should – automatically or manually – react to attacks. We will now discuss several measures and security controls that we can select from, so that they can be applied to our LMS infrastructure in each of the three phases prevention, detection, and reaction:

- *Prevention* includes several generic security controls, such as limiting server access to the absolutely required TCP/IP ports by means of a packet filter firewall, timely or even automatically applying operating system and software updates, encrypting and cryptographically integrity-checking all the data that is transferred and stored/backed up, applying rigorous access control, secure and tamper-evident central logging, and isolating services from each other by running them on dedicated (eventually virtual) machines. LMS-specific preventive measures usually concentrate on application-level access control. As discussed by (Yong, 2007), role-based access control (RBAC) may serve as a base for authorization management, but certain LMS-specific extensions are recommended, which are implemented by the available LMS products to a varying degree. While, for example, learners, instructors, and LMS administrators are typical roles, the access decisions and personalization features must also be based on role parameters, such as the study course a learner is enrolled in. The default settings for access to any data, including course material and user data, should follow a white-listing approach, i.e. each access privilege should have to be explicitly assigned to the relevant roles. This way, extensions made to, for example, the user profiles do not become publically visible to all users unless this is deliberately enabled by a configuration change. Further measures should be taken regarding LMS fraud prevention. For example, email notifications sent to learners should be digitally signed by the LMS: As a benefit for the users, phishing email attacks are less likely to be successful when users can rely on the fact that each official email has a verifiable cryptographic signature. On the other hand, learners cannot manipulate the email's content unnoticed; for example, a learner might receive a notification stating that he has not passed a test, but change the email's content and submit the modified, yet seemingly official email somewhere else, which at the very minimum can cause a lot of confusion.

- *Detection* is the most difficult aspect, since a high detection rate usually cannot be achieved without a high rate of false positives, i.e. unnecessary security alerts, which in turn deteriorate the attention paid to related warning messages. For web-based LMS infrastructures, network-based intrusion detection systems (IDS) can be used. However, in order to improve their reliability, they require configuration adjustments, which are typically performed by letting the IDS learn how the LMS is used regularly; larger discrepancies to this behavior are then considered to be attacks after the training period has finished. While certain attacks are easy to detect, e.g. massive amounts of network traffic as a part of a denial-of-service attempt against the web server, more sophisticated attacks are hard to spot, especially if they aim at only unauthorized reading, but not writing of data. Due to the general lack of equally sophisticated tools for automated LMS monitoring, an administrator with enough time to carefully watch the LMS log files for suspicious behavior on a daily basis is required; unfortunately, this is typically one of the first tasks to be sacrificed when the same administrator has got heaps of other work to do. To a certain degree, usage metrics can be used to detect suspicious activities, for example when the number of logins performed per day by a single user clearly rises above a threshold, or a larger amount of restricted learning material is accessed by the same user within a very short time period. However, getting the thresholds right to avoid too many false positives is difficult, and once attackers are aware of what is being monitored, they will find a way to stay below the radar.
- *Reaction* to attacks takes place during an attack or after an attack that has previously been carried out was unveiled; we discuss automated mechanisms here and will investigate related manual workflows below. Automated reactions are primarily intended to limit further damage to the LMS infrastructure while an attack is going on, especially when no administrator is easily available, such as during night-time and weekends, or when the same type of attack happens so frequently that it would be tedious for administrators to be manually involved each time, e.g. when an attacker makes a larger number of attempts to guess someone else's login password. In its simplest form, an automated reaction is to log the incident for later review, or to escalate the issue, e.g. by sending the administrators an email or ringing their pagers. More sophisticated technical solutions can, for example, dynamically reconfigure the network firewall to block the attackers' IP addresses, or even shut down or temporarily take offline the parts of the LMS infrastructure that are currently under attack. However, such mechanisms must be used with care, because they otherwise could easily be misused themselves for denial-of-service attacks: For example, if too many attempts to login with the wrong password lock the account, an attacker may deliberately use this approach to keep others from working with the system.

Concerning the detection of and the reaction to attacks, the LMS administrators also need to be aware of the possibility that several attacks are carried out in parallel. A massive attack's primary purpose may be to create distraction, so a second attack is not detected. For example, an attacker who somehow manages to acquire administrator privileges may apply several obvious modifications to various courses' material that are easily spotted by hundreds of users, but also manipulate just one of his examination results. Thus, after a central LMS component has been compromised, there first needs to be a fix available for the underlying vulnerability that was exploited, and then a clean new system has to be set up and eventually the most recent reliable backup of the data needs to be restored. Any other approach would be built on the vain assumption that the attacker has not installed a well-hidden backdoor on the compromised system and has not manipulated the LMS data either; re-installing from scratch is the more reliable and often also the faster way to restore order.

## Establishing LMS Security Management Workflows

While the technical security controls contribute to the automation of several operational tasks to a large extent, their use must be orchestrated by means of security management processes, which in turn must be designed with the limits of pure technology in mind. In general, security management is a process that follows Deming's classic Plan-Do-Check-Act life cycle (Deming, 1986), which – in our case – aims at the continuous improvement of the overall security level. We already discussed the design and implementation issues, which correspond to the planning and doing phases, and will now investigate some workflows that are part of the checking and acting phases.

Determining the level of security that is currently achieved is the first step towards identifying weak spots that need to be improved. (Eibl, von Solms, & Schubert, 2006) discuss assessment criteria and the determination of their weights for inclusion in scenario-specific *criteria catalogues*; it can serve as basis for the specification of a check list that is used regularly to evaluate how the security situation is evolving. However, the simplicity of a check list must not mislead to the assumption that redetermining the security level is a quick and easy task; for example, thorough reviews of the permissions assigned to roles and users can be very labor-intensive for large LMS infrastructures.

Security management also includes measures to prevent successful attacks by attempting to detect vulnerabilities in the LMS infrastructure before the attackers do. Usually, security experts will perform *audits* of the system by analyzing the LMS design, implementation, and configuration. *Penetration testing* security tools, such as Nessus, can be used to simulate various types of typical attacks in order to evaluate whether the LMS is vulnerable and how it copes with hostile environments.

Such tests and audits should contribute to and complement *security reports*, which should regularly be created and reviewed by the LMS management as an important part of the overall LMS statistics and reports. Changes in the reported figures should be studied carefully to detect outliers or upcoming trends, since new or updated security controls will become necessary in the course of time to reflect new types of attacks.

Furthermore, an *incident plan* should be developed and updated regularly; it should specify step-by-step workflows regarding the reaction to attacks. In practice, LMS administrators and managers may panic when they are surprised by an acute massive attack or security breach and may take uncoordinated or even wrong decisions and actions. Specifying in advance whom to inform, which components to shut down, and how to react to user inquiries saves precious time and helps to keep one's calm.

Additionally, it should be considered to take *legal steps* to retaliate against attempted or successful attacks. While this can be difficult regarding internet-based attacks because the attackers usually remain anonymous, (Graf, 2002) suggests that there should be a clearly defined escalation mechanism regarding cheating users. The defined penalties should be strictly enforced to avoid the loss of credibility; the policies that already have been established regarding cheating in traditionally written tests and with respect to ICT service abuse can serve as a basis for the LMS acceptable use policy that needs to be specified.

Finally, from the security management perspective some of the *remaining risks* can also be delegated, e.g. to the ICT service provider contracted to the hosting of the LMS infrastructure. For example, security properties, such as the time between the availability of an operating system patch and its deployment on the LMS-related machines, can be specified as a part of the service level agreements (SLAs) between customer and ICT service provider.

## Examples: Implementing Security Policies in LMS Products

Applying the discussed security processes successfully in the real world is still a challenging task. Despite the important contributions of organizational measures and security components like firewalls, the security properties and configuration options of the LMS core decide about the rise and fall of the LMS as a whole. In this section, we review the security-specific administration functionality of two well-known and wide-spread LMS products, OLAT and Moodle, in the context of the goals, risks, and security controls discussed above. We chose these two examples because they take very interesting and rather different approaches towards LMS security management; since both are well-documented open source software packages, you can also compare them with the LMS of your choice easily.

OLAT clearly focuses on LMS application specific security and privacy management, which means that the OLAT administrator is spared from having to deal with server and system security; this actually makes OLAT very attractive for hosting, because the responsibilities for server security on the one hand and LMS application security on the other hand are clearly separated. OLAT distinguishes between several system roles, such as administrator, author and co-author (which equal instructor and tutor in the terminology used above), group manager and user manager.

Concerning account administration, the user managers can manually add new users to the LMS or start batch user imports from Microsoft Excel sheets. While they can also lock user accounts, the deletion of user accounts is usually performed automatically by purging all accounts that have not been used for a specified number of days. While this keeps the LMS user database clean to a certain degree and works fine for small numbers of users, larger installations are typically connected to a local LDAP-based directory service. The LDAP service is not only used for authenticating users, but also to import user records. Therefore, OLAT offers two variants of retrieving LDAP data: Either all the users found in the LDAP server can be imported by means of a bulk load, or each user's data is fetched individually at the first time the user logs into OLAT. Obviously, the second approach is more privacy-preserving because only data about those users is stored in OLAT who actually want to use the LMS. Regularly synchronizing OLAT's user database with the LDAP service also triggers the deletion of OLAT accounts; given that the LDAP server is the organization's authoritative user data source, this efficiently removes retired users' access to the LMS without the need to manually delete accounts.

Groups can be used to assign rights to several users at once; the rights in OLAT are permissions to use LMS features such as file folders, forums, wikis, the course editor, and the assessment tool. As OLAT strives for high usability, rights are only positive and additive, so the group managers and administrators do not have to worry about conflicting positive and negative permissions. OLAT supports basic group privacy management, e.g. by letting the group manager choose whether a member of the group may see who the other members are.

The visibility of learning material can be restricted by start and end dates, by group membership, and by user attributes such as the user's field of study or institution, which exactly matches the instructors' requirements in most practical use cases. The access log files created by OLAT are pseudo-anonymized to preserve the users' privacy: Each user's name is replaced by a randomly assigned number on a per-course basis, so the same user's actions cannot be correlated across multiple courses.

In summary, OLAT implements current best practices regarding security and privacy management, and keeps the related administration tasks as simple as possible.

Moodle takes a more fine-grained approach towards security and privacy configuration. It offers many security and privacy controls that can be configured individually. For example, Moodle includes captcha-based spambot protection, integrates anti-virus software to scan uploaded material, filters HTML code from user-provided content to prevent cross-site scripting, and allows setting up arbitrarily complex policies for local user passwords.

The Moodle documentation also takes the administrator through several concrete security recommendations for the machines the LMS is running on, including security-specific PHP run-time environment configuration, sandboxing, the use of tools to detect hacked machines, and a backup-based recovery workflow in case the LMS has been compromised.

Moodle offers a vast number of ways to integrate the LMS into existing infrastructures: Besides the LDAP integration, which is similar to OLAT, it can also authenticate users by means of RADIUS or IMAP, i.e. checking the user's password against the local email server. Moodle provides a basic set of roles, including administrator, course creator, teacher, grader, student, and guest, and encourages the definition of arbitrary additional roles. To each role, several dozens of very fine-grained permissions can be assigned, such as whether a user who has this role may create new or reply to others' forum posts. There are positive as well as negative permissions that may apply to individual courses, and inhibitory permissions that apply to the LMS as a whole; obviously, creating new roles is complex task which requires diligence and thorough testing.

Moodle can create security reports that warn about potentially problematic system settings, and the users' activities can be monitored closely, e.g. by providing an aggregated view of the user's forum posts, blog entries, and notes. In a nutshell, the Moodle administrator is put in almost full control of system and application security; although this myriad of options can be overwhelming in the beginning, the default settings have been chosen very carefully and allow for a quick start, while there is paved way for fine-tuning security and privacy settings later.

Despite these differences between OLAT and Moodle, security managers must not forget that the LMS itself is just one of the many components of the LMS infrastructure, and thus additional work must be put into securing the other parts because attacks will most often be targeted against the weakest link in the security chain.

## LMS USAGE ACROSS ORGANIZATIONAL BORDERS

Two aspects primarily motivate the use of an LMS infrastructure by institutions other than the one that is operating it. On the one hand, the preparation of high-quality learning material is a time consuming and often expensive process. Thus, many instructors and content authors want to make their work available to more than just their home institution's learners. However, the material should not be completely given away, e.g. by sending a copy to another institution where some other instructor will offer the same course; instead the original authors, instructors, and their home institution shall be put in perspective. This motivation also applies to 3rd party content providers, which typically offer learning material on a commercial basis. On the other hand, the operation of a fully-featured LMS infrastructure may be too expensive for smaller E-Learning projects that want to focus on the content instead of the technology, and for smaller universities that offer only a few online courses. Thus, they often prefer hosted LMS services over setting up their own LMS infrastructure.

For many higher education institutions, the mutual use of LMS infrastructures is also politically motivated. For example, the pan-European Bologna process (European Commission, 2007) encourages students to earn credit points at various universities and apply them to their home university studies. Naturally, this regulation, which was originally designed with respect to physical course attendance, meanwhile has a major impact on the cooperation between E-Learning projects.

A common issue of both aspects, i.e. external learners using a local LMS as well as local learners using an external LMS, is user data and authorization management. We will now discuss how Federated Identity Management (FIM) technology facilitates inter-organizational E-Learning, which new security and privacy challenges arise, and how existing LMS infrastructures can be adapted to support federated access.

## The Federation Concept and Its Benefits for E-Learning

The traditional way to apply for access to internet-based services is self registration. New users have to fill out an HTML form, in which, for example, name, email address, and study course details must be provided. Regarding other universities' LMS infrastructures, often a conformation of enrollment had to be submitted by postal service before access to restricted material was granted. On the one hand, such a long setup phase and the issuance of yet another username and password by each LMS scared away many potential users; on the other hand, the quality of user data on the LMS side often suffered from outdated contact information and even fake registrations.

As a consequence, more user-friendly and reliable ways to manage digital identity information across ICT services in different domains became necessary. After various attempts with proprietary and often complex user data exchange protocols had been made, FIM has meanwhile become a standard approach that is successfully used also for many other higher education services, including electronic library media access, licensed software distribution, and Grid computing.

The basic FIM idea – shown in figure 2 – is to assign each user to a home organization, which is called the user's identity provider (IDP). Whenever the user is accessing an external service, such as an LMS, at an external service provider (SP), the SP can request the required user data from the IDP. There are three types of user information available via FIM:

- *Authentication* assertions: When using FIM, the user does not need a new password for each external service. Instead, the user is redirected to her IDP where she has to login using her home organization's credentials. If she already has logged in there, she is not required to enter her password again, which enables a true single sign-on across organizations' borders. Also, user passwords cannot be intercepted by hostile or compromised external services this way, which improves the overall security situation.
- *Authorization* information: Although in the end the SP has to decide about whether

*Figure 2. Federated Identity Management: basic functionality for LMS infrastructures*

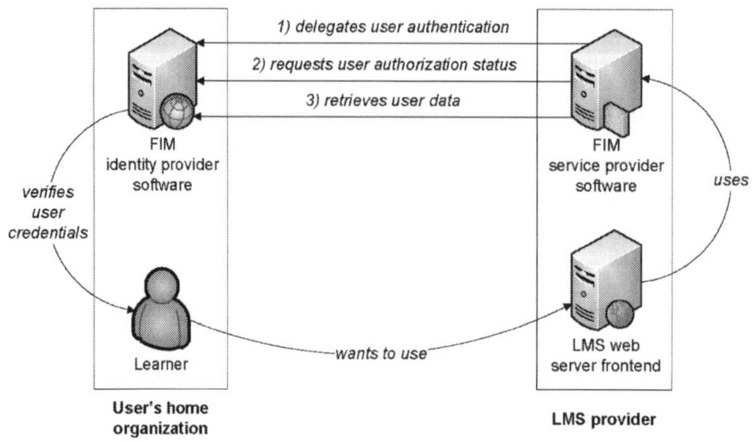


access is granted to the user, the SP can optionally ask the IDP whether the user is authorized for the service or one of its functionalities from the home organization's perspective. This can, for example, be used to let the external LMS user only participate in online classes that fit her study course.

- *User attributes*: Arbitrary user information, e.g. name and email address can be retrieved from the IDP. We will discuss the resulting privacy implications below.

Any number of IDPs and SPs taken together form a so-called federation. Higher education federations have been established in USA (InCommon), Australia (MAMS), most European (e.g. SWITCH-AAI in Switzerland and DFN-AAI in Germany) and several other countries; the number of IDPs and SPs is constantly increasing since several years, demonstrating that FIM is not just a short-lived technology hype.

Using FIM for E-Learning across organizational borders is a successful remedy for the three pain points we discussed above: First, a university's IDP is a reliable data source for the LMS SP, which eliminates the data quality issues of self registration. Second, FIM provides high usability because new users immediately can start to use the external service, without the need to fill out a registration form, send in a confirmation, and manage separate login credentials. Finally, FIM is not a proprietary solution, so existing IDPs can be re-used, and institutions that first have to setup their IDP can use it for many other services without additional efforts and costs.

## Security and Privacy in a Federated LMS

Operating an LMS as SP in a federation has implications which eventually cause two major computer security related changes to the infrastructure. First, participating in a federation typically means offering the service over the Internet. Compared to only campus-internal access, this results in a much larger number of potential external attackers and may necessitate additional security controls; unlike local students, anonymous Internet attackers will unlikely be deterred from hacking or damaging the LMS infrastructure by local policies. Second, FIM requires a trust relationship between the SP and the IDPs regarding the quality of the user data that is provided; this is especially true when some of this data is used to automatically derive authorizations. For example, if certain learning material is restricted to medical personnel and students, we must rely on the IDP to only label digital identities as medical personnel if the users behind these acounts really are. These data quality issues must be asserted on a contractual basis, either bilaterally between IDP and SP, or by using the federation operator as a trust broker.

Privacy protection is even more important in federations than in local deployments, because making full use of FIM technology leads to the IDP-initiated transfer of the user's PII over the Internet to the SP. This obviously necessitates privacy controls to ensure compliance with regulations and data protection acts. The general principle is to minimize the amount of transferred PII by limiting it to the user data that is essentially required on the SP side. Usually, federation-wide policies and contracts between IDPs and SPs will specify the required user data, and the user will have to explicitly approve the transfer. Technical implementations are often based on the business card metaphor as shown in figure 3: Before data is sent by the IDP to the SP, the user will be shown a business card on which all the data is listed that is about to be transferred to the SP, and has to agree to the transfer or has the option to abort this process. For obvious reasons, the users are obliged to carefully study the SP's terms of use and online privacy statement before continuing, because once the data has been sent to the SP, the IDP has no direct control over how the data will be used by the SP and cannot be held responsible for any misuse. However, since the PII is usually the same that would have to be submitted as part of a self registration process, real world practice has demonstrated that the user acceptance of this technology is next to 100%.

## Migrating From Intra- to Inter-Organizational LMS Infrastructures

LMS infrastructures must be FIM-enabled and thus eventually be modified for federated usage. Since FIM is a relatively new technology, only about ten LMS products have yet been adapted (Olshansky & Carmody, 2009); however, a larger number of other vendors and open source projects is currently working on implementing FIM support.

There are primarily two technical changes: On the one hand, user authentication needs to be delegated to the IDP, which requires only minor efforts if the LMS already supports other types of single sign-on systems. On the other hand, user data needs to be retrieved from the IDP using FIM protocols instead of, for example, from local LDAP enterprise directories or self registration forms. This data acquisition approach is facilitated by off-the-shelf FIM modules for popular web server software, such as Apache or Microsoft Internet Information Server. These modules handle all the communication with the IDP and pass the user data to the LMS software, e.g. using additional HTTP headers or environment variables, which are easy to process by the LMS. For example, Shibboleth, which currently is the open source federation software most widely deployed in higher education federations, uses this approach, is available for various operating systems including Microsoft Windows, Linux, and Mac OS, and has successfully been used for FIM-enabling several commercial as well as open source LMS products with favorable development and implementation costs. Among them are the following five wide-spread software packages:

- OLAT can make use of Shibboleth to automatically enroll users when they login for the first time, similarly to its use of LDAP servers, which we discussed above, with the difference that the new users are affiliated with partner organizations and not local. The user's IDP is required to transmit at least the user's email address and a unique identifier. The user's other data, such as first and last name, must be entered manually by the user unless it is sent by the IDP. Additional IDP-sent user attributes, i.e. ones that exceed OLAT's standard user

*Figure 3. The business card metaphor applied to federated LMS user data transfer*

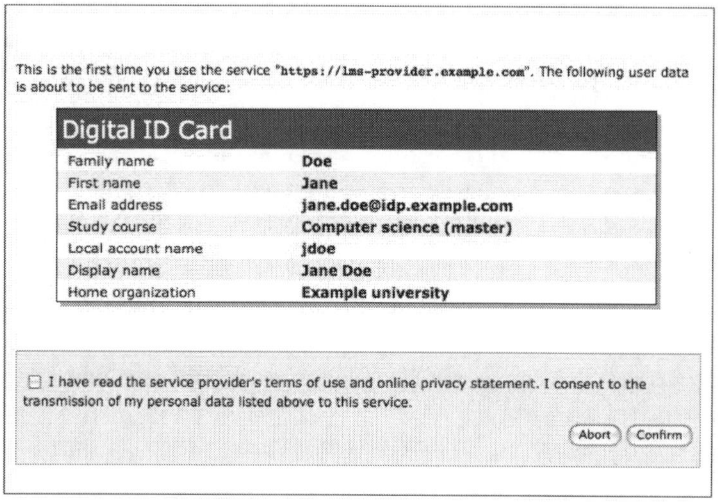


record, can be used in authorization policies, e.g. the ones that control the visibility of learning material, but are not stored persistently on the LMS side.

- Blackboard seamlessly integrates with the FIM modules provided by Shibboleth and thus also can make full use of remote authentication and user attribute transfer features.
- Sakai does not yet support FIM directly, but includes single sign-on features for integration with local enterprise web portals. This interface has been leveraged by the Guanxi portal software, which works as a Shibboleth-enabled frontend to Sakai and takes care of the remote user authentication as well as the aggregation of user attributes for use in Sakai's user profiles.
- Moodle fully supports Shibboleth and has a built-in configuration frontend for administrators, which makes federation setup – including the mapping of federation-specific user attributes sent by the IDP to the user record data fields used by Moodle – quite easy. The login screen can be extended to provide a list of partner organizations, so users are sent to the appropriate IDP.
- IMC CLIX is one of the first commercial LMS products that implemented the underlying FIM communication protocols and thus does not rely on external components like Shibboleth. Before new users who login for the first time are enrolled, CLIX presents the site's terms of use and online privacy statement to the user, which complements the user business card shown by the IDP.

Given the popularity of federated services and the availability of high-quality FIM programming libraries as well as the well-documented underlying FIM standards, we can safely assume that FIM support will become a standard feature of LMS products, just like their LDAP connectivity for local user management, very soon.

## SUMMARY AND OUTLOOK

In this chapter, we have investigated the LMS-specific aspects of security and privacy, which are two essential properties of any ICT service that need to be addressed right from the beginning in order to be effective. We discussed potential threats, attackers, and risks, which lead us to the conclusion that security and privacy must be addressed on the technical as well as on the management level. We then analyzed the security management tasks, such as specifying an LMS security policy and an incident handling workflow. We investigated the security controls that are available for the prevention, detection, and reaction to attacks and how they fit seamlessly in the LMS infrastructure. Finally, we discussed the benefits of federation technology for inter-organizational LMS usage and how its related security and privacy challenges can be addressed.

Since the popularity of E-Learning and the large amount of research and innovation in this field is unbroken, we can safely assume that LMS infrastructures will become even more complex in the future. New LMS features and the growing demand to centrally host these complex infrastructures will cause security and privacy to become even more important and complicated topics. Failing to address them appropriately is not an option for LMS projects with intentions for stable and sustained operation. Luckily, researchers and software producers have started to address topics like privacy-enhanced E-Learning, which puts us on the right track towards an online learning experience without constant worries about the security of our personal data.

## REFERENCES


Anwar, M., Greer, J., & Books, C. (2006). Privacy Enhanced Personalization in E-Learning. *Proceedings of the 2006 International Conference on Privacy Security and Trust (PST 2006)*. ACM.

Borcea, K., Donker, H., Franz, E., Pfitzmann, A., & Wahrig, H. (2006). Towards Privacy-Aware eLearning. *Proceedings of the 2005 Workshop on Privacy Enhancing Technologies (PET 2005)* (LNCS 3856, pp. 167-178). Berlin: Springer.

Deming, W. E. (1986). *Out of the crisis*. Cambridge, MA: MIT, Center for Advanced Engineering Study.

Eibl, C., von Solms, B., & Schubert, S. (2006). A Framework for Evaluating the Information Security of E-Learning Systems. In *Proceedings of 2nd Internation Conference on Information in Secondary Schools: Evolution and Perspectives (ISSEP 2006)* (LNCS 4226). Berlin: Springer.

European Commission. (2007). *The Bologna Process - Towards the European Higher Education Area*. Retrieved April 6, 2009, from http://ec.europa.eu/education/policies/educ/bologna/bologna_en.html

Graf, F. (2002). Providing security for eLearning. *Computers & Graphics*, *26*(2), 355–365. doi:10.1016/S0097-8493(02)00062-6

ISO/IEC. (2005). *ISO/IEC 27001:2005(E): Information technology - Security techniques - Information security management systems - Requirements*. Geneva: ISO Copyright Office.

Martin, B. (2009). *CWE/SANS Top 25 Most Dangerous Programming Errors*. Retrieved April 6, 2009, from http://www.sans.org/top25errors/

Olshansky, S., & Carmody, S. (2009). *Shibboleth Enabled Applications and Services*. Retrieved April 6, 2009, from https://spaces.internet2.edu/display/SHIB2/ShibEnabled

OWASP-Project. (2009). *Open Web Application Security Project*. Retrieved April 6, 2009, from http://www.owasp.org von Solms, B. (2005). Information Security Governance in ICT Based Educational Systems. In *Proceedings of the 4th International Conference on ICT and Higher Education,* Siam University (pp. 109-119).

Yong, J. (2007). Security modelling for e-Learning. In *Proceedings of the 1st International Symposium on Information Technologies and Applications in Education (ISITAE 07)* (pp. 1-5). Washington, DC: IEEE.

Section 2

# Selecting a Suitable Learning Management System:

## Challenges and Solutions

# Chapter 4
# Choosing the Appropriate E-Learning System for a University

**Cerstin Mahlow**
*School of Social Work, University of Applied Sciences Northwestern Switzerland, Switzerland*

## ABSTRACT

*In this chapter the author discuss the introduction of an e-learning system to enhance teaching and learning at a university. The focus is on the decision process choosing a system. Abstract criteria and feature lists are not sufficient for choosing the right e-learning software, even if all stakeholders and their respective requirements are heard. The author argues that "soft" factors should be considered when evaluating e-learning software: (1) The age of the students and their level of education, (2) the pedagogical guidelines and the culture of teaching and learning of the university, and (3) the educational scenarios in lectures and seminars. These factors seem to be only small details and are typically neither mentioned in the requirements nor in the feature lists of e-learning software. Therefore the author proposes that institutions should evaluate prospective systems in real-world scenarios. As a case in point, the author will outline a number of significant differences between two e-learning systems with a focus on pedagogical aspects. The systems can be seen as representative for a certain class of systems; both offer all features that are commonly seen as the most relevant when making a decision for a university e-learning system.*

## INTRODUCTION

Most educational institutions today use e-learning systems to some extent.[1] "Education" and "e-learning" are two generic terms: "Education" on the one hand may refer to primary schools, to universities, or to life-long learning. "E-learning" on the other hand can be simple file upload and download, or the use of elaborate simulations of complex procedures, or full-fledged distance learning courses.

In this chapter, we focus on a university located in Europe and being part of the European tradition

DOI: 10.4018/978-1-61520-853-1.ch004

Copyright © 2010, IGI Global. Copying or distributing in print or electronic forms without written permission of IGI Global is prohibited.

of educational ideals, where it is assumed that learning and teaching benefit from face-to-face interaction of instructors and students. When talking about studying it usually means face-to-face instruction and discussion. E-learning is seen as an additional didactical instrument to enhance teaching and learning and to bring so-called "new media" into the university. Usually an educational institution makes a strategic decision to implement e-learning and then buys a license for a certain software product or someone is tasked with installing and maintaining open source software. From that moment on, the institution claims to "do e-learning."[2]

Depending on the institution, different stakeholders and policies may influence the decision about the software. In this chapter, we will outline the relevant stakeholders and user groups who should be involved in the decision process at a university when the software to use for e-learning is selected. We will assume that implementing an e-learning system from scratch is no option. We will show that today, most e-learning systems offer very similar facilities and features. Thus, decision makers often assume that it doesn't really matter which specific system from a certain class they choose. We will show that this attitude will cause serious problems and we will propose a solution for a more suitable decision process.

This chapter is organized as follows: In section "Background" we give some definitions and elaborate on the general aspects mentioned in this introduction. In section "Stakeholders involved and their requirements for choosing an e-learning system" we briefly describe the stakeholders involved in the decision process for choosing an e-learning system and outline criteria usually considered in this process. The main aspects of this chapter are in section "Don't buy a pig in a poke," where we argue to take into account more fine-grained and less obvious requirements.

## BACKGROUND

In this section, we will define a framework of reference to locate this chapter and give practical definitions of the terms used in this chapter.

### E-Learning

We have already used the term "e-learning" a few times in the introduction of this chapter, but we have not yet given a definition for it. There are thousands of publications in the field (see Conole & Oliver (2007) for a good overview) published in journals[3], at conferences like IADIS International Conference E-Learning (http://www.elearning-conf.org) or in books like this one. However, we have to state that there is no single definition universally accepted by the community.[4] We find definitions emphasizing the learning *process*, defining e-learning as the "process of learning which is supported by the use of ICT (e.g., the Internet, network, standalone computer, interactive whiteboard or portable device)" (JISC, 2006), as well as definitions focusing on the *techniques* used for learning, defining e-learning as the "use of new multimedia technologies and the Internet to improve the quality of learning by facilitating access to resources and services as well as remote exchanges and collaboration."[5] For this chapter we will rely on the definition given by Piotrowski, who defines e-learning as "a general term describing all kinds of computer-mediated and computer-supported learning and teaching" (Piotrowski, 2009, p. 35).

The software used for this support has to offer functionalities for creation, organization, delivery, communication, collaboration, and assessment (Piotrowski, 2009). There are dozens of systems suitable or even explicitly designed for very specific purposes. Here we will use the terms "e-learning software" or "e-learning system" to cover software variously called MLE (managed learning environment), VLE (virtual learning environment), LMS (learning management system), LCMS (learning content management system), CMS (course management system), TELE (technology enhanced learning environments), or LSS (learning support system).

## E-Learning at Educational Institutions

E-learning used by educational institutions has various characteristics. We can open a wide matrix consisting of two axes: one is "educational institution," the other is "use of e-learning systems." Education ranges from primary schools, to universities, to life-long learning. The use of e-learning systems ranges from scenarios using file servers only, to simulations of complex procedures, to full-fledged distance learning courses, where all interactions are exclusively online. Thus, the concrete e-learning at a certain institution can be located in that matrix and can be compared to other scenarios at other kinds of institutions. Figure 1 shows the usage matrix for e-learning systems in educational institutions with three examples: the use of phpBB (http://www.phpbb.com/) for discussions in the foreign language classes in a secondary school, a full-fledged distance course on microbiology at a university and a simulation of ecological processes intended for people interested in further education (or so-called life-long learning). These are only three examples, each course or use of e-learning software can be located within this matrix.

As mentioned in the introduction, we will focus on a university located in Europe and being part of the European tradition. So we can reduce the size of our matrix to a scale, corresponding to the y-axis of figure 1.

## E-Learning at Universities

Three main factors influence e-learning at universities and determine how it should be implemented:

1. All students have completed school and they are at least 18 years old. Some may come directly from school to university, some may have completed an apprenticeship, some may have studied another subject before, or maybe they had a job for some years. Thus, we find a common basis—all students have completed their school education—but the students have different experiences of life. Additionally, even a seemingly homogenous

*Figure 1. Matrix for the use of e-learning systems in educational institutions with three examplary entries*

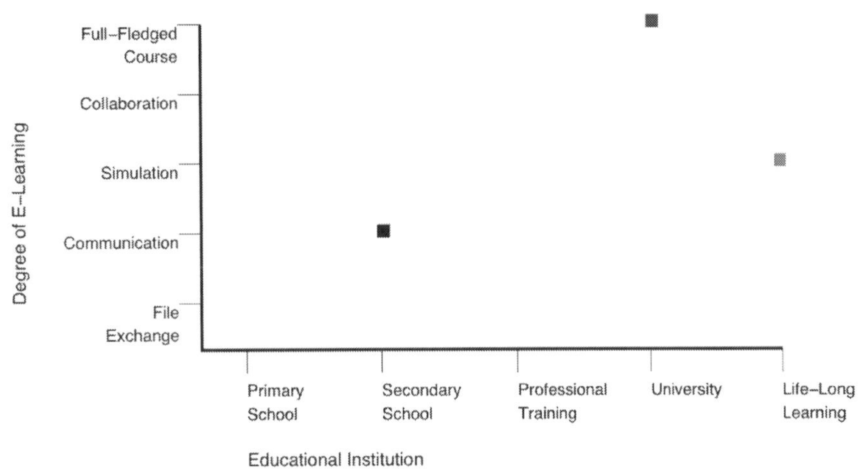

group of students (e.g., almost the same age, coming directly from school, studying the same subject) may have different motivations. They may have chosen this subject for completely different reasons: Some may have chosen it out of pure interest, others may hope to get a certain job, yet others may have wanted to study "something," but have no main interest, or this particular subject may be a prerequisite which has to be completed before being admitted to some other subject.

2. Most universities apply e-learning as an additional pedagogical scenario to enrich instruction and to handle large numbers of students attending a certain class. Thus, e-learning is rather facilitating instruction than replacing traditional teaching. Most universities do not offer distance courses but rather use e-learning tools for so-called *blended learning*, i.e., the didactically motivated combination of face-to-face teaching with different e-learning methods, resources and techniques (Garrison & Kanuka 2004; Wiebcke, 2006).
3. A main factor is the European tradition of teaching and learning—textbooks do not play a big role in European universities. Instructors are typically involved in research as well as in teaching. Instructors also typically create their own materials for teaching and adapt them regularly, so new findings in research are integrated into lectures very quickly. In most cases, e-learning courses are not developed by professional teams, consisting of content experts, instructional designers, and programmers, but by instructors themselves, who try to realize their ideas using the available tools. This has two consequences: (1) Even if a certain subject or a course at one university, e.g., "Introduction to Macroeconomics," is named the same way as a similar course at another university, both courses can be totally different in terms of expected prior knowledge, teaching material used— i.e., slides, examples, reading material, tasks—, exam, ECTS credits, etc. (2) Even if everything is prepared by instructors and readily available, each instructor teaching this particular subject will start from scratch. There is almost no reuse of already existing resources.[6]

## STAKEHOLDERS INVOLVED AND THEIR REQUIREMENTS FOR CHOOSING AN E-LEARNING SYSTEM

As the very first step, a university introducing e-learning usually chooses an e-learning system to serve as the "strategic platform" offered to all instructors and students. Regardless of the emphasis on the learning process or on the techniques (see section "E-Learning"), e-learning as such includes the use of ICT (Information and Communication Technologies). The decision process for an e-learning system involves several stakeholders with different requirements, since e-learning systems can be seen from different perspectives.

### Information and Communication Technology

From a technical point of view, e-learning software is like other software. Therefore the technical staff who maintains the IT infrastructure, Internet and intranet servers should be involved. If the university offers other network applications, such as Web-based e-mail access, a Microsoft Exchange server, or intranet services, the e-learning system should be integrated in the overall user-management structure. Thus, instructors and students should be able to use a single login and password to work with the e-learning system, a principle referred to as "single sign-on:" "an authentication process that lets a user to enter a username and password only once when they log on to a server, yet have access to many applications."[7]

Additionally, depending on the e-learning system, new hardware may have to be installed and should work smoothly with the already existing systems. It may be important that technical staff can benefit from the experiences made by others, available via mailing lists or online forums.

## Administration

For the university administration, the licenses and costs of maintenance are important. There is usually a general budget for software and teaching purposes, but the costs for the introduction and operation of a university-wide e-learning system is likely to exceed this budget. The administration may demand good arguments for extra budgets. If a university has a certain strategy concerning licenses, e.g., only open-source software or software from a certain vendor is allowed, it will influence the decision process as well.

Depending on the organization of the institution, there may also be requirements with respect to the integration of e-learning software with other administrative tools, e.g., user directories for student data. If all processes concerning teaching and learning—e.g., course catalogs, enrollment in courses, exam grades—are handled electronically, the e-learning system should offer interfaces to those information systems. For example, it should be possible that a student, when enrolling for a certain course, is granted access to the corresponding e-learning course automatically, and grades for exams and for tasks solved in the e-learning system should be recorded in the respective student accounts in the user directory automatically.

## Pedagogy

Finally, from a pedagogical point of view, there are requirements concerning the support of certain learning and teaching activities. For the instructor, it should be possible to easily distribute learning material relevant for a course or a single session. Students should be able to access these materials.

Especially in the field of Humanities (in the broadest sense), working with text, i.e., reading, discussing, and writing, is essential. Thus, an e-learning system should provide various tools for communication and collaboration in synchronous and asynchronous settings, e.g., e-mail, forums, chats, wikis, blogs, etc. Studying should also be contextualized, i.e., students should be able to learn and work in environments similar to those they will encounter later in their jobs (Specht, 2008). For some subjects this goal can be achieved by simulations, e.g., training in chemistry labs or microscopy. For subjects like social work, it means working in small groups, collaborating, and discussing various aspects of a task before presenting a solution in the plenum. Therefore an e-learning system should offer tools for groups working on similar or different projects. This requires some project management tools, such as calendars, and collaboration tools, such as file sharing facilities.

With respect to exams, an e-learning system should offer several possibilities for formative and summative assessment (see Cotton (1995) for the differences). Students should be able to take automatically evaluated tests, e.g., multiple-choice quizzes. Instructors should be able to assign tasks—e.g., to prepare an essay or to participate in an online discussion—to all students in a class or to individual students, and to evaluate and grade the results, and to provide feedback.

## Summary

The main concerns of the stakeholders involved in the decision process for e-learning software determine the basic requirements for e-learning software for use at a university. For example, the requirements may be that it is open-source, that it is field-tested, that it works smoothly with other databases used at the university, and that it offers certain functionalities (e.g., file upload and download, discussion forums, multiple-choice tests). These basic requirements will usually be met by several e-learning systems to various degrees.

In the next section, we will present the main factors for the decision process at the author's university. The main focus is on how to find the right tool from a class of similar e-learning systems, so that a system is selected that fulfills the essential requirements of the university as good as possible.

## DON'T BUY A PIG IN A POKE

At the School of Social Work of the University of Applied Sciences Northwestern Switzerland (FHNW)[8] we had to choose an e-learning system. The central ICT services of FHNW offer Moodle for use by the individual universities. However, we did not want to use a system just because it is already available; we rather wanted a system that meets our requirements instead of adapting our teaching to a piece of software. The main criteria—as described in section "Stakeholders involved and their requirements for choosing an e-learning system"—were determined easily: The system should be open-source[9], it should be possible for students and instructors to use their existing user accounts, there should be interfaces to the staff and student administration tools, and it should offer powerful tools for communication, collaboration, and assessment, i.e., e-mail, forums, blogs, chat, wiki, calendar, file exchange, multiple-choice tests, and evaluation tools. Additionally, the system should preferably also be used at other universities to enable the exchange of experiences and collaboration in teaching. Thus, implicitly, a further requirement was that the e-learning system should be field-tested; developing a system from scratch on our own was no option.

These requirements are met by several e-learning systems belonging to a certain class. These systems are open-source, they offer a variety of features for learning activities and there is an active community. Examples are Sakai (http://sakaiproject.org/portal), ILIAS (http://www.ilias.de/), Moodle (http://www.moodle.org), or OLAT (http://www.olat.org). In the next section we will have a closer look at Moodle and OLAT as two representative systems with very similar features and with a very similar history: Both systems were initially developed at the beginning of the 21st century by individuals to solve some concrete user needs and as an alternative to commercial e-learning systems. Both Moodle and OLAT meet all of the requirements mentioned above.

Moodle was initially developed by Martin Dougiamas starting in the late 1990s in Australia to serve as an alternative to WebCT (now owned by Blackboard, http://www.blackboard.com); the first public release was in 2002. OLAT (Online Learning and Training) is a Swiss e-learning system, initially developed by Florian Gnägi, Sabina Jeger, and Franziska Schneider at the University of Zurich starting in the late 1990s to distribute material and facilitate solving tasks in the introductory courses at the Department of Informatics. OLAT has won several awards, e.g., the *Medida Prix*[10] in 2000 and the *IMS Learning Impact Leadership Award* for *Best Open Source Learning Platform in 2009*[11], to name a few. Today both systems benefit from developments done by members of the respective communities. More details of the two systems will be given in the following sections.

If we have two e-learning systems, both claiming to meet the requirements of all involved stakeholders, how to decide which one to use? Does it matter at all? Using Moodle and OLAT as two examples of a class of e-learning systems we will show that the requirements formulated by the main stakeholders are usually not fine-grained enough to serve as a basis for choosing the appropriate software for a specific educational institution.

## It's All in the Marketing Brochures

With respect to the requirements of the ICT and the administration, it doesn't matter whether OLAT or Moodle will be used. Moodle is implemented in the PHP programming language[12] and uses a MySQL[13]database. OLAT was originally implemented using the same technology; however, in 2004 it was reimplemented in Java[14] and the object-relational mapper Hibernate[15], which allows to work with any database, including MySQL, PostgreSQL[16], and Oracle[17].

In Switzerland, the SWITCH consortium[18] offers an authentication and authorization infrastructure (AAI)[19] based on Shibboleth[20] for all Swiss universities, which allows realizing single sign-on for different web applications. Moodle and OLAT both can be configured to support Shibboleth. They can also use LDAP (Lightweight Directory Access Protocol) (see Koutsonikola, 2004) for user management.

Both systems are open-source, and both are used by several educational institutions in Switzerland. The respective communities have very active mailing-lists and forums, which makes it easy to share experiences and to get help from colleagues; answers or solutions can typically be obtained very quickly. Interfaces to user administration databases based on SAP systems[21] are not a general feature of either system, but could be easily implemented. The layout and general visual appearance of both systems can be adapted to conform to the rules of corporate design of our university.

From the pedagogical point of view, we find several similar features to support the educational activities. Table 1 lists selected features according to the six main functionalities described in section "E-Learning." Most of these features can be configured to meet additional requirements. For example, forums are normally readable and writable for all users who have access to it to serve as discussion platforms; however, they can also be configured to be writable for the instructor and read-only for all other users. This allows forums to serve as an announcement board, where discussion is not required and not permitted.

Comparing feature lists and the technologies used, it seems that it would make no difference whether we chose Moodle or OLAT, i.e., we could choose one randomly. Since both systems are open-source and can be installed easily, we could also give both a try and compare them "in action." However, this approach would require substantial resources: Someone from ICT would have to install and configure both systems and create user accounts, and we would need to find some instructors and students willing to create some courses and work on them to simulate real world scenarios. It would also take more time than making a decision based on the comparison of marketing brochures. However, teaching is one of the main activities of our university, and we have quality management processes for teaching in place, so we decided to invest these resources to find the e-learning system which *best* fits our requirements. We will describe our strategy and our experiences in the next section.

## But it isn't that Easy

The features listed in table 1 support certain learning activities, e.g., forums are used for discussions. However, a learning activity cannot be seen abstractly, but its concrete realization is influenced by a number of factors and assumptions depending on characteristics of students, instructors, and the institution, the subject, etc.

For example, there are different types of discussions. The discussion scenario may be "open," i.e., there is no fixed topic or time, and everybody is invited to utter an opinion, to pose a question, or to answer the message of another user. On the other hand, there are also discussions based on an introductory statement or question, with strict rules for how often each participant has to take an active part, and the discussion may be restricted to a certain period of time. There also

*Table 1. Features of Moodle (version 1.9) and OLAT (version 6.3). We use the terminology according to the respective documentation. The feature lists contain elements available out of the box, i.e., no additional plug-ins are considered.*

| Activity | Moodle | OLAT |
|---|---|---|
| Creation | • Editor<br>• Glossary module<br>• Lesson module<br>• Import of content using SCORM | • HTML editor (for single content pages)<br>• QTI editor (for tests, using the IMS QTI standard)<br>• Import and export of content using SCORM, IMS CP and IMS QTI<br>• Glossary |
| Organization | • Lesson module<br>• Access control (enrollment in courses) | • Course editor (create courses consisting of several activities)<br>• Categorize courses and add them to a catalog<br>• Access control (enroll in courses, permit access to certain elements according to certain rules) |
| Delivery | • Delivery of web content<br>• Access control (specify who is allowed when to see and access which elements of a course according to the AAI attributes and date-time specifications)<br>• Play SCORM content<br>• Resource module (upload and Download of material regardless of the format) | • Delivery of web content<br>• Access control (specify who is allowed when to see and access which elements of a course according to the AAI attributes and date-time specifications)<br>• Play SCORM and IMS CP content<br>• Folder (upload and download of material regardless of the format)<br>• Tunneling of content hosted on external servers |
| Communication | • Forum<br>• Chat<br>• Choice (polling)<br>• Calendar | • Send e-mail to other users directly, to group members, or to all participants of a course<br>• Forum<br>• File dialog[22]<br>• Chat<br>• Calendar |
| Collaboration | • Wiki<br>• Glossary module | • Wiki<br>• Special "rooms" for groups, equipped with e-mail, calendar, forum, file sharing, chat, wiki<br>• Glossary |
| Assessment | • Quizzes<br>• Surveys<br>• Assignment module<br>• Workshop module | • Self tests and tests<br>• Questionnaires<br>• Task building block[23] |

exist "mixed" scenarios: For example, each student may be required to post at least five substantial messages, including three postings in reply to messages of other students during the semester, and the postings must be relevant for the topic of the course—if these requirements are satisfied, the student will receive the ECTS credits. Here we have some strict rules (with respect to the number of statements) and some lax ones (the topics of the statements have to be in the subject area of the course, but there is no concrete initial question). In this type of scenario it is important that students are permitted to start a new thread within a forum, since otherwise they could only react to other students.

This simple example illustrates that a statement such as "we need forums for discussions" is not sufficient to exactly determine the scenarios instructors may actually use, and that these scenarios clearly have an impact on the technical abilities of the software in use. We will look closer at two elements essential for teaching in the field of Humanities and Social Work: (1) The support for working in groups and (2) the supported pedagogical principles for setting up courses with emphasis on communication and collaboration.

Afterwards, we will examine usability aspects in section "Usability" and discuss some technical aspects derived from specific didactical scenarios in section "Some more technical aspects." As testbed, we used the OLAT installation at the University of Zurich (which offers access for other educational institutions via AAI) and the Moodle installation of the University of Applied Sciences Northwestern Switzerland, to which the School of Social Work belongs, as mentioned above.

## Working with Groups

Working with groups in higher education has some implicit characteristics: A number of students attending a certain class or course is split up into groups, each consisting of a small number of members. These groups can be created ad hoc, i.e., just for one single task to be solved right now, or they can be created to serve as a "social" structure for an entire course. For the latter scenario, the group members have to work together on a project for some weeks or they have to solve several smaller tasks assigned to them from time to time during the semester. We will here consider groups created to serve as setting for an entire course. The group members will get some tasks they have to solve together using the communication and collaboration tools offered by the e-learning system.

Working with groups plays an important role: (1) In courses with more than 100 students it is not possible to discuss something in depth and to offer everyone the possibility to say something substantial—after some statements there is nothing new to add. Groups with five to 15 students allow fruitful discussions for everyone (Zimmermann et al., 2008). (2) After students completed their studies, they will usually work in teams with less than 100 members. The experience of working in groups helps them to become a valuable teamplayer and to be able to integrate in a team. We can thus fulfill the requirement of contextualization as mentioned above.

### Working with Groups in Moodle

Both Moodle and OLAT allow instructors to work with groups of students. In Moodle, the instructor has to decide which student will be a member of which group. The screenshot in figure 2[24] shows how to configure the settings for a group: The instructor has to decide who of the students enrolled in a course should be member of a certain group. This may seem obvious and easy, but this statement actually makes several implicit assumptions:

- Students have to be enrolled in that particular Moodle course before they can be assigned to a group within this course.
- The instructor decides who will be added to which group.
- The instructor has to add and remove students to groups by hand.

This behavior has two main impacts: Students cannot decide on their own with whom they would like to collaborate, and the instructor is charged with the coordination of the group work. This may be appropriate for a primary school, where teachers have a good overview of the developmental stage of their students, so that they may wish to achieve certain goals by assigning certain kids to certain groups and not letting them decide on their own. However, in a university setting this is quite limiting. University students are usually encouraged to manage their own learning processes, which should also include the selection of partners to work with in a group.

One could of course come up with workarounds, e.g., the students enrolled in a course are asked to form groups and then report the members of each group to the instructor, who then creates corresponding groups in Moodle. However, this is not an attractive option, since the group building has to be done using some other tools or face-to-face, and, moreover, the instructor still has to manually add the students one by one to the respective groups—only the Moodle

*Figure 2. Adding and removing users to groups in Moodle*

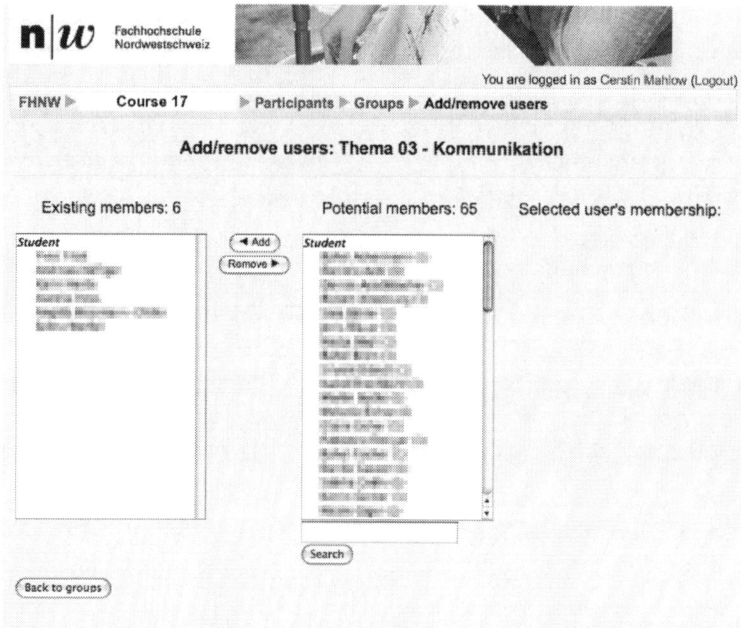


administrator can add lists of students to groups. Moodle offers other group-building procedures: automatic assignment to groups based on different parameters—again, students cannot choose themselves. In the official FAQ you can read this: "Q: Is there any way to enable students to choose a group? A: Not in the standard Moodle 1.9 build, but there are hacks which people have described in the forums."[25]

At our university there are several modules[26] all students of a cohort have to attend. A cohort of Bachelor (BA) students usually has 300 members. In these modules, the students have to work in groups of four to ten people, they can choose there partners themselves. Therefore the instructor has to create 30 up to 70 or more groups and assign the respective members one by one. This will take a lot of time since each student has to be selected—i.e., searched and then marked—from the overall list and assigned to the respective group.

Another aspect to mention is that in Moodle, a group always belongs to a certain course, i.e., it is not possible to create a group without first creating a course.

To sum it up: Moodle does support working in groups, but there are a number of restrictions; in particular, Moodle places the responsibility for the organization on the instructor alone and requires time consuming procedures for creating groups.

**Working with Groups in OLAT**

OLAT offers the same capabilities for working with groups as Moodle: The instructor can add students enrolled in a course to certain groups. However, the instructor can also allow students to decide themselves with whom they want to work together. Thus, OLAT can be used in a way which encourages students to take responsibility for their learning activities. Figure 3[27] shows how to configure a group in OLAT: In this particular case, the group belongs to a certain course, can have at most ten members, and makes use of a waiting list—once ten students have joined this group, further students will be put on the waiting list and will be considered for membership when a group member decides to leave the group.

As in Moodle, the instructor can also manually add or remove students. Unlike Moodle, OLAT

*Figure 3. Creating a group in an OLAT course*

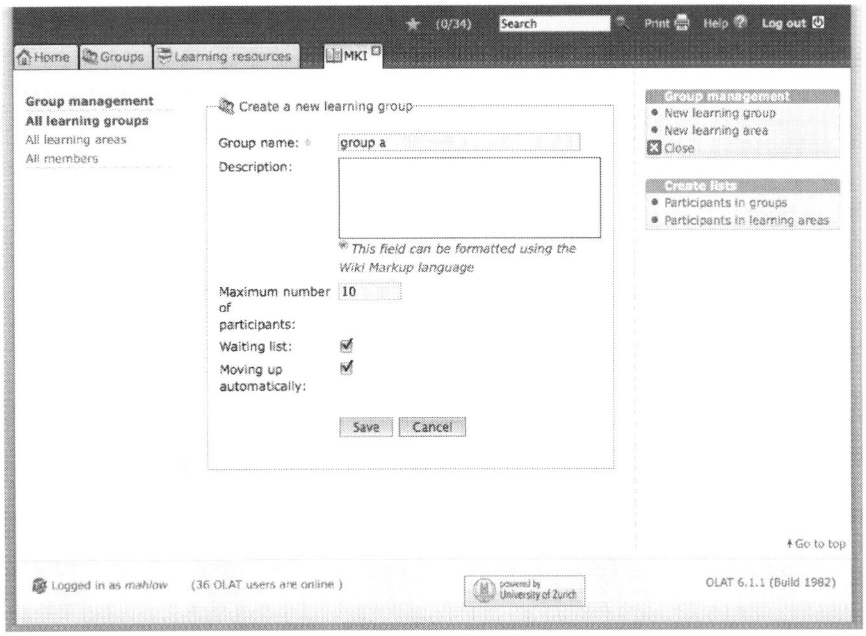


allows uploading a list of names[28] to assign the respective students to a group at once, i.e., there is no need to select every student by hand, which makes this process a lot easier and less time-consuming. However, it is often more attractive to make use of the enrollment building block offered by OLAT. The instructor can configure this building block to be accessible during a certain period of time, e.g., during the first three weeks of a semester, to allow students to join one of the groups belonging to that particular course. Figure 4 shows the screenshot (student view) after enrolling. This particular student has enrolled in "group a" as can be seen from the "Status" column. The student is allowed to leave the group using the link "Cancel" in the "Cancel enrolment" column. Each group may consist of at most 10 members.

What makes OLAT even more attractive than Moodle is that all users—including students—may create groups independent from courses. These "working groups" offer the same features as groups belonging to a course, (i.e., e-mail to members of the group, calendar, file-sharing, forum, chat, wiki). Students may use these groups to organize themselves for group assignments or for exam preparation or even for extra-curricular activities without being controlled by instructors.

To sum it up: OLAT supports several scenarios for working with groups, which range from strictly instructor-controlled groups to completely student-controlled, course-independent groups.

Depending on the institution, the general attitude towards personal responsibility, the educational goals, and the experience of the students, some of the scenarios for working with groups may be more appropriate than others. We think that, in a university, with the aim to teach competences and skills, students should be encouraged to take responsibility for their learning, and thus should be allowed to choose with whom they would like to cooperate. In any case, OLAT supports a wide variety of scenarios, whereas Moodle only has support for a single scenario. Thus, we can state that OLAT offers better facilities for working with groups.

*Figure 4. Enrolling in a group in an OLAT course*

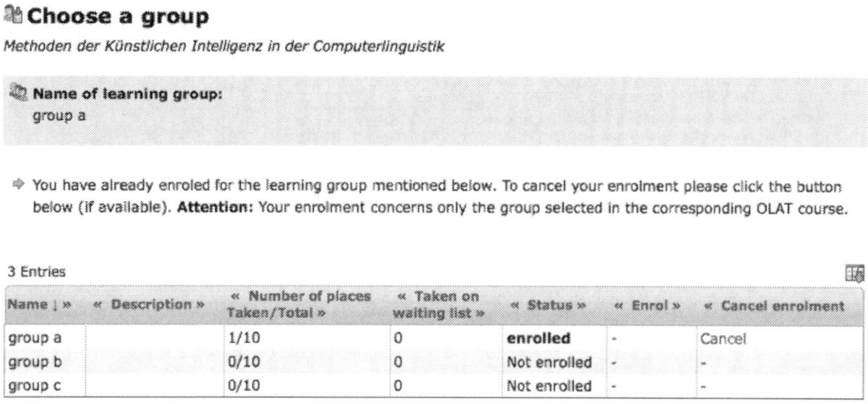


## Pedagogical Principles

Working with groups is one example where the concrete realization of a learning activity depends on pedagogical goals and principles. Moodle claims to be based on "Social Constructionism."[29] The key ideas of the approach are said to be collaboration and communication among students and also among instructors and students.

This seems to allow a wide range of didactical scenarios. In fact, however, the implementation in Moodle restricts the instructor to very few possibilities: Courses have to follow one of three possible structures: (1) by topic, (2) by date, or (3) around an "activity." We wonder: What about hybrid scenarios? What about scenarios following explicit pedagogical approaches, such as Gerson's ECLASS (E = Entry; C = Clarify; L = Look; A = Act; S = Self-Assessment; S = Summary, see Gerson, 2000)? What about a course split into several sections to accommodate large cohorts, where the sections are taught by several instructors on different dates but are on the same subject, the same topics, and have the same assessments? What about offering self-tests only, i.e., no additional material is delivered and no collaboration is intended, to prepare for exams? What about supporting group work on projects only? With Moodle everything has to be inside a course following one of the three structures.

OLAT, on the other hand, does not force instructors to follow predefined recipes, but permits them to construct courses tailored to their particular needs. The "course editor" of OLAT allows instructors to combine so-called building blocks[30] without any restriction in terms of sequences or number of blocks in general, or number of instances of a certain block. This allows reproducing the structures offered by Moodle, as well as any other structure instructors may wish to create. Unlike Moodle, the instructor can set restrictions for building blocks to be visible for certain groups.

Teaching and learning in the field of Humanities or Social Work relies, to a large extent, on discussion and communication. The main value of a course is thus not the material the instructor has prepared beforehand, but the discussions and the material developed collaboratively during the semester. Therefore the instructor may decide to modify the initial setting after some weeks, add learning activities, or skip some. OLAT allows restructuring a course easily at any time.

Another interesting aspect when working with OLAT is the freedom it offers for using features for learning activities. Some resources can be used in several ways: Within a certain course, within several courses at the same time, or not connected to any course at all. This applies to wiki, chat, glossary, and test building block. This allows instructors to make use of a certain element without requiring the overhead of creating a course they do not need.

## Usability

E-learning systems are supposed to be used by two user groups: students and instructors. Instructors prepare courses, including the configuration of elements for learning activities, such as uploading material or assigning tasks, they post to forums, or send messages to the students enrolled in their course. Students will download material, solve tasks and tests, and they are supposed to participate in discussions and to collaborate. The e-learning system should support these activities and make preparing and attending courses as easy as possible. Therefore we will have a look at some usability aspects.

The leading questions when evaluating usability of a website or a web application are (see also Nielson (1994), Helander (1997), Nielson (1997), and Grund et al. (2002)):

- What is the aim of an individual page? Is that aim obvious?
- Is it clear what tasks a user is expected to perform on a given page?
- Is all information required to perform the task available on a given page? Is the information structured clearly?

With respect to navigation, the essential questions are (see Powazek (2006)):

- Where am I?
- Where can I go?
- Where have I been?

Applying these questions or criteria to an e-learning system, we will evaluate how easy it is for students to find relevant courses—"relevant" with respect to the subject they are studying, the courses they have already completed, and the current semester—, i.e., to navigate within the system as such. With respect to a course, the main concern is how to access the different elements of this course, i.e., how to navigate within the course. From the instructor's perspective, one concern is how to navigate to the courses they own and which they may want to configure or maintain—this aspect is similar to the student perspective. The other aspect is the support the system offers to actually setup a course or change some settings.

**Moodle**

Figure 5 shows the homepage of a Moodle installation after login. The left-hand column contains announcements from the administrator. The main part in the middle contains a list of the schools of the University of Applied Sciences Northwestern Switzerland. The right column contains two navigational blocks. The "siteNavigation" displays exactly the same elements listed in the main part. Clicking on an element in the siteNavigation opens the subtree containing further subcategories or actual courses. Clicking on an element in the main part opens the same structure. The "myCourses" block lists the courses a user is enrolled in, using also a folder-like display.

The navigation blocks in the right-hand column are additional modules, developed by the local Moodle administrator and not part of the core Moodle distribution.

It is not possible (except if you program such a module yourself) as a student to see all courses relevant for the current semester. The myCourses block attempts to provide something like this, but it actually shows all courses a student has ever been enrolled in without distinguishing between current and completed courses. It seems very common to completely delete courses at the end

*Figure 5. Home page of the Moodle installation for the University of Applied Sciences Northwestern Switzerland after login, showing mainly navigation elements*

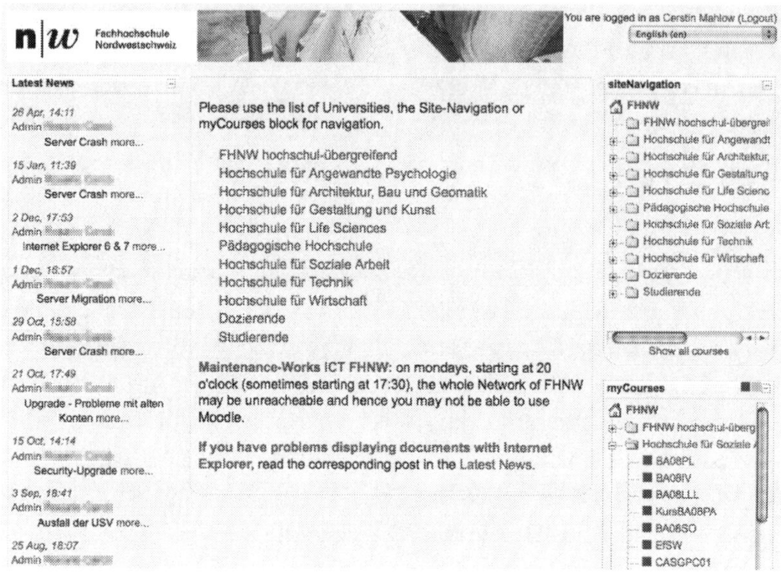


of a semester or to clean them up, i.e., to delete all student-created material and reuse the same course in the next semester. However, as learning is a process, and knowledge and skills should be developed over several semesters, it is counterproductive to throw away everything after a semester ends. Students are generally encouraged to build portfolios and reflect on their studies; therefore they should have access to every course they have attended.

This installation of Moodle serves nine schools, shown in the main part of the screenshot (plus some additional folders). Each school then introduces different subcategories, resulting in a very complex structure. A student looking for a particular course thus has a hard time finding it, unless the instructor provides a URL directly pointing to this course.

Within a course, there is an additional navigation element, as shown in figure 2. Directly under the header of each page is a "breadcrumb path". The path here says: FHNW → Course 17 → Participants → Groups →Add/remove users. A breadcrumb path is normally used to represent the way the user navigated to this particular page; the elements can be used to go back to a certain page. What we find in Moodle, however, is a "breadcrumb path" which effectively displays some arbitrary "path" representing the hierarchy of structural elements a certain page belongs to. In the case shown in figure 2, we navigated from the course (represented by the identifier "Course 17") to the group configuration. We did not travel through any "Participants" page; on the other hand, other elements, such as the school (after "FHNW"), are missing from the path. The path we in fact traveled was: FHNW → School of Social Work → Course 17 → Groups → Group 03 → Add/remove users. When one clicks on a breadcrumb like "Participants", one is directed to a page one has never seen before. There is no possibility to go back to "School of Social Work" to select another course belonging to our university with a single click. We would have to go back to the Moodle start page and then navigate through the hierarchy again.

To sum it up: Moodle is not user-friendly when used to represent hierarchical structures as they often occur at institutions of higher education. It does not allow students or instructors to find and access their courses easily.

In section "Pedagogical principles" we mentioned the structures Moodle offers to create courses: by topic, by date or around activities. Figure 6 shows the first page of a course using the "by date" structure.

At the left column we see some teasers from a forum, the main part is the content of the course. On top of that we see the "breadcrumb path." The first block contains three forums, two MS Word documents, a link and a PDF document. The second block contains a MS Word document and a MS PowerPoint document, etc. We find a structure consisting of several blocks each containing different activities or documents. When we choose one activity, for example the first forum in the first block, we find ourselves on a page as shown in figure 7.

The layout of the page has changed completely. The only familiar element is the "breadcrumb path" on top with its misleading information: We did not travel from the course page to a forums page where we chose this particular forum, but chose this forum directly from the start page of the course. The forum occupies the whole page; there are no right or left columns any more. What is even worse: there is no information about the course structure we could use for navigation. Obviously, Moodle offers certain structures for the course itself shown at the start page of a course, but has also a hidden structure using the type of activities. When accessing a certain activity, this

*Figure 6. First page of a course in Moodle using the "by date" structure*

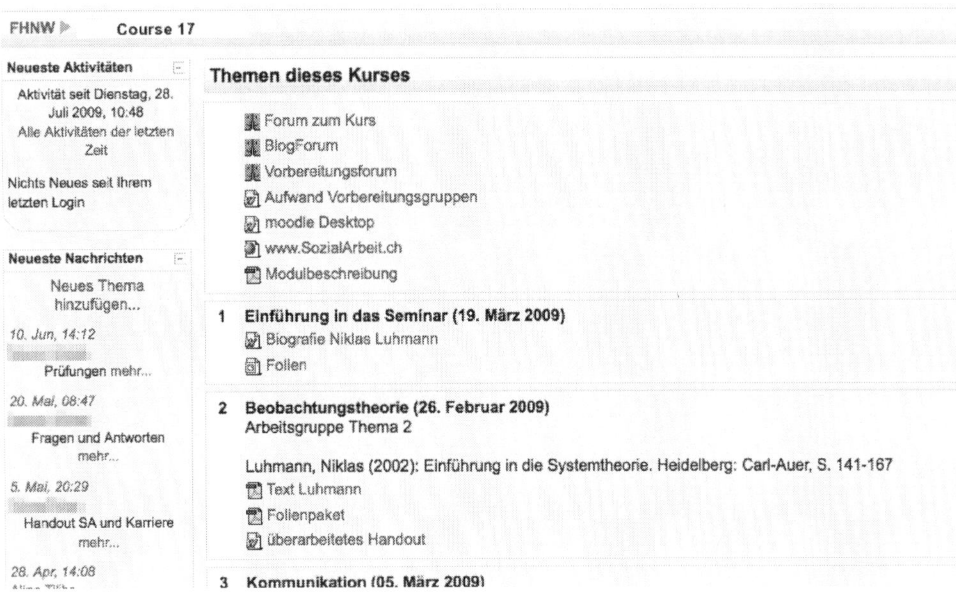


*Figure 7. Screenshot of a single page of a Moodle course showing a forum*

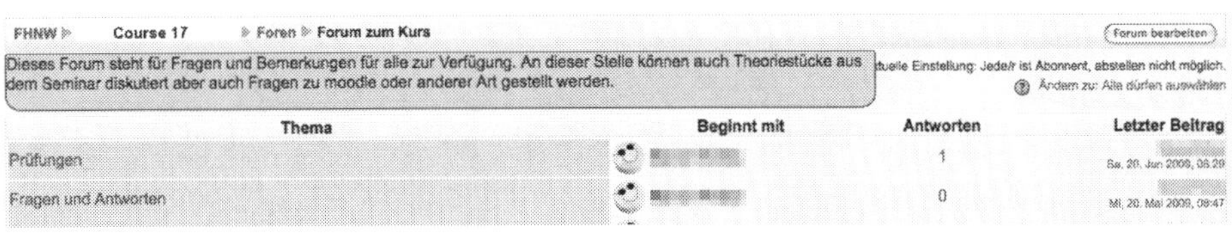

hidden structure becomes visible and is the only one accessible—the original structure of the course is not available for the user unless he goes back to the start page of the course.

To sum it up: Moodle does not fulfill the main principles for usability; its navigation is misleading and confuses the user.

As mentioned above, the instructor in his role of the course author can choose one of the three predefined structures and fill them with activities or documents. There is no possibility to introduce a substructure. Instructors can easily switch between the authoring mode and the student mode to get a feeling how it will look like for students. Blocks and activities can be modified to be visible at a certain date to allow instructors to prepare everything before the semester starts. Of course all elements can be configured, added and deleted at any point in time.

## OLAT

Figure 8 shows the OLAT home page after logging in. On the left are navigation elements for this page. The main part is occupied by several portlets that can be configured individually by each user. The main navigational elements are the "My bookmarks" portlet (listing bookmarks to courses or resources) and the catalog (listing resources according to categories). Users can add a bookmark to a course themselves, and bookmarks can also be removed. Instructors can decide if their courses are listed in the catalog. A course can be listed under several categories. There is also a list of *all* courses and users can search for resources by author, type, or name. Instructors can configure visibility and accessibility of their courses using the AAI attributes. Thus, a particular course may be visible only to students in Geography who have completed the introductory courses and are in their second year.

The overall impression of OLAT is, in our opinion, much more convincing and clearer than that of Moodle. From our experience we can say that students and instructors can easily find the relevant courses. Especially the bookmark function supports fast navigation.

Figure 9 shows the catalog in OLAT. On the left, one can see the navigation; the main part is a list of courses belonging to a certain category of the catalog. Each element consists of the name

*Figure 8. Home page of OLAT after login*

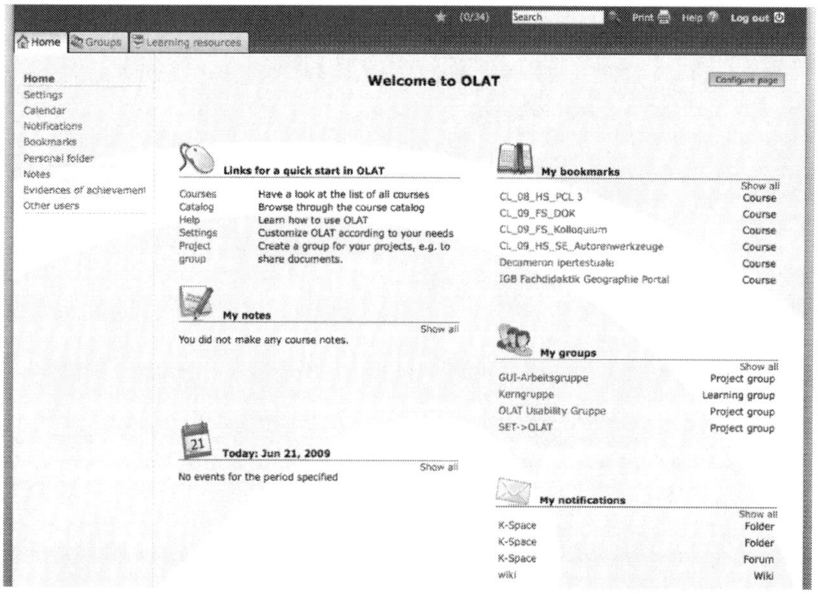

of the course, a short description, and a link to access this course ("Go to content"). In fact, the elements in this list do not have to be courses, but can also be learning resources, i.e., tests, surveys, CP or SCORM elements, wikis, and glossaries. Above the list, one finds the breadcrumb path, which allows for navigation through the catalog and which also represents the context of the current listing. Unlike in Moodle, this breadcrumb path actually represents the path the user travelled without introducing awkward additional stages or suppressing some.

When we enter a course, we get a page like the one shown in figure 10. On the left you see the navigation for the course, on the right you have some tools to make notes, add something to your calendar or set a bookmark. The main place in the middle is occupied by the actual content.

When we enter the forum "Wichtige Infos" ('important news')[31] we see a page as shown in figure 11. We notice important differences when comparing figures 10 and 11 with figures 6 and 7. The navigation of the course remains visible, the user can always see which element of the course

*Figure 9. Catalog in OLAT listing courses relevant for a certain subject*

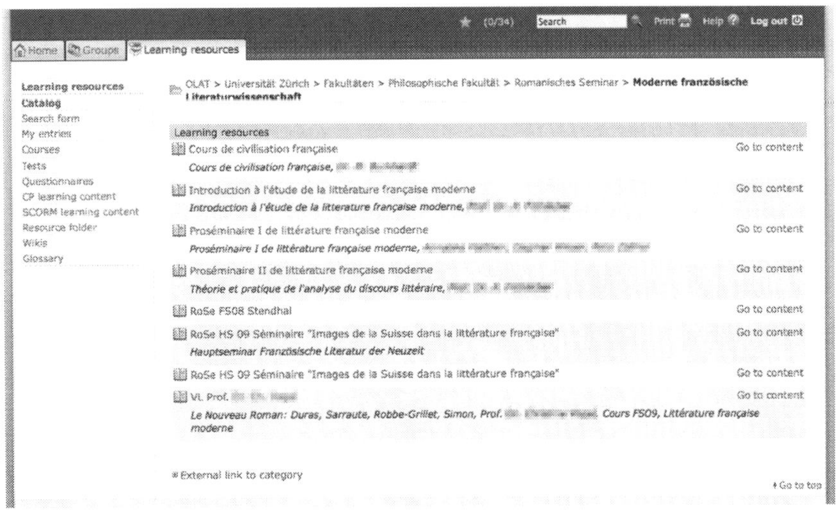


*Figure 10. Start page of an OLAT course*

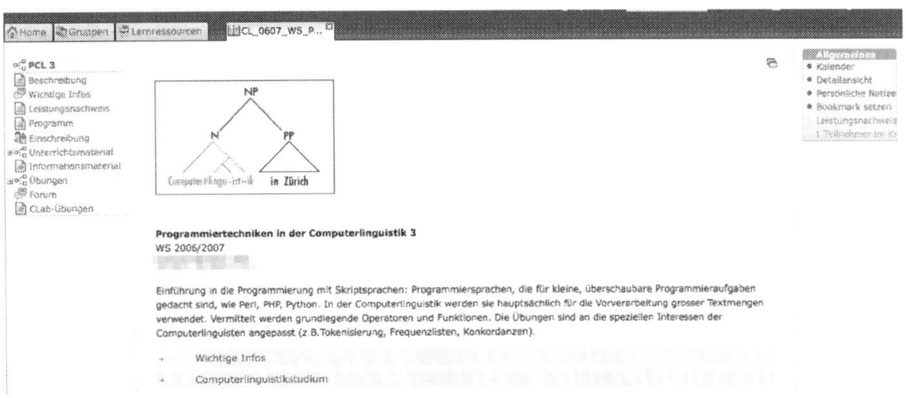

she is accessing (the highlighted one) and which other elements are available. The toolbox at the right column is still available, just the main part of the page changed. There is no need to go back to the start page of the course to choose the next element. There is no hidden structure of acitivities or documents used within a course; the structure or hierarchy presented corresponds to the hierarchy the instructor chose when setting up the course.

To sum it up: The way OLAT presents courses fulfills the demands of usability of websites. Users always now where they are, where they came from, and where they can go. Elements for navigation are always visible at a certain space; users get familiar with the layout very quickly.

From the student's point of view, working with OLAT, i.e., accessing material, communicating, collaborating and solving tasks, is very easy. From the instructor's point of view it is a bit more complex. Since OLAT offers no predefined structures after installing it, instructors have to figure out the structure that best fits their pedagogical needs. However, preparing materials and activities for students and developing appropriate scenarios are a central issues in teaching, so instructors should invest effort anyway. Once instructors have figured out a suitable structure, they can easily use this for other courses by simply copying a course and then customizing it—copying a course creates a new course with the same structure of building blocks, uploaded files and groups as the original one; all groups are empty and all student-created content (e.g., form posts) is removed.

Since the number of elements and their arrangement is not restricted, it is possible to create highly structured courses. As in Moodle, it is possible to define the access time span for each element. Thus, instructors can prepare everything before the course starts or customize the course during the semester. Additionally, certain elements can be made accessible for certain groups only—a feature, not available in Moodle. This allows creating meta-courses to serve as home for several sub-courses. All students can access certain material or activities, other elements are accessible for a certain sub-course and a certain group of students only. Thus, OLAT supports the administration and helps instructors to keep everything together.

## Some More Technical Aspects

At our university, each cohort of bachelor (BA) students has about 300 members. In the first semesters, all students of a cohort are required to attend certain mandatory courses, which means that we have several courses with 300 students. For the face-to-face portion, these courses are split into smaller sections, but all sections have the same content, require students to solve the same tasks, and all students will take the same exam at the end of the semester. An obvious way

*Figure 11. Screenshot of a single page of an OLAT course showing a posting in a forum*

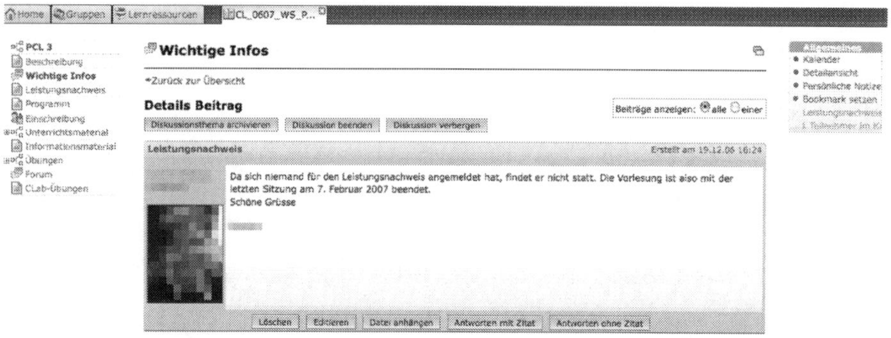

to represent this structure in an e-learning system is to have only one main online course with groups representing the various face-to-face sections. Some discussions and assignments may be done only in some sections, whereas other discussions will be open to all participants—OLAT supports creating those structures as described above. When preparing for the exams, students can do self-tests to assess their knowledge and to get an idea of the tasks to be solved during the exam. Portions of the exams may be automatically evaluated tests, participation in discussions with a very strict setting concerning topic, length of postings and time span for discussion, or essays students have to submit within a week.

Thus, there will be certain days during a semester when all 300 students of a course have to be active on the e-learning system to submit documents, take tests, browse the material for the course, post comments, etc. Of course, each student attends more than one course and we have more than one cohort at a time—the BA is laid out for 6 semesters. The e-learning system thus has to deal with several hundred of concurrent users without any delays.

Although the Moodle documentation states certain hardware requirements—"The general rule of thumb for a single server is that the approx max concurrent users = RAM (GB) X 50 and the approx max browsing users = Approx max concurrent users X 5"[32]—, none of the installations can confirm that these resources are actually sufficient.[33] Apparently, the bottleneck is the MySQL database—even though the web application (the PHP part) can be distributed over several servers, the database has to run on a single server—which means that an increasing number of concurrent users leads to a severe decrease in performance. The version 6.1 (and higher) of OLAT, however, is suitable to be installed using clustering and thus is actually scalable.[34]

In section "It's all in the marketing brochures" we have mentioned the communities for Moodle and OLAT. The members of these communities share experiences on how to use the respective systems. One main aspect is the further development. As both systems are open source, everyone is free to modify the code, to program additional modules and plug-ins. There is such development in both communities, but there is one major difference.

With Moodle, actually everybody is free to publish modules and plug-ins. There is no control loop by the main developer. This is clearly encouraging for programmers. For users, it is very disappointing, as there is no testing or quality control, neither for a specific module as such, nor for the integration and interoperability of a module with others. For example, an instructor wanted to use blogs for several courses. Moodle, as installed out of the box, only offers a site-wide blog, but that would not fit the requirements of this instructor. We did find several third-party blog modules, but it turned out that all of them had bugs and that the installation would impact the overall behavior of our Moodle installation—not to mention what would happen if the core Moodle had to be updated to a newer version. Thus we had to remove these modules and find a workaround using forums instead. The advantage of the participation of everyone at any time is put aside by the disadvantage of very poor software quality.[35]

With OLAT, every user can also modify the code and program additional features. The main developers have added a quality assurance loop for testing such contributions. They regularly publish a reference release including approved modules from other developers. Thus, an additional feature will not be released to the whole community immediately, but once it is released, there is no risk in using it, neither from the point of view of the software, nor from the point of view of instructors and students as end users. Especially the technical staff prefer this approach over that of Moodle.

## The Final Decision

At a first glance, both e-learning systems in question, Moodle and OLAT, seemed very similar, offering basically the same features to support teaching and learning. The technology used seemed to be appropriate and state-of-the-art. When testing both systems with real didactical scenarios, however, it turned out that Moodle is too restrictive to be appropriate for use at a university. Additionally, Moodle exhibited significant shortcomings in terms of usability and technical quality. Therefore, the e-learning systems are not equivalent, as we had initially assumed. After a closer examination, it became clear that only OLAT suits our requirements for an e-learning system for the School of Social Work.

We will thus introduce OLAT as the main e-learning system at our university. If we would have chosen randomly, we may have introduced Moodle, which seemed to be a valid solution at the very beginning. However, when actually using it for all of our courses, we would have faced the disadvantages outlined above. But at that stage, it would have been almost impossible to step back and give the other system a try. Moreover, no one could have guaranteed that the other system would not show similar or other shortcomings. The upfront investment of time and effort into a thorough evaluation of both systems with scenarios accurately representing our actual requirements for teaching and learning in the end prevented us from wasting significant resources for workarounds and third-party solutions.

We therefore encourage stakeholders involved in the decision process for the e-learning system of an educational institution to look beyond the marketing brochures and web pages of the vendor or developer of the candidate systems. It pays off to install a *full working version*—a "test version" may have restricted features and access rights, and thus may give a wrong impression. This should easily be possible for open-source systems. Next, the systems must be tested with scenarios as close as possible to the scenarios the e-learning system is intended to support in daily teaching and learning.

## CONCLUSION

In this chapter, we have shown that abstract criteria and feature lists are not sufficient for choosing the right e-learning software for a university, even when all stakeholders and their respective requirements are heard.

Todays e-learning systems offer very similar feature sets. However, seemingly small details, which are typically neither mentioned in the requirements nor in the feature lists, can make a big difference. Some critical requirements are influenced by the culture of teaching and learning at a specific institution, but these are rarely made explicit. Thus, instructors and students will discover shortcomings of the e-learning system only after its large-scale introduction at the institution, when it is no longer possible to reverse the decision.

We thus propose that institutions must evaluate prospective systems in real-world scenarios. This requires the design of realistic didactical scenarios of the types that are to be supported, taking into account the needs of all involved user groups (i.e., students, instructors, administrative staff, and technical support) before deciding on an e-learning system.

From a pedagogical point of view, we have outlined a number of significant differences between two e-learning systems, namely Moodle and OLAT, which can be seen as representative for open-source systems used at universities, and which both offer all features that are commonly seen as the most relevant when making a decision for a university-wide e-learning system. Generally speaking, we come to the conclusion that Moodle may be suitable for supporting teaching and learning in primary schools, but it is not suitable for hierarchically structured institutions of higher education offering a wide range of courses, seminars, and lectures with different didactical scenarios for educating adults, i.e., universities.

We argue that universities should thus use systems offering both instructors and students a wide range of features, which can be used individually or in combination to meet educational objectives and intentions. OLAT can be regarded as a prototypical e-learning system that meets these requirements.

## REFERENCES

BMBF (Federal Ministry of Education and Research). (2004). *eLearning-Dienste für die Wissenschaft Richtlinien über die Förderung der Entwicklung und Erprobung von Maßnahmen der Strukturentwicklung zur Etablierung von eLearning in der Hochschullehre im Rahmen des Förderschwerpunkts Neue Medien in der Bildung*. Retrieved July 26, 2009, from http://www.bmbf.de/foerderungen/2576.php

Conole, G., & Oliver, M. (Eds.). (2007). *Contemporary Perspectives in E-learning Research. Themes, methods and impact on practice*. London, New York: Routledge.

Cotton, J. (1995). *The theory of assessment. An introduction*. London: Kogan Page.

Garrison, D. R., & Kanuka, H. (2004). Blended learning: Uncovering its transformative potential in higher education. *The Internet and Higher Education*, *7*(2), 95–105. doi:10.1016/j.iheduc.2004.02.001

Gerson, S. M. (2000). E-CLASS: Creating a Guide to Online Course Development for Distance Learning Faculty. *Online Journal of Distance Learning Administration, 3*(4). Retrieved June 21, 2009, from http://www.westga.edu/~distance/ojdla/winter34/gerson34

Grund, S., et al. (2002). Concept for interdisciplinary evaluation of learning tools (CIELT). In F. F. C. Lutz, P. Schulz & L. Cantoni (Eds.), 4th International conference on new educational environments (ICNNE), Manno, Lugano, Switzerland.

Helander, M. (1997). *Handbook of Human-Computer Interaction*. New York: Elsevier Science.

JISC (Joint Information Systems Committee). (2006). *Glossary E-Assessment*. Retrieved July 27, 2009, from http://www.jisc.ac.uk/uploaded_documents/eAssess-Glossary-Extended-v1-01

Kleimann, B., & Wannemacher, K. (2005). E-Learning-Strategien deutscher Universitäten: Fallbeispiele aus der Hochschulpraxis. [Hannover, Germany: Hochschul-Informations-System.]. *HIS-Kurzinformation, B4*, 2005.

Koutsonikola, V., & Vakali, A. (2004). *LDAP: Framework, Practices, and Trends*. Retrieved July 27, 2009, from http://www2.computer.org/portal/web/csdl/doi/10.1109/MIC.2004

Kubicek, H. (2004). *Organisatorische Einbettung von E-Learning an deutschen Hochschulen (Technical report)*. Bremen, Germany: Institut für Informationsmanagement.

Nielson, J. (1994). *Usability Engineering*. San Francisco: Morgan Kaufmann.

Nielson, J. (1997). *Be succinct! (Writing for the Web)*. Retrieved June 21, 2009, from http://www.useit.com/alertbox/9703b.html

Ospelt, T., & Aegerter, D. (2007). *Eine Hochschule—Eine Lernplattform. Eine Anforderungsanalyse*. Diplom Thesis, Zurich University of Applied Sciences, Winterthur, Switzerland.

Piotrowski, M. (2009). *Document-Oriented E-Learning Components*. PhD Thesis, Otto-von-Guericke-University Magdeburg, Germany.

Powazek, D. (2006). *Where Am I?* Retrieved June 21, 2009, from http://www.alistapart.com/articles/whereami

Specht, M. (2008). *Designing Contextualized Learning. Handbook on Information Technologies for Education and Training* (pp. 101–111). Berlin, Germany: Springer. doi:10.1007/978-3-540-74155-8_6

Wiepcke, C. (2006). *Computergestützte Lernkonzepte und deren Evaluation in der Weiterbildung. Blended Learning zur Förderung von Gender Mainstreaming*. Hamburg, Germany: Kovac.

Zimmermann, T., et al. (2008). Dialog mit 200 Studierenden—geht das? Blended Learning in einer Vorlesung mit hoher Teilnehmerzahl. *Das Hochschulwesen, 2008*(6), 179–185. Retrieved July 30, 2009, from https://www.zora.uzh.ch/11297/

## ENDNOTES

1 See Kubicek et al. (2004), or Conole & Oliver (2007). You will find a so-called e-learning strategy for almost any university in Europe. Partly this is due to funding policies; see, for example, the guidelines of the BMBF (Federal Ministry of Education and Research) in Germany (BMBF, 2004).

2 When looking at e-learning strategies published by universities, the e-learning system in place is usually mentioned at the very beginning and as a prerequisite to implement e-learning as a pedagogical option for instructors. See for example e-learning strategies of the Swiss Federal Institute of Technology Zurich (http://www.net.ethz.ch/e-learning-strategie.pdf [accessed 26.7.2009]), the University of Leicester (http://www.le.ac.uk/strategies/elearning/ [accessed 26.7.2009]), the University of Insbruck (http://www.uibk.ac.at/elearning/strategie/ [accessed 26.7.2009]), or Kleimann & Wannemacher (2005). Additionally, when involved as consultant in the process of implementing e-learning at a university—as the author—you will be asked very early which will be the e-learning system to be used.

3 For example eLearning journal, online available at http://www.elearning-journal.de/ [accessed 27.7.2009] or Interdisciplinary Journal of E-Learning and Learning Objects, online available at http://ijklo.org/ [accessed 27.7.2009].

4 In the strategy papers of universities mentioned above, you will find that every institution defines their own understanding of e-learning.

5 http://www.elearningeuropa.info/?page=glossary [accessed 27.7.2009].

6 There is a saying, that university instructors would rather use their colleague's toothbrush than their colleague's teaching materials.

7 http://www.birds-eye.net/definition/s/sso-single_sign_on.shtml [accessed 27.7.2009].

8 The University of Applied Sciences Northwestern Switzerland (FHNW) is a merger of nine universities (the School of Applied Psychology, the School of Architecture, Civil Engineering and Geomatics, the Academy of Art and Design, the School of Life Sciences, the Academy of Music, the School of Teacher Education, the School of Social Work, the School of Engineering, and the School of Business) distributed over four cantons. Their task is to offer and administer degree courses, continuing education, research and development activities and offer consultancy and related services. (http://www.en.fhnw.ch/ [accessed 27.7.2009]).

9 The costs should be as low as possible; open source requires no money for licenses, but for maintenance only.

10 http://www.medidaprix.org [accessed 27.7.2009].

11 http://www.imsglobal.org/pressreleases/pr090514.html [accessed 27.7.2009].

12 http://www.php.net/ [accessed 27.7.2009].

13 http://www.mysql.com/ [accessed 27.7.2009].

14 http://www.java.com/ [accessed 27.7.2009].

15 https://www.hibernate.org/ [accessed 27.7.2009].

16 http://www.postgresql.org/ [accessed 27.7.2009].

17 http://www.oracle.com/ [accessed 27.7.2009].

18 http://www.switch.ch/ [accessed 27.7.2009].

19 http://www.switch.ch/aai/ [accessed 27.7.2009].

20 http://shibboleth.internet2.edu/ [accessed 27.7.2009].

21 http://www.sap.com [accessed 27.7.2009].

22 Discussion based on a document.

23 Instructors assign a task to students; students work on the task and submit their solutions. This document (or the documents) is only visible to the instructor, who will annotate the solution with remarks and give it back to the student. The instructor also will grade the solution handed in by the student. It is possible to allow a resubmission.

24 We will explain the elements and areas of the page in the section "Usability." We have anonymized all screenshots. As the language used for teaching at our university is German, the screenshots showing actual courses contain German content. As the reader of this chapter is not supposed to understand each text shown in the screenshots, but will be directed to certain structural elements, we did not translate these texts.

25 http://docs.moodle.org/en/Groups_FAQ [accessed 27.2.2010].

26 A module is an instruction entity according to the terms used in the so-called Bologna Process (introducing the anglo-saxon system of Bachelor and Master to replace the European degrees of Diplom, Magister and Lizentiat). Unlike lectures or seminars, a module is not restricted to the time span of a single semester, but can include several lectures in several semesters.

27 We will explain the elements and areas of the page in the section "Usability."

28 Instructors usually get lists (as Excel files or as pure text files) from the administration listing the students enrolled in their lecture or seminar. Students have to enroll with the campus management system to ensure the ECTS credits they earn will be registered correctly, etc.

29 http://docs.moodle.org/en/Pedagogy#Social_Constructionism_as_a_Referent [accessed 27.7.2009].

30 Building blocks represent the instances for the features listed in table 1, e.g., forum, test, chat, wiki, and HTML pages.

31 Note that the forum we entered in the Moodle course had the same purpose and was also placed at the beginning of the elements listed in the course.

32 http://docs.moodle.org/en/User_site_capacitics [accessed 30.7.2009]. Interestingly no information about optimal configurations or setups are given.

33 For example, Ospelt & Aegerter (2007) report that the Zurich University of Applied Sciences used 8 GB RAM for their Moodle server. According to the documentation this should be enough to handle 400 concurrent users. However, the system could handle only 25% of the theoretical load. The reasons for this surprising behavior could not be investigated.

34 The University of Zurich uses an installation on three servers with about 2800 courses, over 51000 registered users and over 1000 concurrent users without any performance issues.

35 Even the starting point for Moodle is not very convincing from a computer-science point of view. The word "Moodle" stands for "Modular Object-Oriented Dynamic Learning Environment" (http://docs.moodle.org/en/About_Moodle [accessed 30.7.2009]), clearly refering to object orientation.

However, although PHP supports object-oriented programming, and contrary to the name, it is programmed without using this principle: "Early on I made the decision to avoid using a class-oriented design—again, to keep it simple to understand for novices. Code reuse is instead achieved by libraries of clearly-named functions and consistent layout of script files." (http://docs.moodle.org/en/Moodle_architecture [accessed 30.7.2009]). (The bad impression concerning the quality of the software itself derives thus from the very beginning.)

# Chapter 5
# Technology Enhanced Distance Learning Utilising Sakai CLE and Adobe Connect Pro

**Juley McGourty**
*University of Limerick, Ireland*

**Angelica Risquez**
*University of Limerick, Ireland*

## ABSTRACT

*On-line environments have been incorporated in the Distance learning programmes of the International Equine Institute (IEI) in order to address concerns about streamlining assessment turn-around, distance student attendance at tutorials, providing more detailed and quicker assignment feedback, student peer interaction, student to tutor[1] interaction and, of course, student support. The overriding concern was to provide a more flexible, active learning environment to develop and enhance learning opportunities while, concurrently, integrating more closely the learning activities of the student with the University of Limerick (UL) community. The impetus, therefore, was to make studies convenient and attractive to the location of the distance student, while maintaining educational quality through the provision of pedagogical innovations and at the same time providing a social and interactive environment to support the distance student. In so doing, the IEI uses the collaborative learning environment (CLE) Sakai (www.Sakaiproject.org) to support the distance student and also utilises Adobe Connect Pro™ to deliver on-line synchronous desktop-to-desktop tutorials. This chapter outlines aspects drawn from our experiences with the on-line support and delivery of distance learning programmes. Throughout, various recommendations on enhancing the experiences for students are also presented.*

## INTRODUCTION

The use of on-line technologies[2] has changed contemporary education in ways that were not imaginable in the past. The impact on higher education, and most especially distance learning, has become significant (Kingsley et al., 2009; Spector, 2001). According to Bennett et al., (2004) "Advancements in on-line technologies have facilitated a convergence of distance and campus-based learning". This phenomenon has progressed (Bennett et al.,

DOI: 10.4018/978-1-61520-853-1.ch005

Copyright © 2010, IGI Global. Copying or distributing in print or electronic forms without written permission of IGI Global is prohibited.

2004) to such an extent that programmes offering courses through distance learning have begun involving a mix of reduced face-to-face classes and greater on-line interaction (Lockyer et al., 2006), including on-line access to course materials, on-line virtual tutorials, and collaborative learning environments (CLE), to mention just a few of the many flexible on-line technologies currently available (Concannon et al., 2005).

Universities are also now required to cater to the lifelong learning market by offering greater flexibility in learning opportunities (Robins et al., 1999) as cited in (Eynon, 2008). As a result, flexible on-line learning technologies are becoming widely used in a response to the differing needs of students. Effective integration of these technologies into higher education is becoming an essential proficiency for tutors (Ala-Mutka et al., 2009), especially for those involved in distance learning. Expertise in the use of the technology and competencies in the various learning elements and pedagogy, and a range of social interaction skills, have become necessary to effectively implement and, subsequently, integrate the technology enhanced learning (TEL) into the design of curricula.

The IEI, established in 1993, is part of the Department of Life Sciences in the Faculty of Science and Engineering at the University of Limerick, Ireland. The IEI coordinates the distance learning delivery of Honours Certificate and Diploma in Science (Equine Science) programmes. For further information, see the edited transcription of questions and answers (2009) administered by the IEI to tutors on its programmes (Appendix 1).

## Objectives

The objectives of this chapter are to:

- Discuss the background to the integration of on-line technologies into the IEI programmes
- Evaluate the integration of on-line technologies into the programmes in the context of more traditional learning opportunities and through this, discuss issues, problems and controversies encountered by the IEI
- Discuss solutions and recommendations established by the IEI through their experiences of using on-line technologies on the distance learning programmes

## Background

The concept of lifelong learning has created a greater focus on the provision of open and flexible learning opportunities developed to accommodate the very many people who were unable to commit themselves to full-time education. The Open University, UK enrolled its first students in 1971 and Oscail, based at Dublin City University, begun enrolling distance learning students in 1982. Other institutions across the world, some for many years prior to the 70's and 80's, were also in the business of offering open and distance learning programmes. Increasing numbers of students were looking to access education and training opportunities through use of on-line technologies. Eynon (2008) confirms this logic in universities outside of Ireland. As a result of these initiatives, and based on recognition of a growing demand from potential students and current faculty for wider and more flexible access to programmes, the University of Limerick commissioned a collaborative learning environment (CLE), called Sulis which is based on the SAKAI CLE framework. This CLE reflects the "dual mode package" concept advocated by Enyon, (2008) which caters to students both on and off campus. The Equine Science distance learning initiative, at the University of Limerick, relied on TopClass CLE technology to support delivery of its programmes, until the adoption of the newly commissioned Sulis CLE in 2007.

The need for the utilisation of on-line technologies on the Equine Science distance learning programmes derived, primarily, from the profile of the IEI student population, the facilitation of student accessibility and flexibility in a 'blended–learning' delivery (Saunders et al., 2003), the supplementation of face-to-face and on-line tutorials, and a range of other opportunities, such as the promotion of collaborative learning, on line assessment and greatly reduced costs to the participating students of undertaking studies of a validated programme. The on-line virtual environments of the IEI programmes addressed concerns about student support by facilitating staff–student and student–student interaction, streamlining assessment turn-around, and providing more detailed assessment feedback. In addition, the IEI was acutely aware of the negative feelings resulting from a sense of isolation that a distance-learning student can experience (Buckley et al., 2007). The overarching goal of the adoption of the distance learning paradigm was to provide a more flexible, user-friendly, active learning environment incorporating social, technical and pedagogical solutions. The on-line technologies were additionally developed to enhance learning opportunities, thus "redressing some of the inevitable imbalance" (Dorrian et al., 2009), between students who are able to attend a university campus, and those with life circumstances that make travelling to the physical campus not feasible.

The typical profile of the distance-learning student of the IEI is mature[3] and "part-time". While some students take the course immediately after completing secondary education, a large number of students are already experienced career holders. Students may already be in full-time employment, be on short career breaks and/or have prior family commitments and, therefore, have very little time for attending face-to-face tutorials that could involve travelling distances. Therefore, it was necessary that students be provided with support regardless of their personal circumstances or location in the country. In addition, adoption of the 'new on-line technologies' enabled the IEI to keep pace with other providers of distance learning programmes. Private for-profit higher education institutions are offering through distance learning mode, a wide range of undergraduate and postgraduate programmes.

From a learning perspective, the use of online technologies have also facilitated the ability to directly transfer knowledge from the tutor to the student, encouraging collaborative and self directed learning in a manner that allows students to "process the information so that it becomes personally meaningful to them" (Atkins et al., 2007). This has, therefore, placed pressure on the traditional forms of higher education (Concannon et al., 2005) by offering innovations in the support and delivery of academic material through on-line environments where the design of such environments are interactive, peer-generated and collaborative, and move beyond standard pedagogy and theories of learning. Some observers would agree that such environments can be used to powerfully augment more traditional learning modes and materials (Atkins et al., 2007; Dorrian et al., 2009).

On-line technologies have provided a whole new mechanism for overcoming distance; for accessing information from the convenience of one's home or workplace; for pursuing the accredited learning programmes of institutions such as the IEI, and for interaction with other learners in a virtual environment. On-line technologies can improve the quality of the educational experience by providing rich, exciting and motivating environments for learning (NSF, 2008), and can accelerate positive trends such as increased emphasis on information handling and problem-solving and reduced emphasis on memorising facts (Science, 2000), while encouraging the development of creativity, imagination and self-expression for both students and tutors.

## Delivery and Support

### Using Sulis (SAKAI)

The CLE, Sulis, (Figure 1) utilised by the IEI, is one of the key ICT mechanisms that facilitates the delivery support of the IEI's suite of distance learning programmes. This is built around the framework and the code produced by the SAKAI open source community. The University of Limerick has adopted the title 'Sulis" for its SAKAI CLE platform (http://sulis.ul.ie). The programme of study is divided into courses (modules) calculated to generate a quantum of learning equivalent to approximately 120 hours each. Each module is supported with a customised hard copy learning pack and on-line support material made available over the CLE. The latter support comprises additional information, readings, web and DVD links, on-line library searching, explanations, clarifications and, in some cases, module-specific software, as appropriate to the particular module. In addition, operational information such as schedules of tutorials, information on field visits and on-line discussion opportunities are also provided. Students are enabled to interact with each other, with tutors and with the IEI, and conversely, the IEI and tutors with students. Students can submit their assignments on-line and/or undertake on-line assessments; in both cases students are provided with on-line feedback. In effect, the IEI adopts the view that the CLE, Sulis, operates as the 'virtual classroom'.

The CLE is managed by the IT Division at the University, but customised by the IEI. The IT division, while dictated by the design of SAKAI, moderates the level and range of interactivity possible with the CLE, and this control is evident when navigation and appearance are considered. Each module moderator, in collaboration with the IEI, then populates the content of the CLE to satisfy the needs of their students and/or their curricular and course goals (Atkins et al., 2007). From a pedagogical perspective these sites and tools need to be well designed, relevant to their needs and appropriately embedded into the culture of the course (Conole et al., 2007) to effectively engage with the students' learning processes.

Students are introduced to the use of on-line searching and availability of resources from the outset of the programme and this training, undertaken at the first tutorial (face-to-face), becomes an integral part of their familiarisation with Sulis. This is critical to providing students with some of the tools essential to satisfactorily exploit the research-based, collaborative learning approach promoted throughout the programme modules.

### Using Adobe Connect Pro

The IEI utilises Adobe Connect Pro™ for the delivery of on-line synchronous desktop-to-desktop tutorials. Adobe Acrobat Connect Professional (formerly Macromedia Breeze) is software used to create information and general presentations, on-line training materials, web conferencing, learning modules, and user desktop sharing in a "live classroom" communication environment. IEI student users of Adobe Connect Pro require access to a computer with broadband, a headset (or earphones) and a web camera. Adobe Connect Pro has an intuitive, attractive interface. It runs on the ubiquitous Flash Player across the Internet, and therefore, students and tutors are not required to download any additional software. All virtual meeting rooms have interfaces that are very intuitive and are organized into 'pods'; with each pod performing a specific role.

Some of the features include:

- Virtual meeting rooms
- Audio and visual interaction
- Share pod for presentations, video clips, sound files, and documents
- Polling and testing
- Chatting
- Hands-on labs
- Breakout sessions

*Figure 1. Basic outline of a Sulis page (Powered by Sakai)*

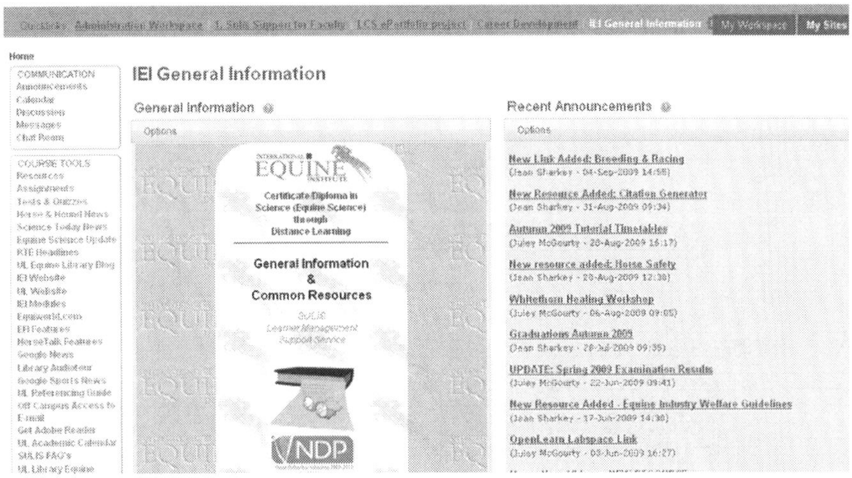


- Recording sessions
- Interactive Whiteboard use

The IEI, after initially piloting Adobe Connect Pro with a group of more experienced Equine Science Diploma distance-learning students, has now adopted it as the on-line mode of delivery for all its modules. The IEI follows the procedure of creating an on-line virtual meeting room on Adobe Connect Pro. Remote students are provided (by email) with an Internet link to the live classroom session, where they enter their name. Once in the virtual room, all members can broadcast live synchronous audio and video material. As long as the students have access to broadband, they can use their own home or an alternative preferred location. Tutors have the option of recording their tutorials and posting these recorded links to the Sulis CLE for access by students at a later date.

Once entered into the room, the students are able to (Figure 2):

- See the tutor and other class members through a web camera
- Communicate with the tutor, and other students through voice protocols or chat rooms
- See the contents of the interactive Whiteboard, PowerPoint presentation or additional file formats in use
- Revisit recorded key lectures and elements of key topics, thus creating a bank of resources
- View pre-recorded practical tasks and procedures for later use.

## Technology Enhanced Learning (TEL)

### Issues, Controversies and Problems

It is generally accepted that any influence of technology will vary as a result of a range of complex and interrelating factors (Eynon, 2008). Implementing a virtual learning environment is not without pedagogic, technological and social impact (Lyons, 2009) on both the student and the tutor. The lack of uptake of technology for educational programmes can be due to a complex mix of inter-related factors including the contrast between a traditional and student-centred approaches to teaching, faculty resistance to change and lack of the necessary digital literacy skills, the fixed and static nature of curriculum and assessment prac-

*Figure 2. Basic outline of Adobe Connect Pro meeting room*

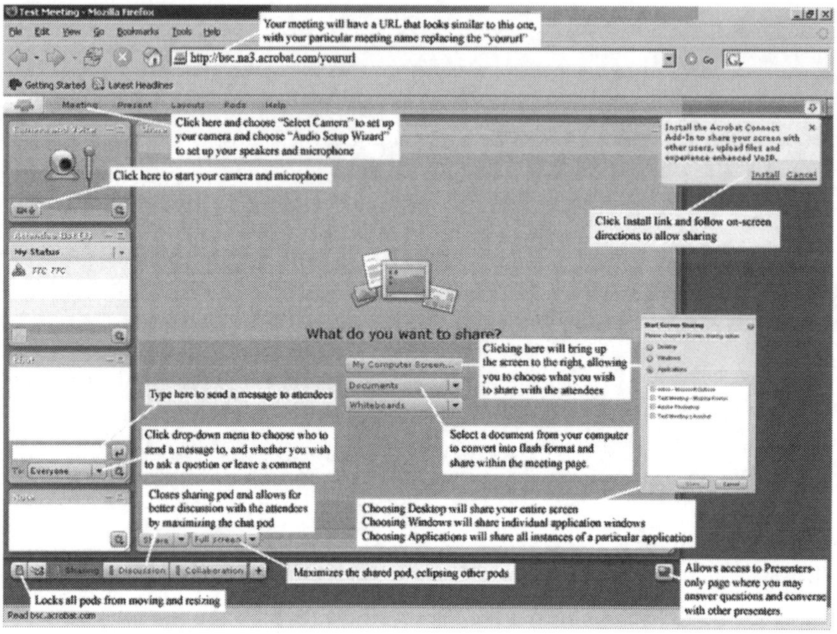


tices, and infrastructural factors such as teacher workload and lack of time (Conol, et al., 2008).

**Pedagogical**

With the rapid development of emerging technologies, the integration of on-line technologies has increasingly attracted the attention of educators (Wang, 2008). What is necessary to consider is that the use of on-line technologies is contingent on not just underpinning traditional methods of teaching and learning but the innovative and interactive use of these technologies to enhance a students academic experience and increase pedagogical quality (Devine, 2005). Eynon, (2008) stated that it is necessary to be concerned about the use of technology for technology's sake and the importance of using the web appropriately (Eynon, 2008). In fact, the argument exists that TEL is not about the technology itself but about the appropriation of technological tools to achieve results (Ala-Mutka et al., 2009; Eynon, 2008).

Careful introduction of on-line techniques into distance learning is therefore necessary, as it may represent a substantive change in learning style for students and for pedagogy on the part of the tutor. Academic module designers need to plan thoughtfully before beginning any on-line technology integration into a module. Learning objectives and contexts need to be identified and the correct CLE tools incorporated. Existing materials and resources may need to be modified or developed for the virtual environment to engage students. While it has been found that equine science students can adjust to web-based learning, the most important predictor of student satisfaction is excellence in pedagogy, irrespective of technology (Dorrian et al., 2009), and in order to achieve this, tutors utilising on-line technologies need to have the appropriate skills. In this context, there is agreement for the need of improved tutor development opportunities focussed on the academic as a skilled pedagogical designer, virtual environment facilitator and academic educator (Devine, 2005; Wang, 2008).

**Technological**

While there is a tendency for Irish universities (Concannon et al., 2005) to utilise the benefits of on-line learning technologies as a mechanism to facilitate and improve learning quality and performance in students both on and off-campus, the self-directed learning techniques such as those typically employed in on-line distance learning are not always met with enthusiasm from both students and tutors (Dorrian et al., 2009). In fact, utilising technologies can often cause anxiety in students and also those tutors who are unfamiliar with on-line environments. This can have a snowball effect in relation to the degree to which tutors encourage and direct students to relevant support sources. Depending on individual departments and tutors, some provide links to recommended sites and CLE tools and resources, while others do not. These findings reflect similar positions reported at the IEI.

The age ranges, educational experiences and academic profiles of the cohorts of the Equine Science distance-learning students are quite disparate. In general, they tend to be older and have been out of formal education for many years. Most would not have previously undertaken any formal third level programme of study. Students belonging to this type of profile are more comfortable with the more traditional face-to-face didactic style of education. Technology is not necessarily a ubiquitous part of their daily environment. Not all students would have an easy access to the Internet nor would the students have the same abilities using on-line technologies. From an educational and, especially, a technology perspective, students may be somewhat uncomfortable with discussing a situation or a topic while not knowing ahead of time what they need to know to succeed. In that sense, they expect to discover or uncover knowledge as they explore a domain, be that a CLE or a book. Many are unfamiliar with the rapid multitasking and accumulation of data that the typical, on-campus student takes for granted; they expect to be told by an authority to read a particular selection of books or manuals and may not have developed the digital or collaborative skills required for utilising on-line technologies prior to enrolling in the IEI programmes.

**Social**

A further significant issue encountered is the inevitable reduction in person-to-person contact when on-line technologies are used; students, especially distance-learning students, currently demand more contact, not less, and a lack of contact is believed to negatively effect students' motivation to learn (Eynon, 2008). Both the student and the tutor have expressed that they miss the face-to-face interaction and the body language that is an integral aspect of such teaching. In fact, in many cases the interpersonal skills of the tutor are almost of as much importance as the reliability and intuitiveness of the pedagogy and technology utilised.

## Solutions and Recommendations

**Pedagogical**

It is important that students are not only comfortable with on-line technologies, but also it is imperative that tutors do not find the change in delivery method to be a barrier to effective teaching. Tutors need to be motivated and prepared to take advantage of the potential opportunities available through the use of on-line technologies. According to Devine (2005), on-line technologies have great potential for radical innovation afforded to distance learning when deployed within a well-developed and robust technical and pedagogical infrastructure (Devine, 2005). Therefore, appropriate selection of interactive tools that are incorporated into the on-line technologies will act as a bridge between good pedagogic practice and effective use of new technologies (Conole et al., 2008). In addition, the interface design of on-line technology support must be attractive so that it can motivate and engage learners (Wang, 2008). Thus it was (and still is) believed that the interface of each module should comprise fewer words, more graphics and much more dynamism or interactivity in a highly structured, more resource-based style of pedagogy when authoring courses for the web (McAndrew et al., 2009).

As a result, IEI tutors are encouraged to adapt the base set of programme materials and resources as suits their individual needs and teaching styles. The lesson plans, materials, resources are stored and organised on-line as required; through username/password authentication the tutors can upload, as well as access, the materials. Students get access to the materials and resources through a similar username/password authentication. In addition, the on-line storage and access of materials and resources, including recorded tutorials, has enabled the creation of a much more coherent and consistent set of programmes.

It is, therefore, most desirable that the pedagogy of on-line tutorials be incorporated into any training sessions relating to the use of on-line technologies. To encourage the use of on-line technologies, tutors, as part of a pedagogy seminar organised by the IEI in conjunction with the Centre for Teaching and Learning at the University of Limerick developed learning activities incorporating on-line technology tools based on the premise that tools were taught to the tutors on a need to know basis to enhance the pedagogical aspects of their modules. Tutors were shown that the technology becomes the link between them and their students.

This seminar had the effect of:

- Creating a much more consistent approach to the development of pedagogical schemes. This arose from bringing all the tutors together, and through discussion, the development of greater communication and the sharing of experiences on learning activities across all modules offered by the IEI.
- Creating a greater engagement of all tutors with the adoption and use of technology; the seminar was timely and enabling in that it coincided with and promoted a more seamless move from the face-to-face tutoring to the use of on-line 'live classroom' technology
- Reducing tutor resistance to the on-line technologies; its uses and potential values were gradually unveiled to them and they became an integral part of the adoption process. Discussion forums are important instructional strategies that have numerous advantages such as promoting critical thinking and knowledge construction and improving relationships (Wang, 2008). These were incorporated as an activity to promote collaborative learning opportunities among the tutors.

Students have now reported satisfaction with the ease of access to the on-line module-specific information and material or pedagogical sources of such material dedicated to their studies. Students felt that this has greatly facilitated advance preparation for on-line tutorials, quizzes, examinations and laboratory sessions. Students have experienced the flexibility of utilising the on-line technologies and can access materials and resources from home and workplace, and can revisit these at any time. In addition, this allows students to manage their time more effectively and encourages self-learning. They can catch up on any missed tutorials or discuss any problems without using the phone or travelling to the University of Limerick campus.

**Technological**

The IEI staff realised early on in its on-line technologies development that when selected tools of the CLE were mandated for use, improvements in students' applications to learning were evident. To assist with the use of these tools, the IEI provided students with specially designed TEL Handbooks offering suggestions on how to study, clarifying the learning objectives, and providing information on how to use the technology as a support tool for the wider task of preparing for the module. The IEI also offered on-demand technical support, assistance and training with aspects of the technology to enable the students to become familiar with and comfortable in the competent use of software and the tools.

When incorporating on-line technologies into the distance learning programmes, the IEI took into account that learners have different approaches to learning. The on-line technologies encourage the students to take responsibility for their own learning and provide them with the choice of when/how they learn. This method instills a responsibility for their learning by maximising their control over its content and delivery (Dorrian et al., 2009). These principles are enabled through a commitment to flexible delivery, providing students with a range of learning resources. In practice, this is achieved through the support of the IEI, which encourages faculty to use a combination of more traditional forms of information transfer and communication (e.g. printed materials, face-to-face lectures, and tutorials and workshops), and innovative methods (e.g. websites, interactive lesson plans, on-line discussion forums, and desktop-to-desktop tutorials). The IEI has also realised that ease of learning is critical for beginners while ease of use becomes more important as users gain experience over time.

Encouraging all tutors to engage with the on-line technologies can be a difficult task for many reasons, but in particular for those who are not comfortable with its use. The response to new learning technology also depends on the effectiveness of the staff development model within the University. An additional barrier to effective delivery of these on-line techniques is the maintenance of excellence in pedagogy, irrespective of the technology used. As mentioned earlier, technical support for the Sulis CLE, on a University campus-wide scale, is provided by the IT Division; however it is the Centre for Teaching and Learning that plays the major role in servicing the pedagogical and training needs of the tutors. This support and training for tutors aids in alleviating feelings of resistance to the technology that might be experienced. These latter needs are serviced through drop-in sessions for students and tutors, seminars, on-demand training, on-line support courses and much more. Encouragement is given through positive reinforcement and changing the way faculty think and respond to the CLE; this has the impact of enabling tutors to further improve their courses by enhancing the quality of the pedagogy used and the support mechanisms employed through the on-line technologies. This creates a positive perspective for the use of CLEs in the University where flexibility and good pedagogical approaches are central to both the student and tutor educational experiences.

Social

The technological component becomes more critical in a technology-enhanced learning environment, where many learning activities are conducted through the support of a computer. On-line learning is often undertaken by the students in their home or place of work in physical isolation from others studying the same material. It is, therefore, necessary that the on-line technologies create and maintain a friendly, interactive environment in which participants feel safe and comfortable enough to interact with one another (Anderson, 2004; White, 2004); it must be available all the time and access must be convenient and fast. The consistency of presentation format for the materials and resources in each module provides familiarity for the students, and makes it easier for them to navigate and access the relevant learning material and develop collaborative and communication skills.

However, despite the general consensus amongst the students that on-line course materials 'were a good thing', the importance of face-to-face contact with tutors was still considered necessary and important. The students interviewed described the benefit of meeting with class colleagues and tutors to discuss work issues. Face-to-face contact was considered vital in building a sense of community or 'belonging' to the class and Equine Science programmes. For many, on-line environments were not a substitute for face-to-face meetings with tutors. To assist with this, the IEI holds the first tutorial of the semester as a face-to-face one, where students can interact with fellow students, engage in essential social networking and gain the training required for additional use of the on-line technologies, and learning skills. Subsequent tutorials are then conducted using Adobe Connect Pro. The use of Adobe Connect Pro has the potential to enhance the pedagogical aspects of the tutorials while assisting in the social interaction preferred by students and tutors alike. Once new users get over the initial shyness of using this technology, their confidence soon grows. Adobe Connect Pro permits the development of a collaborative and communication environment and, in addition, it offers great advantages in areas of communication and savings in terms of time, travel, and interactive learning techniques.

## Discussion

Utilising Sakai and Adobe Connect Pro has the potential to impact on learning processes by fostering new ways of collaborating, in particular by:

- Building on distributed knowledge
- Developing interactive pedagogy
- Enabling peer group learning
- Providing a framework for professional and academic interaction
- Supporting a centralised model of TEL that a larger group of learners can access
- Giving rise to learning communities
- Creating innovative collaborative dynamics
- Allowing learners to generate new learning contexts where reflective learning transforms the very process of learning.

It should be noted, however, that it is the underlying pedagogic design that facilitates this change rather than the technology itself.

The most prevalent theme arising from the pedagogy seminars and workshops conducted with the tutors was the importance of a reward structure that encouraged students to make greater use of and to rely on the on-line resources, without punishing them if they chose not to use them. It was emphasised that students would only fully engage with the on-line technologies when it contributed to achieving an academic grade. In some cases, marks were assigned for involvement in discussion boards that promoted collaborative learning. This is not a new phenomenon. Concannon et al., (2005) also encountered a similar situation and stated, that without adequate reward structures, students were unlikely to access the on-line resources or tasks, despite recognising that they would assist their preparation for the final exam at the end of the semester.

Additional comments from the tutors' seminars included reference to lack of time, lack of interest in using technology, uncomfortable with moving from their current method of delivery, lack of technical skills, lack of technical support, course constraints and role changes.

Therefore, institutional changes incorporating greater incentives for tutors to utilise the on-line technologies in their teaching and module design can be of benefit for those tutors who may be reluctant to change. These incentives can include improved levels of practical and technical support for tutors. Tutors should feel secure in the knowledge that they need only adapt the module in ways that are appropriate to them and their students rather than feel they need to utilise all the tools available. Tutors, for example, responded positively to the assignment submission tool that had been mandated by the IEI. Marking moderation was streamlined, with the IEI gaining at least a week or more in turnaround time. Some tutors were uncomfortable, initially, with the idea of marking on-line, with the technical aspect being the greatest barrier. However, after initial discussion and training, all staff chose to trial on-screen marking and as previously mentioned, turn-around time was reduced from an average of three to four weeks to less than three weeks in all. On-demand support from the IEI was available for any tutors that required it. Almost all tutors expressed satisfaction with the levels of assignment submission, record taking and ease of use that resulted. Initially, resistance was detected from progressing students who felt anxious about using the new on-line tool. IEI staff identified these students early and provided extra support and guidance as required. Students now submit their assignments on-line as standard. There were several initial technical issues. A small number of distance-learning students had unreliable internet connections while others had difficulty in dealing with large files and images. These problems were followed-up and very CLEar instructions and specialised training for students was provided. All student and tutor feedback was considered and, where appropriate and feasible, changes were made and additional planning for subsequent semesters was implemented through an iterative and reflective process.

Students have reported positively on the importance of the ease of navigation in the on-line technologies. Being able to find resources quickly and easily, along with recognising new additions to the site at a glance, was stressed as vital. Students were also positive about the level of interactivity and feedback associated with quizzes, assignments and discussion forums. This was considered especially important when a question in a quiz was answered incorrectly. Students got a fast turn-around time response to their attempts at answering quizzes and on-line assessments. As well as getting an indication of whether they had answered the questions correctly or incorrectly, they were also offered a model answer wherever incorrect responses were given. This process, while of immense educational benefit to the students, also greatly reduces the drain on tutor's time.

In the case of the IEI distance-learning students, much of the support and encouragement is offered through the on-line discussion boards and especially through the chat rooms. These tools are used as a means of communication with tutors, fellow students, and the IEI on task deadlines, requirements, and difficult questions that students could not find an answer to themselves. This is especially true of contact with the IEI, who checks the chat rooms at least twice a day; questions that IEI administrative staff cannot deal with are referred to the relevant moderator or tutor. Additionally, the support element is usually very strong through peer encouragement, and perceived tutor and administration support is crucial in promoting the students' motivation to use the on-line technologies (Ala-Mutka et al., 2009; Concannon et al., 2005). Feedback collected indicates that students feel a sense of support that their queries were dealt with in a timely and professional way. The more critical issue for students was their comfort in the knowledge that technical support was available, if needed. As mentioned, vital support is offered to the IEI students and tutors through training days, TEL Handbooks and then, as an on-demand support, during the business hours of the IEI.

The above findings provided valuable views on the ways that students used the on-line technologies to support their learning, both in terms of how they find and use information and in how they use different communication mechanisms to raise queries and discuss issues with other students, their tutors and the IEI. It also demonstrates how important it is for the initial designers of the modules to be proficient in the pedagogy, technical and social aspects of using on-line technologies. The feedback results suggest that flexible, blended learning innovations of this type may provide a mechanism for improving student responses to education and for maintaining completion rates. This is consistent with the findings of previous researchers who advocate web-based delivery of programmes (Dorrian et al., 2009). In summary, the points listed below are essential requirements, from the point of view of the IEI, for implementing on-line technologies in its distance learning programmes.

- **Communication:** Providing CLEar and detailed instructions and creating appropriate student and tutor expectations of the new technologies.
- **Reducing anxiety and resistance to innovation:** Providing support for students and tutors who may not, initially, feel comfortable operating in the on-line virtual environment.
- **Formal and informal evaluation processes:** Documenting areas of technology that can be improved through reflective iteration.
- **Continued development:** Involving teams of tutors in seminar formats that can provide a high degree of technical and pedagogical support.

## Further Research and Continued Development

The development of these on-line techniques and the pedagogical support required must not remain at a standstill. Innovation, improvements and suggestions from current and future system users are essential. It should also be noted that ensuring the mainstream practice of utilising Sakai and Adobe Connect Pro at the University of Limerick, relies on the quality of educational content placed on the systems. Participants were unanimous in the view that TEL is not only the result of technological innovation, but embracing TEL requires far more than the introduction of new tools. It depends more on pedagogical and organisational innovation than on technology (Ala-Mutka et al., 2009).

The IEI has made the move to technology-enhanced learning and to maintain and deliver the best programmes, it must take advantage of more flexible arrangements and the ability of TEL to improve the delivery and support of its programmes. Taken together, this process has highlighted the importance of open communication, reducing anxiety and resistance to innovation, formal and informal evaluation processes and continued systems development. IEI students are using technologies to support all aspects of learning; directed study, resource discovery, preparation and completion of assignments, communication and collaboration, presentation and reflection. Technology should not, therefore, be simply seen as an 'add-on' for these students. The technologies should provide them with a rich variety of alternatives for interaction and communication in relation to learning and a flexibility of use that enables them to take control of their learning. A new culture must be fostered, where being at school is a motivating, engaging experience, where learners are active stakeholders, where they are owners of tangible learning outcomes (Ala-Mutka et al., 2009).

The pedagogy of on-line tutorials must be incorporated into any training sessions for the software. For IEI tutors, who may have little or no experience in education formats that utilise on-line technologies, the move to on-line teaching involved, in some cases, a drastic change. Access to this new technology involved technical aspects, sufficient skills and adequate understanding of the technology utilised to facilitate education in a virtual environment (Delfino et al., 2007; Kingsley et al., 2009), and the shifting of the primary role of the teacher from dispensing information to facilitating learning (Lockyer et al., 2006). Bennett (2004) also states that adapting to on-line environments has required the development of new skills and changes to teaching practices (Bennett et al., 2004). Currently, the IEI is investing time and effort on a planned continuous professional development (CPD) initiative on the pedagogy of on-line tutorials to assist with these changes. As on-line tutorial development is more challenging, initially, time and technology-wise, the tutorials will be developed over a number of weeks to allow for reflective learning, iterative and reflective design. Tutors will collaborate over time in achieving a pedagogical approach appropriate for on-line delivery. The learning activities in each module will be subjected to an iterative process of reflective observation, acting, sharing and evaluation with colleagues where they will use both on-line "real-world" discussion to build new understandings or to support one another in what will be, for many, a new territory with new pedagogical demands. The evaluation of its impact will be focused on the changes to pedagogy induced by the CPD initiative, the challenges and new opportunities met, and the students' and coordinators' perspectives on the process. Some of the possible questions include these: What new teaching methodologies have been deployed in the on-line tutorials? What competences have tutors acquired that enabled this and what further needs exists? What has been the impact on learners of the methodologies used? Which are the most effective methodologies?

Additional and potential developments might include for consideration the following pedagogical and interactive innovations:

- the use of standalone multimedia to bring immediate access to images, sounds or animations of complex theories
- the development of Flash demos to explain more complex theories
- the guidance of students to Internet resources and the use of, for example, book publishers' accompanying web-based resources, particularly in science and business disciplines – this is a viable alternative to creating "in house" resources
- developing collaborative ventures with OpenLearn universities in other countries.

## Conclusion

Educational technologies are most effective when used properly (Savoy et al., 2009). While many of the same teaching principles apply in face-to-face and on-line tutorials, the latter involves an added layer of complexity and is not comparable in terms of what can be achieved in the same number of hours. On-line tutoring requires time, competence and commitment. Bennett et al., (2004) also propose that enhanced student learning outcomes have been realized when cognitively powerful pedagogical strategies such as collaborative, case-based, problem-based or authentic learning are implemented within technology-supported environments. Through the support of the University of Limerick, tutors are encouraged to use a combination of the more traditional forms of communication and information transfer (e.g. tutorials, labs/workshops, printed material) and the more innovative methods through Adobe Connect Pro and SAKAI (websites, chat rooms, discussion boards, on-line tutorials, on-line assessment, on-line grading/feedback). The IEI aims to enhance the more traditional learning opportunities by a careful integration and mix of new technology-enabled teaching and learning systems.

Complementing traditional educational models with TEL tools is dramatically changing the role of the educator. They are becoming facilitators of the processes of gaining knowledge, in which learners take a far more active part than they have ever done. The tutors' capabilities, enhanced by the availability of TEL resources, enable the creation of motivation in students that encourages them to commit to self-directed learning experiences. It also provides them with structures that keep them focused on the learning objectives and highlighting critical features of the task that they might overlook. This, in turn, facilitates the carrying out of the learning activities so that objectives are met (Ala-Mutka et al., 2009).

It is essential that tutors, who are module content specialists, appreciate the importance of developing necessary pedagogical skills associated with on-line tutoring. Sometimes, tutors may feel constrained by pedagogical issues. In almost all cases, this anxiety and/or resistance is relieved once the technology is used by the individuals and a degree of familiarity is established. It is necessary in all cases that CLEar and detailed instructions are provided regardless of the experience of the user. An established suggestion is to presume the end user has never encountered the system previously. It is also necessary, especially in the case of the tutor, that the appropriate expectations of any new systems are created.

Student learning is now undertaken in a complex and changing environment, with a wide range of technological tools for support. Computer ownership and Internet access is high and students have become accustomed to being able to electronically access information or people on demand (Conole et al., 2007). It is also generally acknowledged that the emerging educational formats of using interactive on-line technologies are more engaging for the learner than the traditional education formats. Engagement and motivation are critical factors for the success of learning experiences and TEL affects the learning processes by engaging participants in learning dynamics that enhance their motivation (Ala-Mutka et al., 2009).

In the higher education programmes offered by the IEI, the utilisation of on-line technologies has become an expected and integral part of the learning process for students. Most participants agreed that the on-line technology integration was also an expected and necessary part of their university experience and of use in their future careers. Major benefits noted included the ease of access to resources, including peer-reviewed journals through the library, and the provision of a central area for students to access and to find information on comprehensive resources pertaining to each module. Negative experiences focused predominately on technical problems that were often due to inexperience in the use of the technology on the part of the student and/or tutor. It is CLEar from this evidence that to look only at the positive and negative factors of technology, is to miss the wider factors impinging on students' use of it as a support mechanism. Age-old problems of students' motivation, peer influence, and study strategy are all as important to the learning process, as are access to technology and computer skills. However, of primary importance to the process were peer encouragement and perceived lecturer and tutor support (Concannon et al., 2005).

In summary, pedagogy, technology and social interaction are critical components of a technology-enhanced learning environment. Technology is more likely to be a basic condition for effective integration of TEL. Sound design of pedagogy or social interaction very much depends on the availability of technological support. Without sufficient support of the technology, many pedagogical and social design activities would be difficult to implement. However, the primary factor that influences the effectiveness of learning is not the availability of technology, but the pedagogical and social design underlying the use of that technology. In general, the IEI found that students consider TEL a valuable support to the learning process. They see it as an additional reinforcement to the traditional face-to-face delivery mechanisms, and they make regular use of the on-line technologies flexible access, and the incentive it provides for ongoing study and continuous assessment in preparation for the final assessments. TEL, based on a solid pedagogic foundation and providing feedback, interaction and access to course materials, is seen as both a benefit, and an improvement in educational delivery and support.

## REFERENCES

Ala-Mutka, K., Bacigalupo, M., Kluzer, S., Pascu, C., Punie, Y., & Redecker, C. (2009). *Review of Learning 2.0 Practices*. IPTS technical report. Retrieved April 18, 2009, from http://ipts.jrc.ec.europa.eu/publications/pub.cfm?id=2139

Atkins, D., Seely Brown, J., & Hammond, A. L. *(2007).* A review of the Open Educational Resource movement: achievements, challenges and new opportunities. *Report to the William and Flora Hewlett Foundation. RetrievedMay5, 2009, from*http://www.oerderves.org/wp-content/uploads/2007/03/a-review-of-the-open-educational-resources-oer-movement_final.pdf

Bennett, S., & Lockyer, L. (2004). *Becoming an On-line Teacher: Adapting to a Changed Environment for Teaching and Learning in Higher Education*. London: Educational Media International, Routledge.

Buckley, W., & Smith, A. (2007). Application of Multimedia Technologies to Enhance Distance learning. *RE:view*, *39*(2), 57–65. doi:10.3200/REVU.39.2.57-65

Concannon, F., Flynn, A., & Campbell, M. (2005). What campus-based students think about the quality and benefits of e-learning. *British Journal of Educational Technology*, *36*(3), 501–512. doi:10.1111/j.1467-8535.2005.00482.x

Conole, G., Brasher, A., Cross, S., Weller, M., Clark, P., & Culver, J. (2008). Visualising learning design to foster and support good practice and creativity. *Educational Media International*, *45*(3), 177–194. doi:10.1080/09523980802284168

Conole, G., de Laat, M., Dillon, T., & Darby, J. *(2007, Mar26-28).* 'Disruptive technologies', 'pedagogical innovation': What's new? Findings from an in-depth study of students' use and perception of technology. *Paper presented at the Biennial Conference on Computer Assisted Learning (CAL 07), Dublin, IRELAND.*

*Delfino, M., &Persico, D. (2007). On-line or face-to-face? Experimenting with different techniques in teacher training.* Journal of Computer Assisted Learning.

Devine, J. *(2005).* Major challenges facing the higher education system in the ICT era. *RetrievedMay26, 2009, from*http://www.elearningeuropa.info/directory/index.php?doc_id=1151&doclng=6&page=doc

Dorrian, J., & Wache, D. (2009). Introduction of an on-line approach to flexible learning for on-campus and distance education students: Lessons learned and ways forward. *Nurse Education Today*, *29*(2), 157–167. doi:10.1016/j.nedt.2008.08.010

Eynon, R. (2008). *The use of the world wide web in learning and teaching in higher education: reality and rhetoric*. Innovations in Education & Teaching International.

Kingsley, K., & Kingsley, K. (2009). A case study for teaching information literacy skills. *BMC Medical Education*, *9*(1), 7. doi:10.1186/1472-6920-9-7

Lockyer, J., Sargeant, J., Curran, V., & Fleet, L. (2006). The transition from face-to-face to on-line CME facilitation. *Medical Teacher*, *28*, 625–630. doi:10.1080/01421590600922909

Lyons, H. (2009). *Case study research methodology for publishing developments in ICT-facilitated learning in higher education - a prescriptive approach*. Innovations in Education & Teaching International.

McAndrew, P., & Santos, A. I. *(2009).* Learning from OpenLearn: Research Report 2006-2008. *RetrievedMay26, 2009, from*http://kn.open.ac.uk/public/getfile.cfm?documentfileid=15028

NSF. *(2008).* Fostering learning in the networked world: learning opportunity and challenge. A 21st Century agenda for the National Science Foundation. *RetrievedMay8, 2009, from*http://www.nsf.gov/publications/pub_summ.jsp?ods_key=nsf08204

Savoy, A., Proctor, R. W., & Salvendy, G. (2009). *Information retention from PowerPoint and traditional lectures*. Computers & Education.

Science, D. E. a. (2000). Learning for life: white paper on adult education.

Spector, M. J. D. T. I. *(2001)*. Competencies for on-line teaching. *Retrieved from*http://purl.access.gpo.gov/GPO/LPS26199

Wang, Q. Y. (2008). A generic model for guiding the integration of ICT into teaching and learning. *Innovations in Education and Teaching International*, *45*(4), 411–419. doi:10.1080/14703290802377307

## ADDITIONAL READING

Anderson, T. (2004). Teaching in an on-line learning context. In Anderson, T., & Elloumi, F. (Eds.), *Theory and practice of on-line learning* (pp. 273–294). Athabasca, CA: Athabasca University.

Babb, K. A., & Ross, C. (2009). The timing of on-line lecture slide availability and its effect on attendance, participation, and exam performance. *Computers & Education*, *52*(4), 868–881. doi:10.1016/j.compedu.2008.12.009

Bouras, A., & Albe, V. (2008). Viewpoints of higher education teachers about technologies. *International Journal of Technology and Design Education*, *18*, 285–305. doi:10.1007/s10798-007-9027-3

Brodie, M., Flournoy, R., Altman, D., Blendon, R., Benson, J., & Rosenbaum, M. (2000). Health information, the Internet, and the digital divide. *Health Affairs*, *19*, 255–265. doi:10.1377/hlthaff.19.6.255

Brown, S., Kirkpatrick, M., Mangum, D., & Avery, J. (2008). A review of narrative pedagogy strategies to transform traditional nursing education. *The Journal of Nursing Education*, *47*, 283–286. doi:10.3928/01484834-20080601-01

Conole, G. ((in press)). New Schemas for Mapping Pedagogies and Technologies. Web 2.0-based e-learning: applying social informatics for tertiary teaching,(56).

Coppus, S., Emparanza, J., Hadley, J., Kulier, R., Weinbrenner, S., & Arvanitis, T. (2007). A clinically integrated curriculum in evidence-based medicine for just-in-time learning through on-the-job training: the EU-EBM project. *BMC Medical Education*, *7*, 46. doi:10.1186/1472-6920-7-46

Das, K., Malick, S., & Khan, K. (2008). Tips for teaching evidence-based medicine in a clinical setting: lessons from adult learning theory. Part one. *Journal of the Royal Society of Medicine*, *101*, 493–500. doi:10.1258/jrsm.2008.080712

DeBourgh, G. A. (2003). Predictors of student satisfaction with distance-delivered graduate nursing courses: what matters most? *Journal of Professional Nursing*, *19*(3), 149–163. doi:10.1016/S8755-7223(03)00072-3

Downes, S. (2004). 2004: The Turning Point [Electronic Version]. Ubiquity, 4 Retrieved 26th May, 2009, from http://www.acm.org/ubiquity/views/v4i46_downes.html

Estabrook, L., Witt, E., & Rainie, L. (2007). Information searches that solve problems: How people use the Internet, libraries, and government agencies when they need help.

Fallows, D. (2008). Search engine use Retrieved 12th May, 2009, from http://www.pewinternet.org/pdfs/PIP_Search_Aug08.pdf

Gillette, J. (2008). Evidence-based dentistry for everyday practice. *The Journal of Evidence-Based Dental Practice*, *8*, 144–148. doi:10.1016/j.jebdp.2008.05.012

Girod, M., & Girod, G. (2008). Simulation and the need for practice in teacher preparation. *Journal of Technology and Teacher Education*, *16*, 307–338.

Gu, X., Zhang, B., Lin, X., & Song, X. (Writer) (2009). Evaluating on-line solutions for experiential support of distance learning by teachers in China [ArtiCLE], Journal of Computer Assisted Learning.

Jackson, L., Zhao, Y., Kolenic, A., Fitzgerald, H., Harold, R., & Von Eye, A. (2008). Race, gender and information technology use: the new digital divide. *Cyberpsychology & Behavior*, *11*, 437–442. doi:10.1089/cpb.2007.0157

Jones, S. (2008). The Internet goes to college: how students are living in the future with today's technology, from http://www.pewinternet.org/report_display.asp?r=71

Khan, K., & Coomarasamy, A. (2006). A hierarchy of effective teaching and learning to acquire competence in evidenced-based medicine. *BMC Medical Education*, *6*, 59. doi:10.1186/1472-6920-6-59

Klem, W., & Weiss, P. (2005). Evidence-based resources and the role of librarians in developing evidence-based practice curricula. *Journal of Professional Nursing*, *21*, 380–387. doi:10.1016/j.profnurs.2005.10.004

Langer, A., & Knefelkamp, L. (2008). College students' technology arc: A model for understanding progress. *Theory into Practice*, *47*, 186–196. doi:10.1080/00405840802153759

Lazar, J., Jones, A., & Shneiderman, B. (2006). Workplace user frustration with computers: an exploratory investigation of the causes and severity. *Behaviour & Information Technology*, *25*(3), 239–251. doi:10.1080/01449290500196963

Levine, A., Bebermeyer, R., Chen, J., Davis, D., & Harty, C. (2008). Development of an interdisciplinary course in information resources and evidence-based dentistry. *Journal of Dental Education*, *72*, 1067–1076.

Littlejohn, A. H. (Writer) (2002). Improving continuing professional development in the use of ICT [ArtiCLE], Journal of Computer Assisted Learning: Blackwell Publishing Limited.

Liu, S.-H., Liao, H.-L., & Pratt, J. A. (Writer) (2009). Impact of media richness and flow on e-learning technology acceptance [ArtiCLE], Computers & Education.

Lorenzetti, J. P. (2007). The Virtual Faculty Lounge: Providing On-line Faculty Development for Adjunct Tutors. *Distance Education Report*, *11*(3), 4–8.

Mackey, T., & Jacobson, T. (2005). Information literacy: A collaborative endeavor. *College Teaching*, *53*(4), 140–144. doi:10.3200/CTCH.53.4.140-144

Madden, M., & Fox, S. (2006). Finding answers on-line in sickness and in health. Retrieved May 12, 2009, from http://www.pewinternet.org/~/media/Files/Reports/2006/PIP_Health_Decisions_2006.pdf.pdf

Metros, S., & Woolsey, K. (2008). (This issue). Digital illiteracies in the age of sight and sound. *Theory into Practice*, *47*, 79–82. doi:10.1080/00405840801992207

Ng'ambi, D., & Brown, I. (2009). Intended and unintended consequences of student use of an on-line questioning environment. *British Journal of Educational Technology*, *40*(2), 316–328. doi:10.1111/j.1467-8535.2008.00899.x

Palmer, S. R., & Holt, D. M. (2009). Examining student satisfaction with wholly on-line learning. *Journal of Computer Assisted Learning*, *25*(2), 101–113. doi:10.1111/j.1365-2729.2008.00294.x

Salmon, G. (2004). *E-Moderating: The key to on-line teaching and learning* (2nd ed.). London: Taylor & Francis.

Shaker, L. (2006). In Google we trust: Information integrity in the digital age. *First Monday*, *11*(4).

Shorten, A., Wallace, M., & Crookes, P. (2001). Developing information literacy: a key to evidence-based nursing. *International Nursing Review*, *48*, 86–92. doi:10.1046/j.1466-7657.2001.00045.x

Singh, H. (2003). Building effective blended learning programs. *Educational Technology*, *43*(6), 51–54.

Skilbeck, M. (2001). *The University Challenged - A Review of International Trends and Issues with Particular Reference to Ireland.* Dublin: Higher Education. Authority and Committee of Heads of Irish Universities.

Sutherland, R., Armstrong, V., Barnes, S., Breeze, N., Gall, M., & Mattthewman, S. (2004). Transforming teaching and learning: embedding ICT into everyday classroom practices. *Journal of Computer Assisted Learning*, *20*, 413–425. doi:10.1111/j.1365-2729.2004.00104.x

Swinglehurst, D., Russell, J., & Greenhalgh, T. (2008). Peer observation of teaching in the on-line environment: an action research approach. *Journal of Computer Assisted Learning*, *24*(5), 383–393. doi:10.1111/j.1365-2729.2007.00274.x

Tillman, H. *(2003).* Evaluating Quality on the Net. *RetrievedMarch26, 2009, from*http://www.hopetillman.com/findqual.html

Varnhagen, C. K., & Wright, D. L. (2008). Learning characteristics of veterinary technology students in a distance-education and an on-campus program. *Journal of Veterinary Medical Education*, *35*(3), 449–455. doi:10.3138/jvme.35.3.449

Wallace, M., Shorten, A., & Crookes, P. (2000). Teaching information literacy skills: an evaluation. *Nurse Education Today*, *10*, 485–489. doi:10.1054/nedt.1999.0439

Wang, C.-H. (2005). Questioning skills facilitate on-line synchronous discussions. *Journal of Computer Assisted Learning*, *21*(4), 303–313. doi:10.1111/j.1365-2729.2005.00138.x

White, N. *(2004).* Facilitating and hosting a virtual community. *RetrievedMay10, 2006, from*http://www.fullcirc.com/community/communityfacilitation.htm

## ENDNOTES

1. In this chapter, tutor refers to the individuals involved in the design, instruction, moderation and delivery of modules in conjunction with the IEI.
2. In this chapter, on-line technologies refer to the use of CLE's and Adobe Connect Pro in conjunction with distance learning.
3. Mature students at the University of Limerick are considered to be over 23 years of age at the time of enrolment.

# APPENDIX 1: TRANSCRIPT OF A PRESENTATION BY THE IEI TO FACULTY AT THE UL

A session of one hour was organised at the beginning of the second semester of 2009 in order to share the experience of the **IEI** with other interested faculty members. The session commenced with a presentation on the use of Sulis and Adobe Connect by the **IEI**, followed by a live demo of the technology and questions and answers session as transcribed below

## Why did you start using this type of technology?

By the very nature of the fact that we are **distance learning**, the **distance learning** students can feel a sense of isolation while studying with us. Our function, and one of our primary functions within the **IEI** is to minimise this sense of isolation and to support the students, and have them integrated into the University community and to make them feel a part of a group as best as we can and obviously technology has helped us an awful lot with this.

## When did you start using Sulis and Adobe Connect?

The whole programme has evolved hugely within the last ten years, particularly with the development of technology, and Sulis is one of those tools that have really helped us along that way. Prior to Sulis we had a different **CLE** called "TopClass"; we had a perpetual licence and the tools that it had were limited. Then Sulis came on board around three years ago, supported by the University of Limerick. With Adobe Connect we started about a year ago, we've done two semesters now, we piloted it first, this time last year and then in September we went forward with a whole stage, so all 5 and 6 stage modules and anyone who signed up for it was desktop to desktop though Adobe and it's worked, and it's working so far very well.

## So you have not had any major problems?

It's worked very well. I suppose it should be pointed out that we're different from the university in that we enrol twice a year, all our modules run twice a year so that gives students the opportunity to complete the programme within their own time and it's a way for us then to provide these services, it's essential then that they have the support mechanism here at the University for that. Our main problems arise from how computer literate the individual student is.

## How do you organise the sessions in Adobe Connect? Do you send them notifications in advance?

Yeah, when you set up your meeting room each student is issued with an email that has a link to unique module room. Adobe also allows us to record the **tutorials** based on the permission of the tutor. So it means that we can link them back into Sulis, and if the tutor is ok with it, we can have links put up on that specific module site if the students could not make the tutorial.

## What time do you normally do it?

Generally, we try to timetable in the evenings, from about six o'clock onwards because most people are working in between the hours of nine and five and **tutorials** are generally scheduled for an hour and a half. They are sent an email for Adobe Connect at least a week before, if not more. They're also sent text reminders to their phone to say, "Don't forget your tutorial", they also receive the timetables at the start of the semester, and they also can log into the calendar on Sulis, if they really forget about it. So, after all that there should be no excuse for them not being available. We provide them with the training at the start of the semester before their first tutorial, so that they can have their computer set up, with their headset working, or camera working. So, they're sent an email, it has a link on it, they click the link, they put their name in the guest box, and then they're into the room. So, you can only access the room if you have the link from the email. We supply them with headsets and web-cameras, we do give them the specs of the ones we use so they do have the opportunity to purchase them themselves or if they have them themselves we let them use that, but they need to have access to a computer and broadband internet.

## What is the main advantage of using these technologies?

The main thing about Sulis and Adobe Connect is that we are trying to encourage as much collaborative learning as possible. One of the methods for Sulis is the use of discussion tools. Students are assessed on how well they argue a point in a discussion, as well with Adobe Connect, the students can see each other, they can see the tutor and they have the opportunity of interacting as they would in a live situation. Some of our students are as far away as Donegal or Antrim, and for them to travel all the way to Limerick can be quite a big deal, so this way we're trying to increase the numbers that attend a tutorial. Most of the **tutorials** are not compulsory and we are trying to show them that if they attend a tutorial that they will learn more, but it is trying to keep them attending the **tutorials** that is the difficultly. There is a lot of opportunity for using Sulis, it has the potential to really increase the teaching ability of the tutor, to improve how a student learns, but I mean there is an awful lot of opportunity for research into that area with regard to Sulis. We're only just touching the surface on how you use Sulis as a teaching mechanism.

## When using quizzes in Sulis, do you have any way of knowing if they're copying?

No, it is a leap of faith, one way of preventing copying is to time the assessment. But there's no guarantee in any assignment that it's the students own either. We try and keep it as secure as possible; we try to prevent plagiarism by submitting assignments through anti-plagiarism software, as much as possible. The on-line quizzes can be quite short or they can be extremely long. If they're extremely long they have a large number of questions and the exact time they require to do it, otherwise you put a really short paper up and they have a very short time, so I mean, they could be getting help, but there's very little you can do about it. But we try and, we trust them as much as we trust a full time on campus student.

## So there's more assessments done, not only on-line quizzes?

Yeah, the assessment, depending on the module, some modules have one or two assignments, they have a full written at the end, same as the full time students, and they either have lab days, practical sessions, or they would have an on-line quiz, depending on the module, it's up to the moderator of the module to determine how they wish to assess the module, same as the full time.

## So there are some outside visits or are they just for lab days?

Yeah, they'd have to, for example anatomy and physiology, come to the University for the full day and they'd do a lab day and they have a small examination at the end of it, but that lab day, I think forms 30% of their overall grade and their assignment would have formed another 30% of their overall grade and they're written examinations would form 40% of their overall grade. We generally do three field visits a semester, and so the three that will run maybe in the Autumn semester are different to the ones that run in Spring, and we try to space them out and vary them year on year, as much as we can. This allows a student coming through the Certificate programme to get a feel for a lot of different activities.

## Do they have to have software downloaded to enter or Adobe Reader?

No, Adobe Connect is delivered across the internet with the ubiquitous Flash Player, so as long as they have latest the flash player, they've no problems with it. It even prompts you at the start to download the latest Flash Player if your computer does not have it.

## What kind of numbers do you have to have in the tutorials?

We've not used Adobe Connect for large numbers, the maximum we've used it for would be about six plus a presenter at a time. A lot of it will depend on the individual's connection to the internet, both student and tutor and on the level of computer literacy of both the student and the tutor, that would be the main factors holding the numbers back. From our current experience, unless the tutor is very experienced using **virtual environments**, smaller numbers in rooms have worked well whereas larger numbers have provided confusion.

## Is there any concern about the difficulty of using Sulis and Adobe?

So far no major concerns. We always try to personalise the meeting rooms a little bit, so that when students do actually click onto something they know that they are in the right place (such as an **IEI** logo). Basically, a tutor will enter and will then activate the meeting room for the students. Once the student enters you can see that their name will pop up on the left hand side in the attendee list pod. Once a tutor is in Adobe and they see a student entering into the room, they can then at that stage activate the students' voice and camera. The tutor has control over the system. They communicate, if there are any problems at that stage with the students; being able to hear the tutor etc., the chat room comes in very useful because they can type their concerns, "Listen, I can see you, but I can't hear you", and then if necessary we go though the troubleshooting that we have just set out for Adobe. Once students and tutors are used to using it and everything has been set up once, they are generally ok. We've had students who wouldn't be hugely IT literate and they've had no problems at all really. They've found it fantastic, and like we've said earlier it means that they can go from the dinner table to the computer to the tutorial and it's worked out really well. The tutor then can share their screen, they can share their documents, if there's something that they are working on, that they're sharing with a student for example, they can actually go to the students' screen to see what they're doing, or what's happening on their screen. Any file format can be preloaded, if a tutor is having a tutorial, they send us their notes the day before. We have them uploaded, if there're any graphics that they want uploaded, any sound files, we can have all that uploaded prior to their tutorial of an evening, so that they can come in and just be able to access it, there's no fuss then that way. There have been some complaints about the time lag in the audio, but this is unfortunately dependant on the individuals' network connection.

### If you wanted to, could you prerecord lectures or tutorials and put them up then? So they could access them anytime they needed if your numbers got too big.

Yeah, you could. We have recorded some **tutorials**, and training films. Tutors can sit in front of their computer and record all that takes place on the screen. You could have your webcam there and record yourself talking and then send the students a link, and say that this is the link to it.

### You've asked them to still their pictures so you're not going to see their mouths going, but you can hear them.

You can turn of their microphones as well during the presentation. If their microphones are on and their cameras are on there's absolute real time communication. There are two options, you can set up that anybody can talk at any time, or that only one person can talk at a time, that's a setting that you set up when you set up the meeting room. We issued them with a document with instructions on Sulis and on Adobe Connect and part of it is what we call etiquette or protocol for using the room, for example we say if you wish to ask a question, you must raise your hand, and then the tutor can say "oh I see John has a question, what is your question?" And then you ask your question. So it doesn't really interrupt the flow of conversation as it is going on in the tutorial.

### Is it completely live?

There is a very slight audio delay due to broadband speed. It is noticeable though if you are asking questions, in any of these desktop to desktop software products like Skype or anything like that, there is a fraction of a delay and we guess it is about two seconds with Adobe Connect, so it means that if somebody asks a questions without raising their virtual hand, there's a two second delay, the tutor is then interrupted talking, and then by the time you get it sorted out there's a bit of a delay the flow of the tutorial is gone, so that's why we say "use these little protocol tools". There's a whole variety them; laugh, disagree, I've stepped away for a moment. You know there is a lot of opportunity for students to interact. You have different looks as well, if you want a discussion look, the presentation goes away the chat area becomes bigger, you can have discussion notes, you can put in a little poll… collaboration views then are different, these are just default settings, so again you have a white board, even divide up your class and discuss a topic, go away and discuss it and come back and you can actually divide up the room into little discussions as well. We haven't used that, but we have looked at the possibility of that. To compensate for the audio time delay we are looking into using Audio Conferencing where Adobe Connect will dial out to a phone number of choice and therefore eliminates the time delay issue.

## When you got the licence, did you figure this out though a user manual or did you have someone show you how to use it?

No, we tend to figure out everything ourselves. The company provided us with great support, there are endless amounts of documentation that you can read, but we did find it was pretty intuitive in that once you got in there and worked through it, it actually was fine.

## How long are those tutorial sessions?

About an hour and a half, sometimes an hour, sometimes two hours; depends on the tutor.

## Tutorials are optional, it's not mandatory to go to those, assessment isn't built into it?

No. Some of them, such as the IT modules and Horsemanship **tutorials** are compulsory, the labs of certain modules are compulsory, but the majority of **tutorials** are entirely optional and because our students tend to have to travel so far we can't say that every tutorial is compulsory. We do say that they are highly valuable and are well worth attending, and the students that do attend tend to get the higher grades than those who didn't, so our way of thinking was trying to find a way that we could get students to attend the **tutorials** without beating them over the head with a stick and **Adobe Connect Pro**vided a way of delivering it to the computer at home.

## And do you combine with face to face, or is it all done on-line?

The first of three **tutorials** beginning in the coming semester will be face-to-face. This face-to-face day will also incorporate technical training in Sulis and Adobe Connect. The additional two **tutorials** will be delivered on-line for the majority of modules.

## And do people, do those that can't take the synchronies version, do many use it afterwards?

They do they have a look at it. We have a centre based in Northern Ireland in Greenmount College, and they actually came in and said "Oh, we can access **tutorials** from previous semesters, this is great for us, for helping the students as well". The genetics tutor up there was able to access the tutorial that we had recorded last Spring. We train the Greenmount tutors as well on Adobe Connect. Our main issues really with any of this technology is, the more the tutors will use the system, the more the students will use the system, and if the students are using the system and are finding that there is nobody at the other end, then they are less inclined to use it again. The whole point of the system is to encourage collaboration, to encourage students, therefore a lot of our work is really trying to get the tutors to use the system. It sometimes takes that little bit longer, but it is well worth the benefit at the end and it seems to be working out well. We've received very positive feedback from all angles and if we were to turn the clock back now I think we would find it very hard, they've come to depend on it. The tutors themselves can be at home, they need never travel, because not all of our tutors are based at UL, some of our Anatomy tutors for example are vets, and they are out and about and for them to come to the University is a little bit more complicated than say being able to sit at home in your living room when you have an hour.

### Is Sulis there for a point, you know where they go to start any of this? So they can reach everything without going somewhere else.

Yeah, they have access to everything available to fulltime students through Sulis. They can access on-line journals, books, and soon video clips and short films to name but a few.

### Is there a notice to say there are lectures here?

There is literally anything we think is of benefit to students on Sulis; access to the library, your blog, podcasts… they've exam timetables, field trip information, and registration forms as well as any extra information we feel the students might need. This semester we've said when you're registering instead of us sending out all the information, go to Sulis, or go to the main website and download it. We're trying to reduce how much we send out, because we use to send out pages of information.

### Are there links to the articles there, do they get linked directly to on-line articles?

They are linked directly to the on-line articles, or sometimes an article link is sent to us from tutors. Sometimes if its articles written by academic staff or people, then we can just post it up and that's perfectly fine, but yeah copyright is a big issue.

# Chapter 6
# Preparing Faculty for a Learning Management System Transition

**Danilo M. Baylen**
*University of West Georgia, USA*

**Mary Hancock**
*University of West Georgia, USA*

**Carol M. Mullen**
*University of West Georgia, USA*

**Mary Angela Coleman**
*University of West Georgia, USA*

## ABSTRACT

*This chapter focuses on the impact of a change in the use of a learning management system (LMS) at one university. Survey data captured faculty members' viewpoints on the transition from one LMS to another, specifically, their dispositions toward technology and change, preparation and prior experiences, need for support, and access to available resources. The inquiry focuses on potential activities and infrastructures that can be established to support the faculty, as LMS users, when a new system is introduced. Also, it explores the types of knowledge, skills, and dispositions that faculty may have or need to effectively and efficiently use the new system to support their work. Finally, strategies are recommended to enhance faculty members' dispositions, preparation, support and access to resources.*

## PREPARING EDUCATION FACULTY FOR A LEARNING MANAGEMENT SYSTEM TRANSITION

The recent acquisition of *ANGEL*, a Learning Management System (LMS) company, by *Blackboard* has sparked the latest discussions on the use of LMSs in higher education. What is an LMS? According to the Office of Information and Instructional Technology (OIIT; 2006), it is a "set of web based tools for teaching, learning, communication and class administration." Ionannou and Hannafin (2008) identified an LMS as a software system designed to manage course content and course activities. Currently the top U.S.

DOI: 10.4018/978-1-61520-853-1.ch006

Copyright © 2010, IGI Global. Copying or distributing in print or electronic forms without written permission of IGI Global is prohibited.

providers of LMSs are *Blackboard, ANGEL,* and *Desire2Learn* (Young, 2009). A recent report of the Campus Computing Survey Project that polled college IT leaders stated 56.8% of colleges who use LMSs run *Blackboard* in the United States (Campus Computing Project, 2008). *Blackboard* provides "easy-to-use tools for designing and managing both web-based and face-to-face courses" (OIIT).

Changes in LMS platforms are fairly common in higher education (Ionannou & Hannafin, 2008; Smart & Meyer, 2005). One of the major reasons for switching to a different provider is increased licensing costs (Smart & Meyer). Another reason is an upgrade to a better, faster, and more robust version of the LMS (Corich, 2005; Ionannou & Hannafin). A third reason identified by Smart and Myer was that some institutions need to have one LMS instead of supporting multiple systems.

Change in a system, especially in higher education, brings about diverse responses and reactions from faculty. Smart and Meyer (2005) reported on how faculty from their university viewed the ease of transition from one LMS to another from a course conversion perspective. The report identified that "parts of the course that did not convert are often time-consuming to reconstruct" (p. 69). This resulted in increased workloads and frustration. However, in the same report, faculty expressed their willingness to convert courses to a new LMS despite the inaccuracy of the course content conversion and the workload involved.

Distance education technologies have presented faculty with the need to adapt to new methods of teaching and learning. Faculty must not only learn the technology but they must also understand the "paradigm shift" in presentation and evaluation of online instruction (Berryhill & Durrington, 2006, p. 52). Oblinger and Hawkins (2006) suggested that most faculty lack sufficient pedagogical and technical expertise to self develop effective online courses, yet Lane (2008) noted few researchers have investigated the effect of LMS design on pedagogy. While a well-designed LMS provides a toolkit for faculty in their development and presentation of courses online, the investment of time is still significant, and the idea of moving weeks or months of work spent in designing and developing a course to a new system often results in a sense of panic on the part of some faculty (Lane, 2008).

This chapter focuses on strategies that can be used to support faculty, when a new system is introduced. What knowledge, skills, and dispositions are needed by faculty members that will encourage effective and efficient utilization of the new system to support their work? What strategies need to be implemented to enhance and provide better support and make resources available and accessible?

## BACKGROUND

A year ago, a member institution of the Georgia statewide university system adopted a newer version of their current LMS in the middle of the academic year. This change was prompted by an initiative of the university systems' technology group to upgrade all campuses from *WebCT Campus Edition* to the *Blackboard Learning System-Vista Enterprise License* (OIIT, 2006). In addition, after this decision was made, *Blackboard* announced it would purchase *WebCT*. Every member institution using the new LMS was allowed to select a name for its specific campus system. For this member institution, University of West Georgia, the new LMS was called *CourseDen*.

A survey was conducted by a group of faculty members (referred as researchers) interested in learning how this transition was managed by the university, and how the faculty perceived such actions. The population surveyed involved the faculty in a College of Education. To develop the survey, the researchers performed a literature search on possible reasons faculty members resist using a new technology-based tool. Very few articles were found that discussed issues specific to the transition from one LMS to another, but considerable literature was located regarding the availability and use of LMSs in institutions of higher education. Like changing textbooks from a previous semester for a particular course, a change in delivery tools, such as, a new LMS translates to a considerable investment of time, especially in course preparation (Chao, 2008; Smart & Meyer, 2005). Chao pointed out that when course content is converted from one system to another, the faculty usually ends up redesigning the course despite the use of templates. However, familiar features of the new LMS interface design could result in its ease of use and contribute to increase in positive adoption (Beatty & Ulasewicz, 2006; Smart & Meyer, 2005). Also, the change in the LMS system required adequate orientation and support for those who may use the new system (Beatty & Ulasewicz, 2006; Chao, 2008; Zsohar & Smith, 2008). Chao further argued that constant communication between faculty and staff involved in the LMS transition process is a key ingredient towards a positive experience.

After reviewing the literature (Mohr, 2007; Moser, 2007; Weaver, 2006), the researchers decided on four possible factors that facilitate a successful transition from one LMS to another: 1) positive disposition towards technology and change, 2) preparation and prior experiences, 3) adequate support, and 4) available resources. These reasons were used as categories to anchor the generation of statements used in the survey.

## DATA COLLECTION

The research project was completed as a pilot study given that the researchers' institution changed LMS three times in the past four years. The researchers were members of the target population and experienced these changes. In developing the tool for data collection, the researchers reviewed several existing surveys focused on technology usage, technology integration practices, and perceptions of online teaching and learning. Items used in this research survey were generated primarily from these resources. Informal conversations between faculty members, staff, and researchers, in both the College of Education and the university at large, provided insights and ideas on how to refine the initial survey items.

After sorting the final survey items, four factors emerged as facilitating a successful transition from one LMS to another: 1) disposition towards technology and change, 2) preparation and prior experiences, 3) adequate, appropriate and timely support, and 4) available and accessible resources. The first two factors, disposition towards technology and change, and preparation and prior experiences, each contained five items. The next two factors, adequate, appropriate and timely support, and available and accessible resources, each had four items. Finally, four items were added to capture faculty members' perceptions and preferences for teaching online.

A twenty-two item Likert-type survey was developed with six choices ranging from strongly agree to strong disagree. The list below identifies the items supporting each factor.

### Disposition Towards Technology and Change

1. When I first heard the news of the change I did not want *CourseDen* to replace *WebCT*.
2. People like me have positive attitudes about *CourseDen*.
3. I was not afraid about this change from *WebCT* to *CourseDen*.
4. Using *CourseDen* is easier than I anticipated.
5. I believe moving to *CourseDen* is a step in the right direction to support online faculty teaching.

### Preparation and Prior Experiences

1. My knowledge and experience with *WebCT* prepared me to use *CourseDen*.

2. I had difficulty finding *CourseDen* features that I am familiar with from *WebCT*.
3. I was alerted sufficiently in advance that *WebCT* would be replaced by *CourseDen*.
4. I have discussed with another faculty/staff how I can effectively use *CourseDen* to teach my courses.
5. The training sessions I attended did prepare me to use *CourseDen*.

### Adequate, Appropriate and Timely Support

1. There is nobody around me that I can ask for support when I have problems working with *CourseDen*.
2. Information I have about *CourseDen* was helpful in getting me started.
3. When I have problems working in *CourseDen*, I feel confident I can get help by calling the Distance Education office.
4. I did not need to attend a *CourseDen* training to inform me how to use it.

### Available and Accessible Resources

1. I have the appropriate software applications to develop my course in *CourseDen*.
2. My computer is updated (or upgraded) so I can work effectively using *CourseDen*.
3. There are people who can help me develop a pedagogically-sound course in *CourseDen*.
4. When I have questions related to working in *CourseDen*, it takes a lot of time to find the answers.

Also, the researchers were interested in faculty members' perceptions of teaching online and preferences if given the choice of teaching their courses online instead of face-to-face. Several survey questions were added as listed below (see also Table 3 in the Appendix):

1. I like to use technology tools to support the classes I teach.
2. I like to teach totally online classes.
3. I believe that teaching online is not as good as teaching face-to-face.
4. If given the opportunity, I would like to teach all my courses online.- How did you ensure reliability and validity of this instrument?

## CONTEXT

The target population comprised of faculty members from the College of Education of a state university. The college has six academic departments with 83 full-time (tenured, tenure-track, and non-tenured) teaching staff during the 2008-2009 academic year. The researchers decided not to include one department, i.e., Media and Instructional Technology (MIT), since all faculty members teach online. With eleven MIT teaching staff excluded, the total pool of survey participants numbered 72. Further, one of the researchers is a faculty member, so it was decided to exclude this individual in the total faculty count resulting in 71 potential survey participants.

Purposive sampling strategies were used in soliciting survey participation and completion. Online courses at this university are identified as being taught at least 95% online (N) or 51% - 94% online (D). The first strategy involved asking current LMS users who teach N and D courses. The second strategy involved asking teaching staff who used the current LMS but had not taught N or D courses. To ensure a response quota of at least 30 percent, the researchers decided to contact target participants directly and ask them to complete the survey after the first and second data collection strategies.

A total of thirty faculty members participated and completed the survey (Table 1 in Appendix). Gender distribution mirrors the composition of the college faculty with twenty-two (73%) female and eight (7%) males. Nine (30%) faculty members identified themselves to be between 31 and 40 years old while 14 (46.67%) were over 50 years old. Each of the five departments was represented (Table 2 in Appendix) with 50% of faculty members identifying themselves as being in their first 5 years of service to the university --- primarily assistant professors and instructors.

Beyond demographic data, the researchers surveyed faculty preferences for online teaching. Almost 97% of the faculty surveyed stated they like to use technology to support the classes they teach. About 60% expressed that they like to teach totally online classes. A little over 43% of survey respondents reported no preference between online and face-to-face teaching. However, only 33% believe that online teaching is as good as teaching face-to-face. Further, only 10% reported they would like to teach all their courses online if given an opportunity while 70% disagreed with such a proposition. These variations in preferences regarding online teaching compared to face-to-face teaching may be related to attitudes about Web-based tools and the use of LMSs.

At the university level, the Distributed and Distance Education Center (DDEC) was given the task of providing training to interested faculty on using the new LMS. These training events were delivered during the end of the fall term and in the beginning of the spring term to help with the transition. Survey participants were asked if they participated in these training events. Two-thirds of the survey respondents (66.7%) reported availing themselves of such training in preparation for the implementation of the new LMS in spring 2009.

## FINDING COMMON GROUND: ISSUES AND STRATEGIES

A review of the literature revealed several factors that contribute to a successful transition. Included in these factors are those that emerged in this case study, such as, a positive disposition towards technology and change; preparation and prior experiences; adequate, appropriate, and timely support; and available and accessible resources. Given these, the researchers identified strategies that university staff can implement to facilitate a smooth and positive LMS transition for faculty members.

### Disposition towards Technology and Change

#### Issues

The researchers chose the word "disposition" to describe the attitudes held by faculty. As they conceptualized it, disposition encompasses the emotional reaction as well as the cognitive beliefs held toward an object or concept, in this case technology or change. This understanding was the basis for creating the survey items.

Table 4 (in Appendix) provides survey data on faculty responses related to disposition towards technology and change. Forty-three percent of the faculty surveyed reported their openness to the idea of replacing *WebCT* with the new LMS in comparison to those who felt strongly against it (27%). A high percentage (77%) expressed a positive attitude to the new system and these faculty members were not afraid of the change from one LMS to another (43%). About half of the faculty surveyed (47%) reported that they found the new LMS easier to use than they had anticipated. Finally, more than half (57%) stated that they "believe moving to the new LMS is a step in the right direction to support online faculty teaching." A deeper look at the response pattern of disposition-related statements revealed a negative reaction to change which is a common human response (Haymes, 2008; Mohr, 2007). This is an indicator that change itself can be an anxiety-producing experience, which can elicit responses of resistance. University staff need to focus their efforts to reduce the fear of change so faculty will develop a more positive attitude about change in general and transition to a new LMS in particular.

### Strategies

It is common to react negatively to change, especially if it will disrupt a routine or put individuals in an unfamiliar or uncomfortable position. The change that comes with new technology tends to be particularly stressful for those required to use it (Haymes, 2008). At this institution, verbal statements were heard from peers during this transition that reflected fearful feelings of what the change would entail. The following list describes strategies that might be helpful to university staff in developing positive dispositions toward technology and change among faculty members in time of transition to a new LMS.

First, it is important to communicate a positive attitude about the transition. Staff involved in helping faculty to learn the new system should acknowledge the differences between the old and the new system and reassure faculty members that help will always be available.

Second, timing is critical to help faculty openly accept the change. Past experiences in learning and managing the implementation of new LMSs taught the researchers that it is a lot easier to sell the idea of transition from the old to a new LMS if it is not implemented in the middle of an academic year.

Third, it is important to have the time necessary to make the change. Moser (2007) notes that investment in educational technologies must reflect recognition of the time faculty spend integrating technology into their teaching. It is important to give ample lead time for faculty and staff to prepare. In this case, the training for the new LMS transition should have started earlier in the fall term schedule. The implementation of the training caught many faculty members between end-of-term activities and preparation for spring courses. The preparation, assessment and evaluation activities, at the beginning and ending of a semester, are time-consuming and stress-inducing events in higher education. Those involved in making decisions as well as implementing them should be aware of the implications of the timing of major changes such as these.

Fourth, training is important during transition, but staff should use various delivery formats to reach various audiences. Diaz, Garrett, Kinley, Moore, Schwartz, & Kohrman (2009) stated, "successful faculty development programs should include … delivery in a variety of on-campus and off-campus formats" including face-to-face, blended, self-initiated, self-paced, anyplace/anytime that accommodates just-in-time needs. Online tutorials are great especially for tech savvy faculty members. However, this is only a small percentage of the faculty population.

Finally, the creation of intrinsic and extrinsic reward systems can facilitate positive dispositions among faculty members in learning to use the new LMS (Diaz et al, 2009; Weaver, 2006). By providing incentives, such as course releases, small grants, stipends, etc., a faculty recipient becomes an ally in helping communicate and articulate the value of using the new system to support teaching and learning activities on a regular basis and on the front lines.

## Preparation and Prior Experiences

### Issues

One's background plays a major part in one's behavior in the face of change (Haymes, 2008; Weaver, 2006). Since a high percentage of university faculty members use some form of technology in their communication, content sharing, consultation and collaboration with colleagues, the researchers believed that such preparation and prior experiences influence how they will function with the introduction of the new LMS. The statements included in the survey asked for prior experiences with the most recent LMS used before the spring semester of 2009, familiarity with features found in old and new systems, and availability of information from others or training sessions.

Table 5 (in Appendix) provides survey data on faculty responses related to preparation and prior experiences. From the survey, more than three-fourths of the faculty surveyed (83.3%) reported that their knowledge and experiences with *WebCT* prepared them to use the new LMS. Half of the survey respondents (50%) reported that they found familiar and similar features from *WebCT* in the new LMS. Other preparation and prior experiences reported by participants that supported a positive transition included advanced notification and awareness of the change (63.3%), exchanges on effective use of the new LMS with others (70%), and relevance and value of the training attended about the new LMS (63.3%).

### Strategies

Faculty from diverse academic disciplines bring with them a wealth of experience as they move from one institution to another. These experiences, in the form of knowledge, skills and dispositions, can influence how they manage their workloads, relationships, and productivity. The following strategies can help staff support faculty in managing changes in the tools they use to deliver content, interact with students, and evaluate student performance in their courses.

First, university staff can survey faculty for "expert knowledge and skills" with the current LMS so identified faculty can be organized to help support others. Also, a survey can capture data on faculty needs, so training opportunities can be tailor-made to ensure they are adequate, appropriate, and timely. Moser (2007) found that it is essential to establish a culture of support for faculty and she recommends "continual needs analysis" as one way of fostering such support.

Second, "frequent" LMS users within departments can be identified. They can be encouraged to become unofficial spokespersons not only in communicating changes like this transition to the new LMS, but also to champion for effective teaching and learning with technology. They can become members of a cadre of informal supporters to assist if similar changes occur in the future.

Third, visually pleasing printed or electronic materials can be created that identify similar features between the old and new systems. Also, differences between old and new systems can be identified with annotations on how to work with them and make them user-friendly to faculty members.

Fourth, university staff and volunteer faculty members can demonstrate how to use the new LMS through micro show-and-tell events to other faculty members. These can become desirable workshops that showcase easily implemented technology innovations (Chism, 2004) – low investment on time, high return on productivity. Also, this means becoming sensitive to faculty work cycles and schedules. In addition, these demonstrations can be delivered as video vignettes that can be accessed easily on the Web.

Finally, it is important to provide faculty with time for reflection (Chism, 2004; Moser, 2007) as they make attempts in learning the new system. Mini-retreats for faculty to think and reflect are also good venues to provide personal encouragement, communicate the intrinsic value of the new tools given the faculty's context and discipline, and stimulate new inquiries on teaching practices.

## Adequate, Appropriate and Timely Support

### Issues

Change, of any kind, always demands adequate, appropriate and timely support. This support comes in many forms including provision of relevant information and access to individuals who can provide assistance. It can also take the form of educational experiences (e.g., workshops, seminars, demonstrations) in small group settings. Weaver (2006) found that characteristics of attendees at workshops on online teaching changes through time. Early adopters have been replaced by the mainstream majority with different course development and training needs. University staff must be sensitive to these needs in order to provide appropriate support.

Table 6 (in Appendix) provides survey data on faculty responses related to adequate, appropriate and timely support. Faculty surveyed (80%) reported that there is always someone they can ask for support when having problems related to working with the new LMS. Others (73%) reported that information provided by the university about the new LMS was helpful in getting them started. Survey responses also indicated that faculty members (73%) feel confident in calling the DDEC on problems related to the use of the new LMS. The DDEC is the sole unit on campus providing support to faculty members in designing and developing their courses using the current LMS. The DDEC conducts workshops and one-on-one consultation on how to use the different LMS features. Finally, less than half of survey respondents (43%) reported that they did attend training on using the new LMS prior to full implementation.

## Strategies

Support is a critical service to faculty members, especially in using and integrating available technology-based tools in their teaching. Diaz et al (2009) found that support programs must be "valuable, relevant, current, and engaging" (p. 55) and they should include best practices in online teaching and learning. Support infrastructures need to be created in developing technology-rich environments for teaching and learning. Without adequate, appropriate, and timely support, faculty could find themselves disadvantaged by non-working tools and inaccessible resources in the delivery of instruction. Several strategies are suggested to avert the occurrence of such a situation.

First, identify university staff who are "experts" on specific features of the new LMS that faculty can call directly for support. With this information, one can create a directory containing names of individuals, areas of expertise, emails, direct phone numbers, and best time and day to call. This provides a convenient and expedient means of support 24/7. Also by having this information published, posted, and distributed across campus and various formats, faculty members always have ready access to resources to meet their just-in-time needs.

Second, university staff can provide models of appropriate web-enhanced and hybrid courses that were developed using the new LMS. By identifying effective course designs, faculty gain ideas that will help them in designing and developing their own courses that integrate available technologies.

Third, the university needs to nurture the creation of faculty learning communities by disciplines or across disciplines. Faculty can be encouraged to form information support networks to promote effective use of the new LMS. Using a faculty teaching faculty model or mentoring to frame the services provided is a good strategy. Availability of peers who are knowledgeable and skilled in using specific LMS features can be an asset when distance and time is limited. The benefits of this model increase when faculty members who share common disciplinary interests or teach in related disciplines network with each other.

Fourth, faculty need to be provided with co-teaching and collaborating opportunities in course development and delivery for effective technology use. As more faculty members take advantage of services provided by university staff, it is critical to rethink how these types of services are provided and who will be involved in providing them. Weaver (2006) stated that ongoing personal support for teaching staff is not sustainable in the long run unless more individuals are involved.

Finally, the university technology office should adopt an update, upgrade, and replacement cycle on technology-based tools and applications for faculty and staff. At the researchers' institution this cycle occurs every 3-4 years given the age of equipment, changes in commonly used software applications, etc. University staff in charge of technology replacement should assist faculty in determining whether they require new equipment based on factors identified above. This should occur on a regular basis.

## Available and Accessible Resources

### Issues

Resources come in different forms: equipment, printed and electronic materials, software applications, server and work space, people, time, and funding. Equipment is the tool of the trade. For faculty members this can take the form of computers, printers, scanners, phones, LCD projectors, etc. Printed and electronic materials come in the form of manuals, books, journals, magazines, CDs, and websites. An example of a software application easily accessible to faculty members is the Microsoft Office Suite containing *Word*, *PowerPoint*, *Excel*, and so on.

In addition to the aforementioned resources, faculty members require space to locate and store their completed projects as well as works-in-progress. These spaces come in various forms, such as, servers, external hard drives, physical space, etc. Server space refers to the capacity or availability of virtual storage space for electronically-based materials such as course websites, electronic presentations, databases, and digital files. Physical or work space is another form of space where one can be productive, such as, offices, laboratories, and classrooms.

People in roles, such as, instructional designers, multimedia developers, trainers, technical support specialists, and instructional technologists are another resource to assist faculty. These individuals are not only helpful in designing and developing pedagogically sound courses, but also in the effective and efficient use of technology-based tools. Finally, time and funding are critical in acquiring quality resources to produce quality products.

Table 7 (in Appendix) provides survey data on faculty responses related to available and accessible resources. Surveyed faculty reported they have the appropriate software applications to develop their courses in the new LMS (73.3%) and their computers were updated (or upgraded) (70%). Survey responses also indicated that faculty members have access to individuals who can help them develop a pedagogically-sound course (80%). Finally, when faculty members have questions related to working in the new LMS, they stated it does not take long to find the answers (60%).

### Strategies

Access to various resources, beyond computer equipment and software applications is critical to successful transition of a new system. Not only is technical support needed, but also required is guidance in designing and developing appropriate infrastructures. Several strategies are listed to help faculty and staff acquire the resources necessary in working with the new LMS and producing quality web-based or hybrid courses.

First, university staff should continuously review new equipment and software applications that effectively support course development activities of faculty members. These reviews should involve faculty members so that they can provide feedback on the appropriateness of these tools in their contexts or disciplines. Further, university staff and individual faculty members should team up to pilot the use of selected instructional tools. Working together in implementing a new instructional tool can facilitate a better understanding of how the tool is used and integrated into the curriculum for both faculty and staff. Also, it will build capacity for problem solving in technology related situations among faculty members.

Second, it is important to increase and enhance the visibility and accessibility of online tutorials. Online tutorials are often buried in layers of web pages that make them difficult to find. University staff need to make online tutorials visible to faculty through different means. For example, university staff can survey faculty on their need for more information in using a specific technology feature. Once a request for information is received, the university staff can embed a link to a web-based resource in an email message that can be sent to specific groups who expressed a common interest. Also, links to online tutorials can be placed in heavy traffic/ highly visited websites or presented in digital video formats to entice curiosity and interest to its readers.

Access to online tutorials can be enhanced by using different delivery formats. For example, a brief in-person introduction to online tutorials can be delivered during regularly scheduled departmental meetings, a streaming video presented over *YouTube*, a "how-to" column in a web-based magazine or other printed material.

Third, university staff needs to develop a repository of good examples of web-enhanced and hybrid models of courses. Chism (2004) discussed creating a database on best practices on teaching and learning with technology. These examples can be enhanced by providing a video commentary, highlighting key elements that make a specific course effective and high-quality.

Fourth, the university needs to fund and establish a teaching and learning unit, if it does not have one already. It can be staffed full-time by at least one person, who possesses a broad understanding of good teaching, effective learning, and integration of technology. This teaching and learning unit can be viewed as a one-stop source of services for faculty and staff. This unit will not only be responsible for providing one-on-one consultation services but also will take the lead in developing learning communities.

Finally, the university can sponsor a technology and learning resource fair so that faculty can share what they have created to support the effective delivery of their courses in any format. These materials, such as *PowerPoint* presentations, digital videos, handouts, and rubrics, can be considered reusable learning objects that other faculty members can adapt to support the teaching of their individual disciplines. Other learning objects can take the form of creative games that can add fun to the learning process, and research literature that provides information and evidence in support of best practices. As part of the fair, faculty participants can vote for the best resources. The selected faculty artifacts will provide recognition not only of faculty efforts to effectively use available resources, but also will celebrate their creativity and resourcefulness. This form of recognition can be intrinsically motivating to faculty and can instill positive attitudes toward the use of current and emerging technologies (Diaz et al, 2009; Weaver, 2006).

## CONCLUSION

Any change, especially involving technology, in any context poses a challenge to those who are implementing it as well as those who are impacted by it. Dispositions, preparation and prior experiences, support, and resources can produce positive or negative responses from those affected by the change. However, it is clear from the survey data that no factor identified by the research team has achieved a 100% agreement from the respondents. Even without 100% agreement, a conclusion may be drawn that faculty dispositions about the transition to the new LMS were generally positive, with nearly one-half of the participants open to change. Further, over half of the participants felt they had good preparation and positive prior experiences; adequate, appropriate and timely support; and resources that were accessible and available.

These results should be framed in the context of the university support infrastructure for technology. For example, the faculty members in the college of education of the researchers' institution receive upgraded hardware every three years. This means that faculty members' desktop equipment is replaced with new machines. Financial resources of other institutions may limit how they implement their replacement policies and these may differ from the policies of the institution used in this case study. This difference in policy implementation can contribute to faculty dissatisfaction of their work environment at these other institutions.

Also, the researchers' university creates an environment in which faculty members have access to software applications provided through the technology office. Further, faculty members at this university have the freedom to install and use non-standard software packages to meet their needs. Again, the policies developed at other universities may differ from this case given their individual and unique needs. This lack of access to new software applications can limit the effectiveness of faculty's teaching practices withstanding the amount of information available and accessible to them through the use of these technology-based tools and applications.

At the researchers' institution, the DDEC personnel provide excellent and responsive technical support for faculty who currently teach, or wish to teach, online. In most institutions, this type of support is usually provided by information technology professionals. Faculty members need this support to successfully navigate a technology-rich teaching environment. However, courses developed and delivered by faculty based simply on technical support often lack the pedagogical rigor expected for excellence in teaching.

Lastly, there are two additional supports for departments and faculty identified in this case study that encourages the use of LMSs for instruction. First, faculty and departments receive monetary incentives for teaching online courses and most of the time the money received is controlled by the faculty in that department. Second, the college of education has an active mentoring program, which provides stipends to support the pairing of an experienced online instructor with a novice faculty member who wants to develop a quality online course. Depending on a specific institution's culture and infrastructure for faculty support, similar incentives may or may not exist at other places.

After reflection on both the data collected and the literature reviewed, the researchers identified strategies that support faculty as LMS users when a new system is introduced. Also, key aspects of knowledge, skills and dispositions were discussed to better understand how they could enhance the effective and efficient use of the new LMS. Finally, strategies on how to provide better support for online teaching and make resources available and accessible to faculty members were discussed.

In this case study, the researchers found that managing change before it happens is critical to a successful LMS transition. It is recommended that the strategies identified and discussed in this study should be further investigated for their applicability in diverse contexts. Finding better ways of motivating and encouraging faculty members to continuously communicate, consult, and collaborate with university staff about teaching and learning with technology is essential especially to a community in transition.

## REFERENCES

Beatty, B., & Ulasewicz, C. (2006). Online teaching and learning in transition: Faculty perspectives on moving from Blackboard to the Moodle learning management system. *TechTrends*, *50*(4), 36–45. doi:10.1007/s11528-006-0036-y

Berryhill, A. H., & Durrington, V. A. (2006). The online course: The development and implementation of training and support. *Distance Learning*, *3*(2), 51–53.

Chao, I. T. (2008). Moving to Moodle: Reflections two years later. *EDUCAUSE Quarterly*, *3*, 46–52.

Chism, N. (2004). Using a framework to engage faculty in instructional technologies. *EDUCAUSE Quarterly*, *2*, 39–45.

Corich, S. (2005). Let's get ready to Moodle. *Bulletin of Applied Computing and Information Technology*, *3*(3). Retrieved from http://www.naccq.ac.nz/bacit/0303/2005Corich_LMS.htm.

Diaz, V., Garrett, P. B., Kinley, E. R., Moore, J. F., Schwartz, C. M., & Kohrman, P. (2009, May/June). Faculty development for the 21st century. *EDUCAUSE Review*, 47–55.

Haymes, T. (2008). The three E-strategy for overcoming resistance to technological change. *EDUCAUSE Quarterly*, *4*, 67–69.

Ioannou, A., & Hannafin, R. D. (2008). Course management systems: Time for users to get what they need. *TechTrends*, *52*(1), 46–50. doi:10.1007/s11528-008-0111-7

Lane, L. M. (2008). Toolbox or trap? Course management systems and pedagogy. *EDUCAUSE Quarterly*, *2*, 4–6.

Mohr, E. (2007, September/October). Incremental innovations: Changing the culture of teaching and learning. *EDUCAUSE Review*, 6–7.

Moser, F. Z. (2007). Faculty adoption of educational technology. *EDUCAUSE Quarterly*, *1*, 66–69.

Oblinger, D. G., & Hawkins, B. L. (2006, January/February). The myth about online course development. *EDUCAUSE Review*, 14–15.

Office of Information and Instructional Technology. (2006). *Implementing USG's online learning management system*. Retrieved from http://www.usg.edu/oiit/about/factsheets/webctvista.pdf

Smart, K. A., & Meyer, K. A. (2005). Changing course management systems: Lesson learned. *EDUCAUSE Quarterly*, *2*, 68–70.

The Campus Computing Project. (2008). *2008 national survey of information technology in higher education*. Retrieved from http://www.campuscomputing.net

Weaver, D. (2006). The challenges facing staff development in promoting quality online teaching. *International Journal on E-Learning*, *5*(2), 275–286.

Young, J. R. (2009). Why Blackboard's plan to buy a rival sparked a campus uproar. *Chronicle of Higher Education*. Retrieved from http://chronicle.com/free/v55/i3/37a00102.htm

Zsohar, H., & Smith, J. A. (2008). Transition from classroom to the web: Successful strategies for teaching online. *Nursing Education Perspectives*, *29*(1), 23–28.

## APPENDIX

*Table 1. Characteristics of population surveyed by age and gender (N=30)*

| Age Range | Male | Female | Total | % |
|---|---|---|---|---|
| 31 - 40 | --- | 9 | 9 | 30.00 |
| 41 - 50 | 2 | 2 | 4 | 13.33 |
| 51 and above | 5 | 9 | 14 | 46.67 |
| No Response | 1 | 2 | 3 | 10.00 |
| **Total** n (%) | 8 (26.7) | 22 (73.3) | 30 (100.0) | |

*Table 2. Characteristics of population surveyed by years of service to the university and departmental affiliation * (N=30)*

| Years of Service | C&I * | CEP* | ELPS* | HPESS* | SLPSE* | Total by Years of Service | % |
|---|---|---|---|---|---|---|---|
| 1-2 years | 3 | 2 | 1 | 1 | 2 | 9 | 30.00 |
| 3-5 years | 2 | --- | --- | 2 | 2 | 6 | 20.00 |
| 6 and over | 6 | 2 | 4 | 1 | 2 | 15 | 50.00 |
| **Total by Department** n (%) | 11 (36.7) | 4 (13.3) | 5 (16.7) | 4 (13.3) | 6 (20.0) | 30 (100) | |

* Department of Curriculum & Instruction (C&I), Department of Counseling and Educational Psychology (CEP), Department of Educational Leadership and Professional Studies (ELPS), Department of Health, Physical Education, and Sports Sciences (HPESS), and Department of Speech-Language Pathology and Special Education (SLPSE)

** Media and Instructional Technology (MIT) department is excluded given that all its degree programs are delivered online.

*Table 3. Pattern of survey responses on statements related to teaching online (N=30)*

| General Statements Related to Teaching Online | Agree n (%) | Middle of Road n (%) | Disagree n (%) | No Response n (%) |
|---|---|---|---|---|
| I like to use technology tools to support the classes I teach. | 29 (96.7) | 1 (3.3) | --- | --- |
| I like to teach totally online classes. | 18 (60.0) | 7 (23.3) | 4 (13.3) | 1 (3.3) |
| I believe that teaching online is not as good as teaching face-to-face. | 7 (23.3) | 13 (43.3) | 10 (33.3) | --- |
| If given the opportunity, I would like to teach all my courses online. | 3 (10.0) | 5 (16.7) | 21 (70.0) | 1 (3.3) |

*Table 4. Survey Responses on Statements Related to Disposition Towards Technology and Change (N=30)*

| Disposition Towards Technology and Change | Agree n (%) | Middle of the Road n (%) | Disagree n (%) |
|---|---|---|---|
| When I first heard the news of the change I did not want *CourseDen* to replace *WebCT*. | 8 (26.7) | 9 (30.0) | 13 (43.3) |
| People like me have positive attitudes about *CourseDen*. | 23 (76.7) | 7 (23.3) | --- |
| I was afraid about this change from *WebCT* to *CourseDen*. | 5 (16.7) | 12 (40.0) | 13 (43.3) |
| Using *CourseDen* is easier than I anticipated. | 14 (46.7) | 12 (40.0) | 4 (13.3) |
| I believe moving to *CourseDen* is a step in the right direction to support online faculty teaching. | 17 (56.7) | 13 (43.3) | --- |

*Table 5. Survey responses related to preparation and prior experiences (N=30)*

| **Preparation and Prior Experiences** | **Agree n (%)** | **Middle of the Road n (%)** | **Disagree n (%)** | **No Response n (%)** |
|---|---|---|---|---|
| My knowledge and experience with *WebCT* prepared me to use *CourseDen.* | 25 (83.3) | 5 (16.7) | --- | --- |
| I had difficulty finding *CourseDen* features that I am familiar with from *WebCT.* | 5 (16.7) | 10 (33.3) | 15 (50.0) | --- |
| I was alerted sufficiently in advance that *WebCT* would be replaced by *CourseDen.* | 19 (63.3) | 6 (20.0) | 4 (13.3) | 1 (3.3) |
| I have discussed with another faculty/staff how I can effectively use CourseDen to teach my courses. | 21 (70.0) | 5 (16.7) | 4 (13.3) | --- |
| The training sessions I attended did not prepare me to use *CourseDen.* | 1 (3.3) | 8 (26.7) | 19 (63.3) | 2 (6.7) |

*Table 6. Survey responses related to adequate, appropriate and timely support (N=30)*

| **Adequate, Appropriate and Timely Support** | **Agree n (%)** | **Middle of the Road n (%)** | **Disagree n (%)** |
|---|---|---|---|
| There is nobody around me that I can ask for support when I have problems working with *CourseDen.* | 1 (3.33) | 5 (16.7) | 24 (80.0) |
| Information I have about *CourseDen* was helpful in getting me started. | 22 (73.3) | 8 (26.7) | --- |
| When I have problems working in *CourseDen*, I feel confident I can get help by calling the Distance Education office. | 22 (73.3) | 5 (16.7) | 3 (10.0) |
| I did not need to attend a *CourseDe*n training to inform me how to use it. | 11 (36.7) | 6 (20.0) | 13 (43.3) |

*Table 7. Survey responses related to available and accessible resources (N=30)*

| **Available and Accessible Resources** | **Agree n (%)** | **Middle of the Road n (%)** | **Disagree n (%)** |
|---|---|---|---|
| I have the appropriate software applications to develop my course in *CourseDen.* | 22 (73.3) | 7 (23.3) | 1 (3.33) |
| My computer is updated (or upgraded) so I can work effectively using *CourseDen.* | 21 (70.0) | 5 (16.7) | 4 (13.3) |
| There are people who can help me develop a pedagogically-sound course in *CourseDen.* | 24 (80.0) | 5 (16.7) | 1 (3.33) |
| When I have questions related to working in *CourseDen*, it takes a lot of time to find the answers. | 2 (6.7) | 10 (33.3) | 18 (60.0) |

# Section 3
# Supporting Technologies for Student Tracking, Evaluation, and Synchronous Course Delivery

# Chapter 7
# Plagiarism Detection Tools in Learning Management Systems

**Sergey Butakov**
*Solbridge International School of Business, South Korea*

**Vladislav Shcherbinin**
*American University of Nigeria, Nigeria*

## ABSTRACT

*The main objectives of this chapter are to review the state-of-the art in plagiarism detection methods, discuss the most popular software tools available on the market and describe the new open architecture for plagiarism detection tools. The proposed architecture emphasizes the extensibility feature that allows it to be easily adapted for handling new types of assignments in the future. This chapter shows how the proposed architecture was implemented in a desktop application and a server-side plug-in for the Moodle course management system. An extended set of user trials is provided to support the proposed solutions. This set includes extensive tests for intra-corpal and internet plagiarism searches, tests with non-English assignments and promising results on cross language plagiarism detection.*

## INTRODUCTION

The problem of digital plagiarism arose along with development of digital technologies. In the last two decades, another dimension developed as the Internet became the most popular tool for information access. Digital plagiarism appears not only in education but it also exists in different forms in industry (Nitterhouse, 2003), web design (Bailey, 2006), and research publications (Boisvert & Irwin, 2006). Although from this point forward this chapter primarily focuses on digital plagiarism in education, many of the concepts discussed here can be extended to plagiarism detection on websites and in research papers and other sources of information.

The level of plagiarism has remained high all over the world for the last two decades (Austin and Brown, 1999; Hart and Friesner, 2004). For example, one study estimated the proportion of students in American high schools engaged in different kinds of plagiarism as up to 90% (Jensen, Arnett, Feldman, and Cauffman, 2002). Another study has also indicated a high level of plagiarism in higher education institutions in Ethiopia (Teferra, 2001). In this

DOI: 10.4018/978-1-61520-853-1.ch007

Copyright © 2010, IGI Global. Copying or distributing in print or electronic forms without written permission of IGI Global is prohibited.

chapter under student submission, we mean a file that contains virtually any kind of information such as text or a picture, sound file or motion picture. This chapter primarily concentrates on plagiarism detection tools with regards to text submissions but it will be highlighted if the approach can be extended to other kinds of submissions.

Software tools for uncovering digital plagiarism are evolving along with information technology, moving from simple single-user desktop applications in the early 1980's to the Internet-based plagiarism detection services of today. Plagiarism detection tools uncover plagiarism using two main approaches: (i) comparing the submission with other submissions by searching for similarities in content or (ii) comparing the submission with previous work done by a student and looking for unusual style.

This study is focused on the search of similarities between the submitted work and other documents. For purpose of detection, each case of digital plagiarism can be categorized into two types: intra-corpal (local) and extra-corpal (global) plagiarism. Extra-corpal plagiarism occurs if parts of a paper were obtained from a source outside the learning community (university, school, or learning center) and used without an appropriate reference. Such sources for global plagiarism include the Internet, books, periodicals, CDs, and P2P networks (Underwood and Szabo, 2003). The second type of plagiarism is intra-corpal plagiarism, where the original source of the plagiarized submission is located inside the learning community. The source can be from the same class, another section of the course, or the same course offered in a previous semester. As many studies have found, local plagiarism is very common all over the world (Austin and Brown, 1999), especially in developing countries where Internet access is limited or expensive (Odinma, Butakov, Grakhov, and Bollou, 2008). For example, Teferra (2001) shows that in Ethiopia, copying from classmates may account for as much as 80% of academic misbehavior.

The evolution of plagiarism detection tools brought up a lot of advances in both algorithms and implementations of anti-plagiarism methods. The following section provides the background information on the available plagiarism detection methods and tools. The subsequent sections discuss the open architecture of plagiarism detection tools and provide extensive descriptions of user trials.

## BACKGROUND

This section briefs the reader on methods for plagiarism detection and the available software solutions. We start with detection methods because advances in algorithms and information technologies in general are exactly the main driving force for software development. Since most commercial software tool vendors do not reveal the algorithms they use, we hope that a general description of methods and available tools will give the reader an understanding of state-of-the art technology in this field.

### Plagiarism Detection Methods

As mentioned in the introduction, methods for plagiarism detection in texts can be separated into two major groups: authorship detection methods and text comparison methods. The methods from the first group (also known as stylometry or intrinsic plagiarism detection (Eissen & Stein, 2006)), detect anomalies in the style of writing. This is very similar to the natural process of detection used by many professors, if, for example, they encounter a nearly perfect English paper submitted by a student who is not a native speaker of English. Such a paper would look suspicious and might be compared with possible sources of plagiarism. These methods require some background information about the student: in most cases, examples of previous writing. Some of the methods from this category can work without background information, instead checking style consistency within a single paper, but they might be vulnerable to 100% plagiarized text, as it can be consistent in style. In this study, we concentrate on the second type of methods: the comparison methods.

Comparison methods are based on the idea of comparing submitted work with other works from the same field or domain. The comparison task can be split into two major steps: access documents that might be similar to the submission and then proceed with a detailed comparison. The first step defines the scope for plagiarism detection: if the text is being compared with other texts submitted in the same institution, then it can be considered as local (intra-corpal) plagiarism detection. Learning management systems play the vital role of data source in intra-corpal plagiarism detection, as they store all the digital submissions from within the learning community (Romero, Ventura, & Garcia, 2007). If such a system also has access to submissions outside the learning community, then it can be used for global plagiarism detection. In the selection phase, the system should preselect from the outside source a set of documents—the set may be very large—that might be similar to the submission being processed.

Detailed comparison of documents should be precise enough to find similarities within documents and yet be fast enough to compare a submission with a large set of other submissions. There are a number of well-tested algorithms for such a comparison that are suitable for plagiarism detection (Clough, 2000; Maurer, Kappe, Zaka, 2006). These algorithms can be categorized into two large groups: feature extraction-based algorithms and document fingerprinting algorithms. The former are based on calculations of certain attributes in the text in question. For example, they may count the average number of characters in a string, the average number of words in a sentence, and so forth. As the result of feature extraction, each document turns out to be represented by a vector of *n* elements that can be used for comparison with other documents. To perform the comparison the algorithms will then calculate the distance between these vectors using Euclidian or more complicated metrics (Jones, 2001). Other algorithms from this category use a variety of pattern recognition techniques to find similarities, such as those in neural networks that are naturally designed for pattern recognition (Engels, Lakshmanan, and Craig, 2007). Feature extraction algorithms show good performance in one-to-many comparison, as each document is represented by a limited number of features: a vector with *n* elements. However, these algorithms can be affected by minor changes in text style and/or they may require intensive pre-processing, such as the feature selection itself, to be performed before they can be implemented.

The algorithms forming the second category process a text as a sequence of symbols and look for similarities with another text—another sequence. These algorithms do not require much preprocessing and they are suited for all types of texts, such as essays, source code, worksheets with answers and interview transcripts, to name a few. To avoid the computationally expensive comparison of string sequences, these algorithms substitute the text with a numeric representation, also called a fingerprint. A fingerprint is much smaller than the original document, and thus the fingerprints of two different documents might be compared to each other very quickly to determine the level of similarity between them.

One of the most well-known algorithms for document fingerprinting is the Winnowing algorithm by Schleimer, Wilkerson, and Aiken (2003). The main advantages of the Winnowing algorithm are as follows:

1. It does not depend on a specific structure of input. Plain text (e.g., source code, essay, or podcast script) can be used as an input.
2. It works quickly on large sets of data. Scaling the grammar size allows it to perform a quick check followed by a detailed comparison.

3. The fingerprint of a document worksheet/template, if any, can easily be removed from the student submission.

The documents' fingerprints, compiled by the fingerprinting algorithm, are used as a basis for comparing documents to each other. Figure 1 shows a simplified example of a one. vs. all comparison. While compiling the fingerprint, the algorithm translates the submitted text into the set of hashes: the four-digit numbers in Figure 1. After compilation, the hashes obtained are to be compared with those from system storage. In our example, the fingerprint of the submitted Document X includes three hashes with the same values as the hashes belonging to Document 01 in system storage. It means that the level of similarity in this pair-wise comparison is 60%, or the number of hashes with the same values divided by the total number of hashes in Document X.

There are many practical realizations of plagiarism detection algorithms. The following section provides information about the major players in the anti-plagiarism software market, starting with the giant service Turnitin and then moving to applications from relatively small startups.

## Anti-Plagiarism Software Solutions Available on the Market

There are a number of plagiarism detection systems on the market that can be used to check whether a paper was plagiarized "globally" or "locally". One of the most well-known is the Turnitin system developed by the US company iParadigms LLC (Turnitin, 2009). The algorithms used by Turnitin for document search and matching are not available to the public. An institution can subscribe to the service and use it with the direct access or through plug-ins for the most popular course management systems (CMS), including Blackboard, WebCT, ANGEL, Desire2Learn and Moodle (Turnitin, 2009). Turnitin compares the submitted assignment with those from its own database and provides a report that indicates the level of similarity between the submitted and stored documents. Turnitin is certainly the most popular service, well-known for its large databases (10 billion crawled pages from the Internet, over 70 million student papers, and over 10 thousand major periodicals as of May 2009). Although its large database is considered an advantage of the service, storing student papers in a third-party database may cause some legal issues with copyrights (Foster, 2002).

*Figure 1. Simplified example of one vs. many comparison*

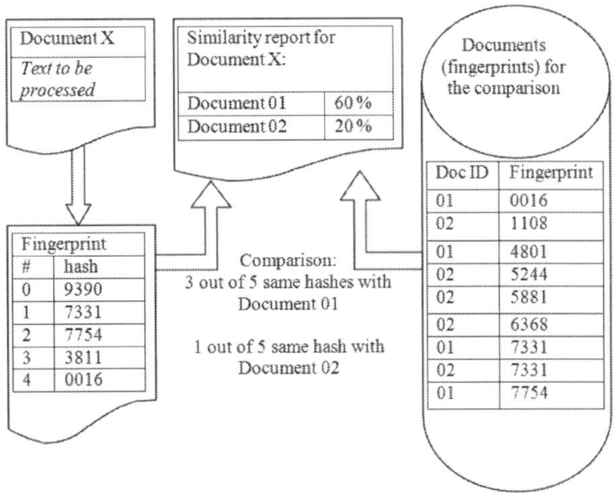

Some services use Turnitin at the back-end, adding their own interface with extra services (see JISC, 2009 as an example). Other popular global search services include MyDropBox, recently converted to Blackboard's Safe Assign service (Safe Assign, 2009). It may be used either through Blackboard or as a standalone service that compares the submission not only with publicly available pages on the Internet but also with documents from some subscription-based libraries. A service called Antiplagiat is being used by many universities and national education bodies in Russia (Antiplagiat, 2009). It provides similar services as Turnitin, extending the search to some national research databases. The recently started Collaborative Plagiarism Detection Network (CPDN) (Kulathuramaiyer, Zaka, and Helic, 2008) is based on a distributed architecture. CPDN employs publicly available search facilities provided by major search engines and supports the exchange of information across a network of servers to prevent cross-institutional plagiarism.

There are software products on the market that can help a professor speed up the search for locally plagiarized papers. These products include the CopyCatch software suite, which uses linguistics attributes to detect similarities in documents (see CopyCatch, 2009); Plague Doctor, which is based on feature extraction technology coupled with neural networks (Engels, *et. al.,* 2007); and the SNITCH software suite, which is based on such metrics as the average length of words in the text to detect potential plagiarism (Niezgoda and Way, 2006). The recently developed PAIRwise system (PAIRwise, 2009) is an open-source tool that provides both global search and comparison of locally submitted documents. Also there are a few specialized systems that operate on specially formatted texts (mostly program source codes). Measure of Software Similarity, or MOSS (Schleimer, *et. al.*, 2003), is one of the most well-known tools of this sort; it is implemented through the web as a non-commercial service. Another service for detection of similarities in computer programs is JPlag (Prechelt, Malpohl and Philippsen, 2002). The Picapica project (Picapica, 2009), which supports local and global searches, combined with stylometry (Eissen and Stein, 2006) and employs heuristics algorithms to avoid highlighting of properly cited text.

Among the abovementioned systems, such major players on the market as Turnitin and Safe-Assign can be integrated with a university-wide CMS such as BlackBoard, WebCT or Moodle. CMS databases are a very valuable source of information on student academic activities (Romero, *et al.* 2007; Romero and Ventura, 2007). These databases store digital copies of student assignments in a well-structured form, providing an ideal platform for local plagiarism detection tools by giving access to a complete set of assignments submitted within the learning community. Although the Turnitin service can be used as an add-on for many CMS's, two requirements may be considered as an obstacle to its adoption. First, the institution must pay for the appropriate Turnitin subscription, which can be a problem for small educational organizations. Second, there must be a stable, 24/7 network connection between the CMS and the Turnitin servers. This requirement might prove to be a major obstacle for educational institutions in developing nations, where Internet access is both expensive and unstable (Odinma, *et. al.* 2008).

The abovementioned problems with current plagiarism detection systems, along with the need for expandable tools that could cope with new types of digital submissions, inspired us to propose a new architecture for plagiarism detection tools and to develop software based on this architecture.

## PLAGIARISM DETECTION TOOLS: ARCHITECTURE AND USER TRIALS

This section starts with some details on the software tools developed, followed by descriptions of experiments that were performed to test these tools.

## Extendible Architecture for Plagiarism Detection Tools and Its Implementations

This subsection presents a new architectural solution for plagiarism detection tools. It starts with the general description of an open architecture and then proceeds with the details of its implementation on desktop and server platforms.

### An Extendible Architecture for Plagiarism Detection Tools

Software architecture affects such important characteristics of a system as performance, usability and extendibility. In this project, we have focused on the extendibility of the system architecture as the key for future development of the system. The general architecture for a plagiarism detection tool is presented in Figure 2. The architecture is assumed to be extendible as it can easily be adapted for the processing of new types of submissions (this will be discussed later in this section). As can be seen from Figure 2, the local search is performed in four main steps:

1. the system tokenizes the document and removes meaningless symbols (punctuation characters, whitespace, etc.).
2. the system compiles the fingerprints of submitted documents and stores the results in its database.
3. if required, the worksheet fingerprint is removed from the student submission.
4. the system compares the fingerprint of the submitted document with fingerprints of "local" documents.

The global search adds four more steps to this sequence:

*Figure 2. The architecture of plagiarism detection module*

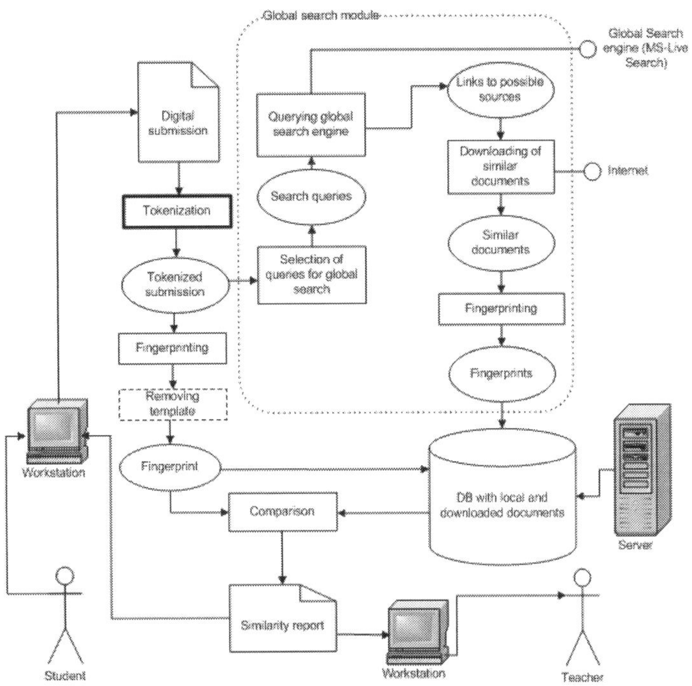

5. the system prepares search queries. At this stage the system may use different methods to prepare text search queries that will be used later on. For example, the selection process may be adapted to a specific language's grammar, or the random selection of the sliding window position may be implemented.
6. the system queries a global search engine to find documents on the Internet that may be similar to the text within the student submission. Some of the existing systems, like Turnitin (Turnitin, 2009), use their own web search facilities by maintaining huge archives of documents the system has crawled.
7. the system downloads documents from the top search results.
8. it compiles fingerprints for documents from the top search results.

Three main features make the proposed architecture extendible: (i) it can easily be coupled with any search engine; (ii) it can adopt any method to calculate fingerprint; and the most important feature, (iii) the Tokenization step, allows the architecture to adapt to new submissions as soon as methods to convert them to text are available.

For adaptation purposes, tokenization becomes the most vital part of data processing. It may be trivial if the system works with text submissions, as tokenization only removes unnecessary information from the text, providing the fingerprinting stage with pure text without spaces and punctuation. For example, tokenization of the first six words in the previous sentence would give the following result: FORADAPTATIONPURPOSESTOKENIZATIONBECOMES. Of course, tokenization becomes more complicated and interesting if we want to address non-textual plagiarism or cross-language plagiarism. One of the main concepts to be introduced in this project is to think about *tokenization as the preprocessing of a student submission*. In this case, *the goal of tokenization is to get plain text from any type of submission*. This is a simple procedure if the system works with a text document, whether formatted or unformatted. However, extending the system to process other kinds of submissions makes tokenization more complex. For example, speech recognition software might be used to obtain a text script from an audio podcast submitted by a student. Once the system has translated the submission from an audio file into plain text, it can look for similarities across the web and check whether the student has simply read some text found on the Internet. Or, if the submission is suspected as being plagiarized from language X and submitted in language Y, then it can be translated from Y back to X in the tokenization step. After tokenization, the anti-plagiarism tool no longer cares about the format or language of the original submission because it is provided with the corresponding plain text. Such extensive preprocessing in the tokenization step brings a plagiarism detection tool to a new advanced level and makes it possible to extend it to allow processing of new types of submissions and to address cross-language plagiarism issues. From the teacher's point of view, such architecture eliminates concerns about the type of assignment that needs to be processed: text essay, source code or, potentially, audio podcasts.

The proposed architecture was implemented in two software packages: a desktop application for plagiarism detection and, on the server side, an anti-plagiarism plug-in for the Moodle CMS. The following subsections brief the reader on some details of these software tools.

## Architecture of Stand-Alone Plagiarism Detection Software

The standalone desktop version of plagiarism detection software targets teachers that do not use the Moodle CMS, as well as any other user who needs to perform plagiarism checkups for texts. This group of users may include web masters, conference reviewers, editors, etc. The software can work on any desktop with Microsoft Windows XP or above and employs Microsoft SQL Server (including its free Express edition) to manage data. The desktop version is based on the architecture described in the previous subsection and has the following distinctive features:

- it provides a teacher with a compilation report that finds a minimum spanning set of documents that covers the plagiarized text. This is especially useful when the suspicious document was compiled from many sources.
- the desktop version can handle obfuscated text— such as text that has been intentionally altered with spelling mistakes or minor changes in syntax. This is a very useful feature for the plagiarism detection market and is based on the implementation of the following ideas: (a) fuzzy string comparison based on Levinstein's editing distance and (b) automatic spelling correction in search queries implemented out-of-the-box in Bing, the newest search service from Microsoft.
- it uses multiple search/downloading threads to speed up global search and downloading of potential sources.
- document compression is being implemented in the system to reduce the space consumed by the system database on the user's desktop.

Most of the experimental results recorded in the user trial subsection of this chapter were performed with the desktop version of the software.

## Detailed Architecture of the Plug-in for the Moodle CMS

The architecture described in the beginning of this section was also used to implement a plagiarism detection module for Moodle—one of the most popular open source CMS's. Development was planned with the following features in mind: (i) the plug-in should incorporate the advantages of local and global search; (ii) it should allow asynchronous global search to be performed at scheduled (off-peak) times; and (iii) it should be an open-source contribution to the Moodle community.

The document tokenization stage shown in Figure 2 requires plain text to be extracted from any proprietary document formats, such as Microsoft Word documents or Adobe PDF files. This is a simple operation on the Windows platform and can be performed with Microsoft Filter Pack or Adobe PDF filter, but on the LAMP (Linux+Apache+MySQL+PHP) platform it requires third party open source tools. These tools sometimes do not have all the required features and proper documentation.

Moodle CMS is well known for its easy-to-use plug-in implementation layer that allows the administrator to extend the functionality of the system easily. Once the anti-plagiarism plug-in is embedded into the server, it provides an easy-to-use interface for a teacher (Figure 3). Most of the settings such as *n*-gram size or query length are hidden from the teacher and left for administrator access only, as server performance and the results of the search are very sensitive to them and changing these settings is not recommended for non-experienced users. For example, changing of the *n*-gram size can make newly uploaded documents incomparable with those submissions that are already in the database.

Using the simple interface, the teacher—or other user—needs only to select the type of search to be performed (either local or global) and the assignments to process (Figure 3a). After processing, the user can access the report that provides the level of similarity among the submissions, as displayed in Figure 3b. Of course, this report can show a high level of similarity between the submission and a properly cited on-line refer-

*Figure 3. Teacher's interface for anti-plagiarism plug-in: (a) settings and (b) results report*

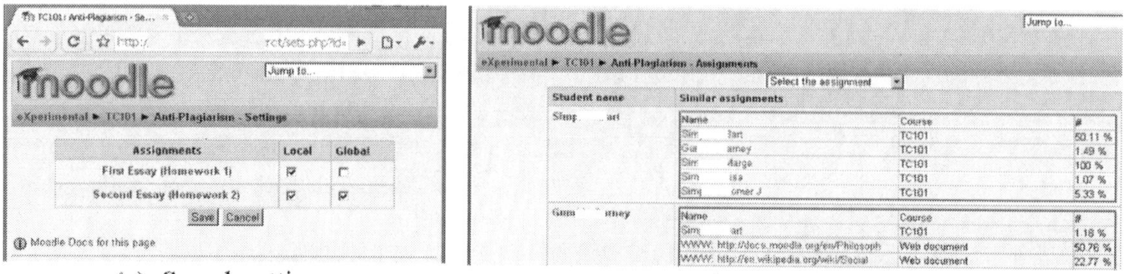


(a): Search settings (b): Search results

ence. In the last stage, the manual side-by-side comparison by the teacher is required to confirm the plagiarism.

## User Trials

Different kinds of real and simulated student submissions were processed to evaluate the performance and usability of the software. We started with submissions from two different courses that were taught at the American University of Nigeria (AUN) in 2006 – 2008. The first course, named "Introduction to Computer Science," has been offered since Spring 2006 to computer science and software engineering majors. The course was taught with active use of the Moodle CMS: students submitted all the assignments through the system and professors input all grades and comments into the system as well. The submissions from this course were used to check the system's performance in detecting intra-corpal plagiarism. The submissions from the second course ("University Writing I") were used to explore the characteristics of the global search. Both search experiments were performed using the desktop version of the software.

### User Trials on Intra-Corpal Plagiarism

The results of searches for local plagiarism are shown in Table 1. The course was observed for four semesters and involved the four professors teaching the course. In the first semester (Spring 2006), the course was taught by two professors (referred as A and B in Table 1). They taught it together using the same course slot in Moodle, so the results were aggregated, and the two professors' results were not able to be separated for analysis. During the second semester, the course taught by professor B in Spring 2007 was observed, and the third and fourth classes to be observed were courses taught by professors B and C in separate course slots in Fall 2007. Finally, the last courses to be observed were those of professors C and D, teaching in Spring 2008, also in separate slots. The teaching in separate course slots gave us the opportunity to observe results from different professors separately.

Each student in the course was required to submit eleven homework assignments. In each assignment, a student had to insert some information from the course-specific software simulator, as well as his/her own conclusions from the simulation's results, into a worksheet. The worksheet was a Microsoft Word file, which the students were expected to fill in and submit through the Moodle CMS. All students were expected to do the same assignment.

Using the data described above, the search for plagiarized papers was performed with manual cross-checking for confirmation. From Table 1 we can see that the detected level of plagiarism

*Table 1. Summary of local plagiarism cases in four semesters*

| | Total number of submitted assignments | Total number of plagiarized assignments | | Number of assignments plagiarized from another section | | Found with manual grading | | Found by software | |
|---|---|---|---|---|---|---|---|---|---|
| | | # | % | # | % | # | % | # | % |
| Teachers A and B; Spring 2006 | 292 | 85 | 29% | - | - | 22 | 26% | 85 | 100% |
| Teacher B; Spring 2007 | 237 | 44 | 18% | 17 | 38% | 5 | 11% | 44 | 100% |
| Teacher B; Fall 2007 | 76 | 13 | 17% | 13 | 100% | 4 | 30% | 13 | 100% |
| Teacher C; Fall 2007 | 263 | 59 | 22% | 24 | 40% | 21 | 36% | 59 | 100% |
| Teacher C; Spring 2008 | 93 | 12 | 13% | 5 | 42% | - [a] | - [a] | 12 | 100% |
| Teacher D; Spring 2008 | 62 | 10 | 16% | 6 | 60% | - [a] | - [a] | 10 | 100% |
| Total | 1023 | 223 | 22% | - | - | 52 | 26% | 223 | 100% |

[a] not applicable (professors did not perform a manual search because the system was running).

varied from 17% to 29% and that cross-section plagiarism varied from 38% to 100%. It is natural for students to assume that professors would not compare worksheets submitted in different sections, or work uploaded in previous semesters. It is important to note that all the professors involved in the survey stated that it was extremely time-consuming to search for the original source of the suspicious work. They estimated that search sometimes took up to half of the total grading time. The level of plagiarism was noticeably higher before the system implementation in Spring 2008. In Spring 2008, the students were aware that all the submissions would be processed by the system, but some apparently did not believe that it would be effective. This observation agrees with the results by Underwood and Szabo (2004), showing the lack of students' belief in teachers' ability to identify plagiarism. Students who submitted copied assignments for the first homework assignment received very poor grades, and professors also demonstrated the system during class time on a set of anonymized assignments. This demonstration has basically stopped plagiarism in this course, with only a few minor exceptions.

This experiment provided very promising results for intra-corpal plagiarism detection using a set of 1000+ student submissions. It is very important to note that these submissions were template-based and thus hard to analyze with the most of the tools available on the market

## User Trials on Internet Plagiarism Detection

The data set that was used to check system performance on global searches consists of 155 essays submitted in the "University Writing I" course. Thirty-four students were enrolled in the course and were assigned five essays, each about two pages long. As can be seen from the results presented in Table 2, the level of plagiarism detected in these submissions varied from 10% to 20%. To check each submission, the system had to retrieve between 17 and 32 documents from the Internet. Traffic consumption per assignment was relatively low: about 3.4 MB at maximum. Table 2 also shows that the number of downloaded documents is correlated with the number of words in the essay: the longer the submission, the more documents the system had to download from the web. On the other hand, there was no correlation between the number of words and generated traffic. This effect can be explained by the fact that traffic depends on the type of resource to be downloaded. A good example is the food essay: it is very likely that web documents describing food include a lot of pictures, which leads to higher traffic generation and a small correlation coefficient.

As mentioned earlier in the chapter, the global search is based on excerpts from student submissions. The system selects short sequences of words starting from the beginning of document, and the essential parameter at this stage is the number of consecutive words to be selected for a query. For example, if the number of words is two, the previous sentence would be broken up into queries of two words: "The system", "system selects", "selects short", etc. If the number is three, then the set of queries will be: "The system selects", "system selects short", "selects short sequences", etc. The length of the search query affects the efficiency of global search, and how many words should actually be used for querying the search engine requires some investigation. It is obvious that the fewer words the system selects, the less relevant the search results will be. On the other hand, if the system uses long queries, then the search results will be more relevant but vulnerable to minor changes within this selection.

To determine the optimal number of words for a global search, we repeated the search, setting the number of consecutive words to values from 5 to 10. The results of this experiment are shown in Table 3. It can be seen that, if the system uses more words in the query, each search takes less time and consumes less traffic, but the results become less precise and system can more easily miss the plagiarized paper. Thus, increasing the number of words in the query could decrease the search accuracy of the system. On the other hand, using short queries (six or fewer words) increases traffic and leads to many wrong hits, or false positives. Based on this experiment, seven-word queries were used for the global search presented in Table 2.

As we can see from Tables 1 and 2, the most promising result of the experiments is that all plagiarized submissions were uncovered without any wrong hits. Thus, the proposed technology

*Table 2. Results of global plagiarism search for five essays*

| Essay name / Essay Characteristic | Compare & Contrast | Cause & Effect | Anne Frank | Food | Process |
|---|---|---|---|---|---|
| Number of submissions | 29 | 34 | 32 | 29 | 31 |
| Number of plagiarized papers | 3 | 5 | 4 | 4 | 6 |
| Avg. number of words in the assignment | 522 | 520 | 536 | 485 | 401 |
| Avg. number of downloaded documents per assignment | 24 | 30 | 32 | 18 | 17 |
| Average traffic per assignment, MB | 2.2 | 2.6 | 3.4 | 1.1 | 1.2 |
| Average number of downloaded docs per 100 words in the assignment | 4.7 | 5.7 | 6.0 | 3.8 | 4.3 |
| Average traffic per 100 words of assignment, MB | 0.4 | 0.5 | 0.6 | 0.2 | 0.3 |
| Correlation coefficient for number of words and number of downloaded documents | 0.5 | 0.7 | 0.8 | 0.5 | 0.8 |
| Correlation coefficient for number of words and traffic | 0.3 | 0.7 | 0.5 | 0.1 | 0.7 |

*Table 3. Search results with varying numbers of consecutive words as query phrases*

| Number of consecutive words as the search query | Average number of downloaded documents | Average traffic, MB | Plagiarized papers detected | Plagiarized papers not detected | False positive |
|---|---|---|---|---|---|
| 5 | 124 | 7.54 | 6 | 0 | 2 |
| 6 | 45 | 3.14 | 6 | 0 | 2 |
| 7 | 17 | 1.21 | 5 | 0 | 0 |
| 8 | 6 | 0.49 | 4 | 1 | 0 |
| 9 | 3 | 0.24 | 4 | 1 | 0 |
| 10 | 3 | 0.28 | 4 | 1 | 0 |

helps effectively to uncover cases of "local" and "global" cheating and to improve course outcomes. The results of on-site implementation of the system in Spring 2008 show that once students see the anti-plagiarism system in action, they change their attitude and put more effort into the course.

From a technical point of view, our software solutions have four main advantages: (i) the system can be used in asynchronous mode to perform global search at off-peak times, (ii) the system "searches globally but acts locally"—meaning that student essays are not being transferred to a third party and thus the system helps to avoid copyright issues other systems have (Foster, 2002), (iii) the system uses only free third-party services, and (iv) the system can be extended to handle new types of submissions by extending the tokenization phase. After summarizing the results of system implementation for intra-corpal and extra-corpal plagiarism detection on the large archive of real student submissions, we can conclude that it was very successful.

To run the detection process properly, the user needs to adjust a number of different search settings. For the proposed software solutions most targeted users, such as educators and course administrators, are not professionals in the field of digital plagiarism detection. Of course, a plug-and-play tool to check student submissions would be the most appropriate solution for the majority of users. To give these tools a more like plug-and-play feel, we had to find default quasi-optimal values for search algorithm parameters. These parameters should solicit the reasonable amount of time to perform the search while not compromising on the quality of the search. The rest of this subsection reflects on some experiments performed while tuning the default search values for the software.

To locate the possible sources of submitted documents on the Internet, the algorithm uses the sliding window approach. It takes each sequence of *l* words from the text and uses these sequences to query the search engine. Some researchers argue that such an approach may cause a very large number of search queries to be sent (Knight, Almeroth, & Bimber, 2004), and they propose to use words from the beginning of each sentence. Although this narrows down the number of search queries from nearly the number of words to the number of sentences in a document, it might be vulnerable to minor changes in the text. If a plagiarist changes the beginning of most of the text's sentences, the search algorithm may fail to locate the source. Also, the performance of such a selection method can be affected by the structure of different natural languages. The revival of random selection was proposed as the simplest and most language-independent method to narrow the number of search queries.

The total time required to find the possible source of plagiarized document on the web consists of the following parts:

- time to send queries to the search engine and receive the results, or query time;

- time to download the documents from the Web and perform preprocessing and compile fingerprints;
- time to compare the fingerprints of the submitted and downloaded documents.

The time required for the last two operations depends mostly on the computational power and network bandwidth allocated for the service. Time to query the search engine depends on the number of queries to be sent and on the bandwidth. Since queries and replies are not that large, we may say that query time mostly depends on the number of search queries to be sent. Random selection of search queries can significantly reduce the number of search queries and of course reduce the total time required to find the source of a plagiarized paper.

To test the random selection we performed a global search on five simulated "copy-and-paste" plagiarized papers, each of about 550 words: a paper consisting of 50% plagiarized text (Paper *A*); a paper containing 100% plagiarized text taken from an article in Wikipedia (Paper *W*); and another 100% plagiarized paper—Paper *M*—taken from one of the internet paper mills, www.schoolsucks.com. Partially plagiarized paper *A* was altered in three ways: Paper *A.A* consisted of plagiarized text followed by original text, paper *A.B* was a mixture of original and plagiarized paragraphs—each original paragraph was followed by a plagiarized one—and *A.C* was a mixture of sentences—each sentence from the original was followed by a plagiarized one.

The test was performed using a sliding window of seven words to query the search engine. The length of seven words showed good performance on documents in English in our experiment, although the optimal number may be different for other languages. The query algorithm processes a document following three major steps. First of all, the algorithm creates the complete set ($X$) of search queries, selecting sequences of seven words from the beginning of the document. The total number of search queries is $|X| = w - 7$, where $w$ is the number of words in a document. In our cases the total number of queries for all documents was about 550. In the second step the algorithm randomly selects a certain number of queries from the $X$ set: 5%, 10%, ... 95% and finally 100% for a query subset. This subset is used to query the Microsoft Live Search service. After sending selected queries to the search engine, the algorithm takes top links, whose rank is based on the number of times the link appeared in the top results returned by the search engine, and forwards these links to the downloading step.

The aggregated results of this experiment are presented in the Table 1. The most promising outcome is that even with a small number of randomly selected search queries—as low as 5%—the source of a plagiarized document appears in the top search results.

Such positive results obtained with random selection of search queries suggest that processing time and traffic workload can be reduced while looking for the source for a plagiarized paper. This is especially important for institutions in developing nations where Internet connectivity may be expensive and unstable.

## User Trials with Non-English Texts

Plagiarism detection in non-English languages can be performed using the architecture presented above with no changes. The major challenges may appear in global plagiarism detection, and they are related to the encoding of web-pages. It is very common for web texts on non-English websites to fail to provide the correct information about the encoding that is being used, in the body of HTML pages. Therefore, the text extraction from the web page may include incorrect information or suffer from incorrect auto-detection of the encoding table. Although the trend to use Unicode is on the rise, many national web pages still use old-style one byte encoding.

*Table 4. Results of the search experiments with selection of search queries from 5% to 100%*

| | Source is on the top | Avg. query time for 5% selection | Avg. query time for 50% selection | Avg. query time for the full search (100%) |
|---|---|---|---|---|
| Doc. A.A | 100% | 8 s | 81 s. | 160 s. |
| Doc. A.B | 100% | 8 s. | 81 s. | 157 s. |
| Doc. A.C | 100% | 7 s. | 80 s. | 162 c. |
| Doc. W | 100% | 6 s. | 69 s. | 147 s. |
| Doc. M | 87% | 6 s. | 80 s. | 155 s. |

To check the performance of the desktop solution, a set of 42 Russian documents was processed. All these essays were deliberately copy-pasted from paper mills with minimal changes. Some of the essays were compiled from two or three sources. A global search was performed with the following parameters: 8 words in excerpt (query); 100% queries were used for the search. The result was the system was able to find all the sources for all the papers indicating the remarkable ability of the multi-lingual search. The total number of downloaded documents was 6,827 which means that about 162 documents were downloaded to check each assignment. As it was indicated in the description of the intra-corpal plagiarism detection, the number of downloaded documents to check essays in English was much lower (see Table 2). This fact can be explained by the fact that each of the Russian essays was about 10 pages long - significantly longer than the ones that we used in the English language search. The experiment proves that regardless of the language, public search engines provide excellent results for plagiarism detection in different languages, with zero cost. As we know from publications, (Head, 2007) students use public search engines as the first tool to look for the relevant information on the web. Based on the results provided above, we can assume that plagiarism detection based on facilities provided by public search engines will also give good results with other languages and assignments.

## Cross-Language Plagiarism Detection

The issue of cross-lingual plagiarism is being reported in many fields such as journalism, education and even poetry (Breytenbach, 2005). The number of research projects focused on cross-lingual plagiarism detection is quite limited, and they are concentrated mostly in the development of statistical models for detection of such cases (Pinto, Civera, Barrón-Cedeòo, Juan, and Rosso, 2009; Lee, Wu, and Yang, 2008). As cross-lingual plagiarism detection is a new area, there are no established tools on the market that would provide out-of-the-box solutions for end users.

The approach that was suggested in the architecture section of this paper (section 3.1.1) is simple in terms of implementation because it does not require implementation of new models or methods to work with plagiarized text. It requires one additional step to be implemented in the tokenization phase (Figure 2): the suspicious submission should be translated from the source language to the language of the possible original (target language).

To check our assumption, we ran a small experiment with five newspaper articles manually translated by professionals, from English (target language) to Russian (source language). English articles were taken from the newspapers available on-line such as Guardian, LA Times, Foreign Policy and Reuters. These translations simulated cross-language plagiarism with original papers available on-line in English. To perform

*Table 5. Results of the search experiments with selection of search queries from 5% to 100%*

| | Number of similar articles found | Number of actual copies of the original article found | The highest and lowest similarity levels for the original source (including copies) |
|---|---|---|---|
| Article A (Guardian) | 15 | 13 | 2.47% / 2.23% |
| Article B (Guardian) | 3 | 2 | 3.94% / 3.15% |
| Article C (Foreign policy) | 17 | 12 | 3.75% / 1.62% |
| Article D (LA Times) | 4 | 4 | 1.77% / 1.09% |
| Article E (Reuters) | 8 | 7 | 2.55% / 0.39% |

automated translation back to the target language, the free multi-language translation tool available in Microsoft Live (TM) web services was used.

The quality of automated translations was not very close to the original articles in the newspapers, but surprisingly, it was good enough to locate the source material on the Internet. The detailed results of the experiment are shown in Table 5. As we can see in all five cases, the originals along with their copies on different web sites were found with a low level of wrong hits. For example, for the first article, only two out of fifteen suspicious texts were not copies of the original article in Guardian.

Although we had to explicitly define the target language, the most promising result that we got in this experiment is that for large texts plagiarized from another language. Even with the current state of machine translation tools, it is possible to embed automated translation features into plagiarism detection tools and find the source. Such a result gives a good foundation for future work in this field.

## FUTURE RESEARCH DIRECTIONS

The plagiarism detection problem in e-learning is very similar in its persistence to security problems in information systems at large. Just as hackers are trying to overcome the security measures, plagiarists are trying to misrepresent their efforts by submitting copied works or ideas. As the chapter shows, with the current state-of-the art tools and technologies, the detection of copy-and-paste plagiarism in texts can be considered a technically solved problem. Despite the many improvements that are still to be made in such systems, the following open problems require a lot of efforts from researchers and software developers:

- Text obfuscation: the level of obfuscation in plagiarized texts can vary from minor changes in spelling to complete rewriting of the text using stolen ideas. The latter case is the more difficult to detect and requires a high level of "understanding" of text semantics by the machine. This can be addressed by employing Semantic Web technologies for text comparison in plagiarism detection.
- Plagiarism in non-text submissions. Although the architecture proposed in this chapter is capable of handling non-text submissions by converting them to text on the tokenization phase, the industrial implementation of this technology requires that a very good job be done on the tokenization phase. For example, the processing of audio podcasts would require high quality speech recognition to be implemented during this stage. The solutions for this problem will evolve along with technologies for sound and motion picture processing.
- "Deep" web search. Although many plagiarism detection systems claim to have access to periodicals and e-libraries, it is

still an open problem to detect content plagiarized from such sources, as well as content written by someone other than the plagiarist. Authorship detection methods should be very helpful in addressing this problem, providing complimentary service to document comparison systems.
- Integration with existing CMS and LMS software packages. A lot of work will have to be done by software developers to equip existing systems with plagiarism detection tools.

These open problems create a lot of opportunities for exciting and rewarding research in the field of learning management systems and plagiarism detection.

## CONCLUSION

Plagiarism remains one of the painful problems in many fields including education, business, journalism, research, etc. This issue is being addressed in many ways depending on the industry: in education, apart from the development and implementation of institutional policies on cheating, software tools provide a great assistance to teachers helping to prevent plagiarism in courses they run. It is a very visible trend in learning management systems development to equip them with plagiarism detection tools. Such tools at large are based either on detection of anomalies in texts (authorship detection) or on text comparison. The latest provide good evidence to a teacher if plagiarism has been detected, and over the last decade, they were proven to be an efficient solution for plagiarism detection in education.

The chapter provided the description of available methods and software solutions for plagiarism detection. Based on this review, we proposed an open architecture for an anti-plagiarism tool. The suggested architecture includes a preprocessing stage to convert virtually any kind of digital submission into text form and processing stages that use this text to look for similarities among submissions. The chapter also described two software solutions based on the proposed architecture and provided extensive user trials to show the efficiency of these solutions.

The promising results of user trials include intra-corpal plagiarism detection with template-based submissions, extra-corpal plagiarism detection in English and Russian and finally extra-corpal cross-language plagiarism detection. These promising results also support the opinion that the public search engines can serve as an external storage for Internet plagiarism detection and moreover such complimentary to search services as translation can be used for cross-lingual plagiarism uncovering.

An important part of learning management systems, plagiarism detection tools still require a lot of work be done to address such issues as text obfuscation, handling of non-text submissions, and including "deep" web into plagiarism search. These open problems provide a lot of opportunities for future research in the field of plagiarism detection, and hopefully will provide more assistance to educators in the near future.

The software tools described in the chapter are accessible for download from www.siberiasoft.info.

## ACKNOWLEDGMENT

The authors want to acknowledge productive discussions with Prof. Cheryl Pavlik and Prof. Aghileh Djafari Marbini on the issue of plagiarism in freshmen courses. We also want to express our appreciation to Dr. Alexey Mikhailov from Altai State Technical University for providing us archive with essays in Russian language.

## REFERENCES


Antiplagiat. (2009). *Antiplagiat plagiarism detection service*. Retrieved May 20, 2009, from http://www.antiplagiat.ru

Austin, M. J., & Brown, L. D. (1999). Internet Plagiarism: Developing Strategies to Curb Student Academic Dishonesty. *The Internet and Higher Education*, *2*(1), 21–33. doi:10.1016/S1096-7516(99)00004-4

Bailey, J. (2006). Web Design Plagiarism. *Plagiarism Today*. Retrieved June 12, 2009, from http://http://www.plagiarismtoday.com/2006/08/02/web-design-plagiarism/

Boisvert, R. F., & Irwin, M. J. (2006). Plagiarism on the rise. *Communications of the ACM*, *49*(6), 23–24. doi:10.1145/1132469.1132487

Breytenbach, K. (2005, August 12). Poet accused of plagiarism gives back her prize. *Cape Times*. Retrieved June 16, 2009, from http://www.capetimes.co.za/general/print_article.php?fArticleId=2834085

CDPNet. (2009). *Collaborative plagiarism detection network*. Retrieved June 3, 2009, from http://www.cpdnet.org

Clough, P. (2000). *Plagiarism in Natural and Programming Languages: An Overview of Current Tools and Technologies. Research Memoranda, CS-00-05*. UK: Department of Computer Science, University of Sheffield.

CopyCatch. (2009). *CopyCatch Software Suite*. Retrieved May 20, 2009, from http://www.copycatchgold.com/

Eissen, S. M., & Stein, D. (2006). Intrinsic Plagiarism Detection. *Lecture Notes in Computer Science*, *3936*, 565–569. doi:10.1007/11735106_66

Engels, S., Lakshmanan, V., & Craig, M. (2007). Plagiarism detection using feature-based neural networks. In *Proceedings of the 38th SIGCSE technical symposium on computer science education* (pp. 34–38).

Foster, A. L. (2002). Plagiarism-detection tool creates legal quandary. *The Chronicle of Higher Education*, , A37.

Hart, M., & Friesner, T. (2004, December). Plagiarism and poor academic practice – A threat to the extension of e-learning in higher education? *The Electronic Journal of e-Learning, 2*(2). Retrieved June 8, 2009, from http://www.ejel.org/volume-2/vol2-issue1/issue1-art25-hart-friesner.pdf

Head, A. J. (2007) Beyond Google: How do students conduct academic research? *First Monday, 12*(8). Retrieved June 16, 2009, from http://firstmonday.org/htbin/cgiwrap/bin/ojs/index.php/fm/article/view/1998/1873

Jensen, L. A., Arnett, J. J., Feldman, S. S., & Cauffman, E. (2002). It's wrong, but everybody does it: Academic dishonesty among high school students. *Contemporary Educational Psychology*, *27*(2), 209–228. doi:10.1006/ceps.2001.1088

JISC. (2009). *JISC plagiarism detection service*. Retrieved June 11, 2009, from http://www.submit.ac.uk

Jones, E. L. (2001). Plagiarism monitoring and detection - towards an open discussion. In *Proceedings of the Seventh Annual Consortium For Computing in Small Colleges Central Plains Conference on the Journal of Computing in Small Colleges* (pp. 229-236).

Knight, A., Almeroth, K., & Bimber, B. (2004). An Automated System for Plagiarism Detection Using the Internet. In L. Cantoni & C. McLoughlin (Eds.), *Proceedings of World Conference on Educational Multimedia, Hypermedia and Telecommunications 2004* (pp. 3619-3625). Chesapeake, VA: AACE.

Kulathuramaiyer, N., Zaka, B., & Helic, D. (2008). Integrated Copy-Paste Checking: Design and Services. In *Proceedings of the 2008 Eighth IEEE international Conference on Advanced Learning Technologies* (pp. 519-523). Washington, DC: IEEE Computer Society. Retrieved from DOI= http://dx.doi.org/10.1109/ICALT.2008.312

Lee, C., Wu, C., & Yang, H. (2008). A Platform Framework for Cross-Lingual Text Relatedness Evaluation and Plagiarism Detection. In *Proceedings of the 2008 3rd international Conference on innovative Computing information and Control*. Washington, DC: IEEE Computer Society. Retrieved from DOI= http://dx.doi.org/10.1109/ICICIC.2008.76

Maurer, H., Kappe, F., & Zaka, B. (2006). Plagiarism – A Survey. *Journal of Universal Computer Sciences*, *12*(8), 1050–1084.

Niezgoda, S., & Way, T. P. (2006). SNITCH: A Software Tool for Detecting Cut and Paste Plagiarism. In *Proceedings of SIGCSE '06 conference* (pp. 51 – 55).

Nitterhouse, D. (2003). Plagiarism: Not just an academic problem. *Teaching Business Ethics*, *7*, 215–227. doi:10.1023/A:1025017019246

Odinma, A. C., Butakov, S., Grakhov, E., & Bollou, F. (2008). Planning, designing and implementing an enterprise network in a developing nation. *International of Journal of Enterprise Network Management*, *2*(3), 301–317. doi:10.1504/IJENM.2008.018783

PAIRwise. (2009). *The PAIRWise system*. Retrieved May 20, 2009, from http://www.pairwise.cits.ucsb.edu/system.htm

Picapica. (2009). *Picapica plagiarism detection service*. Retrieved from http://www.picapica.net

Pinto, D., Civera, J., Barrón-Cedeòo, A., Juan, A., & Rosso, P. (2009). A statistical approach to crosslingual natural language tasks. *J. Algorithms*, *64*(1), 51-60. Retrieved from DOI= http://dx.doi.org/10.1016/j.jalgor.2009.02.005

Prechelt, L., Malpohl, G., & Philippsen, M. (2002). Finding Plagiarisms among a set of programs with JPlag. *Journal of Universal Computer Science*, *8*(11), 1016–1038.

Romero, C., & Ventura, S. (2007). Educational data mining: A survey from 1995 to 2005. *Expert Systems with Applications*, *33*(1), 135–146. doi:10.1016/j.eswa.2006.04.005

Romero, C., & Ventura, S., & Garcia, E. (2007). Data mining in course management systems: Moodle case. *Computers & Education*. Retrieved from doi:10.1016/j.compedu.2007.05.016.

Safe Assign. (2009). *Safe assign anti-plagiarism service from Blackboard*. Retrieved May 14, 2009, from http://www.safeassign.com

Schleimer, S., Wilkerson, D., & Aiken, A. (2003 June). Winnowing: Local Algorithms for Document Fingerprinting. In *Proceedings of the ACM SIGMOD International Conference on Management of Data* (pp. 76-85).

Teferra, D. (2001). Academic dishonesty in African universities-trends, challenges, and repercussions an Ethiopian case study. *International Journal of Educational Development*, *21*(2), 163–178. doi:10.1016/S0738-0593(00)00037-7

Turnitin. (2009). *Turnitin brochure*. Retrieved May 14, 2009, from http://turnitin.com/static/pdf/Turnitin_brochure.pdf

Underwood, J., & Szabo, A. (2003). Academic offences and e-learning: individual propensities in cheating. *British Journal of Educational Technology*, *34*(4), 467–477. doi:10.1111/1467-8535.00343

# Chapter 8
# Using a Learning Management System to Facilitate Learning Outcomes Assessment

**Steven F. Tello**
*University of Massachusetts Lowell, USA*

**Luvai Motiwalla**
*University of Massachusetts Lowell, USA*

## ABSTRACT

*Increased demands for accountability among state and federal policy makers require that colleges and universities improve the process of measuring student learning outcomes. Despite a growing need, there has been limited development of integrated, electronic processes and tools that facilitate assessment of student progress toward program-level learning outcomes. Collecting student course materials, classifying by program and course-level objectives and reporting the results remains a tedious and labor-intensive task. This project demonstrates how course-level assessment data from a learning management system (LMS) can be utilized for program-level outcomes assessment. A pilot system was developed to integrate data from a LMS to provide continuous reporting of program and course-level assessment with minimal additional effort from faculty and students. This chapter shares the authors' outcomes assessment system development approach, faculty development approach, and the lessons learned from their project, including the challenges confronted during system implementation.*

## INTRODUCTION

Over the past decade institutions of higher education have focused increased attention on the assessment of student learning outcomes. In part, this is a reaction to increased calls for accountability in higher education as expressed by various political and competitive forces, encouraged by the debate surrounding the reauthorization of the Higher Education Act of 1965 (Lovett, 2004; Shulock, 2004). However, it also reflects a growing understanding on the part of faculty and administrators regarding the role outcomes assessment plays in both shaping the student learning experience as well as shaping the academic programs and curriculum from which students graduate. Both regional and professional accreditation agencies acknowledge this growing call for accountability and have integrated assess-

DOI: 10.4018/978-1-61520-853-1.ch008

Copyright © 2010, IGI Global. Copying or distributing in print or electronic forms without written permission of IGI Global is prohibited.

ment standards into the accreditation process (AASCB, 2005). Measurement of these broad program-level outcomes requires program administrators, faculty, advisory boards and other constituent groups to determine what metrics constitute mastery or demonstration of a particular skill or accomplishment.

This expanded interest in outcomes assessment, while important and welcomed, creates several challenges for college faculty and administrators. For example, the manual process of monitoring the integration and achievement of learning outcomes within courses and degree programs is typically a labor intensive, paper-based process often driven by accreditation visits and timelines. Faculty and administrators are asked to report which outcomes are addressed by which course activities and how student achievement is assessed. This information, along with sample student work, is then collected by department chairs and deans who organize and present it to visiting accreditation teams. Therefore, there is a need for a systematic and automated approach to linking course level activities and assignments with program goals and institutional mission for assessment purposes.

While assessment as a tool for improving student learning and educational programs offers great promise, existing processes for presenting outcomes assessment and collecting and summarizing student achievement are somewhat limited. In a 2008 report, Eduventures found that 76% of institutions surveyed identified difficulty in collecting and analyzing outcomes data as a "top five challenge" in implementing comprehensive assessment programs. There is a need for a more systematic process and an easier method for linking course level activities and assignments with program and institutional-level learning outcomes (Maki, 2004). Some institutions are experimenting with the use of ePortfolio systems to collect and organize student learning artifacts over the course of their tenure with an institution, and others use dedicated electronic outcomes assessment systems to collect and report on student achievement toward course, program and institutional objectives. However most institutions surveyed by Eduventures had not yet purchased an institutional software system for managing the outcomes assessment process

Although most universities have not purchased dedicated outcomes assessment systems, a growing number already own or license a Learning Management System (LMS) system. Today's LMS have the ability to capture and store every course activity, whether an exam, assignment, project, or discussion, along with the grades and evaluation of each of these activities. This capability allows the LMS to be used to facilitate the collection, analysis, organization, and reporting of course-level assessment data. The authors of this chapter initiated the use of a LMS system for outcomes assessment in an effort to improve the learning environment of students, to enhance the communication between program administrators and faculty, and to satisfy the reporting needs of their professional accreditation agency. This chapter:

- examines the role of the LMS in facilitating the measurement of student learning outcomes,
- discusses the technical approach used in adapting an LMS to measure student outcomes assessment,
- discusses strategies for facilitating faculty adoption and use of the system,
- reviews the challenges confronted and lessons learned from this case study,
- discusses future directions in the use of an LMS for outcomes assessment.

## ELECTRONIC OUTCOMES ASSESSMENT

There is some confusion regarding the terminology associated with outcomes assessment. Terms such as learning outcomes, learning objectives and assessment are sometimes used interchangeably, and incorrectly. In addition, outcomes assessment is often interpreted differently by the various stakeholders in the academic environment. A student typically considers a course examination or project a form of evaluation or assessment of his or her individual course performance. An instructor or faculty considers these measures along with student interactions and course evaluations, typically in the broader context of his or her course. A dean considers enrollment, retention, and completion rates as indicators of faculty and program effectiveness while a policy maker might consider the impact graduates make to their respective communities as evidence of program effectiveness and value. Each of these stakeholder groups considers outcomes assessment from their various perspectives. While each set of needs is important, it does make it difficult to implement one process or method for collecting, analyzing and reporting assessment data and activities. Therefore, the following section first defines our concept of outcomes assessment and then discusses the rationale for using an LMS for outcomes assessment.

## Outcomes Assessment

Otter defines learning outcomes as typically broad statements describing "what a learner knows or can do as a result of learning" (Otter, 1992, p. 2). Allan (1996) distinguishes between learning outcomes (broad consequences of learning) and learning objectives, which are more focused and measurable statements of future accomplishment typically subsumed under subject- or discipline-based outcomes. The Higher Education Quality Council (1993) supports this distinction, encouraging institutions to include institutional review policies that include and differentiate between learning objectives and learning outcomes. Palomba and Banta (1999) help to focus our definition on outcome measures by stating "Assessment is the systematic collection, review, and use of information about educational programs undertaken for the purpose of improving student learning and development." This definition of assessment is broad enough to incorporate varying institutional perspectives and focuses on the collection and review of various types of course data evidence with the intent of ultimately improving student learning.

A deeper understanding of the role outcomes assessment plays within higher education, along with accreditation requirements which specify the establishment of program-level outcomes assessment (AASCB, 2005; NCATE, 2002), have resulted in considerable discussion regarding how to best define, measure and achieve outcomes assessment. Typically, professional accreditation agencies such as the Association to Advance Collegiate Schools of Business (AACSB), the National Council for the Accreditation of Teacher Education (NCATE) and Accreditation Board for Engineering and Technology (ABET) require evidence of student progress towards meeting program-level learning outcomes. This evidence can be provided through various measures, including knowledge-acquisition measures (Nicholoson, Barnett & Dascher, 2005), achievement measures (Miller, Seay & Chamberlain, 1991), student perception measures (Glynn & Rajendran, 1993), and employer and alumni perceptions (Nicholson, et. al., 2005). However the various measures and sources of data available present a challenge to colleges and universities as they determine which data to collect, how to collect it and how to report their findings. These factors have led to the inconsistent use of technology for outcomes assessment purposes (Conole and Warburton 2005).

Outcomes assessment requires accumulation of course activities from a range of courses over a period of time, then matching these activities to specific learning objectives for a specific degree program. This process is outlined in Figure 1.

Theoretically, the process starts with the development of program-level learning outcomes, which clearly articulate the mission of the academic program and institution and the consequences of program participation and learning for both

*Figure 1. Linking program-level learning outcomes to course-level assessment activities*

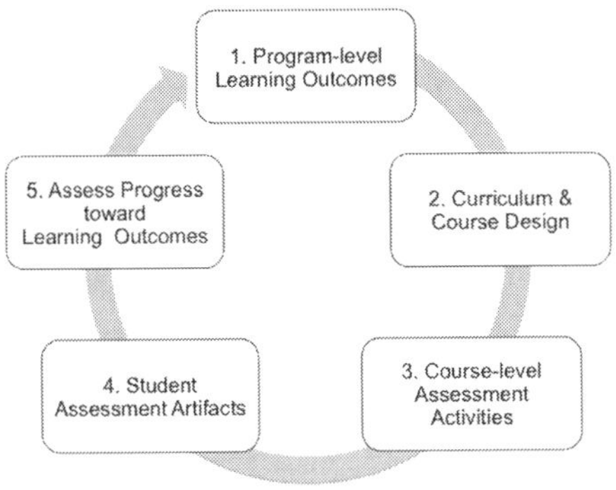


the learner and the greater society or discipline. Course selection and curriculum design then focus on constructing a learning experience that supports the predetermined learning outcomes. The design of individual courses and assessment activities follow, the point where faculty make choices regarding course content (what is taught in each course) along with how they will assess student comprehension and mastery. Ideally, this work should begin with program-level outcomes in mind, asking how course-level activities contribute to achievement of program-level learning outcomes. Next, sources of artifacts used to assess student progress toward program-level learning outcomes should be collected, organized and archived. These assessment artifacts could include written papers and projects, quizzes and exams, homework assignments along with the notes and grades assigned by their various professors in their individual courses. While it is not necessary to collect every assignment or exam each student submits, when collected and organized by course, program or outcome, they do provide potentially powerful insight into the abilities and understanding forming among students in a particular course and program.

Of course here lies the challenge. Prior to the introduction of electronic LMS and e-portfolio systems, the presentation of student work was a manual, paper-based process, requiring the physical collection of assessment artifacts, the organization and storage of these artifacts until a reviewer traveled to the site and sifted through notebooks or boxes of course material. Unless one is schooled in the techniques of document review, the ability to discern trends, strengths and challenges is limited by the papers submitted and the accreditor's ability to sift through these numerous materials. The broad adoption of LMS in particular means that many course assignments and activities are now occurring in an electronic format. If student course materials are submitted and stored in electronic formats, these assessment artifacts and information regarding corresponding assignments can be retrieved and displayed in electronic format as part of the outcomes assessment process.

## Outcomes Assessment and Learning Management Systems

Using a LMS for outcomes assessment is still in the early stages both from a technical and organizational perspective. A LMS can assist in the collection, organization and reporting of assessment artifacts, while promoting new online approaches towards the assessment of student learning. However, LMS have not yet included a set of tools tailored to the evolving requirements of outcomes assessment. The current technology does not aggregate data from multiple courses, departments or programs in a manner that aids reporting of student progress across courses and programs. While some common tagging methods for identifying and exchanging LMS content between systems exist (e.g., IMS standard), LMS tend to use their own data types and methods for the coding and storage of student assignments, complicating efforts to extract and repurpose student assessment artifacts. Currently, multiple systems are required to extract, classify and report on LMS assessment data.

Document management systems like Word and Adobe are used for collecting, presenting and sharing documents, reporting tools like Cognos are used for the extraction, transfer and loading of data from LMS databases to outcomes assessment systems, additional curriculum planning tools like LiveText and Task Stream are used for planning and defining program goals, and rubrics development tools are used for the evaluation of student work. These solutions require educational institutions to hire expert IT staff to manage, integrate, and maintain a suite of software tools, and require complex training and faculty oversight. The technology required to conduct electronic outcomes assessment is thus expensive, confusing and tedious.

A fully integrated LMS with outcomes assessment tools could simplify this task as well as make most activities transparent to administrators, faculty and students. What is needed, according to Neil Allison, is a set of systems that can embed assessment in curriculum design by providing a set of tools to facilitate both summative assessment (after), and formative assessment (during) as part of regular course activity (Hampson, 2008). These future LMS should be able to extract student artifacts like assignments and exams from their databases and rubric-based evaluation forms and link them to program learning outcomes. These knowledge-driven intelligent automated systems would improve assessment quality without significant time involvement from the faculty and/or administrators.

Several commercial outcomes assessment tools have entered the market over the last few years (see Appendix A). These systems target outcomes assessment activities such as planning, accountability, learning goals matrix development and reporting. They include four common components: 1) *Planning Tools* – which provide environment for strategic planning, creating learning goals and criteria for measurement; 2) *Tracking and Analysis Tools* – which provide customizable templates for defining learning objectives, rubrics, surveys, course evaluations and portfolios; 3) *Reporting Tools*– which provide customizable templates for reports, dashboards and institutional accountability; 4) *Collaboration Tools* – which include email, calendaring, document sharing and management, and conferencing capabilities for assessment teams to communicate and collaborate effectively. Some provide limited integration abilities with LMS while others have no integration. Most of the vendors were at the early stages of developing outcomes assessment software in 2006 when we started our eOutcomes project.

The key weakness of these systems is that they currently do not fully integrate with learning management systems (LMS) or other campus-wide student registration and reporting systems. The only exception to this is BlackBoard Learn, a relatively new system, which claims to provide integration with their Blackboard LMS data. Another problem with these tools is that they are not yet independently validated with faculty, staff and students at higher education institutions. No data is available on how successful these systems have been in measuring outcomes or in improving the quality of assessment in academic programs.

## THE eOUTCOMES CASE STUDY

The goal of our outcomes assessment project was to design and develop a systematic and automated approach to collect, organize and report learning outcomes electronically with LMS integration. A pilot study with twelve on-campus web-enhanced courses in business and general education was conducted on our campus over two years. Our vision was to implement an outcomes information system that would allow faculty and administrators to click on a program-level learning outcome, see a list of courses that addressed this outcome, then see the assessment activities and student assessment artifacts associated with each learning outcome (see Figure 2 for a screenshot of the eOutcomes System).

*Figure 2. eOutcomes system as viewed by faculty*

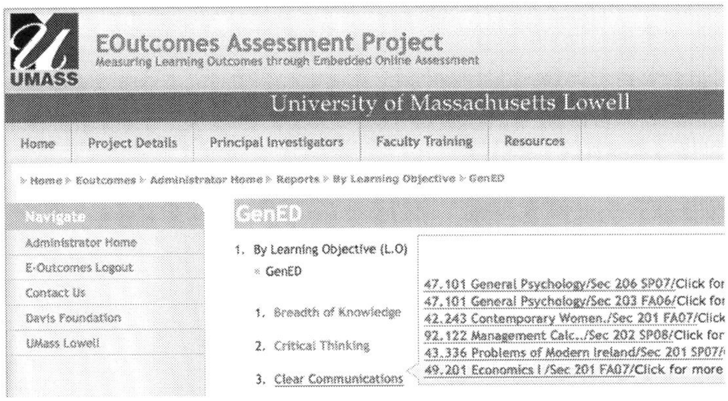


## Method and Approach

The integration of an LMS into an on-campus course without a reduction in class meeting time is typically referred to as web-enhanced or hybrid learning (Allen and Seaman, 2006; Ko and Rossen, 2001). The intent of this modality is to enhance on-campus course activities with online resources, assignments, communication tools and assessment activities. In order to collect, classify and report course-level student assessment activities to program-level learning outcomes, it was necessary for the course-level assessment activities to be conducted and recorded electronically. In this regard, use of the LMS provided increased student access to course materials and assessment activities while also serving to electronically store and record individual student activities and their performance on these activities. Prior to the start of research activities, the proposal was reviewed and approved by the University's Institutional Review Board which required the informed, written consent of all students and faculty participating in this study.

This project utilized an action research methodology as described by Kemmis and McTaggart (1988) and Stringer (1999), where the researchers worked with the subjects to change practice in regard to outcomes assessment. Action research was first used in industry settings in an effort to improve the efficiency of business processes and work teams (Glesne, 2006; Whyte, Greenwood & Lazes, 1991). This qualitative method engages the knowledge and experience of major stakeholders in the research process, incorporating their comments and feedback and modifying the treatment as required. The researchers serve as facilitators, introducing the treatment, problem or challenge, to the participants; facilitating a discussion around the topic and recording suggestions and potential solutions. The researchers assist the participants in implementing and evaluating the effectiveness of the solutions deployed. This cycle continues for the duration of the research project.

The major stakeholders included the faculty and students participating in the study, administrative department chairs and deans from the respective academic programs and the campus information technology unit that hosts the University LMS. The business and general education curricula were selected because each program had developed an extensive set of program-level outcomes which included several common elements (see Figure 3).

The researchers served as consultants knowledgeable in the design and deployment of web-enhanced courses; colleagues engaged in the teaching and assessment process and as facilitators, assisting stakeholders to articulate their respective needs in relationship to the development of an electronic outcomes assessment process. The researchers designed and modified both the

*Figure 3. Comparison of business school and general education learning outcomes*

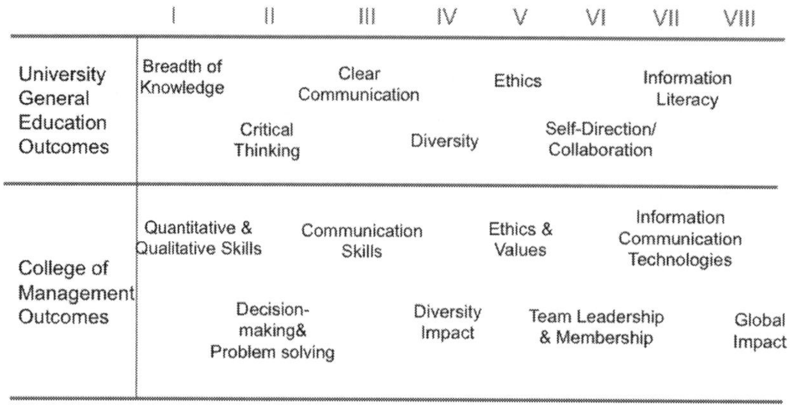


electronic outcomes assessment process and the faculty training based on their respective expert knowledge and numerous conversations with study participants.

The data summarized in this paper were collected at the conclusion of the Fall 2006 and Fall 2007 semesters. Table 1 list the courses and student enrollments of those participating in this two-year project.

During the first year of this two-year project five faculty participated, two from the business school and three from the General Education program. Each faculty member participated in a summer training program and then developed a web-enhanced course site for the 2006–2007 academic year. During the second year of the project, an additional seven faculty participated in the project, four from the business school and three from the General Education program. The second year faculty also participated in a summer training program and then developed and offered a web-enhanced course during the 2007-2008 academic years.

## The eOutcomes System Life Cycle

While developing the eOutcomes system we used the system development life cycle (SDLC) process (outlined in Figure 4). This allowed us to follow a systematic process of identifying, designing & implementing, and validating eOutcomes at our university. For user-centric systems, like eOutcomes, it is often better to have a structured methodology to avoid missteps and coordinate the design and development tasks properly amongst the members of a large design team.

SDLC uses a *systems approach* for problem-solving which basically states that complex problems need to be broken-up into smaller manageable problems using a systems' hierarchy and then the team develops a solution for each problem within the hierarchy of any enterprise system (Motiwalla and Thompson 2009). The structured or phased approach is designed to catch problems at an early stage before they become a major risk to the system implementation process. The SDLC process requires both technical and non-technical problem solving skills; therefore, design and implementation of systems like the LMS, must consider technology in context of the organization's business processes, culture, and people.

### Identification Phase

*Identification of Program-level Outcomes Assessment Needs* – this involved meeting with academic deans and department chairs to identify the evidence required to assess student progress

*Table 1. Project course offerings And enrollments*

| Course Name | Enroll | Discipline | Sem. | Program |
|---|---|---|---|---|
| Contemporary Women Writers | 17 | English | FA06 | General Education |
| History of Crime & Social Control | 29 | History | FA06 | General Education |
| General Psychology | 35 | Psychology | FA06 | General Education |
| Business Finance | 155 | Finance | FA06 | Business Education |
| Marketing Principles | 98 | Marketing | FA06 | Business Education |
| Exploring the Universe | 103 | Physics | FA07 | General Education |
| Economics I | 39 | Economics | FA07 | General Education |
| Management Calculus (2 sections) | 71 | Math | FA07 | General Education |
| Organizational Behavior | 150 | Management | FA07 | Business Education |
| Operations Management (2 sections) | 67 | Operations | FA07 | Business Education |
| Financial Accounting (2 sections) | 217 | Accounting | FA07 | Business Education |
| Operations Management | 30 | Operations | FA07 | Business Education |
| **Total Enrollments** | **1011** | | | |

*Figure 4. eOutcomes project system life cycle process*

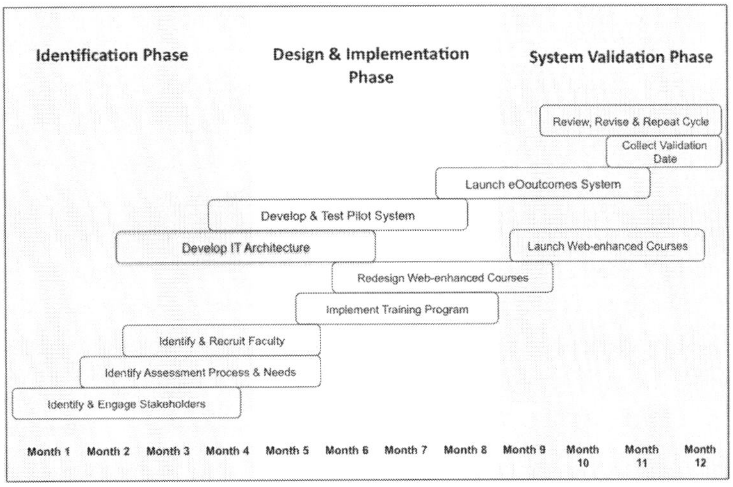


to program-level learning outcomes. The business school outcomes assessment required evidence that demonstrated student ability to apply communications skills, quantitative and qualitative analytical skills, information technology skills, team leadership and membership skills along with an understanding of ethics and the impact of diversity and globalization on business organizations. The General Education program required evidence that demonstrated student ability to apply communications skills, information literacy skills, critical thinking skills, and self-direction and collaboration skills along with an understanding of ethics, the impact of diversity and familiarity with several different areas of knowledge. The evidence typically collected at the program level included student case studies, papers, projects and exam performance.

*Identification of Course-level Assessment Activities* – this involved meeting with participating faculty to identify the various types of assessment activities each used in their respective courses. This phase also required a discussion with faculty regarding how these course-level assessment activities contributed to program-level learning outcomes. The point of this conversation was two-fold. First, it assisted the researchers in considering how to best translate traditional classroom assessment activities (e.g., papers, exams, presentations) to electronic assessment activities in the LMS. Second, it assisted faculty participants to think about how these classroom activities aligned with program-level outcomes assessment. These conversations occurred at the annual faculty training sessions and in smaller, one on one discussion with the researchers.

*Identification of Technical Design Requirements and Challenges* – this involved meeting with IT staff who operated the University LMS. These staff were familiar with both the functional operation and use of several LMS course tools and the ability to collect and extract data from the LMS. While the researchers were both experienced online instructors, familiar with the various LMS course tools, the ability to collect and extract data from the LMS was dependent on the IT staff's knowledge, ability and willingness to provide root level access to this data. This latter issue, the ability to find and extract data from the LMS, proved challenging and resulted in the decision to design an external, electronic outcomes assessment system for the classification and reporting of assessment data.

## Design and Implementation

The design and implementation of eOutcomes was an ongoing iterative process during both the first and second years of the project. The design of the system was followed each year by the summer faculty training program and the deployment of web-enhanced courses in the subsequent Fall semesters. Revisions to our electronic outcomes assessment system, approach to training and deployment of web-enhanced courses occurred along a similar timeline.

### Database Development

During the first year we created an initial design and development of the electronic reporting process and system. Working closely with IT staff, the researchers developed the eOutcomes architecture, as shown in Figure 5. While the researchers initially planned to classify course-level assessment activities in the LMS, the LMS design at the time, along with concerns that this work might impact other users of the LMS, precluded the modification of LMS code to accommodate this approach. Additionally, the manner in which the LMS coded and stored student assignments prevented the researchers from extracting individual student assignments (as evidence of student progress toward learning outcomes) from the LMS. This meant that an external, electronic outcomes assessment database and a method for extracting course-level data from the LMS would have to be developed.

An external database was developed using MySQL, and a web-interface for faculty and administrators facilitated the classification and reporting of student, course and program level assessment activities. A batch process was executed daily to extract student and course level data from the LMS, using the Gradebook Report available in Blackboard Vista. This report included student performance data along with a record of all course assignments. This daily data extract was parsed using a custom Perl script and then imported into the eOutcomes system. If any graded activity (e.g., exam, quiz, project, assignment, discussion thread) in the LMS changed, it was updated in the eOutcomes system during the next batch process. The eOutcomes system was then accessed by faculty to classify course-level assessment activities according to program-level outcomes. Faculty also uploaded electronic samples of student work

*Figure 5. eOutcomes database architecture*

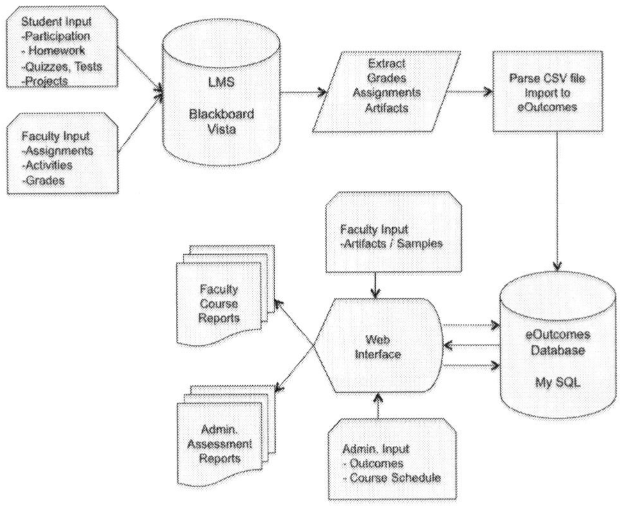


to eOutcomes, providing additional evidence of student progress toward program-level learning outcomes.

**Faculty Development**

To ensure successful deployment, we conducted a two-day summer training workshop for each set of participants. The curriculum examined the development and assessment of learning outcomes, an introduction to the LMS course tools, and training in how to develop web-enhanced course materials and electronic assessment materials (available online at http://eoutcomes.uml.edu/eoutcomes/resources.html). Participants were asked to bring their course syllabus and related materials to the training and received hands-on assistance from project staff. During each training session, participants raised new questions and challenges in regard to moving their specific course into a web-enhanced format:

- how do I manage a high-enrollment lecture class online?
- how do I reduce cheating on online exams?
- do students have to submit all materials electronically?
- how do students submit Math problems online?
- if I put my materials online, won't the students skip my class?

These questions and others had to be answered in order to facilitate faculty transition to a web-enhanced model, thereby increasing the opportunity for electronic outcomes assessment. Questions like these *posed technical challenges* (e.g., What format should student papers be submitted in?), *pedagogical challenges* (e.g., How do I design and manage web-enhanced course activities?) and *commitment challenges* (e.g., Is electronic outcomes assessment worth all this extra work?). In response to faculty requests, we expanded the training program to a third day, later in the summer, when faculty could come back to campus to discuss the challenges they encountered while preparing web-enhanced course materials. We provided technical support throughout the year and worked with the LMS/IT staff to answer the more difficult technical questions.

During these training discussions one experienced, tenured faculty member shared a frank and surprising observation, he said "You know,

I hadn't really thought a lot about outcomes when I designed my past courses…I know what I need to cover, I structure the course around the content, then design a test and assignments to make sure the students are learning the material…but I don't really think about programmatic outcomes." Perhaps this comment isn't really that surprising. In fact it underlines the challenge in moving higher education toward outcomes assessment. For the most part, faculty focus on the courses they teach and the students in these courses, while department chairs, deans and administrators take responsibility for organizing and conducting programmatic assessment activities. In the traditional, paper-based assessment system, administrators may request copies of course materials once a year or less, limiting faculty engagement in the process. This participant's comment suggests that the redesign effort required to move paper-based assessment materials into an electronic, web-enhanced format as part of this project helped him to think more deeply about the relationship between course-level assessment activities and program-level learning outcomes.

### System Validation

While several commercial electronic outcomes assessment tools exist on the market, no vendor independent study examines their acceptance and success rate among college faculty and administrators. Clearly, college administrators who are responsible for outcomes assessment reporting can recognize the value of a system that improves and facilitates the current manual process. However faculty who are somewhat removed and uninvolved in outcomes assessment may not see the need or value in an electronic process. Since successful adoption of an electronic outcomes assessment tool requires the support and participation of both faculty and administrators, we examined faculty experiences and perceptions regarding the use of an LMS for outcomes assessment. Three research questions guided our validation study:

1. Which LMS tools do faculty use for student assessment purposes in web enhanced courses?
2. Do faculty perceive a benefit in the use of a LMS for linking course-level and program-level assessment activities?
3. Does electronic outcomes assessment provide some type of cost benefit over traditional, paper-based methods for measuring outcomes?

Question one and two were examined through an online survey of participating faculty and through discussions with faculty after deployment of web-enhanced course activities. The researchers interviewed all faculty participating in the project at least once outside of the training sessions to discuss their individual use of the LMS for outcomes assessment activities. Question three examined whether or not the embedding of outcomes assessment within the LMS facilitates efficient classification and reporting of program-level assessment data for accreditation purposes. The researchers interviewed participating faculty and academic administrators to develop a sense of the faculty usage and interest in outcomes assessment system and the economic consequences of implementing an electronic system.

## FINDINGS

Faculty participants were surveyed regarding use of the LMS in their respective web-enhanced courses (at the end of each semester) and their use of the eOutcomes system (toward the end of the study). This data is summarized in Appendix C and discussed in more detail below. Faculty feedback regarding how each used the LMS assisted us to better understand how assessment artifacts are collected and stored in the LMS. Use or non-use of an LMS tool (e.g., discussion forum, quiz or exam tool) on the part of the faculty plays a significant role in determining how effectively the LMS can serve in the collection and organizing function of the outcomes assessment process. Faculty feedback regarding their perceptions of the utility and functionality of the eOutcomes system helped us to refine the systems interface and functionality. Feedback regarding faculty use of both the LMS and eOutcomes system helped us to understand the time commitment required by faculty when involving them in the outcomes assessment process.

## LMS Use by Faculty

The LMS in use at the time of this study was Blackboard Vista 3.x. Like many LMS, Vista has a common set of course tools including Assessment tools (quizzes, exams, surveys), Assignment tools (for posting/collecting homework assignments), Communication tools (discussion forum, chat, email), Content tool (for presenting/sharing lecture notes, presentations and related material), a Gradebook tool, an Outline/Schedule tool, a Syllabus tool, and a Team tool (for grouping students for course related activities). All participants in this study received Vista training prior to participation in this project. Project training activities focused on examining how faculty might use these various LMS tools in a web-enhanced course to facilitate the collection and organization of student assessment artifacts (e.g., papers, exams, presentations, homework). Participants were asked to introduce a minimum of three LMS tools into their first semester web-enhanced course. A majority of participating faculty (60%) placed more than 50% of their course material and activities online in the LMS.

As demonstrated in Figure 6, nearly all courses used the Outline/Schedule, Lecture Notes, Assignment and Syllabus tools, followed by use of the Gradebook, Discussion, Assessment and Group/Teams tools. This is consistent with an earlier University of Wisconsin study which found that faculty use LMS primarily as an administrative tool to distribute course materials and manage quiz administration (Morgan, 2003).

The range of tools used by faculty participants suggested that the LMS could be used as a reliable source of student assessment artifacts, evidence of student progress toward course-level learning objectives. The use of the Gradebook tool in 9 of the 12 courses supported our ability to extract student performance data from the LMS into the eOutcomes database. However use of a tool in the LMS by faculty, did not necessarily mean student work would be stored in a universal electronic format that could be extracted from the LMS and then shared as evidence of student progress. Further conversations with faculty indicated that

*Figure 6. Use of online course tools within participating courses (N=12)*

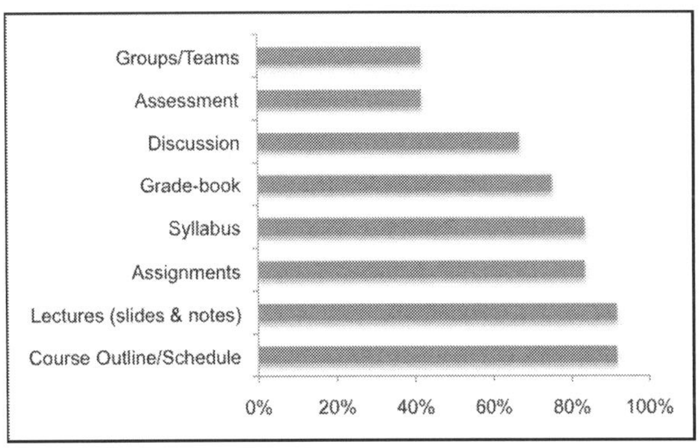

a range of assignment types were being used by faculty in the LMS. Assignment types ranged from homework responses posted in a discussion forum, to online quiz responses, to the posting of online papers and presentations. While collection and organization of assessment activities in the LMS was critical to our approach, we also needed a universal format for displaying student-created content as evidence of student progress. Since the LMS stores student artifacts in different formats depending on the manner of submission (e.g., as a discussion thread or as an attachment), we ultimately determined that faculty would need to save sample student work as Adobe PDF files (using the "Print to PDF" function) in an external database.

In addition to understanding which tools faculty used in the LMS, we also examined how frequently they used these tools (Figure 7). A review of faculty usage of LMS tools quickly indicates that some LMS tools were used more frequently than others. The Lecture and Syllabus tools were used on a weekly or daily basis in a majority of courses while 60% to 70% of participating courses never used the Exam or Quiz tool. This finding has important implications for electronic outcomes assessment – if assessment activities do not occur in the LMS, you cannot extract evidence of student progress from the LMS.

The major obstacles identified by faculty toward using the LMS for exam and quizzes were concern over cheating, lack of sufficient computer lab space for large section classes and difficulty in presenting or recording content electronically (e.g., showing student work on math exams). In most web-enhanced courses, faculty were reluctant to offer unmonitored online exams. While the LMS software supported some online exam controls (e.g., random-draw questions, password protection, time limits, lock-down web browser) faculty generally required students to take exams on campus using paper and pen.

Exams and quizzes are however only one type of assessment activity. Homework assignments, discussion postings, student projects and other activities also assist faculty and programs in assessing student progress. Over 80% of the courses used the Assignment tool to collect and manage

*Figure 7. Frequency of LMS tools as used by faculty in each web-enhanced course (N=12)*

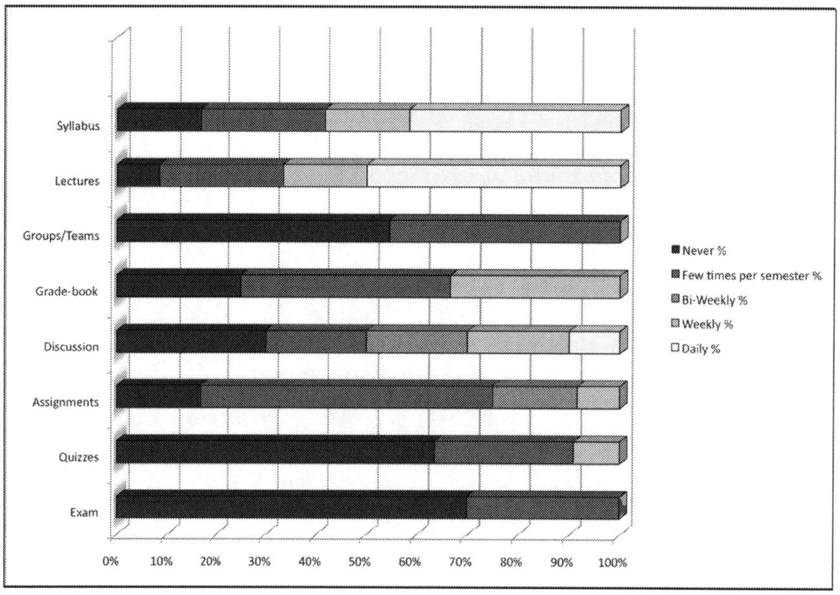

student assignments, 70% used the Discussion tool and 45% used the Groups/Teams tool. These faculty reported a generally favorable reaction from students regarding the use of these online tools in their web-enhanced courses and a majority of faculty (81%) agreed or strongly agreed that use of the LMS contributed to the success of their students in participating courses. Ultimately, the use of these tools in our web-enhanced courses has helped in collecting and organizing assessment materials for faculty and provided electronic resources for our outcomes assessment process.

## Faculty Perceptions Regarding eOutcomes

In regard to the use of the eOutcomes System, the majority of faculty indicated that the system was easy to use (80%) and that the system made the process of capturing and measuring outcomes assessment easier (73%). However, despite their participation in the project and general agreement that the eOutcomes system facilitated the outcomes assessment process, only 54% indicated that the system was a useful tool for capturing and measuring learning outcomes. Our sense of this discrepancy is that some faculty did not recognize the administrative value of consolidating course-level data within one system for program-level analysis. Despite this one discrepancy, 63% of the faculty indicated they would use the system in place of manual, paper-based outcomes assessment and 62% agreed that other faculty would find it useful. We conclude that faculty generally agreed with the project investigators that this project leveraged technology and an appreciation for outcomes assessment in order to strengthen the teaching and learning practices at the University.

## Cost Impact

Our discussion of cost impact and savings is somewhat limited by the technical challenges faced in attempting to integrate with the LMS (and the accompanying need to develop an external eOutcomes system) along with the effort required to train faculty in both the use of the LMS for web-enhanced courses and in the eOutcomes process. If we had fulfilled our original goal of conducting electronic outcomes assessment (including the organization, classification and reporting of student assessment artifacts) within the LMS, we would not have spent technical time on the design and development of an external eOutcomes system. Additionally, faculty would only be trained in the use of the LMS and would not require additional training in the use of the eOutcomes process. This was the streamlined process we originally envisioned.

Unfortunately, the design of the LMS required an external eOutcomes process which necessitated additional technical and faculty development effort. Despite the development challenges encountered, participating faculty generally found the electronics outcomes process agreeable and preferred to a manual process, where they would have to collect, copy and organize physical assessment artifacts. A conversation with the Dean of the Business School and the Director of Undergraduate Education suggests that there are cost savings in terms of the time spent preparing paper-based assessment materials. The Director reports that he spends days reviewing and organizing the physical assessment artifacts handed in by faculty, if they are actually handed in. He also reports the need to "remind" faculty regularly that these materials are due (or overdue, as the case may be). From their administrative perspective, there is a significant savings in time associated with the electronic batch acquisition of assessment materials. In addition, the dashboard and graphic reporting tools built into the eOutcomes system eliminated several days of manual data entry as the Director transferred each courses' coversheet data into his own Excel spreadsheet. Faculty also reported additional areas of savings, including:

- Supply savings – less Scantron sheets and paper for exams, less duplicating of course materials
- Proctor savings – less time spent scanning, reviewing and reporting exam grades (less room for error)
- Faculty time savings due to:
- Online grading exams, many are auto-scored
- Online grading quizzes, many are auto-scored
- No longer required to fill out assessment spreadsheets
- No longer required to collect samples of student work making copies
- Overall reduction in usage of paper, data re-entry time and recording errors

## CONCLUSION

The demands of learning outcomes assessment increasingly impacts all institutions of higher education, however emerging information technology tools can facilitate the process of collecting, organizing and reporting campus progress toward meeting program learning outcomes. We have shared this case study example in the hope that other institutions will find our approach and methods beneficial and adaptable to their institutions. Based on this research effort, we share the following lessons critical to supporting implementation of an electronic outcomes system:

- *Develop a clear understanding of your outcomes assessment process*: as discussed earlier in this chapter, students, faculty, administrators and external stakeholders each have different perspectives regarding outcomes assessment. It is critically important that program stakeholders have a clear understanding of expected program outcomes, accreditation requirements and the relationship between course-level assessment activities and program-level outcomes assessment. While the assessment cycle shared in Figure 1 is intuitive, our experience suggests faculty do not typically integrate program-level concerns into their course-level activities. Before any technology plan is developed, faculty need to understand and value outcomes assessment.
- *Sustainable commitment from senior administration*: in order to secure faculty involvement, administrative staff and technical support the executive administration must not only support the project verbally but also make a sustainable financial commitment in terms of equipment and staff time. Electronic outcomes assessment crosses institutional boundaries, requiring faculty to work with IT staff, curriculum staff and administrators. Cross-boundary collaboration typically requires the encouragement and support of senior administrators.
- *Faculty commitment & support*: we found that in order to realize the efficiencies of an electronic outcomes assessment process, faculty had to make a concentrated commitment to using the LMS for the organization, collection and grading of the majority of course assessment activities. This meant time developing electronic course materials, changing how they graded course materials and time to rethink past assessment practices. In addition, their willingness to learn new processes and to take risks with their courses and peers is important. Certainly, the incentives the grant supported (i.e., Tablet PC and a development stipend for each participant) helped encourage faculty to take this risk with us.
- *Technical Support for faculty*: the process of converting course content into online format, especially beyond posting lecture notes, is complex and difficult. Creating and assessing online assignments, quizzes and exams requires considerable technical

support in the early stages of the project as well as over the long-term. When last minute problems with exams and grades occur, faculty demand immediate resolution. Technical support staff must be in place to assist in resolving these technical issues or both students and faculty will be disillusioned with the project.

- *LMS, IT Services support & involvement*: one of the key reasons this project succeeded was due to the support and commitment of the UMass Online technical staff and administration. They were willing to take risks in providing access to the online course data on an ongoing basis. There were several challenges in the early stages of the project but we were able to overcome and streamline the data transfer from Vista to the eOutcomes system.

The development and use of web-based assessment activities to facilitate program-level outcomes assessment requires an additional time commitment on the part of faculty and, possibly, students. Faculty must learn to use the various LMS tools for web-based assessment activities and may also need to redesign assessment activities to facilitate electronic submission. Additionally, while typically focused on course-level teaching and assessment, program-level assessment requires faculty to think about how course-level activities relate to higher level, program outcomes. In some instances this may also lead to the redesign of course syllabi and course activities, a commitment of additional time and effort. However trends in online education suggest that more and more faculty are developing web-enhanced courses, despite the effort, in order to increase student access to course materials. If this effort is expanded to include the electronic collection of student assessment activities and LMS developers integrate outcomes assessment tools into their systems, the ability to provide real-time reporting on student progress toward program and institutional outcomes can be realized while ultimately saving faculty and administrative staff time.

## ACKNOWLEDGMENT

This project was funded by a grant received from the Davis Educational Foundation established by Stanton and Elisabeth Davis after his retirement as chairman of Shaw's Supermarkets, Inc.

## REFERENCES

AASCB. (2005). Overview of Assessment. *Assessment Resource Center, AACSB International*. Retrieved January 6, 2006, from http://www.aacsb.edu/resource_centers/assessment/overview-why.asp

Allen, E., & Seaman, J. (2006). *Making the grade: Online education in the United States, 2006*. Needham, MA: Sloan Center for Online Education. Retrieved September 20, 2008 from http:/./www.sloan-c.org

Conole, G., & Warburton, B. (2005). A review of computer-assisted assessment. *ALT-J*, *13*(1), 17–31. doi:10.1080/0968776042000339772

Eduventures, Inc. (2008). *Challenges and opportunities in postsecondary outcomes assessment – benchmarking institutional progress*. Boston: Eduventures, Inc., Retrieved March 22, 2009, from http://www.eduventures.com/

Glesne, C. (2006). *Becoming qualitative researchers* (3rd ed.). Boston, MA: Pearson Education.

Glynn, K. A., & Rajendran, K. N. (1993, September). Perceptual-based student outcomes assessment process in the marketing curriculum. *Journal of Education for Business*, *69*, 11–17.

Hampson, K. (2008). Neil Allison: Education, Outcome Assessment, and Technology. *Higher Education Management*. Retrieved May 22, 2009, from http://highereducationmanagement.wordpress.com/2008/12/03/neil-allison-outcome-assessment-and-technology/

Higher Education Quality Council. (1993). *Student Assessment and the Classification of Awards*. London: Higher Education Quality Council.

Kemmis, S., & McTaggart, R. (Eds.). (1988). *The Action Planner* (3rd ed.). Geelong, Australia: Deakin University Press.

Ko, S., & Rossen, S. (2001). *Teaching online: a practical guide*. Boston: Houghton Mifflin Company.

Lovett, C., & Robert, T. (2004). We Need an Honest Conversation. *The Chronicle of Higher Education*, 51. Retrieved January 6, 2006, from http://chronicle.com/weekly/v51/i02/02b00601.htm

Maki, P. (2004). *Assessing for Learning: Building a Sustainable Commitment Across the Institution*. Sterling, VA: Stylus Publications.

Miller, F., Seay, R., & Chamberlain, D. (1991). The current status of outcomes assessment in marketing. *Journal of the Academy of Marketing Science*, *19*(Fall), 353–362. doi:10.1007/BF02726511

Motiwalla, L., & Thompson, J. (2009). *Enterprise Systems for Management*. Upper Saddle River, NJ: Prentice Hall.

NCATE. (2002). *Professional Standards for the Accreditation of Schools, Colleges, and Departments of Education*. Washington, DC: National Council for the Accreditation of Teacher Education.

Nicholoson, C., Barnett, S., & Dascher, P. (2005). Curriculum assessment in marketing programs: current status and examination of AACSB core standards at the program level. *Marketing Education Review*, *15*(2), 13–26.

Otter, S. (1992). *Outcomes assessment in Higher Education*. London: UDACE.

Palomba, C. A., & Banta, T. W. (1999). *Assessment Essentials*. San Francisco, CA: Jossey-Bass.

Shulock, N. (2004). No Less Than a Cultural Shift. *The Chronicle of Higher Education, 51*. Retrieved January 6, 2006, from http://chronicle.com/weekly/v51/i02/02b00601.htm

Stringer, E. (1999). *Action research: A handbook for practitioners* (2nd ed.). Thousand Oaks, CA: Sage.

Whyte, W. F., Greewood, D. J., & Lazes, P. (1991). Participatory action research: Through practice to science in social research. In W. F. White (ed.), Participatory Action Research (pp. 19-55). New York: Sage.

## APPENDIX A

The following is a list of key software vendors in the outcomes assessment area today.

- *LiveText*: LiveText has three sets of tools 1) Evidence -- which gathers student data from courses or programs, 2) Assessment – tools like rubrics, surveys and peer-evaluation forms for evaluating performance against learning outcomes, and 3) Analysis – report generation tools to present data in different formats for various stakeholders. Supports three categories of stakeholders 1) Students – allowing them to create e-portfolios, look at instructor feedback and collaborate in teams, 2) Instructors – allowing them to create course templates, assess student learning and develop course and e-portfolios, 3) Administrators – allowing them to create assessment reports, self-studies and program assessment plans and manage other accreditation requirements. For more information visit http://www.livetext.com.
- *Tk20 CampusTools:* Provides a process for data collection, assessment planning, and comparison tools to link course activity to learning goals, ability to generate a variety of reports for compliance, analysis, and program improvement. Tk20 also provides survey and assessment rubric tools, faculty profiling tools, customizable reporting tools.For more information visit http://products.tk20.com.
- *Task Stream:* A learning achievement and accountability management system provides Web-based tools for capturing and analyzing institutional processes, student learning data from classes, programs and institutions and provides analysis and reporting tools. Accountability Management focuses on administrators to aid in measuring institutional effectiveness while the Learning Achievement Tools focus on faculty and students ability to develop course artifacts, rubrics, surveys and e-portfolios linked to learning objectives. Provide reporting tools which aggregate and present data analysis for administrators and faculty. For more information visit http://www.taskstream.com.
- *TracDat:* Enterprise-wide system for managing assessment process. Provides tracking tools that integrate data at program and course levels. Aids alignment among learning goals, course and student activities with tools for curriculum mapping, rubrics and surveys. Includes reporting tools to assess progress toward course and program goals. For more information visit http://www.nuventive.com/.
- *Blackboard Learn:* The Learn platform is from Blackboard, Inc. the same company that provides LMS software for e-learning. A relatively new software that provides support for outcomes assessment with multiple data collection measures on student progress. It provides tools for planning, assessment, accreditation and reporting. Includes collaboration tools for managing a document repository. For more information visit http://www.blackboard.com/learn.

## APPENDIX B

*Table 2. Faculty survey responses regarding use of the Vista LMS and the eOutcomes system (N=12)*

| Faculty Use of the Learner Management System (LMS): | Faculty Participants' Responses |
|---|---|
| | |
| Total course materials placed online | 60% posted more than 50% of materials online |
| Tools used most frequently by faculty | Syllabus, Lecture Notes, Grade-book & Discussion Forums |
| Tools used regularly by faculty | Assignments & Discussion Forums |
| Tools used in-frequently by faculty | Assessments & Teamsv |
| Tools never used by faculty | Chat Room |

| Faculty Perceptions of Student Use and Impact of the LMS: | Faculty Participants' Responses |
|---|---|
| Tools used frequently by students | Syllabus, Lecture Notes, |
| Tools used regularly by students | Assignments Grade-book, & Discussion Forums |
| Tools used in-frequently by students | Teams |
| Tools never used by students | Assessments & Chat Room |
| Student reaction to use of Syllabus, Lecture Notes and Grade-book | Favorable or Highly Favorable |
| Contributed to student success | 81% agreed LMS contributed to student success |

| Faculty Perceptions regarding use of eOutcomes System: | Faculty Participants' Responses |
|---|---|
| Overall usage | 84% used eOutcomes System more than once/semester |
| User interface | 80% agreed that interface was easy to navigate |
| User prompts | 82% agreed that user prompts were easy to understand |
| Process of capturing learning outcomes | 60% agreed that it met their expectations |
| Process of capturing learning outcomes | 72% agreed that the time it took to accomplish tasks in the system was acceptable |
| Process of capturing/measuring learning outcomes | 73% agreed that it made the process easier |
| Usefulness of system | 54% agreed that the system was a useful application for capturing/measuring outcomes |
| Instead of manually recording learning outcomes | 63% agreed they would use eOutcomes system |
| Future use in other colleges | 62% agreed other faculty would find it useful |

# Chapter 9
# eUreka:
## A Campus-Wide Project Work Management System to Support Constructivism, Reflection and Collaborative Learning

**Daniel T.H. Tan**
*Nanyang Technological University, Singapore*

**Adrian D.H. Lu**
*Nanyang Technological University, Singapore*

**Sheryl E. Wong**
*Nanyang Technological University, Singapore*

## ABSTRACT

*Project work is an established learning activity for students. It is a learner effort-based endeavour towards the higher learning objectives or Bloom's taxonomic outcomes of beyond application into analysis, evaluation and creation. Its many forms include the final year (or capstone) project, term or mini-projects, group project work, field trips and studies, etc. It can include and extends beyond problem-based learning. The nature of such project work might be investigative, analytical, case study, information gathering, research or development. Nanyang Technological University (NTU) has explored an extension of eLearning from a means of knowledge transfer and delivery coupled with pedagogy to that of knowledge creation and discovery. Via eUreka, this web-based project work management system facilitates the constructivist learning approach of students to be more effectively managed, supervised and mentored. Web 2.0 tools such as weblogs facilitates opportunities to participate, deliberate, feedback, comment and collaborate in the process of learner-centric knowledge creation and discovery.*

## INTRODUCTION

Project work is an endeavour towards a learning objective or outcome. It involves goal setting, planning, execution, monitoring and evaluation. Project work is anchored in the view that education is "....rooted in reality. Reality is in making things and doing things, and to do this, certain qualities of mind and character need cultivation" (Armstrong, 1950). Project work comprises learning activities that involve "an in-depth study of a particular topic,

DOI: 10.4018/978-1-61520-853-1.ch009

Copyright © 2010, IGI Global. Copying or distributing in print or electronic forms without written permission of IGI Global is prohibited.

usually undertaken by a whole class....small groups....and occasionally by an individual" (Katz and Chard, 2000).

Project-based Learning (PjBL), thus, is an instructional strategy that organizes learning around project work. Students are usually organized into multiple project groups to solve, typically, real world complex tasks and challenging questions based on an authentic (and usually work-related) scenario. In working towards the solution or outcome of the task, the students need to apply their existing knowledge and thinking skills to make meaning of the issues at hand to construct new knowledge. Some benefits derived from project-based learning include increased motivation, development of problem-solving and higher order thinking skills (Thomas, 2000; Bransford, Brown, & Conking, 2000). In addition, the need to communicate, share, interact and work as a member of a team will provide students with the learning opportunities to inculcate life skills, negotiation skills, people skills, critical-thinking skills, communication skills and problem-solving skills.

Project-based Learning (PjBL)[1] is a constructivist pedagogy to encourage deep learning by providing learning opportunities for students to adopt an inquiry based approach in dealing with issues and questions/problems presented in the project scope and expected outcome. This in turn, will prepare students to cope with real and relevant situations in their own lives. Students are required to be independent, and sometimes inter-dependent to be accountable for their contribution and effort in the project team. The students take ownership of the problem/task set in the project and learn to manage, set goals, and strive to meet project expectations. They can tap on either skill sets they already have, or struggle to develop them while doing the project.

As students strive to solve problems and discover the consequences of their actions - through reflecting on past and immediate experiences - they construct their own understanding. In this constructivist approach, learning is thus an active process that requires a change in the learner.

Making meaning of the context of learning content and the environment, and construction of shared knowledge through interdependent social participation are some areas of involvement for project-based learning.

In Nanyang Technological University (NTU), Project-Based Learning is a key pedagogical approach that is adopted by faculty members in the assessment of student capability, independence and ability to apply knowledge. NTU students are assessed not only on what they know of the subject matter through in-class tests, individual assignments and term examinations, but also their teamwork skills such as the ability to contribute productively to group projects and assignments. The objective of this approach is to equip students with the necessary life skills that will prepare them for the workforce.

In September 2004, eUreka, an online Project Work Management System (PWMS) was introduced to establish a web-based platform for the centralized management of Final Year Projects (FYPs). The FYP – similar to the capstone project - is a major formal assessment activity for graduating students in the University. The system is aptly named after Archimedes' *"I have found it!"* to capture the students' adventurous spirit and insight in the pursuit and discovery of new knowledge and passion for innovation.

Since its maiden launch in 2004, eUreka has been gradually adopted as a PWMS for other project-based activities such as the Industrial Attachment (IA) Program (student internship program), faculty post-graduate research projects and student study groups as well as planning of non-academic projects, such as activities by student clubs, seminars and conferences organized for the staff in the University.

The launch of eUreka at NTU also coincided with an important chapter in the history of the Internet – the Web 2.0 revolution where the Web has evolved into a participatory environment. In line with the philosophical underpinnings of Web

2.0, eUreka seeks to empower student and staff with a more participatory experience such as, using weblogs in an institutional/campus-wide system.

Advocators for the use of computers see it as a tool to supplement and transform traditional classroom content (Cuban, 1986). Cuban also recognized the use of information and communication technology (ICT) in education but stressed the need for teachers and students to understand both the technology and the social practices that make it a powerful force. The success for the adoption of technology in education will thus require clear goals and recognition of the role of end users to determine their needs in the planning, design and implementation of any web-enabled system.

The use of ICT was explored and deliberated in the design and development of eUreka, working towards a web-based project work management system to provide a centrally managed platform for the documentation, sharing and collaborative application of knowledge among students and professor-supervisors. Its design was led by a team of professors, with the processes of project supervision defined as the project requirements.

The use of eUreka was not intended to replace the face-to-face supervision for the formal projects but was designed to support generic project work process and documentation. With a good grasp of the project progress, face-to-face discussions have become more focused, agenda specific, and generally more productive. The resulting outcome was a web-based tool that enhanced the collaborative and supervisory processes of project work. Being online, eUreka also permits any time, any place access.

The purpose of this paper is to describe the pedagogical underpinnings which the system is based on – social constructivism, cooperative learning, and regulations of learning. The paper will also cover the scope for use of web technology (vis-à-vis Web 2.0 tools where appropriate) to better align the student learning experience to the expected workflow and management of projects. Lastly, the paper will present the usage data since its implementation, focusing in particular, feedback received from faculty and student users.

## CONSIDERATIONS LEADING TO THE DESIGN AND DEVELOPMENT OF EUREKA

What is and Why NTU embarked on the design and development of eUreka? This section covers the background leading to the need for an institutionalised platform to address issues/concerns on the management of project work in NTU.

Before eUreka, faculty members managed their projects individually by getting students to record their work in a logbook and provide progress status via e-mail and face-to-face meetings. This physical logbook, as hardcopy material, was subject to contamination, stains as well as wear-and-tear. Also, they have been known to have been misplaced or lost. Access to such logbook entries was inconvenient as at any one time, only either the student or faculty (as project supervisor) could access it to write/update/review. The supervisor holding on to this log-book could disrupt recording of on-site observations or findings during that time. While a physical logbook could provide valuable documentation for project work, a web-based system has more affordances that are not available in a hard-copy version.

Keeping documents and records online could reduce the prospects of students or project supervisors losing (in parts or its entirety) the physical logbook. In addition, the ability of the supervisor to keep track of the project progress online would also enhance the face-to-face sessions with students. A preview of the students' work progress by the faculty would result in more productive and focused discussions with the students.

From an academic perspective, there is also the need to enhance the documentation of knowledge creation and learning experiences that are unique to students' involvement in project work. Knowledge assets discovery and creation are critical to the project work flow process and appropriate documentation is essential to capture such innovative ideas, theories, principles and models.

The need for computer-mediated tools (such as online forums and weblogs in eUreka) in conjunction to face to face meetings will better facilitate communication and collaboration among the project team and supervising faculty. These tools will also support learning reflections and assessment of project work management.

Laffey *et al.* (1998) advocated that there is the need to consider instructional and learning processes when looking into the design of project-based learning support system. The instructional processes include scaffolding and coaching via activity-timeline-deliverable planning. The learning processes cover personal management, resourcefulness, knowledge representation, communication, problem solving, collaboration and reflection.

At the time of eUreka's inception in early 2004, there was no available centralized academic platform in NTU to track and manage the project work process and organize the created knowledge assets by students. The lack of key documentation, effective and convenient communication channels, and proper monitoring processes for projects resulted in lost opportunities of good ideas being ineffectively communicated or documented for implementation.

Recognizing the value of knowledge-asset creation and management, NTU thus embarked on the design and development of a web-based platform to create a centralized system for better management of project work outcomes for the learning community.

eUreka Project Work Management System was developed using Java and JSP. It operates on resin web services on a load-balanced configuration of Windows 2003 Enterprise servers, coupled to fault-tolerant Microsoft SQL 2000 servers.

## THE PEDAGOGICAL UNDERPINNINGS BEHIND THE DESIGN OF EUREKA

In his review on Project-Based Learning (PjBL) research, Thomas (2000) listed five criteria that drive a PjBL assignment: centrality, driving question, constructive investigations, autonomy, and realism:

1. "PjBL projects are central, not peripheral to the curriculum." In PjBL, the project is the main teaching strategy where students learn the central concepts of the subject matter through working in their projects.
2. The "driving question" (Blumenfeld et al., 1991) or an "ill-defined problem", as coined by Stepien & Gallagher (1993) in project-based approach is the question(s) that students strive to answer or find a solution to.
3. *Projects involve students in a constructive investigation* whereby there will be project activities that allow them to hone their decision-making, problem-solving and knowledge building skills.
4. Autonomy refers to projects that are student-driven and not supervisor-led. The supervisor's role is to provide the necessary scaffolds and advice to the students, and not to spoon-feed them with information.
5. Realism refers to projects that are authentic and reflect real problems or issues, and the solutions and prototypes created by students that can be real-world applied.

Generally, project work in NTU is carried out with the above 5 main criteria in place. Project work in NTU, regardless of whether they are Final Year Projects (FYPs), Industrial Attachment (IA) projects, undergraduate or post-graduate course-work projects, has a significant bearing/impact in the assessment component of the overall program. Indeed, in the case of the FYPs and IA projects, project work is "central to the curriculum" and students are assessed primarily on the process and product/solution they have devised when assessed at the end of the project work. Students are empowered to take ownership of the project work from selecting the investigation topic to initiating their consultation time with their tutors and IA supervisors.

This paper will discuss the use of features in eUreka in relation to the teaching of the language module 'The Art of Academic Writing' and the use of the system to facilitate the monitoring and management of Industrial Attachment programme (IA) for the NTU students.

The design of eUreka is primarily to support a student-centric learning experience that allows students, tutors and supervisors to collaborate in their journey of knowledge discovery Unlike a typical Learning Management System (LMS) in an institution that is essentially supervisor-centric, eUreka allows students to initiate, manage and own their project sites. With the exception of the Assessment module, students have both 'read and write' access to all the other main modules in eUreka system. It facilitates the freedom and space to participate, deliberate and collaborate in the process of knowledge creation. Given the scope for student control and ownership to the project site, the faculty would still need to be mindful of their role to monitor, direct and assess the students in their project process and derivation of expected outcomes.

In project work, as in constructivism, students learn by knowledge application and discovery. They apply prior knowledge and understanding to work in real world or scenario-based problems. Students – either individually or in a team – consider the project requirements or then systematically approach the project execution for a successful project outcome.

eUreka allows students to take ownership of the project. In fact, all their efforts, decisions, planning and design efforts are documented in their respective project sites. Even in a team project, each member has his/her own space while they share a repository, discussion forum, and project Gantt Chart in the Activities Module to plan, manage and adhere to an agreed projected timeline. This Gantt Chart has proven to be a powerful scaffolding tool, as the students with their mentor/supervisor build up the timeline, activities, expectations and deliverables (refer to Figure 1). If required, the faculty/supervisor can choose to send email alerts to the students should they digress from the project objectives or timeline. This helps the students to stay on course in their project work.

From the student's perspective, the process of planning using the easy-to-understand Gantt chart

*Figure 1. The familiar and easy to use Gantt Chart allows students to plan their project milestones and assign tasks to team members*

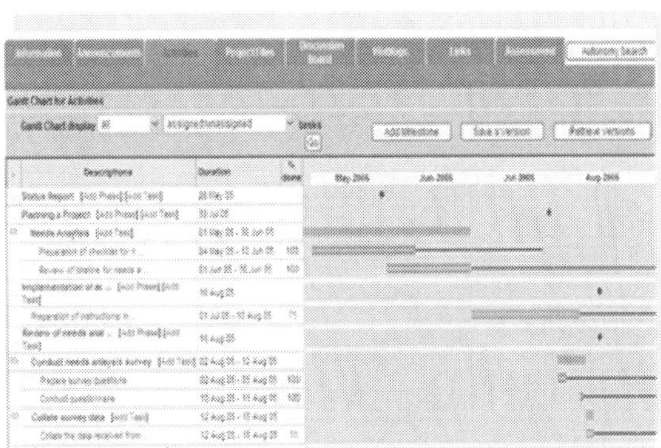

tool provides powerful constructivist scaffolding for knowledge construction, discovery, application and project management for milestones, deliverables, and task scheduling and phase completion.

In addition, the versioning feature of the 'Project Files' Module allows the upload of documents to be time-stamped to better keep track of project deliverables and deadlines (see Figure 2). Multiple versions will be archived and tracked.

The use of Project Files Module was extended by faculty tutors to apply regulatory learning for the academic writing course ***'The Art of Academic Writing'***. This project-based course emphasizes on collaborative and recursive writing where students peer-review each other's drafts for their writing assignment. Tutors in the course have found the versioning feature in the Project Files Module to be useful for keeping track of the different versions of draft submitted for their review. The time-stamp on the documents allows the tutors to track the history of draft uploads.

eUreka also provides the affordances for learning to take place through self-regulation, others-regulation and object-regulation (Vygotsky, in Chaiklin, 2003). According to Vygotsky, learning is a social-psychological process that takes place when learners learn through self-reflection, from their peers and artifacts such as model text. In the language module, the comparison of the different versions of the draft allows students to reflect on their weaknesses in previous work, and make improvement to the current draft. Learning through regulation by their peers takes place when students use the feedback to improve their subsequent drafts. Students can also learn from other class-mates' materials and exemplary work uploaded and shared by the tutors.

Through this collaborative exercise in peer-review and commenting on each other's writing, the students also learn important social graces and skills in the use of language. One example is in using acceptable etiquette language to clearly critique their peers' work. They will, thus, exercise diplomatic and constructive feedback in their analysis. Moreover, learning through such collaboration will allow them to pick up important verbal and written communication skills that will benefit them when they join the workforce.

In summary, eUreka seeks to put in place an online platform to facilitate constructivist learning theory in that learning is an active process and students will learn best by making sense of the content and context with the faculty providing scaffolding support (Brooks & Brooks, 1993). The scaffolding support in the use of version tracking in Project Files, peer review, learning through regulation and learner control in the management of the project site are clear

*Figure 2. Time-stamp in the Project Files Module allows tutors to keep track of students' writing drafts*

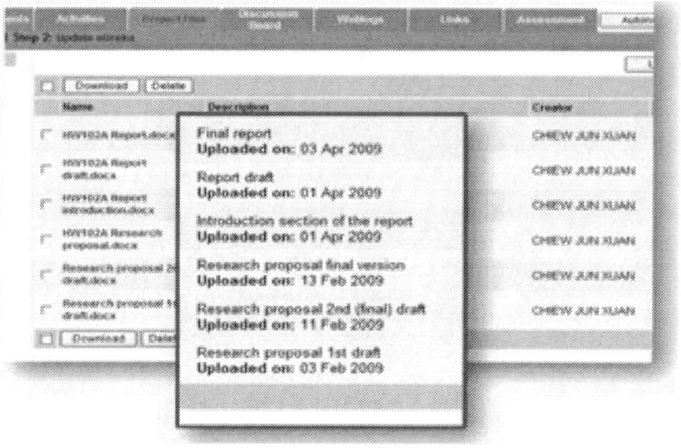

*Table 1. Tools in the eUreka project work management system*

| Module | Description |
|---|---|
| Information | A summarized overview of a selected project site. The user will be able to access the latest announcements, project files uploaded and updates to work flow process for the project. |
| Announcements | This section provides announcements, news and administrative updates. It can also function as a form of reminder for project members. |
| Activities (making meaning of project scope, monitoring & managing work flow process and working towards project outcome) | Users can create milestones and tasks to monitor and manage the work flow processes and development for the project. The Gantt Chart generated will provide a graphical representation of the overall plan and work flow process. Version tracking keeps history of the project work flow process (milestones and tasks).<br>**Milestone** marks a scheduled event in the project duration that signifies the completion of a major deliverable or a set of related deliverables.<br>**Task** refers to the specific assignment in relation to the work scope for the project goals and outcome. |
| Project Files | This is the document repository in eUreka system. It includes a customisable document structure which allows uploaded files to be stored under multiple folders. Files can be uploaded to the repository, for the project material, reports and documentation of the project work flow process. |
| Discussion Forum (Active involvement in project issues, striving towards feasible solutions) | The online forum facilities group project discussion. While it has limited benefit for individual projects, in group project – members and supervisors can raise issues for discussion. |
| Weblog (Web 2.0 application for collaborative and constructivist learning approach) | Weblog is the online journal or diary to record individual member's progress, observations, thoughts, insights, responses, learning experiences, etc. It acts like the personal journal/ log book for students to document their work. Members can share and allow others access to their weblogs. |
| Links | This is a repository of useful web links for reference in the project work. The user can create his/her own folders for better management of relevant and useful project links. |
| Assessment (for Faculty Supervisors only) | This section provides tools for assessment rubric creation and customizable assessment templates. Scores can be weighted, updated and eventually exported out to other systems (such as, in NTU, the Marks Entry Systems). |

examples of elements in constructivist approach in the design of eUreka.

## EUREKA: FEATURES AND ROLE IN NTU

Today, eUreka is widely used by NTU students for their Final Year Projects and postgraduate projects. It provides a one-stop resource repository and work flow process management system that facilitates the application of project management skills, albeit in real-world projects. A summary on the features/modules in the system is given in Table 1.

Since July 2008, the adoption of eUreka has been extended to ***Industrial Attachment (IA) programme***. As part of the undergraduate curriculum, NTU students are required to undergo a period of industrial attachment (IA), varying from 8 to 24 weeks. Students on IA are assigned to work in private or public organizations, both locally or overseas, during this period.

Traditionally, IA project work at NTU centered on the use of a physical logbook or journal (Figure 3) that was maintained by the student documenting his observations and reflections for the attachment.

To better facilitate the supervision for IA programme, the physical logbook was replaced by the use of weblog in eUreka – the online access provided convenience to the faculty to access and monitor the work progress of students on attachment. Graphs of data results, photographs of samples or experimental set-up, video of live

*Figure 3. Sample log book entries*

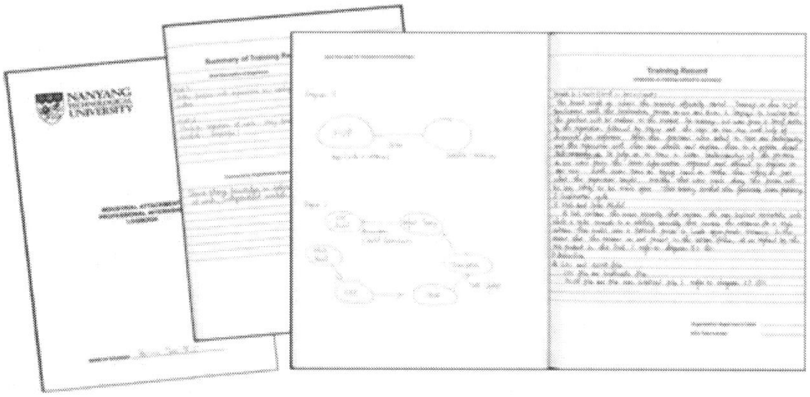

specimens, etc can be attached as a file for reference - this is beyond what the paper log book can do. This feature, thus, better facilitates the documentation of learning experiences for the students on their attachment program as well as provides easy online access anytime anywhere for the supervisor and tutor.

eUreka also provides an effective mentoring and communication tool for the IA supervisor, NTU tutor and internship student. The expectations and assessment criteria for the IA assignment can be clearly communicated to students using the Gantt Chart in the Activities Module. As students spend most of their time at the off campus IA site, effective timely communication between the NTU tutor and the student via email alerts and weblogs, also meant a fairer and more open grading system is established. In addition, the tutor will, likewise, be in a better position to moderate the students' performance.

Faculty has provided feedback that Gantt Chart in eUreka has helped students to better manage their project timeline and provide early alert prompters for need to re-direct or guide the students in their project tasks/goals. Majority of the Students (68.8% of the survey respondents) found the weblogs useful for documentation of their reflections and learning experiences.

Based on the feedback from the students and faculty involved in the pilot launch for the IA porgramme from July to December 2008, the use of eUreka for IA programme has been further explored and enhancements to the features include intermittent auto-save for weblog compositions, better work flow process to alert/prompt supervisors on uploaded documents for review and system generated reminders to better manage project timeline and deliverables.

## WEB TECHNOLOGY AND EUREKA

The Web 2.0 revolution has undermined the exclusive rights of webmasters to create and disseminate information on the Internet and extended the rights of content creation to the common web users, empowering him with a voice on the worldwide web. It propagated the egalitarian ideals of "folksonomy" that encourages the collaborative practice of creating and managing content through "tag" annotations. A term coined by information architect Thomas Vander Wal, (Mathes, 2004), this revolutionary "bottom-up" way of content classification allows web users to create and classify content through "social tagging" as opposed to the traditional subject indexing approach practiced by web experts. Mathes (2004) describes this as "author creation" rather than "professional creation" of data. This meant that web-users with common interest in a subject matter can collaboratively create, comment and edit each other's work on the Internet through a Web 2.0 application such as a blog or wiki. In addition, the contributor need not be a subject matter expert as information he gives will be verified by other contributors using the comment or edit functions.

In essence, the web itself has not changed (technically), but it is the way which web application developers design the applications for the web and how web users interact with the web that have changed. The Internet has basically been transformed from a medium where people look for information to a platform where people participate and contribute towards personal insights and knowledge creation. These key changes that took place in the Web 2.0 age can be summarized below:

1. Allowing users to have ownership of the content which they upload on the web
2. Fostering a climate of participation where web users contribute and collaboratively work towards knowledge creation
3. Creating a level playing field whereby both experts and novices can contribute equally on any subject matter, thereby forging friendships amongst people of diverse backgrounds but with common interest
4. Offering a more user friendly and intuitive interface for users to create and upload content

In many ways, eUreka is a system that displays the characteristics of web 2.0 application in that it provides for:

- user ownership of content,
- participation and collaboration towards knowledge creation
- learning opportunities to develop life skills to meet the challenges of the working world

## User Ownership of Content

eUreka project sites are owned by the project initiator, which can be both faculty and/or students. A student who creates a project site in eUreka will automatically be appointed as the project owner and can enroll other students or faculty members into his project site. He can also assign these new members roles as Supervisor, Member or Guest (with read only access). Subsequently, the students can upgrade members as supervisor to help him co-manage the project site. If there is a change in group membership, the supervisor can also un-enroll members from a project site, or enroll new members (see Figure 4).

With the exception of the Assessment Module (available only for faculty to configure their assessment rubrics), students can access and use the other modules in the project site (see Figure 5). These range from putting up a message to all project members using the Announcement Module to sharing bookmarks in the Links Module.

The Project Files Module allows students to use it as a repository where they upload, organize and share their resources such as project documentation, graphic and multimedia files with their fellow project members. Essentially, it gives students a sense of ownership that they are the ones who have researched and contributed to the repository of resources. At the end of the project cycle, students can download and keep a copy of their project site (in html page format). This can be used as part of their portfolio to be showcased to prospective employers or post-graduate supervisors.

### Encouraging Participation and Fostering Camaraderie

In the Activities module, tasks can be assigned to each project member and email reminders sent as a due date approaches. Students can also track the different versions of the Gantt Chart (in graphical format) using the 'Save a Version' feature so as to

Figure 4. Project owners can easily add new members or edit the membership status by clicking on the "add project members", "edit" and "remove" buttons

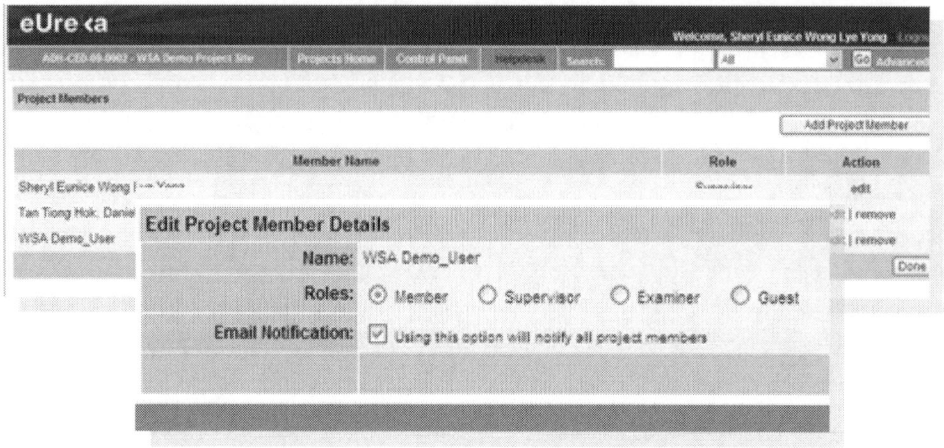


Figure 5. The 8 main modules in eUreka are located on the top frame of the system and easily accessible by users

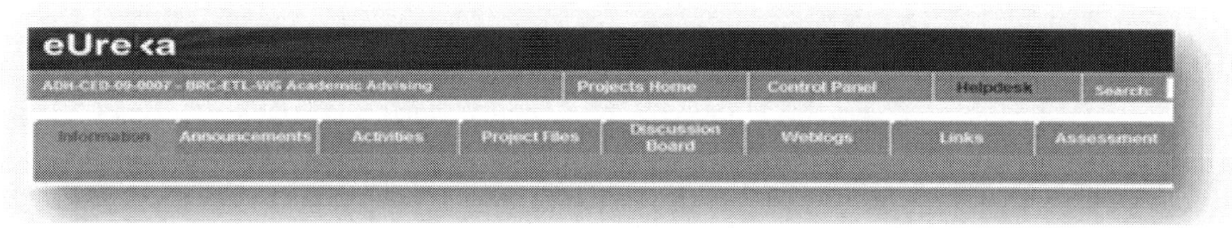


document changes made to previous milestones and tasks. Unlike a typical project management system whereby the project manager takes the driver's seat and sets the pace of the project, the Activities feature allows students to plan and negotiate the milestones, activities, tasks and project deliverables with the project supervisor and fellow team members. Planning of the project schedule is a collaborative scaffolding exercise whereby students and their project supervisor negotiate project deliverable with timelines. It is not a one way communication channel whereby all project deliverables are dictated by the supervisor, thus making project management process more participative.

The Discussion Board module facilitates communication and discussion among project members. Using the intuitive scaffolding icons, members can categorize their discussion threads as "Ask a question", "Identify a problem", "Discuss a topic" or "Suggest improvements" to better facilitate discussion and brainstorming amongst project members. In this discussion forum, all project members are empowered with a voice to articulate their ideas and suggestions regardless of their depth of understanding in the subject matter. This helps students to generate and share ideas such as problem-solving suggestions. It also encourages and fosters camaraderie between project members as they work collaboratively in seeking a solution for their project work. Members can also record their learning reflections on a given discussion topic using the "My Reflections" function that is available in the Discussion Board module. These

*Figure 6. The Gantt Chart used for project planning with versioning functions*

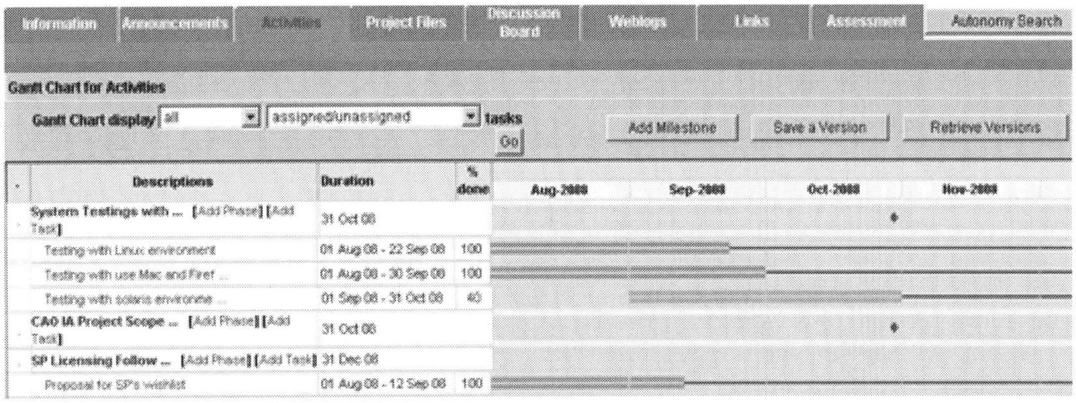


*Figure 7. "My Reflections" function allows students to record their learning reflections from a discussion thread. Scaffolding icons used to organise threads and better facilitate discussion.*

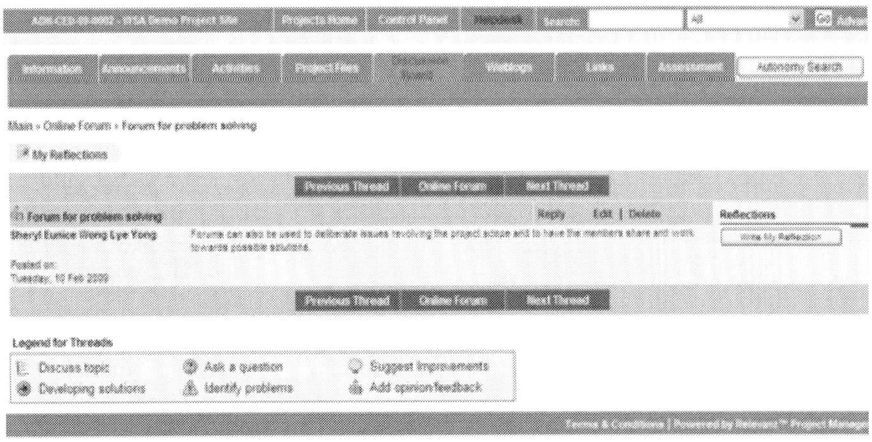


reflection entries will serve as testimonies on "what had worked" and more importantly "what failed" in the project work so that lessons from the mistakes can be learnt.

## Weblogs and the Learning of Life Skills

Reading, writing and critical thinking skills are important literacy skills which students need to develop. In the information age, it has been argued that digital literacy is another pre-requisite skill which students need to acquire for sociability, lifelong learning and employment opportunities (Resnick, 2002). Web 2.0 tools like weblogs can be valuable educational technology tools that have a two-fold advantage of helping students improve their reading and writing skills as well as gain the necessary digital literacy skills to meet challenges in the information age.

The Weblog tool in eUreka has an added advantage when compared to many other blog sites – security and privacy. Weblog entries are not accessible to the public, except to project team members. This facilitates accountability of individual contributions, hence making assessment transparent and fair.

Each project member will have his/her own weblog space and they can grant access and share these weblog entries with their fellow project mates and supervisor(s). Weblog entries can be 'whispered' to selected project members. Weblogs promote self-expression and is a place where students can develop highly personalized content to reflect their thoughts and feelings. Yet weblogs connect students and supervisors in the project groups where members can comment and give feedback to one another on issues related to the project. Weblogs provide students with a platform where ideas are shared, questions are asked and answered, and consensus valued.

Faculty members have remarked on a different experience when using eUreka to supervise students vis-à-vis the traditional periodic face-to-face meetings and updating themselves reading their physical log book. Contact time for the latter tend to be rather limited, typically only during such meetings, besides sporadic emails and sometimes, phone calls. However, with eUreka, faculty supervisor feels a greater awareness of the student's work by reading the easily accessible weblogs. They are able to provide more timely comments, have more touch points and contact with the students, albeit online. Students perceive greater involvement by the supervisor, which translates to a higher sense of welfare. This leads to greater motivation in sustainable self-directed effort and support in project work outcomes. In other words, eUreka system provides effective support to enable Bloom's higher domains of application, analysis, synthesis and evaluation in student learning and performance.

## EUREKA SINCE ITS MAIDEN LAUNCH IN SEPTEMBER 2004

Interestingly, it was observed that users in Phase I (maiden launch in September 2004 to present)

*Figure 8. Weblog entries from a student eUreka Project site on academic writing: documentation of a student's learning journey through the module.*

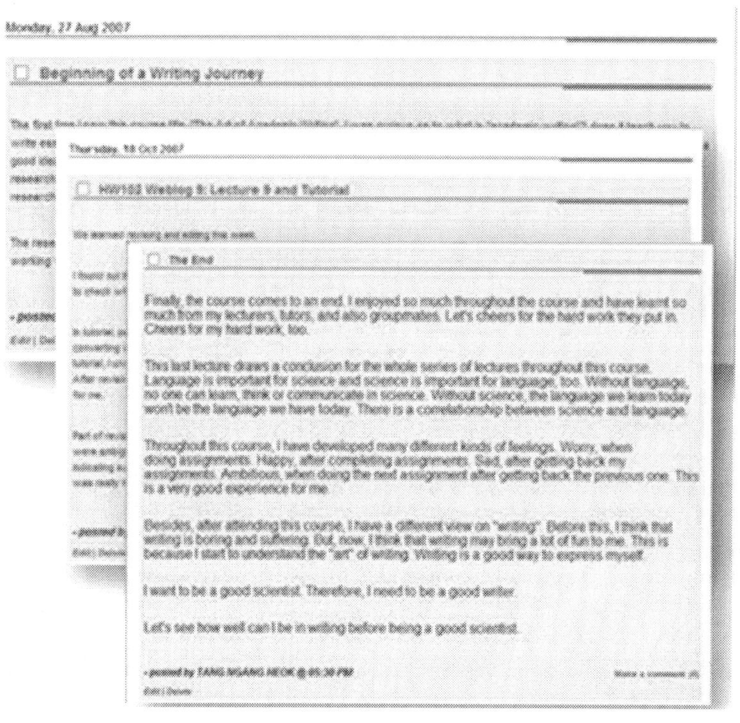

first viewed the system primarily as a useful repository for resource upload and documentation of their work. The monitoring process in milestones and tasks provided an efficient organizational tool for the various activities and tasks needed to successfully complete the project. The use of a weblog for educational purposes was initially less enthusiastically received by faculty and some students.

Since then, there has been steady growth in the number of project sites and usage of its tools. For the period 2008/09, there are 2,303 active project sites (see Figure 9). Active project sites include activities (creation of milestones and tasks to monitor project process/progress); creation of discussion forums/weblogs or upload of files. The number of files uploaded has grown from 593 in 2004/05 to the present 7,101 files in 2008/09 (see Figure 10).

More significantly, there has been a steady increase in the use of reflection and collaboration tools via weblogs and online forums. From the usage data, one can suggest that faculty and students appreciate the use of weblogs and forums for personal sharing. This sets the stage for greater user ownership, self-directed and student-centred learning experiences. Students can now put on records their learning journey, the instances of successes, failures, frustrations and "now I know" experiences – and when their project is referenced for use by future students, its complete diary of the journey will help ramp-start the new project under a new project team to the next level of the research initiative.

The number of weblogs created has increased from 124 in 2004/05 to 4,612 in 2008/09, an increase of 37.19 times (Table 2). Discussion forums saw an increase of 14 times in the number of forums created (from 68 forums in 2004/05 to 958 forums in 2008/09) to facilitate deliberation, sharing, review of postings and contributions towards feasible solution to issues surrounding the project scope and content.

## NEXT STEPS FORWARD FOR EUREKA

eUreka is easy to use and faculty are extending the use of the features such as File Versioning for report drafts in regulatory learning, reflections and collaborative writing skills. The versioning also keeps record of the history of the draft towards the final output for the group assignment. The

*Figure 9. Active Project Sites in eUreka*

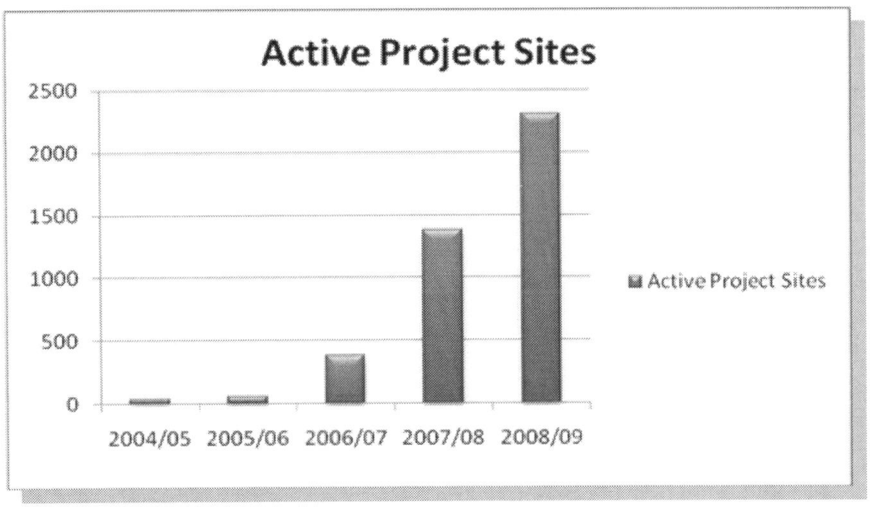

*Figure 10. Files upload to the repository in eUreka*

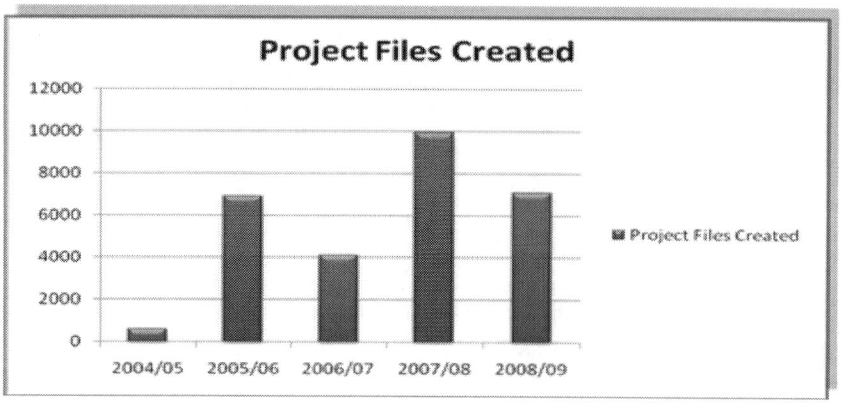


*Table 2. Number of new weblogs and forums created*

| Year | 2004/05 | 2005/06 | 2006/07 | 2007 /08 | 2008/09 |
|---|---|---|---|---|---|
| Weblogs Created | 124 | 138 | 416 | 2,002 | 4,612 |
| Discussion Forums Created | 68 | 229 | 343 | 745 | 958 |

use of weblogs can be seen as an online journal to replace the physical logbook, but it can also serve as sharing of relevant experiences in the management of projects with team members and faculty to comment and provide useful feedback for steps forward. eUreka was built and enhanced with feedback from the student and faculty communities. User feedback has provided invaluable inputs for further system enhancements.

Future development should see eUreka providing online 'crawlers' to relevant external web resource sites that relate to the project contents, as well as digital rights management to cope with the increasing volume of uploaded files/resources. To better cater to research projects, the eUreka team is considering the addition of basic financial module to keep track of disbursement of research expenses.

Other possible areas for extension of use for eUreka can cover cross project sites sharing of resources. A wiki tool might be useful for some types of team work, though its assessment of individual contributions might be challenging.

## CONCLUSION

This paper provides a historical narrative to the evolution of eUreka project work management system in NTU, describing the Institution's aim to automate the project work management. In addition, it also discusses how eUreka empowers students by allowing them to take greater ownership and accountability of their work.

Constructivist and discovery learning is well facilitated with good channels of communication and supervision incorporated in the design of eUreka. The Gantt chart facilitates scaffolding processes and comments from weblogs provide a powerful channel to guide learning and doing. Supervisory time is more effectively used when crucial stages are reached; supervisors keeping track of the progress can make prior arrangements to be present during such times.

eUreka also provides affordances (namely through its participatory modules and functions) for students and staff members to work collaboratively in their quest for knowledge discovery.

It has served as an effective organizational tool to capture and manage the valuable knowledge assets that could be created with the project work process.

Building on the constructivism learning theory, learners construct knowledge by working together interdependently, in communities of knowledgeable peers. Knowledge is not absolute, it is local and historically changing – we construct and re-construct knowledge time after time (Bruffee, 1995). This favours active learning over passive learning. To this end, eUreka seeks to put in place a web-based networked learning community for students to interact, negotiate, discover, manage, construct and build up knowledge assets pertinent to the project objectives and expected outcomes. (5,776 words)

## REFERENCES


Armstrong, J. M. A. (1950). *Projects and Their Place in Education*. London: Sir Issac Pitman & Sons, Ltd.

Blumenfeld, P. C., Soloway, E., Marx, R. W., Krajcik, J. S., Guzdial, M., & Palinscar, A. (1991). Motivating project-based learning: Sustaining the doing, supporting the learning. *Educational Psychologist*, *26*(3-4), 369–398. doi:10.1207/s15326985ep2603&4_8

Boss, S., & Krauss, J. (2007). *Reinventing Project-Based Learning: Your Field Guide to Real-World Projects in the Digital Age*. New York: ISTE.

Bransford, Brown & Conking. (2000). *Designing Effective Projects: Characteristics of Projects Benefits of Project-Based Learning*. Retrieved June 17, 2009, from http://download.intel.com/education/Common/en/Resources/DEP/project-design/DEP_pbl_research.doc

Brooks, J., & Brooks, M. (1993). *In Search of Understanding: The Case for Constructivist Classrooms*. Upper Saddle River, NJ: Prentice Hall.

Bruffee, K. A. (1995). *Collaborative Learning: Higher Education, Interdependence, and Authority of Knowledge*. Baltimore, MD: John Hopkins University Press.

Chaiklin, S. (2003). The Zone of Proximal Development in Vygotsky's Analysis of Learning and Instruction. In Kozulin, A., Gindis, B., Ageyev, V., & Miller, S. (Eds.), *Vygotsky's Educational Theory and Practice in Cultural Context*. Cambridge, UK: Cambridge University Presents.

Cuban, L. (1986). *Teachers and Machines: The Classroom Use of Technology Since 1920*. New York: Teachers College Press.

Cuban, L. (2001). *Oversold and Underused: Computers in the Classroom*. Cambridge, MA: Harvard University.

Katz, L., & Chad, S. (2000). *Engaging Children's Minds: The Project Approach* (2nd ed.). Norwood, NJ: Ablex Publishing Corp.

Laffey, J., Tupper, T., Musser, D., & Wedman, J. (1998). A computer-mediated support system for project-based learning. *Educational Technology Research and Development*, *46*(1), 73–86. doi:10.1007/BF02299830

Mathes, A. (2004). *Folksonomies - Cooperative Classification and Communication Through Shared Metadata Computer Mediated Communication*. Retrieved June 17, 2009, from http://www.adammathes.com/academic/computer-mediated-communication/folksonomies.html

Resnick, M. (2002). Rethinking learning in the digital age. In Kirkman, G. (Ed.), *The global information technology report: Readiness for the networked world*. Oxford, UK: Oxford University Press.

Stepien, W., & Gallagher, S. (1993). Problem-based learning: As authentic as it gets. *Educational Leadership*, *50*, 25–28.

Stern, C. (2002). *Information literacy unplugged: teaching information literacy without technology.* White paper prepared for UNESCO, the US NCLIS and National Forum for Information Literacy. Retrieved June 17, 2009, from http://www.nclis.gov/libinter/

Thomas, J. W. (2000). *A Review of Research on Project-based Learning.* Retrieved June 17, 2009, from http://www.bobpearlman.org/BestPractices/PBL_Research.pdf

Wikipedia. (n.d.). *Project-based learning.* Retrieved June 17, 2009, from http://en.wikipedia.org/wiki/Project-based_Learning

## ENDNOTE

1 http://en.wikipedia.org/wiki/Project-based_Learning

# Chapter 10
# Improving the Tracking of Student Participation and Effort in Online Learning

**Ian Douglas**
*Florida State University, USA*

## ABSTRACT

*Much research into educational technology is focused on tools for supporting teaching and learning. In contrast to this work, relatively little research is conducted into technology that tracks student participation and effort. No matter how good the educational technology, learning is dependent on a sufficient input of effort from the student. Most Learning Management Systems have some tools for tracking students, but they are currently difficult to use and underused by instructors. This chapter examines the importance of tracking in student management, reviews attempts to improve the quality of tracking tools, and suggests paths for future research based on the deficiencies in current tools.*

## INTRODUCTION

The focus on educational technology has mostly been to create better tools for teaching and learning. The emphasis is on supporting the teacher's design and operation of a course while assuming the student comes to the course with the right set of attitudes and behaviors. There is another side to the formula leading to successful learning and that is the student participation and effort. No matter how well a course is designed, and no matter how good the content or technology, students will not succeed unless they put effort into their learning. For reasons out of the control of the instructor, a student may not be in the correct psychological frame of mind for learning. They may not be aware of the level of effort required, or may have too many other commitments and distractions to provide the right amount of effort.

Traditionally, one of the main tools for measuring participation and effort has been class attendance. Studies of class attendance show that it is strongly correlated with class performance. For example, Romer (1993) compared student performance and attendance in an intermediate undergraduate economics class and found a statistically significant

DOI: 10.4018/978-1-61520-853-1.ch010

Copyright © 2010, IGI Global. Copying or distributing in print or electronic forms without written permission of IGI Global is prohibited.

relationship, with the performance of students who attended all lectures on average a full letter grade higher than that of students who attended a quarter of the lectures. Devadoss and Foltz (1996) reported a similar relationship in a study of 400 students from four universities over three semesters. Chen and Lin (2008), quantify the effect as corresponding to between a 9.4 and 18% improvement in exam performance for those who attend classes.

The findings related to attendance can lead to class attendance being required, or at least a requirement for monitoring attendance. At primary and secondary educational levels there may even be a policy of notifying parents when a student fails to attend a class. This only deals with the problem on a physical level, given that a student can physically attend but cognitively be absent from the proceedings in class. Although there is a long established literature on student motivation (Keller, 1983), and there are certain teaching techniques that can be employed to encourage cognitive participation and keep students' attention (e.g. breakout groups, random questioning), there are usually opportunities for students to tune out of what is going on in class. There is an element of self-discipline involved in learning and when this is lacking it needs to be identified and the student provided with appropriate feedback and counseling.

The potential for daydreaming in class has been supplemented by the potential distractions provided by technology in the classroom. Allowing laptops and mobile devices in class may have educational benefits, but it also allows attention to be divided between text messaging and Internet browsing during class time. There is a myth that the modern student has developed a greater ability to multitask; this has no basis in scientific research. Dividing attention generally impairs performance, as has been shown by the distracting effects of driving and talking on the phone.

Online classes present both a problem and an opportunity in relation to participation and effort. The problem is that attendance at synchronous events is often not required, and when it is, students cannot usually be seen. In a face-to-face class, a teacher can to some extent monitor the level of cognitive attendance by looking at the facial expressions and gazes of the students in class. In an online class this feedback is absent and although a student may appear to be present, they could be watching TV at the same time as undertaking a class. However, in online classes technology used presents new opportunities for automatically tracking students and reporting the data to both them and their instructors.

This chapter will review attempts to develop better technology for tracking students' participation and effort on courses. This chapter will begin by considering research into the usefulness of tracking data. In the following section we will describe the typical tracking tools found in current learning management systems. We will then review the research efforts that have been aimed at improving the quality of tracking tools, providing some examples of tools that demonstrate the variety of concerns related to tracking tools (e.g. improving data visualization). Finally we will consider what research is needed in future, not only to improve the tools, but also to better integrate them into course management practices.

## STUDENT TRACKING RESEARCH

There have been a number of researchers who have looked into the factors that may cause some students to not fully participate in an online class. Mason (1994) classified online participants into three groups: active participants, lurkers who process activity but do not contribute and those who do not take part. Taylor (2002) came up with a similar classification of workers, lurkers and shirkers. In the context of online participation, Vonderwell and Zachariah (2005) define participation as "taking part in a dialogue for engaged and active learning". They identified five factors that are important in effecting participation: technology and interface, content area experience, student roles, instructional tasks and information overload.

Given that participation is important, and not easily equated to taking class attendance online, it is important to have some means of monitoring students, measuring levels of participation and effort, and identifying situations equivalent to non-attendance, or low effort. These are situations that might benefit from an instructor's intervention. Many systems used to manage online learning such as WebCT, Blackboard and Moodle collect interaction data and provide some basic tracking tools for monitoring student activity.

It should be noted, that there is some debate on the correct terminology for such systems, which are often referred to in the literature as Learning Management Systems (LMS). Watson and Watson (2007) argue that an LMS relates to more general concept with a richer set of features, and these systems are better referred to as Course Management Systems. However, with each new version these systems add more features. They often allow plug-ins and add-ons, and can be used to support non-course based learning. The division lines between what is an LMS and what is a CMS are becoming blurred just as what once were referred to as authoring tools are no longer easily distinguishable from general web development tools.

Douglas & Alemanne (2007a,b) conducted two studies that used data from an LMS to evaluate the measures of participation and effort in an online course and their relationship with performance. They analyzed discussion posts, synchronous class speaking activity, email contacts to the instructors, click counts on the course web site and combined score for all activity. Only the click count was provided automatically through the LMS system used (Blackboard). The rest of the items had to be counted manually.

In one study with forty students (Douglas & Alemanne, 2007a), the click count ($p=0.0002$), the number of discussion posts per student ($p=0.1$) and synchronous class attendance ($p=0.035$) were found to be significant predictors of student success in the course. In a second study with thirty students (Douglas & Alemanne, 2007b), although individual measure of all the emails sent to the instructor, all the discussion posts and all the incidents of voluntarily contributing in synchronous classes (e.g. asking a questions) were not found significant, a combined score for these measures was significant ($p=0.014$). In the same study the number of clicks accumulated on the course web site over a semester was also a significant predictor of success ($p=0.032$).

Despite being a simple measure, the click count may be a useful indicator of student effort that predicts student success on a course. In using the click measure it may be possible to identify problem students within the first few weeks of any course. Despite the evidence that tracking tools provide useful information, they are not generally used by institutions in formal monitoring, or by individual instructors. Douglas (2008) reports on an informal survey of the 35 faculty members in a College where online teaching is common. Only two people reported using the tracking facilities built into the university's implementation of Blackboard. Both had only recently used it and reported difficulty in learning how to interpret the data provided. There seems to be a general lack of awareness of both the tracking tools and their potential usefulness in monitoring students.

## EXISTING STUDENT TRACKING TECHNOLOGY

Figure 1 illustrates a screen shot from the main tracking tool in the Blackboard LMS, one of the most widely used proprietary systems (Bradford et al, 2007), showing a student's click count over a period of nineteen days. It shows which days the student logged into the course web site, how many clicks were registered on those days and the accumulated click count for the period selected.

In addition to being able to track the clicking activity of individuals, the system can show the aggregated data for the whole class. The user may select to look at a specific date range, but the default is to show all the data captured since the course became active.

This tool is only available to the instructor; the students cannot see their own activity or compare it with the rest of the class. It is also a crude measure in that you cannot assume that more clicks equal more learning effort (discussion inputs and content access). They can just as easily equate to incorrect interactions due to a student not being familiar with the user interface. In order to measure whether a student has accessed an individual piece of learning material, or contributed to an online discussion, the instructor has to go to the effort of setting up tracking for individual content items, assuming they have the knowledge of how to do this.

Tracking of individual content items is accessed by going to the content area and changing to an edit mode (first screen on figure 2), rather than using the main tracking tool, which is in the instructors' control panel under the label "statistics". After changing to the edit mode the user must select the "Manage" button, which displays a list of attributes that includes "statistical tracking". Statistical tracking is disabled by default on every content item. There is no means of globally changing this default. The instructor must click on "enable tracking" to begin recording student activity with a given content item (second screen on figure 2). The instructor also has the opportunity to enable tracking when first adding an item of content. Once it is enabled, the instructor can go through the same sequence of switching to edit view, selecting manage, and then selecting "View statistics", which brings up a form in which individual students and/or date ranges can be selected (third screen on figure 2). Clicking on the "Submit" button on this form will provide a display similar to figure 1 for this individual content item showing when and how many times a student accessed the content. These operations must be repeated for every content item an instructor wishes to track and view statistics on. Setting and viewing statistics for individual

*Figure 1. A screen from the Blackboard course statistics-tracking tool*

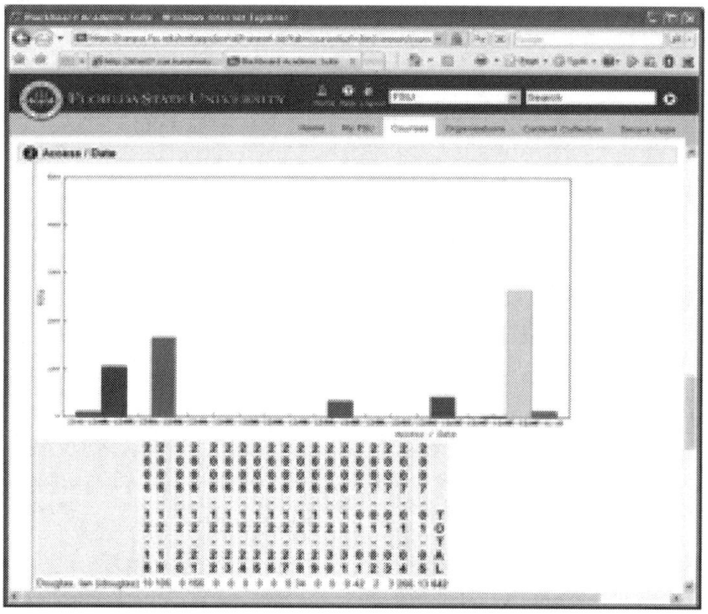

content items can be a time consuming operation when a course contains a large number of content items.

Tracking of communications such as contributions to threaded discussions are not covered in the statistical tracking feature. Contributions to asynchronous discussion are not differentiated in the main tracking tool separate from access to content. Another problem with the tracking tool is that it only works for asynchronous activities. If a student participates in a synchronous class using chat or voice systems, their activity must be tracked manually.

There are some clear usability problems with the tracking tools provided in current LMS that may contribute to them being underused. It is also not easy to process the information, particularly with large class sizes (Dringus and Ellis, 2005). There is no central dashboard type function where an instructor can see which content items a student has or has not viewed. The reports on click activity could be made more useful by filtering out navigation clicks from clicks on content. An increasing number of researchers are now investigating these issues and a number of third party tools and add-on software have been developed to address them.

## RESEARCH ON IMPROVING STUDENT TRACKING TECHNOLOGY

Romero, Ventura and Garcia (2007) identify a new area of research called educational data mining,

*Figure 2. The sequence for setting up and viewing individual content statistics*

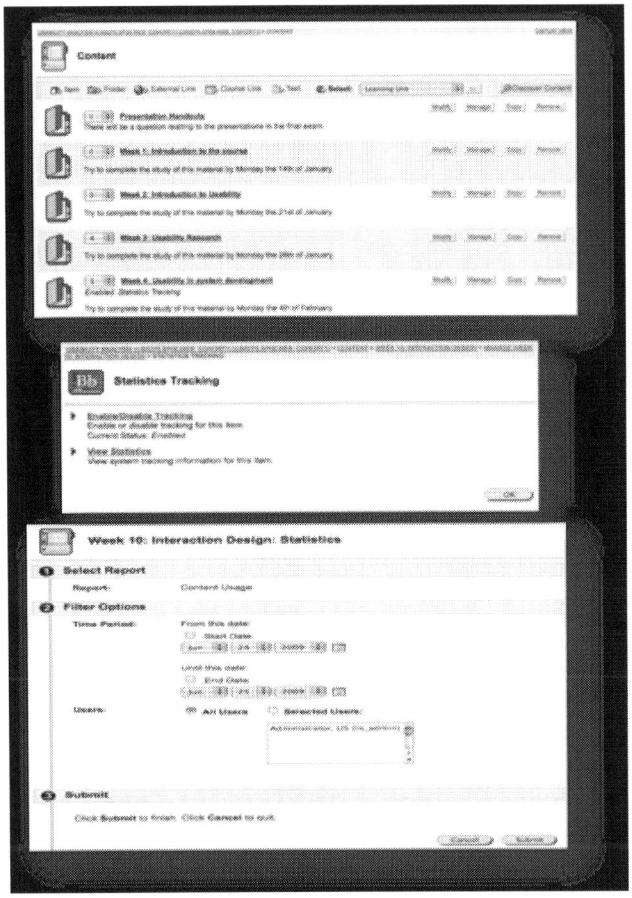

deriving from the area of computer science that investigates algorithms for discovery of patterns across large data sets (Klosgen & Zytkow, 2002). They present a comprehensive review of the techniques, the potential applications, and describe a case study using Moodle. These techniques are not only aimed at student tracking, they can also assist in evaluating the structure and effectiveness of course content and organization. The log files and databases of online tools such as LMSs present the raw data for educational data mining. The process of data mining involves collecting data from the educational tool, doing some preprocessing (such as converting it to an XML format), applying data mining algorithms and then interpreting the results. There are a number of data mining techniques. The ones most relevant to the tracking of participation and effort include:

1. **Statistics:** This is the level of tools already including in systems like Blackboard, showing items such as the relative number of clicks on the course web site on different days of the week (Zaïane, Xin, & Han, 1998).
2. **Information visualization:** These techniques attempt to present data and its relationships in more sophisticated and interactive ways that enable instructors to visually identify patterns (Mazza and Dimitrova, 2005).
3. **Clustering:** This is used to identify grouping of related data. Romero, Ventura and Garcia (2008) describe using the Weka (Weka, 2007) system to categorize students into three clusters: very active, active and non-active.
4. **Classification:** This involves labeling new patterns that emerge from the data analysis. For example, Cocea & Weibelzahl (2006) report identifying learners with low motivation who are likely to drop out.

Information Visualization (IV) (Card, Mackinlay, & Shneiderman, 1999) is an area of research that promises improvements in tracking tools. IV provides visual representations of data that enhance understanding of its meaning. An example here is CourseVis, which uses the data already collected by the WebCT system. Mazza and Dimitrova (2005) argue that the tabular format used for presenting data in existing LMS tools (see figure 1) is poorly organized and difficult to interpret and this is the reason the data is rarely used. CourseVis was composed of a number of visualization tools. The "cognitive matrix" has the student name along the x-axis and the main concepts covered in the course along the y-axis. A performance score indicator is shown at the intersection. The performance score indicator is derived from quiz questions relating to the concept and is color-coded from red (poor score) to green (good score) with non-attempts coded as black. The color-coding allows instructors to immediately identify students with problems, or concepts that seem to be causing trouble for the whole class (indicated by a mostly red and black horizontal line).

An informal evaluation of the CourseVis tools suggested that they enable instructors to quickly and more accurately obtain information about their students. In particular, instructors noted that it would enable the early identification and potential prevention of problems in online courses such as non-participating students who might drop-out.

CourseVis forms the basis for a more recent tool called GISMO (available at http://gismo.sourceforge.net/). GISMO adds visualization to Moodle, the most commonly used open source course management system. Figure 3, illustrates a screen from GISMO which provides a single view of the access to every resource item for every student on a course. The names of the students are represented along one axis and the resources along the other. The intensity of the blue dot at the intersection represents the number of times the student has accessed a particular resource. If there is no blue dot the student has not accessed the resource. There are similar screens available for tracking student attendance and participation in discussions. Mazza & Botturi (2007) report that the access visualization features of GISMO allowed the identification of two slackers on the course. This led to a more detailed investigation of the students concerned, and why their participation patterns differed from the rest of the class.

Participation in online classes is not just about accessing learning materials posted online, it also involves various forms of communication. Asynchronous threaded discussions are an important part of many e-learning courses. Research into communication via technologies, such as threaded discussions, has focused on what factors promote better participation in discussions. Hewitt and Teplovs (1999) found activity and growth in discussion threads were more likely to occur when replies were posted within a day of the most recent posting. It is suggested that as time passes and threads and posts grow in number there is increased competition for students' attention (Hewitt, 2005).

As discussed in Kienle and Ritterskamp (2007), an important role of the instructor is to instigate and encourage quality interaction in such discussions. However, it should be noted that some research shows that more knowledge sharing, collaboration and community building occurs in learner-learner communication (Schrire, 2006). Analysis tools providing feedback on what is occurring in these discussions are potentially useful in monitoring and improving the value of the discussions and the participation levels of different students. Jeong, (2004) argues that the qualitative nature of posts is very important in determining the responses, with critiques and particular types of argumentative exchanges producing higher response rates. Having students pre-classify their contributions to discussions using a predetermined set of message/response categories is seen as helpful in getting students to reflect on their contributions.

In the process of conducting research into the factors involved in promoting good discussions, researchers have developed analysis tools that can augment an LMS. Jeong, (2003) describes the discussion analysis tool (DAT), a stand-alone tool that takes threaded discussions exported from Blackboard and performs a sequential analysis to produce quantitative descriptions of the interactions occurring. The DAT is primarily useful as a research tool, but in recognizing the need for practical tools Joeng also created Forum Manager (available for download at: http://mailer.fsu.edu/~ajeong/ForumManager/) for instructors to obtain statistical information on posts to their

*Figure 3. Gismo allow tracking of student access to all course content on one display*[1]

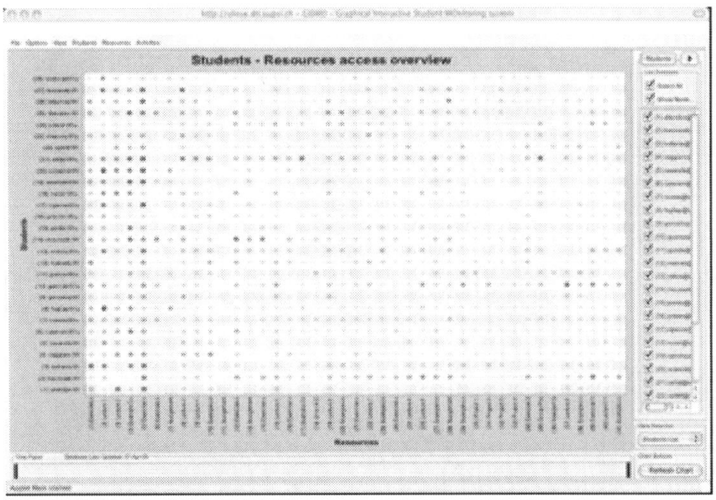

discussion boards. The tool uses data downloaded from Blackboard into an Excel spreadsheet and allows the user to navigate around and read the threaded messages offline (see figure 4), and use the "Count postings" and "Performance Analysis" tools to analyze each discussion and students' performance in the discussion. The Forum Manager counts postings per student, number of threads, average length of discussion threads, percent of messages posted as replies, number of replies elicited by messages posted by the student, number of times the student reciprocates a reply, number of different days the student posted to the discussion, percentage of messages posted with user-specified keywords and phrases.

Jyothi et al (2007) describe a visualization tool for displaying the interaction patterns of asynchronous discussions. The tool can be embedded into the Moodle open source LMS. It is based on Schrire's framework (2002), which includes a graphical representation of the social presence and management of the online communication patterns. The tool uses a radial tree layout for visualizing the analysis of a threaded discussion. Each participant is represented as a small color-coded circle and the arrows between circles represent the dialogue between participants. The connected circles radiate out from a central point in chronological order. In addition to showing the flow of conversation activity, the visual chains help determine the active participants and the social presence of different students in the class. The instructor can click on the node to get a detailed view of the interactions of a particular student.

Group work is a particular problem in relation to participation in online learning. When students participate equally, they achieve a good learning experience (Webb, 1995). However, there is the opportunity for what is generally referred to as

*Figure 4. A screen shot of forum manager*[2]

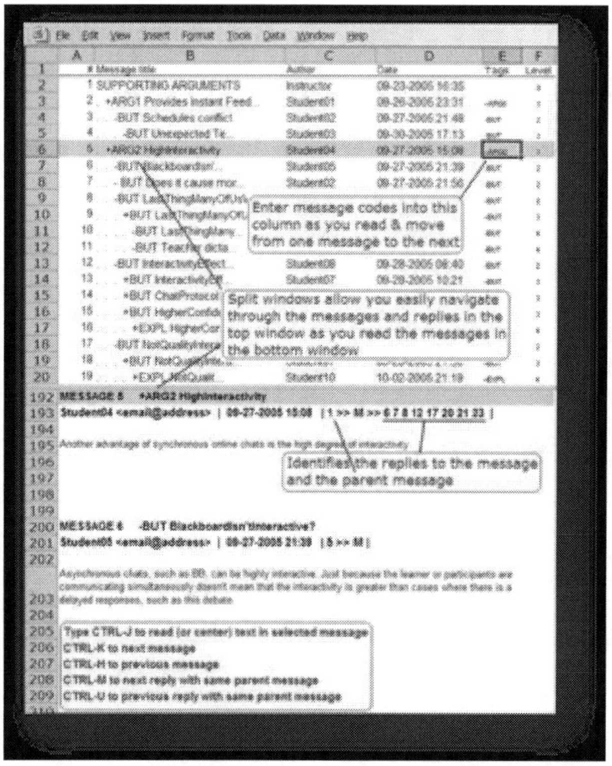

social loafing, where students invest less effort in group work compared to their individual work and rely on the efforts of others to achieve a satisfactory group result. Group work in online courses has been reported to have a number of problems. For example, Fjermestad (2004) reports that online groups need more time to complete tasks than face–to-face groups, and Baltes et al (2002) reports lower levels of satisfaction. The use of tools that enable social comparison and regulation may provide a means to increase participation and reduce social loafing (Michinov & Primois, 2005).

Janssen et al (2007) studied the effect of a Participation Tool (PT) added into an online chat system used during student project work. The tool (see figure 5) presents a visualization of the activity of different group members relevant to their use of a shared messaging system. The system displays a number of different sized spheres that show the measures of individual students compared with the group. The measures are quantitative rather than qualitative and include the number of messages sent and the average length of the messages. Students can zoom in to see their own group profile or zoom out to see how their group compares with others in the class.

An evaluation study was conducted on the participation tool with 52 students using the tool and a control group of 17 not using the tool. It was found that students with access to the tool sent more long messages and had a greater equality of participation. Students using the tool also reported that they knew better when a group member was not working hard.

Mazza and Dimitrova (2005) include a "discussion plot" tool in their CourseVis collection of tools. Discussion plot uses data from the LMS discussion board to create a 3D scatter plot using the variables: originating student of a thread, date, topic and follow-ups. The instructor can rotate among these variables, with three variables on the dimension and the fourth represented by spheres in the 3 dimensional space. For example, an instructor can view which students initiated threads about a topic at a given time, while the number of follow-ups to the thread is indicated by the size of the sphere positioned at the intersection of the three dimensions of originating student, date and topic. The instructor can use this tool to see what topics are popular in discussion compared to ones that are neglected, and identify students who are not initiating any discussions. This would allow the instructor to give encouragement to students who are not initiating discussions and gentle restraint to students who may be spending too much time initiating new discussions. The author of this

*Figure 5. A screenshot of the participation tool*[3]

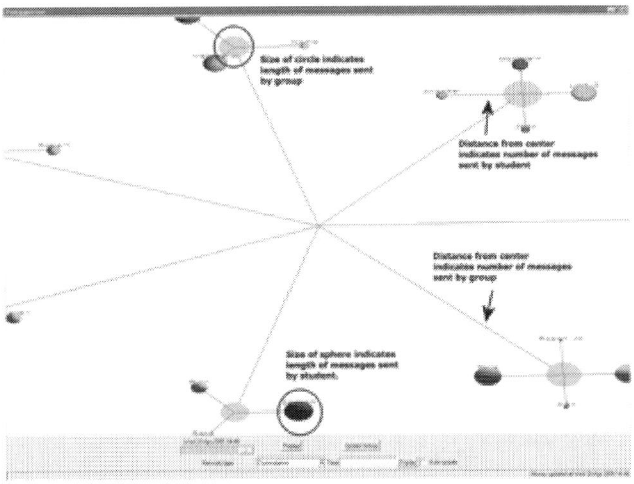

chapter identified a student in a previous course who was a high contributor to the discussion forum, yet a poor performer on class assessments. The student was neglecting the study of the content in order to spend more time engaged in discussion, which he found more enjoyable.

## FUTURE DIRECTIONS

Relative to other areas of educational technology research, there has been little focus on tracking tools and their potential impact on student success. In addition to creating new tracking tools, the following five areas are where more research is needed.

### Making Tracking Tool Use More Seamless and Improving Their Interfaces

A current theme in a lot of the research into new tools is improving the interfaces. A lot of the data required is already automatically collected. It is making it easily available and viewable that seems to be the problem. The vendors of LMS could make this process easier by investing more in supporting tools for tracking and better visualization of the results. Usability is a general problem with such systems (Kakasevski et al, 2008), which have tended to add more and more features and in the process make the systems more difficult to navigate. This may be a factor in why the existing tracking tools are not used by most instructors.

### Promoting Awareness of the Need to Use the Tools

Compared to studies that show the effectiveness of various instructional techniques, there are relatively few studies that show the effectiveness of student tracking. More studies would result in a weight of evidence that may influence education administrators to invest more in tracking. In the same way that class attendance has been made mandatory in many face-to-face courses (or at least taking roster has), some form of automated tracking could be encouraged by administrators if there was an accumulation of evidence supporting its use. More studies such as those by Douglas & Alemanne (2007a,b) showing tracking data is predictive of success would be helpful. Studies showing that interventions based on the data provided help in supporting student success would be even more compelling.

### Providing Tracking Data to the Students

Research in tracking systems is primarily focused on providing reports to instructors. Students are provided with feedback on the outcome of their effort in the form of grades in an online grade book, but little feedback on their relative input. Students may think they are putting reasonable effort into a course only to be disillusioned when they get their assessment scores. Investigating the effect of students being able to see their own participation measures on an online course is a useful area of research. Providing feedback for the student may provide a useful way of encouraging better participation. If a student saw that their scores on participation measures were below the class averages it may cause them to reflect on their effort.

Students often have to adjust their participation and effort levels to different demands and it helps to have a guidepost to doing this. Moving from different course levels, from face-to-face to online instruction and between different subjects and instructors with different demands requires adjustments of effort. It is possible given the amount of material online that students often miss things. Getting access to a tracking system would be a useful way of reminding students of what they have not yet covered in the learning content, instructions and assignments that are posted online.

### Considering Ethical Issues in Tracking Student Data

Yeaman (2004) notes that the ethical issues in the use of e-learning tools are a neglected area of research. In a study of ethical concerns among 20 professional educational technologists, Lin (2007) noted that there was a "keen awareness of the impact of technology on individual privacy" and that one subject specifically raised the concern of "if we do track students, how should we responsibly and ethically use that information?" Prior to the advent of such tools this was not a concern for instructors.

Students may regard tracking of all their online course activities by instructors and/or institutions as a "big brother" approach. There are potential legal (e.g. data protection) and ethical issues (e.g. basing grades on tracking data) that need consideration by relevant researchers. There is a need to consider policy issues in the use of tracking tools. At the very least students should be made aware that tracking is part of the systems they are using and that instructors have the ability to check on their course related activities.

### System Wide Tracking of Learner, Rather Than Just Course Based Tracking

Tracking tools tend to have a narrow course-based view, but effective tracking should be system wide. System wide tracking could be argued to be a defining feature of a Learning Management System compared to a Course Management System. Students should not only be able to see their level of participation compared to their peers on one course, they should be able to see an aggregation of tracking across courses. It may be that exerting greater than average effort in one of their courses is leading them to neglect the others. This information should also be available to their instructors and advisors.

Given that there is some evidence to suggest that participation data is important, there are implications for student advising. The aggregation of participation data across all courses for each student into a personal participation profile, would allow quick identification of students whose participation falls well below the average across all their courses and provides the ability for both advisors and instructors to be proactive in offering remedial assistance and advice. Students may need specific guidance in adjusting their effort when transitioning between different levels (high school to college) and modes (face-to-face to online) of learning.

Improvements in cross course tracking would provide useful data for educational planning, curriculum design and educational research. Are there courses with significantly higher participation rates? Is there an imbalance in demands of different courses? Are instructional techniques used in some courses improving motivation? Can evidence of individual student problems be identified early from individual participation profiles?

Such benefits can only arise if LMS research and development focuses more on the tracking as a core feature providing value, rather than a periphery feature to the instructional tools. The evidence of a few studies carried out into the importance of tracking data suggests a need for this focus. Even the most sophisticated course will not engender learning without a sufficient level of participation and effort from the student. What constitutes a good measure of "sufficient participation" is a key research question.

## CONCLUSION

In traditional forms of instruction, participation has primarily been measured by class attendance. There are relatively few studies considering an equivalent (or improved) measure in online learning. This paper has presented a review of studies concerning measures of students' participation and effort in online classes. The importance of even crude measures such as click count has been demonstrated in the few evaluation studies done, and the variety of tools available that track various kinds of data for instructors is increasing. The basic level of tools provided in current Learning Management Systems is inadequate and they do not seem to be used much by instructors. Partly to blame for the fact that these tools are underutilized is that there is no clear strategy for using them, the value of the data is not known and there are notable design deficiencies in their usability.

Although there are researchers that have developed prototypes of more sophisticated tools, this is a relatively neglected area of research. There is still a need to further develop these tools to allow the easier automation and customization of participation data collection and visualization. The data should be provided to students and advisors, not just instructors. There is also a great need for research into the creation of technologies that can monitor participation in a more qualitative way.

The research carried out to date has implications for those who build learning management tools, suggesting the need to improve and publicize the usefulness of the student tracking features. If student inputs (effort and participation) can be tracked in the same way as student outputs (assessment scores), instructors would have a useful mechanism to identify students who need assistance and students would have a useful tool for self-evaluation.

## REFERENCES

Baltes, B. B., Dickson, M. W., Sherman, M. P., Bauer, C. C., & LaGanke, J. S. (2002). Computer-mediated communication and group decision making: A meta-analysis. *Organizational Behavior and Human Decision Processes*, *87*(1), 156–179. doi:10.1006/obhd.2001.2961

Bradford, P., Porciello, M., Balkon, N., & Backus, D. (2007). The Blackboard Learning System: the be all and end all in educational instruction? *Journal of Educational Technology Systems*, *35*(3), 301–314. doi:10.2190/X137-X73L-5261-5656

Card, S., Mackinlay, J., & Shneiderman, B. (1999). *Readings in Information Visualization - using vision to think*. San Francisco, CA: Morgan Kaufmann.

Chen, J., & Lin, T.-F. (2008). Class attendance and exam performance: a randomized experiment. *The Journal of Economic Education*, *39*(3), 213–227. doi:10.3200/JECE.39.3.213-227

Cocea, M., & Weibelzahl, S. (2006). Can log files analysis estimate learner's level of motivation? In Schaaf, M., & Althoff, K.-D. (Eds.), *Lernen-Wissensentdeckung* (pp. 32–35). Adaptivität.

Devadoss, S., & Foltz, J. (1996). Evaluation of factors influencing student class attendance and performance. *American Journal of Agricultural Economics*, *78*, 499–507. doi:10.2307/1243268

Douglas, I. (2008). Measuring participation in Internet supported courses. In. *Proceedings of the International Conference on Computer Science and Software Engineering*, *5*, 714–718.

Douglas, I., & Alemanne, N. D. (2007a). Monitoring participation in online courses. In *Proceedings of the 8th International Conference on Information Technology Based Higher Education and Training* (ITHET 2007), Kumamoto, Japan.

Douglas, I., & Alemanne, N. D. (2007b). Monitoring student participation and effort. In *Proceedings of the IADIS International Conference on Cognition and Exploratory Learning in Digital Age* (CELDA 2007), Algarve, Portugal (pp. 299-302).

Dringus, L., & Ellis, T. (2005). Using data mining as a strategy for assessing asynchronous discussion forums. *Computer & Education Journal*, *45*, 141–160. doi:10.1016/j.compedu.2004.05.003

Fjermestad, J. (2004). An analysis of communication mode in group support systems research. *Decision Support Systems*, *37*(2), 239–263.

Hewitt, J. (2005). Toward an understanding of how threads die in asynchronous computer conferences. *Journal of the Learning Sciences*, *14*(4), 567–589. doi:10.1207/s15327809jls1404_4

Hewitt, J., & Teplovs, C. (1999). An analysis of growth patterns in computer conferencing threads. In C. Hoadley & J. Roschelle (Eds.), *Proceedings of the computer support for collaborative learning (CSCL) 1999 Conference,* Dec. 12-15, Stanford University, Palo Alto, California. Mahwah, NJ: Erlbaum.

Janssen, J., Erkens, G., Kanselaar, G., & Jaspers, J. (2007). Visualization of participation: Does it contribute to successful computer-supported collaborative learning? *Computers & Education*, *49*(4), 1037–1065. doi:10.1016/j.compedu.2006.01.004

Jeong, A. (2003). Sequential analysis of group interaction and critical thinking in online threaded discussions. *American Journal of Distance Education*, *17*(1), 25–43. doi:10.1207/S15389286AJDE1701_3

Jeong, A. (2004). The combined effects of response time and message content on growth patterns of discussion threads in computer-supported collaborative argumentation. *Journal of Distance Education*, *19*(1), 36–53.

Jyothi, S., McAvinia, C., & Keating, J. G. (2007). An interaction visualisation tool for a learning management system. In *Proceedings of the Conference of the Center For Advanced Studies on Collaborative Research,* Richmond Hill, Ontario, Canada, October 22 – 25 (pp. 326-331). New York: ACM.

Kakasevski, G., Mihajlov, M., Arsenovski, S., & Chungurski, S. (2008). Evaluating usability in Learning Management System Moodle. In *Proceedings of the ITI 2008 30th International Conference on Information Technology Interfaces*, June 23-26, Cavtat, Croatia (pp. 612-618).

Keller, J. M. (1983). Motivational design of instruction. In Reigeluth, C. M. (Ed.), *Instructional-design theories and models: an overview of their current status* (pp. 383–434). Hillsdale, NJ: Lawrence Erlbaum Associates.

Kienle, A., & Ritterskamp, C. (2007). Facilitating asynchronous discussions in learning communities: the impact of moderation strategies. *Behaviour & Information Technology*, *26*, 73–80. doi:10.1080/01449290600811594

Klosgen, W., & Zytkow, J. (2002). *Handbook of data mining and knowledge discovery*. New York: Oxford University Press.

Lin, H. (2007). The ethics of instructional technology: Issues and coping strategies experienced by professional technologists in design and training situations in higher education. *Educational Technology Research and Development*, *55*(5), 411–437. doi:10.1007/s11423-006-9029-y

Mason, R. (1994). *Using communication media in open and flexible learning*. London: Kogan Page.

Mazza, R., & Botturi, L. (2007). Monitoring an online course with the GISMO tool: a Case study. *Journal of Interactive Learning Research*, *18*(2), 251–265.

Mazza, R., & Dimitrova, V. (2005). Generation of graphical representations of student tracking data in Course Management Systems. In *9th IEEE International Conference on Information Visualisation*, London (pp. 253-258).

Michinov, N., & Primois, C. (2005). Improving productivity and creativity in on-line groups through social comparison process: new evidence for asynchronous electronic brainstorming. *Computers in Human Behavior*, *21*(1), 11–28. doi:10.1016/j.chb.2004.02.004

Romer, D. (1993). Do students go to class? Should they? *The Journal of Economic Perspectives*, *7*(3), 167–174.

Romero, C., Ventura, S., & Garcia, E. (2007). Data mining in course management systems: Moodle case study and tutorial. *Computers & Education*, *49*, 1–17.

Schrire, S. (2002). Interaction and cognition in asynchronous computer conferencing. *Instructional Science*, *32*, 475–502. doi:10.1007/s11251-004-2518-7

Schrire, S. (2006). Knowledge building in asynchronous discussion groups: going beyond quantitative analysis. *Computers & Education*, *46*(1), 49–70. doi:10.1016/j.compedu.2005.04.006

Taylor, J. C. (2002). Teaching and learning online: The workers, the lurkers and the shirkers. *Conference on Research in Distance & Adult Learning in Asia*. Retrieved May 14, 2009, from http://www.ouhk.edu.hk/CRIDAL/cridala2002/speeches/taylor.pdf

Vonderwell, S., & Zachariah, S. (2005). Factors that influence participation in online learning. *Journal of Research on Technology in Education*, *38*(2), 213–230.

Watson, W. R., & Watson, S. L. (2007). An argument for clarity: what are Learning Management Systems, what are they not, and what should they become? *TechTrends*, *51*(2), 28–34. doi:10.1007/s11528-007-0023-y

Webb, N. M. (1995). Group collaboration in assessment: multiple objectives, processes, and outcomes. *Educational Evaluation and Policy Analysis*, *17*(2), 239–261.

Weka. (2007). Retrieved from http://www.cs.waikato.ac.nz/ml/weka/

Yeaman, A. R. (2004). Professional ethics: The misuse of technology. *TechTrends, 48*(5), 16-18.

Zaïane, O. R., Xin, M., & Han, J. (1998). Discovering web access patterns and trends by applying OLAP and data mining technology on web logs. In *Proceedings of the Advances in Digital Libraries Conference* (April 22 - 24). Washington, DC: IEEE Computer Society.

## ENDNOTES

1. Image obtained with permission of the author from http://gismo.sourceforge.net/
2. Image obtained with permission of the author from http://mailer.fsu.edu/~ajeong
3. Reprinted from Janssen et al (2007) with permission from Elsevier.

# Chapter 11
# Open Synchronicity for Online Class Support

**Clark Shah-Nelson**
*SUNY Delhi, USA*

## ABSTRACT

*Instant messaging and text chat, online collaborative whiteboards, web conferencing and other synchronous Web 2.0 tools are increasingly finding their way into higher education and are available in both commercially-branded and open source varieties. This chapter describes best practices and challenges in using these new tools, focusing on free and open source software for synchronous course delivery, collaboration, learning activities, and technical support, based on the author's experience in online teaching and online-teaching support. Synchronous tools can provide immediate and efficient communication for instructors, learners and support staff, fostering community and establishing a heightened sense of social presence. An increasing number of practitioners in the field of distance learning are using synchronous tools to reach their learning and support objectives (Murphy and Rodríguez Manzanares, 2008). Today, free and open-source tools offer similar functionality as costly enterprise systems. This chapter will describe these new open source tools, the types of needs that drive their use, and strategies for effective use and implementation.*

## INTRODUCTION AND BACKGROUND

In the past, distance learning was by definition asynchronous, relying primarily on correspondence via mail (Prewitt, 1998). By the early 1990s, teaching methods in distance learning became more synchronous, with the use of videoconferencing via cable, microwave, and satellite and audio conferencing, predominantly via telephone. By the mid-to-late 1990s, however, asynchronous tools grew in popularity once again, with the rise of Learning Management Systems (LMS) and Course Management Systems (CMS). In recent years, thanks in large part to the increasing availability of high-speed broadband Internet access, new and highly effective synchronous tools with

DOI: 10.4018/978-1-61520-853-1.ch011

Copyright © 2010, IGI Global. Copying or distributing in print or electronic forms without written permission of IGI Global is prohibited.

great promise for distance learning have begun to emerge.

The trend in distance and online education is toward an increased use of synchronous tools for effective communication between teachers and learners. Many of these tools have been experiencing convergence over the past decade. First, instant messaging (IM), which unlike electronic mail is predominantly used synchronously for relatively brief messages, was coupled with audio and then video chatting capability. Now, in 2009, there are many possible software solutions, including open source and enterprise, which incorporate videoconferencing, webinars, and collaborative whiteboards into a single package.

## INTRODUCTION TO SYNCHRONOUS TOOLS

A decade and a half ago, when the author began working in the field of distance learning, email was, for him, quite new. Back then, instant messaging barely existed as we know it today. There were chat rooms, Bulletin Boards and Internet Relay Chat (IRC). Virtual worlds existed in the form of Multiple User Dungeons (MUDs) and Multi-User Dungeons Object Oriented (MOOs) (Sanchez, 2009). But the "instant messenger" as we know it had yet to be born. Videoconferencing, shared whiteboards, and webinars were still in their early infancy. Today, instant messaging has become ubiquitous. When (a room full of college students was) recently asked who had used an instant messenger, every student's hand was in the air. The same was true of text messaging. Visit the offices of SUNY Delhi's Campus Information Systems (CIS) on any given day and you will see instant messaging and web conferencing in use as people coordinate, get quick answers to pressing questions, support, and inform each other of their whereabouts. Clearly, instant messaging has become useful and necessary to people in a variety of contexts. Young people, students, office workers, grandparents, friends, you name it, all of these people have started using instant messaging because it is instant, convenient and easy.

## INSTANT MESSAGING AND CHAT FOR THE INSTRUCTOR

One of the primary reasons the author began using instant messaging with students while teaching online was to use a communication system students frequently used and knew to get immediate responses, and to know that the student had read the message. Whereas with email, one can send a message and wonder if the other person has received or read it, IM offers direct instant feedback that leaves little doubt as to whether the other person has read the communication. An online instructor can use IM to catch up with students who have not recently completed work, or to give a specific student immediate feedback about grades, questions, performance, etc.

Using the tools provided within the LMS was not enough to effectively communicate with students. Announcements and email were often ignored, but when students could be contacted via IM, specific questions could be asked to guide the students and get direct answers immediately. While there was a "chat" tool within the LMS, it never seemed to be used because one would have to launch it separately (without knowing if anyone else was in the chat room or available), and the chat room would quit automatically if inactive for a certain period of time. It turned out that adopting a certain IM system and informing students of an IM name, as well as getting their IM names to add them as "buddies" was most effective.

Students would often use IM to strike up a conversation, to check and see if an assignment had been graded, ask a question on their homework, get help with an activity or get technical help. If a student was working in the course and had a question, it was extremely easy to make contact via IM, as the student could see availability and initiate contact. However, many high schools at that time did not allow students to use instant messengers, and actually may have blocked certain ports on the network from being used. The students could still use IM to make contact when they were at home or somewhere else. Additionally, online students may lack an equivalent to hallways or traditional spaces for socializing on campus (Hrastinski, 2006) and IM can make up for this, adding a "place" for social interactions with other students.

One of the aspects of online education that can be frustrating for students and instructors alike is the sense of isolation. Another frustrating aspect can be the amount of time asynchronicity can take, when participants are pressed for time. The feeling of a live connection between instructor and students, of the instructor "being there" – available and able to help students when they need it, is vital. Murphy and Rodriguez Manzanares (2008) and Münzer (2003) both found that IM increased the "social presence" in elearning. However, Hrastinski (2005) found that students who adopt the use of an IM system had a higher degree of participation than those who did not, and that the system was primarily used for collaboration and course information, rather than socialization or emotional support.

## FOR HELPDESK: USER SUPPORT

Technical help services such as a helpdesk or an IT department are an excellent venue for the use of IM tools in an online education environment, and this benefits the end users. Think of it this way – if you were ordering something online and had some questions, would you rather leave the computer, find the phone (which might be a search in and of itself), pick up the phone, call, and then maneuver through several layers of menus on the phone before finally getting to speak to a human many minutes later? Or would you prefer to click on "live chat now" and get a human being on the other end almost immediately? A recent visit to the classroom showed that the vast majority of students would rather strike up a text chat with a live person rather than wade through a set of phone prompts. Moore (2003) asserted that learner support in distance education is "one of the most critical elements in determining the success of a distance education program" (p. 141), and that "highest priority should be given to informing every student that an easily contactable learner support specialist is available" (p. 143). So for the end user, IM offers a fast simple way to get help, and an alternative to using the phone or email.

There are distinct benefits of using IM for the IT technician as well. Since what the end user tells the IT person is most likely going to end up in text anyway (to enter contact information, nature of issue, platform, system, etc. into a ticket tracking system); it ends up being more accurate to have this information typed out rather than spelled out over the phone. As a technician on the other end, the possibility of misspellings and miscommunications is far greater over the phone than via IM, where either person can catch any misspellings or corrections immediately. This is also a good use case for giving information back to the student or faculty member, such as links to more information online. This is most easily done in an instant messenger, since a link can be pasted in and the other person can simply click on the link to go to the resource. Furthermore, if it becomes necessary to do some sort of screen sharing, so that the technician can see what is happening on the student's computer screen or vice versa, having an IM session going with the student will make this much easier to provide a URL and/or instructions for connecting to such a system.

In order to support our online teachers and learners, the author has embedded an IM widget into our open source LMS, Moodle, on the system front page and in every course in the system, so that users can seek help by simply typing in a box and connecting with a technician. The widget consists of some simple JavaScript code generated from Wimzi.com – which connects an AOL Instant Messenger (AIM) account to the widget. One or more technicians at various times/days who know the login can then share the AIM account. The advantage of using such a widget is that end users do not necessarily need to even have an IM account, or launch any other application – it's built right into the web browser, and they can just enter a question or comment to get started. Other departments within the university are using this same tool and technique, including the library and career services department.

In the SUNY Delhi CIS office, IM is used on a daily basis for multiple purposes:

- get a quick answer to a question
- see if the person is in his/her office to drop by (perhaps send a quick – "Are you around? Can I drop by?" note)
- remind each other of meetings
- alert each other of current issues
- communicate with each other about incoming phone calls prior to forwarding
- send files to one another (easier than email attachments)
- inform others of whereabouts and/or availability
- instigate a help ticket

There are many other reasons why one might use IM instead of email. Whereas email lends itself better to more formal, official, lengthy communications that may need to be accessed again in the future, or which are carbon copied to others, IM is a better bet for quick back-and-forth communications and coordination. If a typical IM conversation took place via email, one would end up with up to 20 messages or more clogging up the inbox. When sending files via IM, one can just drag a file or set of files onto a user name to send them – thus avoiding the extra clicks and hassle of attachments.

## CHOOSING THE RIGHT IM PROTOCOL/SYSTEM

One of the most important factors of using IM is choosing which system or protocol to use so that all participants can easily reach each other. Since not all systems play together well, it is important to choose one and stick to it. In our office, we've chosen AIM as the common protocol to use, which simplifies the process of connecting to other users. AIM is also compatible with other applications such as iChat on the Mac, and GoogleTalk. Up until recently, AIM users could not easily IM with Yahoo users, MSN users, etc. As of 2009, this is changing somewhat, as some of these systems open up to communicate with others out of competition. The open source Jabber/Extensible Messaging and Presence Protocol (XMPP), used by GoogleTalk, further opened up the field as users could create transports or gateways so that they could chat with Yahoo, MSN, and AIM users.

Originally, most instant messaging was done with small programs downloaded from providers, such as AIM, Yahoo Instant Messenger, etc. Soon, this technology incorporated web browser-based IM, so that users would not be required to download separate software. Next, multi-system clients appeared on the scene that could handle IM sessions for many various IM systems simultaneously. They connect users via Yahoo, MSN, Jabber, (GoogleTalk), AIM, and ICQ, among other protocols. Some of these are downloadable client applications and some are web-based ones. Popular multi-system clients include Pidgin, Trillian, and Adium, for downloadable clients, and Meebo.com for web-based. Of course, the only way IM can really work well is if users on both ends keep the application open, so that others can see if that person is online or not and start up a conversation. Naturally, an instructor might want to only do so during particular hours (online office hours).

One hurdle some instructors face if the institution does not have a standard IM system of its own is that students may be spread out among various

IM providers – Yahoo, MSN, AIM, Jabber, etc. In such a case, one of the recommended strategies for an instructor is to encourage (or require) all students to get an account on a system such as Gmail /GoogleTalk or Meebo.com, which is compatible with many of the other systems. The instructor can also embed a widget into the course so that students can see if the instructor is online and make immediate contact. (This does not necessarily help the instructor who needs to get into contact with the student, however!)

## LOCATION, LOCATION, LOCATION

Looking at popular Learning Management Systems, most of them today have started incorporating some sort of instant messaging system so that users can chat without having to download separate software. This can be quite handy for users (the fewer programs online instructors and students have to download, install, and support, the better!) However, one potential issue is that some systems deal with the implementation of IM better than others. One popular LMS, Angel, for example, has an instant messaging system, but users have to specifically go to that link and start it up, rather than just have it working as users enter the course. This may limit use of the tool due to the extra effort users must put in to access it. In Moodle, a list of "online users" shows up on the page automatically in the course if the instructor chooses, and users can then see who is online and strike up a conversation. This particular system has a bit of lag between sending and receiving messages (unlike most IM client sessions). Nevertheless, this still imparts a sense that "somebody is here," breaking participants' isolation, and as such is a step in the right direction.

Since the built-in IM capabilities of many LMS's may not quite be up to par with other IM programs, some instructors or support staffers may want to utilize widgets to bring IM into an LMS environment. As mentioned earlier, one can set up a widget using Wimzi.com – which is part of AIM. Meebo.com and Plugoo.com are other sites with similar functionality. After pasting in the Wimzi widget into an HTML block in a course, a participant can see if the instructor is online or not, put their name into a blank, and send the instructor an instant message immediately. (Figure 1) This means we do not have to worry about whether or not we have "friended" each other (added each other as contacts in that system). This is a highly effective way to offer users technical support, or students a direct link to the instructor during office hours. While not a good way to connect a whole bunch of users all together, or to see if a set of users in a course is online, it is useful for several people to find out if one person is online and communicate with that person. Technical and course support are the best uses.

Another location for IM that has become more commonplace in the last several years is in email systems. Google's Gmail, for example, has built-in chat right in the same browser window as the email, and does not require other software downloads. If course members are set up with Gmail accounts

*Figure 1. IM widget embedded in Moodle for live student support. (Adapted from Alejandra M. Pickett's "Intro to Online Teaching" course, State University at Albany, NY. Used with permission.)*

at the beginning of a course, all could quite easily communicate using IM. Also, most IM systems such as GoogleTalk, AIM, iChat, and others are now able to support sending and receiving Short Message Service (SMS) messages, which makes this all interoperable with cellular telephones.

## CHAT

The author will differentiate "chat" from "IM" by defining chat as "multi-person" vs. IM as "one person" communication.

What are some of the ways teachers and students are using live chat in the online classroom?

- Live discussions (more lively and engaging than threaded discussions, but also tougher to moderate)
- Practice (foreign languages, debates, etc.)
- Homework help and immediate feedback

For most effective use, chats need to be curriculum and objective-driven and assessed with pre-scheduled times/dates defined in the syllabus. Alternatively, if there's an available chat area for students to wander in at any time, the system should show immediately upon entry to the course homepage that someone is in the chat room so that users can know to open it. Without scheduled sessions or overt information on other users in the chat room, students often wander in to chat rooms, say "Hi" or the like, and then wander out when they find no-one else is there. If there is no way to tell if other people are in the chat room before entering, except by clicking to open the chat, waiting for it to load, and then seeing if anyone else is there, the chat fails as a tool for spontaneous communication. Finally, if chats are scheduled, it is probably most effective to make them a core part of the defined curriculum; driven by a learning objective and they should be assessed with a rubric to define levels of participation. Attendance at scheduled live chats that are not incorporated into the course often have poor attendance.

Many LMS's, including Blackboard, Angel, Moodle, eCollege, and Desire2Learn, have built-in chat systems so that instructors can provide live areas for chat, although some suffer from limited functionality, such as the absence of a whiteboard or specialized tools for math, or they impede spontaneous ad hoc chats. As for other features, these do vary widely, but some feature transcripts of any past chat, whiteboards, tools for math equations, audio, video, or any/all of the above. It is possible to use third party software or web sites such as Meebo.com for establishing chat rooms in a course. Meebo offers a way to embed a chat room, just like it does for IM. And most IM systems will allow one to add other users to a conversation, start up chats, and ultimately do the job of both chat and IM.

## WEB CONFERENCING, VIDEO CONFERENCING, WHITEBOARDING, AND SCREEN SHARING

In the 1990s many schools invested in the infrastructure for synchronous audio/video-based distance learning (Baggaley, 2008). This often included classrooms equipped with cameras, VCRs, LaserDisc players, computers, monitors, and microphones that were connected to other similar classrooms via cable, satellite, microwave, Ethernet, and/or, the bandwidth mothership, fiber optics. Setting up such an environment could be quite a costly endeavor, particularly if the school did not have its own network and had to pay for the bandwidth needed to send and receive all those packets of video and audio data over the network.

These days, however, all of this in some form or another can be accomplished using relatively inexpensive software and using typical broadband such as digital subscriber line (DSL) or a cable modem offered by most telecom companies. Web browser-based and Java-based software that is easily downloadable and capable of facilitating various uses for audio, video, screen sharing, whiteboarding, webinars, office hours, live support, etc. are now available for as little as nothing and as much as hundreds of thousands of dollars. Some universities have enterprise contracts for software such as Wimba, Elluminate, WebEx, Adobe Acrobat Connect, among others. This, we could call the "high level of commitment and budget" group.

For institutions that do not use these systems, free, open source, and lower cost options can achieve some of the same functionality.

## DIMDIM CONFERENCING SOFTWARE

DimDim is an open source conferencing tool that has been making giant strides in this realm in the last year or so, and has built an integration pack for Moodle. One can sign up for a free DimDim account and use their servers or download the source code and host an open source version. DimDim features audio, video, screen sharing, whiteboarding, recording, conference call bridging, text chat, and is Flash-based so it is extremely easy to get started using. Of the various tools the author has used, it is one of the easiest and most full-featured.

The first step in getting to know DimDim is to visit the web site and sign up for an account. There, users can take a video tour of the various aspects of hosting a DimDim meeting and taking advantage of this software. Once logged in, the primary choices are: "Host Meeting" or "Join Meeting."

After clicking on "Host Meeting" one must set a number of options for the setup of the web meeting, such as the room name, when to start/schedule the meeting, and what types of functionality should be included in the meeting. Starting on the "General" tab, one can either "Start Now" or "Schedule" the meeting for a specified time in the future. By clicking on "Options," another important feature is the ability to email participants an agenda, as well as an invitation with easy link to get into the meeting. Also, if the meeting is intended to be private, a "Meeting Key" or password that attendees are required to type in to enter the meeting can be set that prevents viewers who do not have the correct key from viewing. (Figure 2)

Next, using the "Features" and "Dial-In" tabs, users can choose whether to include video, audio, a telephone conference call number and chat. Webinars, or seminars that take place entirely online have become quite popular and can easily take place using DimDim. Generally, webinars involve a number of participants calling into a conference call line and/or listening to audio on the computer while watching a presentation, Importantly, when setting up a webinar in DimDim, where the other participants do not need audio (and where, in fact, the audio from other participants can be a distraction if enabled), the "Microphone" and "Auto Handsfree" settings should be set to "Off" as in Figure 3. It is also very important to decide whether or not to include the dial-in number. If you do not plan to have people call into the conference number, it's better to keep that set to "Off" so that people are not confused and trying to dial the phone number. Once you have selected all the settings for the meeting, simply click "Start" to begin the meeting.

From a meeting hosting standpoint, DimDim is extremely easy and quick to set up meetings, whether they are pre-scheduled or impromptu. DimDim's use of a "room name" makes it simple, with one permanent URL to send users to your DimDim room. When in a meeting, clicking on the room name at the top will copy the URL into the clipboard for easy pasting into an email or instant message.

From the participant standpoint, DimDim is also extremely easy, as once the link is clicked, the browser goes right to a browser check page that makes sure the user has both an acceptable browser for DimDim and version of Adobe Flash

*Figure 2. DimDim.com meeting scheduler*

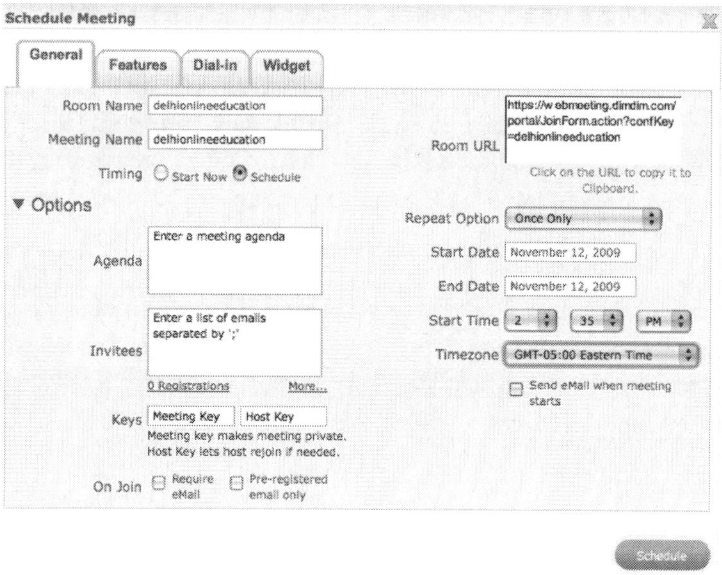


*Figure 3. DimDim.com features tab*

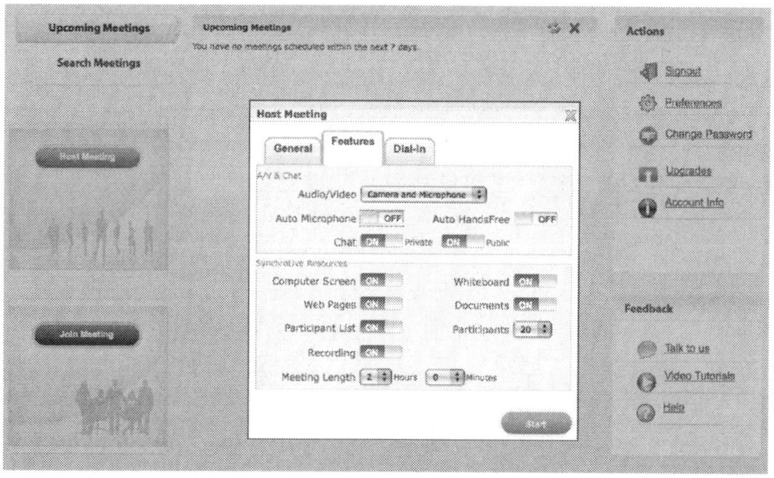


player/plugin. If those check out, the user is taken right into the web meeting without further delay. Other systems such as Elluminate frequently require Java and other downloads that delay the access into the room.

Once in the meeting, the host can assign a certain number of microphones and cameras to users to speak using the Voice Over IP (VOIP) protocol. This is perhaps one of the most annoying aspects – rather than giving everyone a microphone and ability to speak, or rotating based on people "raising their hands" to speak, the presenter must manage who has the microphones and/or video cameras. (For those who would like a smoother system for meetings with multiple people, the FM Live Communication option described later in this chapter is recommended.) The presenter can also assign any other user to be the presenter. As for audio, a great feature is that users have the ability to change the audio level of other participants next to their names, in case someone's microphone is coming through too loudly or not loudly enough. So each user has some control over how loudly/softly the others come through.

Beyond the audio/video capability, there is a chat area on the right side of the screen, and users may have to open that chat manually, particularly if the presenter is showing a document or web page or switches between the various "inputs" that can be shared. These inputs include: documents such as PowerPoint presentations (PPT, PPTX) and PDF files; web pages which are entered by URL, shared desktop screens, and a whiteboard, upon which participants can scribble, draw, and flesh out ideas.

DimDim offers other functionality in its preferences, such as the ability to set up the features you want to use as defaults for each meeting you start to host, whether or not to use a "Lobby" (so that the host has to let people in as they arrive, otherwise, they must wait in the "lobby"), widgets that uses can click on to go immediately to a room, and personalization such as meeting titles, custom watermark images, collaboration space URL, and return URL, to which users are forwarded at the end of the meeting.

In the author's experience, DimDim seems best suited for one-way audio webinars, especially if they need to be recorded, as well as for end user support when screen sharing is needed. When using DimDim's VOIP system for audio, it is highly recommended that all participants use headsets for both listening and speaking. When using speakers instead, there is a great likelihood of the existence of echo and other audio annoyances.

## FM LIVE COMMUNICATION

Open University's FM Live Communication is a free Flash-based fully web-based videoconferencing system. One can sign up for a free account on the Open University OpenLearn or LabSpace Moodle servers (http://labspace.open.ac.uk/course/view.php?id=2951) and then book meetings using their excellent and elegant system. Again, it's Flash-based and very easy to start and use. It does not offer screen sharing, however, so it is better suited for meetings than for helpdesk support uses. For distant colleagues collaborating on projects, FM Live is an invaluable communication tool.

After registering for an account, confirming the email address, and logging in, one can visit the FM course at the address above, and click on "FM Live Communication" in the "Learning Tools" block in the upper left (Figure 4), and then click on "Book a Meeting."

The next stop is the FM OpenLearn server, where the date and time of the meeting are chosen. By default, the current date and time are selected, and the time is always given in the server's cur-

*Figure 4. The FM live communication link is in the learning tools block to the left*

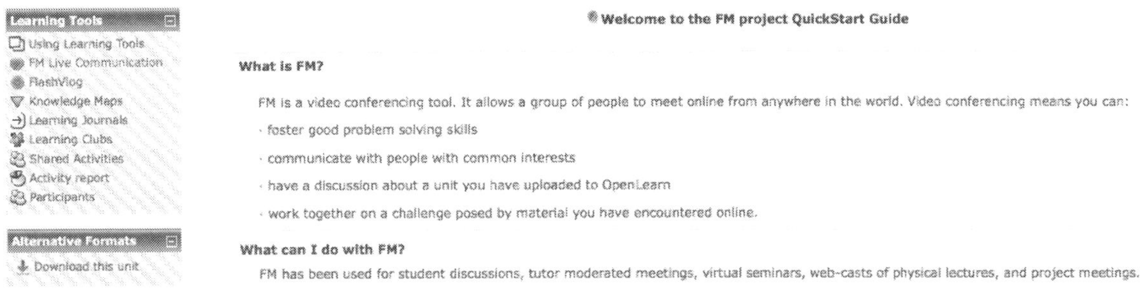

rent time, which is Greenwich Mean Time +0100 (Figure 5).

The system will then email a meeting invite, and forward to the My Events page, which shows all past and present meetings. This page will have a URL link to the meeting, which is what all attendees will need to join (Figure 6). That link can be emailed or posted in a discussion forum. Once the meeting has begun, users can click the URL to join the meeting. If guests are allowed when setting up the meeting, other users will be able to join without setting up an account in advance. Otherwise, users will need to set up accounts and log in to join the meeting. Upon entering, Flash will ask permission to use the microphone and/or web camera, and users can enter their name and click "ENTER" to join.

In the meeting space, users can see either a set of auto-updating still images of the other participants, or a list of names in the upper right. Below that frame is a set of tabs for various functions: prefs, xtra, vote, urls, chat, and FM. Perhaps the most

*Figure 5. The Flashmeeting booking screen is shown*

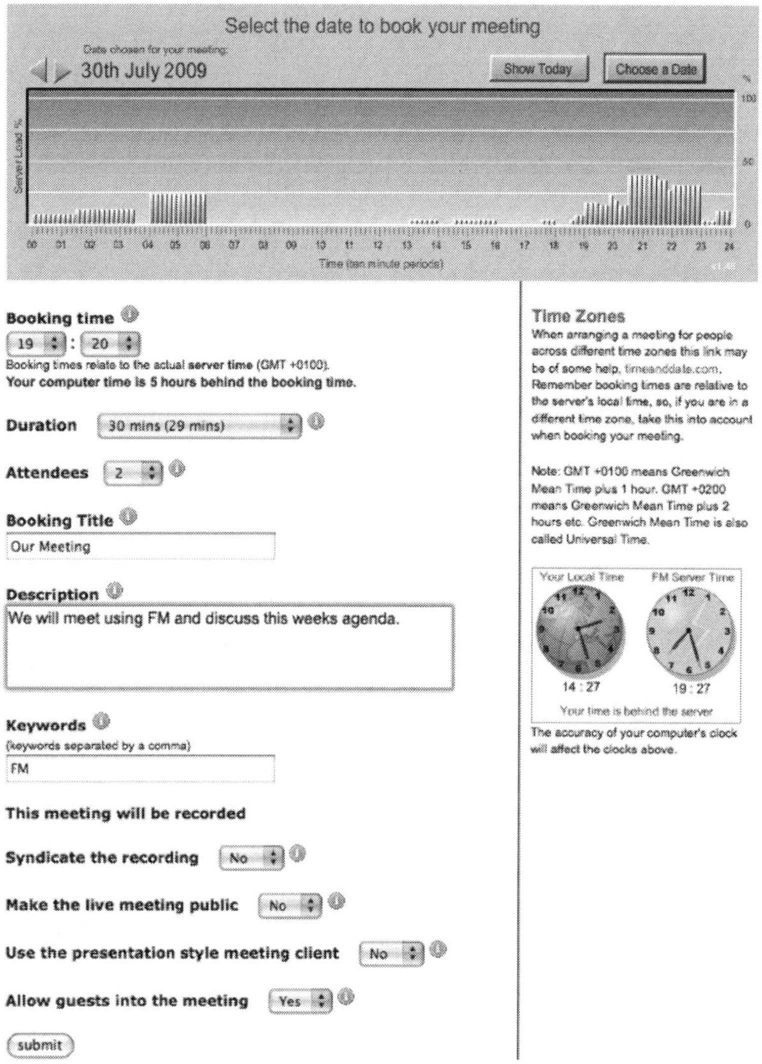

*Figure 6. The My Events page showing the link to the upcoming meeting and all past meetings*

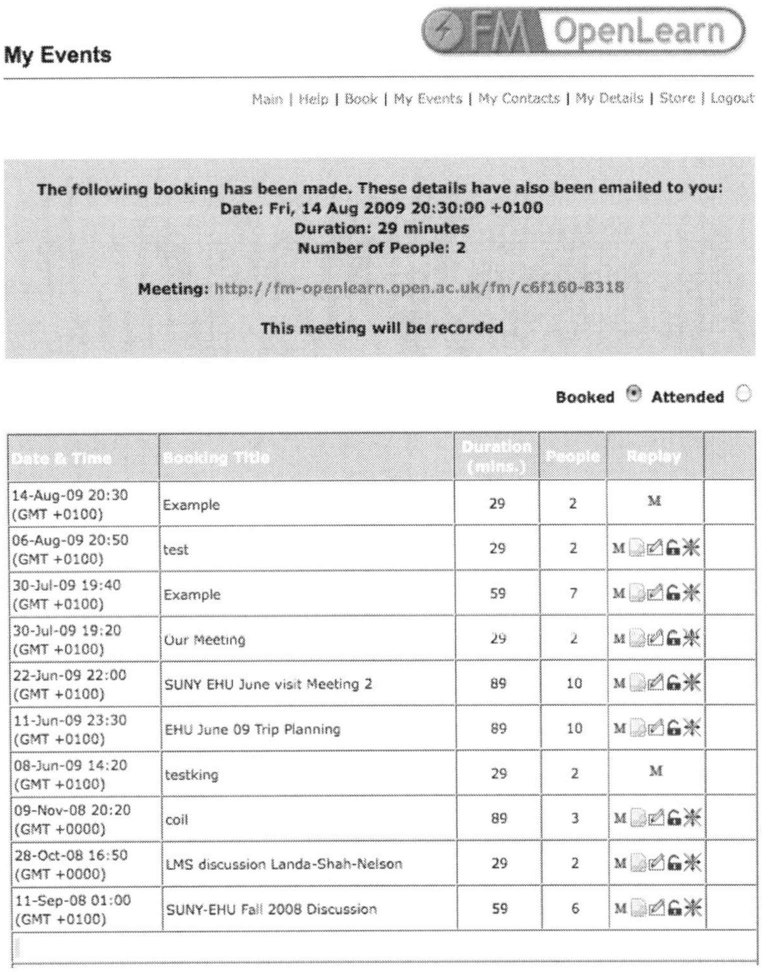
**My Events**

FM OpenLearn

Main | Help | Book | My Events | My Contacts | My Details | Store | Logout

**The following booking has been made. These details have also been emailed to you:**
**Date: Fri, 14 Aug 2009 20:30:00 +0100**
**Duration: 29 minutes**
**Number of People: 2**

**Meeting:** http://fm-openlearn.open.ac.uk/fm/c6f160-8318

**This meeting will be recorded**

**Booked** ◉ **Attended** ○

| Date & Time | Booking Title | Duration (mins.) | People | Replay | |
|---|---|---|---|---|---|
| 14-Aug-09 20:30 (GMT +0100) | Example | 29 | 2 | M | |
| 06-Aug-09 20:50 (GMT +0100) | test | 29 | 2 | M | |
| 30-Jul-09 19:40 (GMT +0100) | Example | 59 | 7 | M | |
| 30-Jul-09 19:20 (GMT +0100) | Our Meeting | 29 | 2 | M | |
| 22-Jun-09 22:00 (GMT +0100) | SUNY EHU June visit Meeting 2 | 89 | 10 | M | |
| 11-Jun-09 23:30 (GMT +0100) | EHU June 09 Trip Planning | 89 | 10 | M | |
| 08-Jun-09 14:20 (GMT +0100) | testking | 29 | 2 | M | |
| 09-Nov-08 20:20 (GMT +0000) | coil | 89 | 3 | M | |
| 28-Oct-08 16:50 (GMT +0000) | LMS discussion Landa-Shah-Nelson | 29 | 2 | M | |
| 11-Sep-08 01:00 (GMT +0100) | SUNY-EHU Fall 2008 Discussion | 59 | 6 | M | |

important of these are chat – where text-based chat happens, prefs – where audio levels, inputs and camera inputs can be changed, and xtra – where files are uploaded and shared and a whiteboard can be added to the meeting if need be (Figure 7).

FM regulates who is "broadcasting" by keeping it to a stream of one person at a time. This helps prevent people from talking over one another, and keeps things quite orderly in the meeting. Once someone has clicked "Start Broadcasting" their audio/video will be broadcast to the others. If another person wants to speak, he or she can click on "Join Queue" and FM will keep track of who is next. Then, when the current speaker clicks "Stop Broadcasting," the next person in the queue will pop right up and broadcast. For those who don't want to wait, there is a handy and appropriately named "Interrupt" icon. In this way, FM creates a seamless flow between attendees, and they can chat with each other in text in the chat box in the lower left, or share URLs when broadcasting, as well as use the 'vote' tab to show support or lack thereof for questions at hand.

Following the meeting, (which ends abruptly at the allotted time with no way to extend it), participants can visit the same URL to view the recorded version of the meeting. This is a great feature for reviewing what happened, taking secretarial min-

*Figure 7. The view from within a Flashmeeting*

*Figure 8. The recorded Flashmeeting, which can be viewed after the meeting has finished*

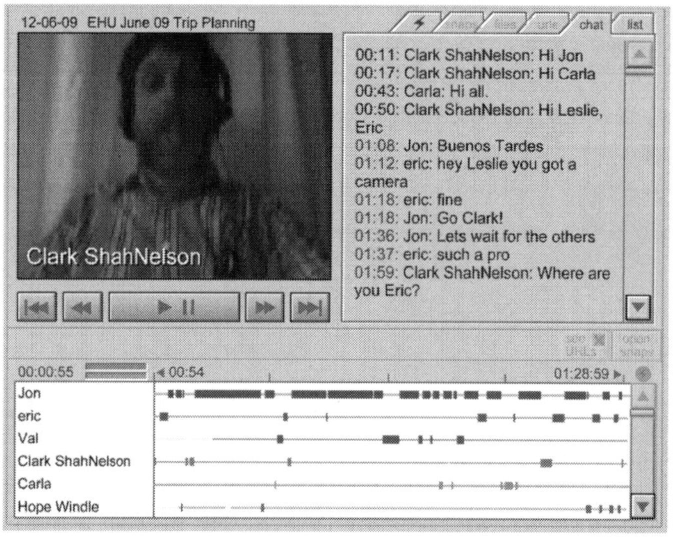


utes or notes, or for those who missed the meeting or a portion thereof. The recording features a color-coded timeline that shows who spoke, when, and one can easily skip forward through the meeting to certain points using the forward arrows. The chat is also recorded and accessible in the chat tab at the top, as are URLs, files, and snapshots from the meeting (Figure 8). The archived meetings seem to be available on the FM server indefinitely.

What FM does not offer is the ability to share desktops or display a presentation as DimDim does, moving along with the presenter. It also does not offer any kind of telephone bridge for users to call in via a standard phone. What it does offer is an easy way for groups to meet online in an orderly elegant system.

## WHY AND WHEN CONFERENCING TOOLS BECOME IMPORTANT

The author started using these tools after experiencing hours of discomfort while attempting to describe to users how to do certain tasks in a software application, or trying to fix problems that they can not very accurately describe in words. The ability to show one's computer screen to students and faculty was needed, and for them to show the support technician their screen while helping them with issues, questions, and training. An easy link or widget that users could click on to join in a screen sharing session was also needed, as well as a free/cheap way to connect with remote users for online meetings, discussions, and collaborative work, and a way to share on-campus training and workshops with remotely working faculty. Such synchronous tools have proved highly effective for supporting users, and have rapidly become essential for a variety of tasks.

These types of software can also help instructors effectively meet their educational needs and objectives. For example, if the course objectives include some sort of activity that requires remote/distant learners to collaborate, either in small groups or as a class, for subjects that rely heavily on live manipulation of visual items (such as math), or for office hours that offer the most complete set of tools that can be used to interact with students. An instructor might find, for example, that having live chat or instant messaging office hours but neither voice/audio nor the ability to share screens is too limiting. To effectively offer office hours, students should be able to contact and work with the instructor as easily as possible – by IM, phone, and/or using one of the tools listed above, such as DimDim – which offers many multiple ways for users to communicate with one another. Bennett and Lockyer (2004) found that students might be able to take advantage of just-in-time support since teachers can be more easily reachable and available. (Bennett & Lockyer, 2004) An instructor could set up a DimDim office, for example, and simply place a link in email or in the course to the "virtual office." Students can then click on the link, type in their name, join the instructor, and then choose how to communicate – from text chat, to VOIP audio, to video, to telephone or any combination of those. The instructor also benefits from being able to demonstrate ideas and concepts in multiple ways to the student – using a whiteboard, using files on the computer, web pages, or sharing what is on his or her screen.

Similarly, students can set up their own accounts and/or "rooms" to work collaboratively on projects, have discussion groups, and study groups. Additionally, researchers found that the ability to have chat communication between instructor and students was desirable among students in elearning (Atan, Rahman, & Idrus, 2004).

To get started using the above listed tools, the first step is to simply go to the web site and sign up for a free account. All of these sites offer excellent resources for getting started including videos, FAQs, knowledge bases, and in most cases, user forums where one can post questions and interact with both other users and with technical support staff for the product. These are important, because most users may not receive support from their institutions' IT departments, unless there are issues related to bandwidth or ports. Rather, many will need to use the support system for each product itself for technical issues. This may require a slightly new skillset among users that are used to just calling the Helpdesk and getting assistance. Users may need to get used to:

- setting up accounts in the user forums for the product
- searching for relevant topics in the user forums
- posting questions and interacting with other users (who may or may not be very friendly)
- perusing FAQ documents
- finding the appropriate channels to report issues directly to the software company

Effectively incorporating these tools into courses can be accomplished in several ways. As was mentioned earlier, to set up a DimDim meeting, for example, an instructor could send an email invitation to students; insert a link into the course; embed a little widget into the course that students click on at the appropriate time to join; or all of the above. In general, the author would encourage users to do all of the above.

## ESTABLISHING GROUND RULES SURROUNDING THE USE OF SYNCHRONOUS TOOLS IN THE COURSE

Along with choosing the tool itself, whether it is IM, FM, Chat or DimDim, instructors should offer students a variety of other kinds of information to support effective use of the tools. These include documentation on how to get started; lists of required hardware, such as web cams, headsets, etc., along with links to recommended items; links to any download and install any required software (Flash, Java, Yugma, etc.); and links to any browser tests or other pre-entry sites. Instructors should also consider providing a a description of rules, regulations and expected netiquette surrounding communications between students and instructor (and consequences if these are breached) Hines and Pearl (2004) compiled an excellent example of "chat protocol" for class chats that includes using an exclamation point (!) for students to signal that they have a question, and having students practice the chat protocol while asking general questions about the course (Hines & Pearl, 2004).

Rubrics establishing the assessment of student work, attendance and participation requirements (including info for students who view the archived session if that is permitted for credit) are also recommended, as well as initial "tech sessions" for users to test the system in a low stress situation and get used to using the common tools within the system. If there are any differences in expectations for attendance, participation, post-participation, etc. they should be spelled out in the syllabus or activity information page. It is a great idea to have a second person with the instructor during the first several sessions, so that the second person can be available to assist students who are experiencing any technical issues. This way, the instructor can carry on with the other students and not lose valuable time dealing with technical issues of a couple people. The "assistant" should be available via IM, phone and/or Skype or the like and be knowledgeable about the system and getting students set up with software, microphones, etc.

Finally, instructors should provide documentation on what to expect afterwards, including the timing and availability of archive links.

Immediately after the sessions, instructors can improve effective use by getting some quick feedback from the students using polls or surveys, with questions such as:

- What was most beneficial about this session?
- What was lacking?
- What would you do differently in the future?
- What helped and what hindered your learning process?

## FUTURE RESEARCH DIRECTIONS

This chapter includes personal experiences and some anecdotal evidence of using a variety of synchronous tools to facilitate online education support, and it is only a jumping off point for a more in depth study of end users' experiences, preferences, attitudes, and overall satisfaction. As these tools become more frequently used, and as new synchronous tools are created such as Google Wave, Google Voice, Etherpad.com and so on, the need for more documented use cases and data concerning their use arises. While there has been research indicating that more immediacy of feedback, ease of communication, and increased interaction can be benefits of using various synchronous tools, questions remain:

Do course participants with synchronous tools in use:

- feel well supported?
- drop out of courses less?
- receive support in a shorter timeframe?
- have an increased sense of social, cognitive or teacher presence?
- report increased overall satisfaction with their eLearning experience?

## CONCLUSION

Synchronous communication tools can help instructors and students, and the support technicians who assist them communicate more efficiently and effectively. Easier than picking up a phone, more precise than typing what is heard, IM and chats are fulfilling a need for users to get more immediate, direct feedback. The synchronous tools that are found in most of the LMS's today are less efficient and less functionally satisfactory than specialty third-party tools that can be used and embedded into the LMS, such as AIM, GoogleTalk, Meebo, Wimzi, etc.

Whereas videoconferencing was traditionally a costly endeavor, both in terms of equipment and bandwidth/technology, Web 2.0 software tools today can do much of the same functionality at a fraction of the cost. These same tools offer instructors the ability to have webinars, live online classes and office hours, and offer support personnel the ability to help users by means of screen sharing.

Implementing and using synchronous tools, and in particular free/open source tools, can greatly increase the effectiveness of online interactions. The synchronous tools available today make it possible for instructors, students and all types of office workers to connect, collaborate, receive training, and effectively work, even if separated by great physical distance. Thus, these tools function to support and foster learning communities.

## REFERENCES

Atan, H., Rahman, Z., & Idrus, R. (2004, June). Characteristics of the Web-Based Learning Environment in Distance Education: Students' Perceptions of Their Learning Needs. *Educational Media International, 41*(2), 103-110. Retrieved May 13, 2009, from doi:10.1080/09523980410001678557

Baggaley, J. (2008, May). Where did distance education go wrong? *Distance Education, 29*(1), 39-51. Retrieved July 30, 2009, from doi:10.1080/01587910802004837

Hines, R., & Pearl, C. (2004, Spring). Increasing Interaction in Web-based Instruction: Using Synchronous Chats and Asynchronous Discussions. [from Academic Search Premier database.]. *Rural Special Education Quarterly, 23*(2), 33–36. Retrieved May 13, 2009.

Hrastinski, S. (2005). *Introducing a synchronous medium in a computer-mediated distance learning course: Towards understanding how student participation is affected* [PDF Document]. Retrieved from http://www.ascilite.org.au/conferences/brisbane05/confblogs/index.php?blog=143

Hrastinski, S. (2006, August). The relationship between adopting a synchronous medium and participation in online group work: An explorative study. *Interactive Learning Environments, 14*(2), 137-152. Retrieved May 13, 2009, from doi:10.1080/10494820600800240

Moore, M. (2003, September). Learner Support. [from Academic Search Premier database.]. *American Journal of Distance Education*, 17(3), 141–143. Retrieved May 13, 2009. doi:10.1207/S15389286AJDE1703_1

Murphy, E., & Manzanares, M. (2008, March). Instant Messaging in a context of virtual schooling: balancing the affordances and challenges. *Educational Media International, 45*(1), 47-58. Retrieved May 29, 2009, from doi:10.1080/09523980701847180

Prewitt, T. (1998, July). The Development of Distance Learning Delivery Systems. [from Academic Search Premier database.]. *Higher Education in Europe, 23*(2), 187. Retrieved July 23, 2009. doi:10.1080/0379772980230205

Sanchez, J. (2009, February). A Social History of Virtual Worlds. [from Academic Search Premier database.]. *Library Technology Reports, 45*(2), 9–12. Retrieved July 23, 2009.


## ADDITIONAL READING


Bennett, S., & Lockyer, L. (2004, September). Becoming an Online Teacher: Adapting to a Changed Environment for Teaching and Learning in Higher Education. *Educational Media International, 41*(3), 231-244. Retrieved May 13, 2009, from doi:10.1080/0952398041000168084 2

Burnett, C. (2003, April). Learning to Chat: tutor participation in synchronous online chat. [from Academic Search Premier database.]. *Teaching in Higher Education, 8*(2), 247. Retrieved May 13, 2009. doi:10.1080/1356251032000052474

Im, Y., & Lee, O. (2003, Winter). Pedagogical Implications of Online Discussion for Preservice Teacher Training. [from Academic Search Premier database.]. *Journal of Research on Technology in Education, 36*(2), 155–170. Retrieved May 13, 2009.

Johnson, G. (2008, January). The relative learning benefits of synchronous and asynchronous text-based discussion. *British Journal of Educational Technology, 39*(1), 166-169. Retrieved May 13, 2009, from doi:10.1111/j.1467-8535.2007.00739.x

Lee, Y., & Sun, Y. (2009, Spring). Using instant messaging to enhance the interpersonal relationships of Taiwanese adolescents: evidence from quantile regression analysis. [from Academic Search Premier database.]. *Adolescence, 44*(173), 199–208. Retrieved May 13, 2009.

MacDonald, L., & Caverly, D. (2000, Spring). Techtalk: Synchronous distance developmental education. [from Academic Search Premier database.]. *Journal of Developmental Education, 23*(3), 38. Retrieved May 13, 2009.

Munzer, S. (2003, July). An Evaluation of Synchronous Co-operative Distance Learning in the Field: The Importance of Instructional Design. [from Academic Search Premier database.]. *Educational Media International, 40*(1/2), 91. Retrieved May 13, 2009.

Tuovinen, J. (2000, March). Multimedia Distance Education Interactions. *Educational Media International, 37*(1), 16-24. Retrieved May 13, 2009, from doi:10.1080/095239800361473


## KEY TERMS AND DEFINITIONS

**Asynchronous:** Not at the same time. Learning activities are generally denoted as being either synchronous or asynchronous, depending on whether the participants are required to be in a certain place (virtual or not) at a certain set time.

**Chat:** A system for people to communicate with one another using text (as well as other modalities) in a group setting.

**Instant Messaging (IM):** Instant messaging is a system for people to send instant text communications to one another via the computer and a network.

**Learning Management System (LMS):** A computer-based system which functions as an online classroom, complete with grade book, discussion forums, a way to share files and web pages with students, and any number of other functions depending on the software and instructor.

**Screen Sharing:** Sharing one's computer desktop with someone else over a network. Using screen sharing one can show another person what is on the computer screen even though the other person is in another location.

**Synchronous:** At the same time, but not necessarily the same place.

**Videoconferencing:** Conferencing that involves the sending of video via a network.

**Virtual/Personal Learning Environment (VLE/PLE):** New memes to replace LMS, with the emphasis on "environment" – and less on management.

**Voice-Over-IP (VOIP):** Live audio via the web, commonly used for people to communicate in a way similar to a phone, but entirely online.

**Whiteboard(ing):** A whiteboard is a collaborative workspace where users can draw and write. It is commonly used in conjunction with videoconferencing, audio conferencing and/or chat.

# Chapter 12
# Leading Toward Improved Collaboration

**Vickie Cook**
*Greenville College, USA*

**Kara L. McElwrath**
*University of Illinois Springfield, USA*

## ABSTRACT

*As more and more learning is adapted to the Web 2.0 environment, it becomes imperative that faculty and students have the ability to collaborate through instant file sharing in a secure environment. Faculty, staff, and students of colleges and universities are more likely to click on or choose 'File-Save' than to file work in a physical filing cabinet today. Processing, managing, and storing these electronic files is much less structured than filing systems of the past. There is a critical need to manage the explosion of e-mail, documents, chat, and other content. Document creation to the ultimate disposition and storage of documents in both the function of teaching and learning, as well as the administrative functions within a university setting, requires process management that will provide a secure solution. The University of Illinois Springfield (UIS) chose to purchase a Xythos product branded as eDocs to the university community, which provided a basic content management solution. The ABACUS (central guiding priorities) method of critique was used to determine the best possible solution. In this chapter, the authors will discuss how the Xythos product was chosen and how this choice impacted the pedagogy of online learning at UIS.*

## INTRODUCTION

Teaching and learning creates a plethora of paper and electronic documents. As more and more learning is adapted to the Web 2.0 environment, faculty and students need the ability to collaborate through instant file sharing in a secure environment. Managing these documents through electronic storage that integrates with a Learning Management System can create an environment where teaching and learning can flourish (ScanSoft Productivity Applications, 2004). Regulations that guide compliance issues are most likely to be in place for processing, managing, and storing paper documentation. Since 1998,

DOI: 10.4018/978-1-61520-853-1.ch012

Copyright © 2010, IGI Global. Copying or distributing in print or electronic forms without written permission of IGI Global is prohibited.

guidelines for electronic file keeping have been established. The U.S. Department of Education Institute of Education Sciences (2009) maintains that legal requirements for paper documents are equally applicable to electronic documents thus making the retention and management of electronic documents a priority for educational institutions. Some universities, such as the University of California, Irvine, Boston College, and the University of Texas at Austin have found solutions to the amount of storage space required to retain paper files, reduction of staff time in manipulating paper documents used in the classroom, and the management of archived documents (Bourque, Franklin, & Updegrove, 2005). Employees may feel a false sense of confidence in their IT staff regarding the ability of IT staff to reliably retain, manage, and safeguard electronic files. Documents are only useful if users can find information when they need it. According to Records Management, State of Tennessee, "workers can waste up to two hours a day looking for misplaced paperwork – at a total of 500 hours (62.5 days) per year" (Miller, M. personal communication, July 15, 2009). Today, there is a critical need to manage the explosion of e-mail, documents, chat, and other unstructured content from its creation to its ultimate disposition and storage.

The University of Illinois Springfield (UIS) is a small public liberal arts university. UIS currently enrolls approximately 4,900 students which includes about 2,000 graduate students. UIS is organized in four colleges including Business and Management, Education and Human Services, Liberal Arts and Sciences, and Public Affairs and Administration with a base of 220 full-time faculty. There are currently 22 bachelor's degree programs and 25 minors, 20 master's degree programs, and one doctoral degree offered. UIS is a teaching institution that focuses on student engagement and small class sizes with a ratio of 1:12 faculty to full-time student equivalent. UIS was established in 1969 as Sangamon State University by the Illinois General Assembly. On July 1, 1995, the institution became a campus of the University of Illinois, and inaugurated as the University of Illinois Springfield.

UIS has become noted as an award winning leader in online education. Awards include 2008 Excellence in Online Reputation Management from the Society for New Communications Research; 2008 Ralph E. Gomory Aware for Quality Online Education; and 2007 SLOAN-C Excellence in Institution-Wide Online Teaching and Learning Programming. Quality online learning represents a significant investment by faculty and administrators at the Springfield campus in time, effort, and funding (University of Illinois Springfield, 2009). As UIS continues to expand its online offerings to a Web 2.0 environment, there is a continued need for faculty and students to have the ability to collaborate through instant file sharing in a secure environment.

## Background

Even with all the technological success realized at UIS, there remained a need for secure file sharing and storage options by faculty, staff, and students. In an interview with key members of the UIS implementation team, the following items were identified as driving the decision to seek a secure file storage and sharing system that would address needs of both the administrative and instructional programs on campus (Cook & McElwrath, 2009). The need for a secure collaborative learning environment was identified as a priority in 2005 by the Information Technology Services (ITS) staff through their interactions with various stakeholders including faculty, students, academic professionals, and administrative staff across campus. Constituents providing a variety of functions across campus noted the limitations of e-mail and other desktop tools for sharing and managing content. Additionally, concerns by both the users and ITS were discussed regarding the proliferation of department servers and files being shared between a variety of users. Other concerns voiced through a series of discussions with ITS included the loss of valuable data, decreased staff productivity, redundancy in uploading files to the learning management system, and concern for management of electronic files for compliance with legal requirements.

More and more work occurs outside of the confines of an office or classroom thus making anytime/anywhere access to information a critical part of increasing productivity. With increased emphasis on collaboration, information frequently needs to be shared with non-employees without comprising security. An example of this would be the collaboration of two faculty from different universities who might work together on a research project. The sharing of research documents and versions of the project would need to be accessible to both individuals, one of which would not be a part of the internal networked system. "The increasing reliance on digital tools for communication and collaboration has given rise to a new set of concerns in the areas of privacy, data protection, electronic discovery, and information security." (Xythos Software, n.d., p.3).

E-mail is too often used to share important/confidential files. This creates a host of issues, including the inability to manage revisions, control of who has access to the files, and access to historical records of who has viewed/modified the files and when. Without a user-friendly file storing and sharing system, employees continue to use e-mail to share files, thus continuing to expose organizations to additional risks and inefficiencies (PricewaterhouseCoopers, 2008). Universities worldwide have attempted to both find solutions to the issues of file storage and sharing, as well as educating their employees and students about e-mail security risks. The Chinese University of Hong Kong has developed a graphical explanation that provides users with a map of how e-mail attachments travel from user to user and security considerations that transpire (Instructional Technology Services Centre, 2004). Thus, an investigation of an Enterprise Content Management (ECM) system began. This ECM would provide a solution for document storage and file sharing system within a secure environment. An ECM is one type of commercially produced software system that combines technologies, strategies, methods, and tools to capture, manage, store, preserve, and deliver content and documents in a variety of formats to end-users.

Because of the complexity of the components needed for an ECM, it is often difficult for individual users to understand and use on a regular basis. Users of varying skill-levels and comfort must embrace such a system in order for it to be effective. "In fact, industry analysts have reported that less than half of recent ECM deployments have succeeded primarily due to the lack of user acceptance." (Xythos Software, n.d., p.2).

As we (the authors) reviewed the considerations used in determining the ECM system to be implemented at UIS, we created the ABACUS method of critique to find a system that would provide:

- Affordability
- Blackboard Academic Suite ™ functionality to support teaching and learning
- Ample document storage
- Collaboration between internal and external constituents
- User-friendliness
- Secure storage and sharing of documents

Additionally, this chosen system needed to work with the existing technology infrastructure and software which include the standard Windows and Mac operating systems, Microsoft Office Suite, software related to research data collection including SPSS, and other specialized software related to specific fields of study. Many IT professionals thought the solution to the 'File-Save' era was an ECM system. And for many large organizations with specialized requirements, an ECM system was the solution. Such a system addressed document and record management requirements, but came with a high cost both in terms of budgetary considerations and employee staff time to implement, train, and use the system. Even if the ECM resolved the issues in theory, the costly investment would be wasted if the users did not embrace the system. Many companies (Xythos Software, n.d.) have found that an ECM meets the specialized needs of just a few employees. The majority of employees need just a fundamental set of tools, allowing them to securely share files, track revisions made to those files, and access the files at a convenient time and place. A system that can meet these needs would minimize work duplication, allow users to easily transfer and reuse knowledge, and support the establishment of best practices.

## Deploying a Solution

For organizations such as UIS, which has many users with a wide variety of technical skills and comfort levels, a limited budget, multiple platforms, and a host of unstructured electronic files, ECM was not a viable option because of cost and difficulty of use for a variety of skill levels of the end-users. We needed a Basic Content Services (BCS) system that would allow users (comprised of faculty, administrators, staff, and students) a solution.

After a comprehensive review of what was needed, which included a variety of end-user interviews, discussions with other campuses both inside and outside the University of Illinois system, and reviewing current literature regarding BCS systems, the ITS group worked on writing a Request for Proposal (RFP) and began investigating possible solutions. Xythos provided a series of questions on their website that the ITS group utilized as they discussed solutions with constituents across campus and during the evaluation of possible solutions:

- Is the application easy to use?
- Will the system be available and accessible to all knowledge workers in the organization?
- Are systems compatible with existing technologies?
- Can records management functionality be applied transparently? In other words, will users be able to work with their files in the same or a similar manner as before? Content is more valuable when it is accessed in context. (Xythos Software, n.d.)

In addition, the ideal solution would be cross platform, not browser-specific, and without a dedicated client that had to be downloaded. This would provide end-users with the most autonomy in utilizing the system and ensure that it would be used through increased user-friendliness.

A Request for Proposal (RFP) was prepared and six software vendors responded. A separate RFP was prepared for the hardware storage component. Of the presentations, Xythos was determined to provide the best match for the defined needs of the university. Its handling of Learning Management Systems (LMS), including Blackboard, which UIS has used for a number of years, was a critical piece in the decision. Xythos allowed linking files in a variety of LMSs, thus allowing for changes to files to be made just once by faculty. These changes would be reflected in all linked files throughout the LMS. This flexibility not only made document management easy for teaching faculty but also paved the way for a smooth transition for LMS vendors if a change was needed in the future. In addition, Xythos was willing to provide on-going support and install the integration component to the UIS Blackboard LMS. This made their offer very attractive, particularly the inclusion of ongoing support in the implementation of the product for the greater UIS community.

Xythos also came with the flexibility for even better customization. Through recommendations from Xythos representatives, and the Blackboard Users Group, ITS staff found common goals in implementing the Xythos product with Northwestern University (NU). Northwestern is located in Evanston, Illinois. Due to shared goals, shared use of Blackboard, and close proximity, UIS partnered with Northwestern University (NU). NU IT staff were willing to share innovations that they had created in their deployment of the Xythos product on their campus. Northwestern's staff had created a bridge allowing for integration with the Blackboard Active Directory. This bridge simplified account creation and the log in process for use in online education. In addition, UIS IT staff developed a program to map a drive to specific user accounts, thus making access to files even simpler for the user. Having a partner institution to share solutions and integrate software use was a tremendous consideration in deploying this vendor choice. Collaboration among IT staff can be one of the most critical components in building a strong IT support system. This collaboration between NU and UIS has continued to provide strong support for both institutions.

UIS is not unlike many universities with mobile faculty and staff, online students who may never be on campus, increased need for collaboration with other universities, businesses, and organizations, and legal and compliance requirements that must be met and maintained. Xythos products offered UIS many benefits that would meet the need for a basic content management system. Using the ABACUS method of critique, the Xythos solution was implemented. A safe and secure web access and file storage for faculty who would be using the system to supplement their research efforts, to provide significant file sharing with students, and for integration of files in online classrooms via the Blackboard LMS were all components of the ABACUS critique.

Students needed a solution that would hold course files, assignments and homework, and group work collaborations in a safe and secure environment to eliminate problems with home computer equipment, the possible destruction or loss of a portable drive. The ability to collaborate with other students and with their teaching faculty in a secure environment was a valuable consideration.

Easy deployment and support by the Information Technology Services staff for end-users was important. Faculty and staff development that could be accomplished through short training sessions was desired. Training included implementation of a series of workshops including the following topics:

- Introduction to eDocs
- An alternative to e-mail attachments
- Blackboard Integration
- eDocs: A collaboration tool
- Using eDocs in specific platforms (Windows, Mac)
- Basic eDocs functionalities (Information Technology Services, 2009)

Additionally, UIS users were comprised of both Mac and PC devotees. Currently, approximately 50 faculty use Mac products but there are no estimates related to student usage. PC users were in the decided majority during the eDocs implementation; however, more and more faculty and students appear to be choosing to move toward Mac options. A system that allowed either platform user to integrate their files was a valuable asset. Finally, the system needed to be scalable, as uses and users were added to the university.

Improved academic collaboration through a versioning system adds value for faculty grading, teaching faculty, research faculty, administrating committees, and students collaborating on a variety of group documents. Documents that were changed by various collaborators could be stored and utilized by the group. Previous versions could be compared to current versions. Through the versioning option, it is not necessary to continuously wonder if the version being used is the most current. The software managed the versioning of the documents.

Additionally, subscriptions to notifications of file usages and changes, as well as a commenting feature added overall worth to the final product. Administrative committees could utilize this function to allow users to choose to receive e-mail notification when changes to committee documents were completed. This proved extremely valuable in working in groups to collaborate on hiring decisions, committee review of processes and policies, and for faculty and students to be notified of updates to shared files.

Teaching faculty and students needed to be assured that a file storage and sharing system would compliment current existing online technologies. Files that could be uploaded into the Blackboard LMS and updated by a single change provided a streamlined effort in the collaboration between faculty who shared course sections and programs that shared course-room documents, such as rubrics and assessments. Sharing files with students through bridges and links made the system a valuable repository of learning objects for specific courses, as well as program consistency.

In any economic climate, but particularly in today's economy, reduced IT costs are a necessity (Nelson, 2007; Wellman, 2006). Finding solutions that eliminate overhead costs and create efficiencies in education speak to the concerns of the Secretary of Education, Future of Higher Education Commission (2006) that are voiced in *Costs, prices and affordability: A background paper for the secretary's commission on the future of higher education concerning the rising costs of higher education*. Cost reduction through the elimination of unnecessary file storage to additional servers and media for course documents, student files, and administrative committee files provided a strong financial impetus for locating the correct system.

According to experts universities and colleges are experiencing cost savings using basic content management systems (Miller, 2009, personal communication). Jay Rozgonyi, Director of Computing and Network Services at Fairfield University (2008), stated "Our preliminary estimates are to save about $500,000 over the next two years in hardware related costs alone. Xythos is helping turn file management and storage essentially into self-service procedure, allowing our network administrators to pursue other tasks." Similarly, Paul Bishop, Vice President of IT at Santa Barbara City College (2007) estimated the campus will be saving more than $160,000 a year; not just from deploying Xythos but by streamlining inefficient processes and providing a more feature-rich document management experience. Bishop (2007) stated "Almost half our campus population was underserved because we did not offer a web-enabled document collaboration service. With Xythos, we've leveled the playing field. Xythos helps us to better support our high-mobile community with a solution everyone can use."

Large file storage and transfers had created bottlenecks of data inside the Blackboard course space or LMS in past semesters. Rather than purchasing new equipment to supply larger servers, the BCS system provided the answer for these bottlenecks. Additionally, a system that allowed users to create their own specific work spaces without direct involvement from IT staff would provide faculty, staff, and students with ease of use and allow IT staff to be focused on higher productivity with other work projects. This self-directed use of the system provided efficiencies for the staff who were managing the product. Xythos was the vendor who provided the solutions and options for additional value, while meeting the technological needs of the campus.

Once the RFP from Xythos was accepted, an ITS working group was formed to begin planning for the successful launch and implementation of the document management system. The ITS team met regularly for nearly a year before the anticipated launch of Xythos to the campus. The team was responsible for:

- Branding the Xythos solution
- Marketing the product to the UIS community
- Selecting a group of Early Adopters to help test the system

- Writing policies for usage, including storage quotas, appropriate use, and backup maintenance procedures
- Developing a roll-out plan, including marketing and promotion, and communications with key administrative committees
- Designing training and support materials, including a webpage, printed documentation, live training sessions, recorded training sessions, and how-to videos

First and foremost, the branding needed to be unique, easy to remember, self-explanatory (not just another acronym), and catchy. After much debate among the ITS working group, 'eDocs' was chosen. From that point on, Xythos became branded as eDocs on the UIS campus.

For the campus to embrace eDocs as a solution, it was imperative that it be marketed in a way that clearly showed how it was better in terms of collaboration than current solutions. Toward this effort, notices and publications that explained the benefits of this storage and solution system were released to the campus community six months prior to the launch. These internal information sharing sessions not only described the features of eDocs but also compared eDocs to current systems and practices to assist with better understanding by the end-users.

UIS hosts a Technology Day event for faculty, staff, and students annually in February. eDocs was introduced to the campus during the 2008 Technology Day via a poster display. Faculty, students, and staff volunteered to become early adopters, and a group of about 50 faculty, staff, and students were quickly given eDocs accounts to begin experimenting. These early adopters were provided support through workshops, documentation, how-to videos, and one-on-one assistance by IT staff in setting up and using their eDocs accounts. They were given access to begin storing and sharing files both internally within the UIS community and to collaborate with external users. Some faculty and students began to use eDocs to share files and build links in the Blackboard LMS as well as in classrooms across campus.

The next step was to provide ongoing training and support. Bi-weekly training sessions, which were open to the entire campus community, were offered for the first six months of using the eDocs system. These hands-on workshops were detailed how-to sessions that included printed documentation that users were given for assistance as they began using the system. Online workshops were also offered to meet the needs of our online students and adjunct faculty who were not available during the day to attend on-campus workshops. In addition, workshops were recorded, archived, and posted on an eDocs webpage along with accompanying printable hand-outs, allowing users to review the training at their convenience (Information Technology Services, 2009). The Technology Support Center offered technical support via e-mail, phone, and in-person.

The eDocs webpage evolved into more than merely a launching point for users to log into the system. A series of short, one to four minute videos highlighting each of the features was made available on the webpage. In addition, a 'Best Practices' section was added to showcase concrete examples of how eDocs might be effectively used. A 'Known Issues' section was also included, identifying features that were not currently working as desired.

eDocs became and continues to be a service that is explained during student orientation, showcased in newsletters, presented during conferences, included in new employee orientations, and more. The eDocs system was effectively born.

## UIS Survey of Users

In preparation to writing this chapter and to more fully understand how eDocs was being used by the greater UIS community, we invited the community of faculty, staff, and students to tell us what eDocs features they used, and where they accessed their files using eDocs. Within 6 months of launching eDocs to the campus community, 1854 users had logged into eDocs and over half of those (930) had data stored in their accounts.

E-mail invitations to participate in a survey were sent to 321 Civil Service staff, 275 Academic Professional staff, and 384 full-time and part-time Faculty. 241 respondents participated. 73 faculty and 102 staff participated. 66 students chose to participate by completing a survey. Finally, some respondents sent unsolicited e-mail testimony of how they used eDocs, while others who were identified as heavy users of the system were asked specifically if their stories could be shared. The results of those personal interviews are found below.

Table 1 reflects how often individuals utilized eDocs and for what purposes. Although respondents indicated that they most often utilized eDocs to store files and share files with others members of the UIS community, some respondents also indicated that files were shared with colleagues outside of the UIS community, to host websites, and to integrate files with Blackboard (Figure 1.)

Figure 1 depicts the self-reported use of eDocs by faculty, staff, and students. Five primary functions were given as choices. Storing files was the most often chosen response (n=118) by the three groups followed by 103 respondents who indicated that they used eDocs to share files with other UIS colleagues.

To further understand where eDocs was being utilized, respondents were asked to identify their location of use. We found that students (n=70)

*Table 1. Utilization of eDocs*

| | Daily | At least once a week | Several times a week | Occasionally throughout a month | Only when asked (for example, when someone shares a file with you) |
|---|---|---|---|---|---|
| Faculty | 7 | 8 | 5 | 15 | 7 |
| Staff | 11 | 11 | 7 | 14 | 6 |
| Students | 6 | 12 | 8 | 14 | 40 |
| | | | | | |

(Cook & McElwrath, 2009)

*Figure 1. Faculty, staff, and students use eDocs in a variety of ways. (Cook & McElwrath, 2009)*

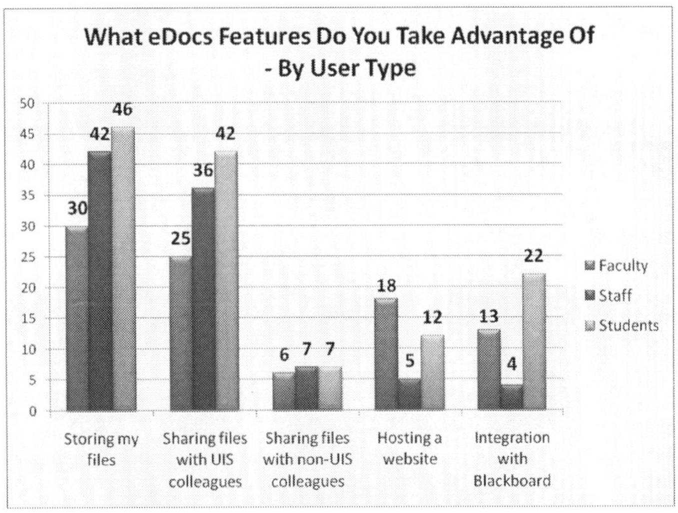

were the number one user group of eDocs in their homes. Staff and faculty (n=85) were most likely to use eDocs in their offices on campus (Figure 2).

## Follow Up Interviews

The following stories were shared with the authors by faculty and staff who have given their permission to serve as examples of how eDocs is used in the process of teaching and learning.

Dr. Brian Jackson, professor in English 101 and 102, shared information regarding his use of eDocs to improve student writing. He creates a separate folder for each student in his eDocs space. Each student uploads all drafts of papers and all in-class pre-writing to his or her own folder for the instructor to review. The instructor, in turn, uploads annotated papers containing feedback in each student's folder. The course is completely paperless. At the end of the semester, students assemble a portfolio of their best papers from the work in their folder (B. Jackson, personal communication, May 27, 2009).

Dr. Jackson also serves as a clinical instructor in writing for the Center for Teaching and Learning (CTL), where he is charged with supervising writing tutoring. Graduate assistants in the CTL can access the assignments from the English Department's eDocs space, so they are prepared when working with individual students (B. Jackson, personal communication, May 27, 2009).

Dr. Beverly Bunch, an Associate Professor in the Public Administration program, with a joint appointment in the Center for State Policy and Leadership, shared her use of eDocs. She uses eDocs extensively in the courses that she teaches (personal communication, May 27, 2009). In the online section of Intergovernmental Relations, the students prepare a presentation (PowerPoint with narration) and upload the file to their eDocs area. Students post the link in the discussion forum in Blackboard. Other students in class may review the presentations of fellow learners in their small group of three students and give feedback. Students revise their presentations based upon the feedback and submit for grading and viewing by the whole class.

In the blended section of the Nonprofit and Governmental Public Financial Management course, the students post an audio recording (made using Audacity) in their eDocs area and then put the link in the discussion board, so others students may listen to the recordings. The students also work in small groups to prepare training materials to be put on a Facebook page. The students use their individual eDocs area to host some of the files that they link to their Face-

*Figure 2. Identified locations where eDocs is currently used. (Cook & McElwrath, 2009)*

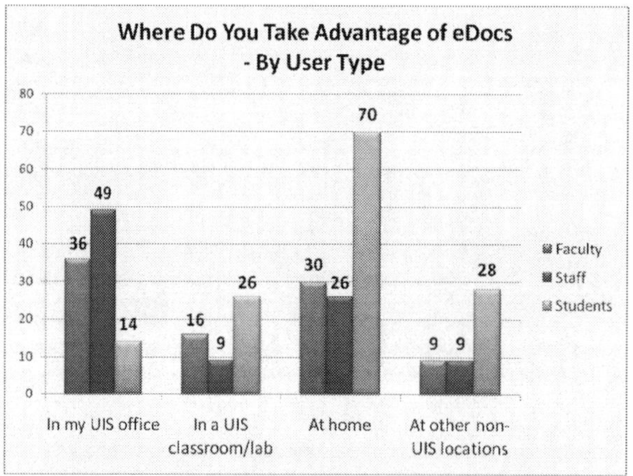

book pages (B. Bunch, personal communication, May 27, 2009).

As another example, Dr. Keith Miller, a professor in the Department of Computer Science at UIS, University Scholar, and the Schewe Professor in Liberal Arts and Sciences, uses eDocs to distribute lectures which are recorded using Articulate in his online and on-campus courses in computer science, computer ethics, and philosophy. Examples of Dr. Miller's use of eDocs for lectures are available at: https://edocs.uis.edu/kmill2/www/PHI442L1/player.html; or https://edocs.uis.edu/kmill2/www/lecture8csc540/player.html (K. Miller, personal communication, May 27, 2009).

Kim Rutherford, Learning Specialist with the UIS Office of Disability Services (ODS), describes her use of eDocs in providing material in an alternative format to students. "eDocs has revolutionized our alternate format delivery process. Before, we had to either e-mail huge files to students, have them come in and save it on their flash drives, or other methods which were time consuming for both student and staff. Now, when the material is ready for the student, all we have to do is upload to eDocs and send them a link, instead of several e-mails or setting up a time to meet to transfer material to a flash drive. I could praise eDocs all day long for what it has done for our students and text conversion; we have been waiting a long time for something this simple and efficient. eDocs is simple, fast, effective, and has truly revolutionized our alternate format process. We are able to process material at a much faster rate, and deliver it so quickly. My students have told me how much of a difference eDocs has made to them because they receive their material easily and quickly, from any computer at any time. Simply stated… eDocs is a dream come true for ODS." (K. Rutherford, personal communication, May 27, 2009).

Other comments that were shared by various respondents to the original survey included these uses to support student learning through practical applications within the university:

- Using eDocs to bolster intern placement through sharing information among and between students, faculty, and selected internship mentors.
- The use of eDocs as a portfolio tool that can be implemented through internship placement, or used as a learning documentation tool to provide evidence of assessment for state and regional accreditation bodies.
- Faculty and students may share student portfolios easily without overtaxing e-mail space or sharing portable file drives and devices.
- Departmental portfolios can be created to address industry standards. Students may then upload specific evidentiary support for having met specific standards. A link is shared with the grading faculty who can insert feedback and comments on the students' portfolios. The student receives a notice through a subscription choice in eDocs that the file was reviewed by the professor and notes made to the file.
- These processes are effective and efficient as learning and assessment tools and create a practical application for using and sharing files between students and faculty.

The UIS Associate Provost for Information Technology/CIO, Information Technology Services, Farokh Eslahi, shared how eDocs has provided a cost savings for the ITS Department. "Since the implementation of eDocs, we have seen cost savings in two areas: technology support and server and storage hardware and software. Prior to eDocs, ITS was managing a growing list of file server "shares" for individuals and departments. In addition, ITS was dealing with issues of large e-mail attachments that ended up in all mailboxes, and duplicate content in Blackboard. Adding and removing users for the shares, changing permission, backups and restores, etc. were mostly managed centrally by the Technology Support

Team. With a self-managed document management system, users can perform those functions themselves, therefore freeing up our technology support staff to focus on other responsibilities. Blackboard integration has certainly been another big plus – allowing users to store content only once in a secure, central location. Also, ITS was able to decommission the file server used for the shares. Another saving in terms of hardware and software is that there is no longer a need to maintain a separate web server for faculty, staff, and student personal web pages given the web publishing support in eDocs. Another saving is the amount of storage we now need for the workstations in labs and classrooms. All lab and classroom PCs are configured to automatically connect to a user's eDocs space upon login allowing users to store their documents." (F. Eslahi, personal communication, July 7, 2009)

## Supporting Teaching and Learning: Building Collaboration

Much of the success of eDocs may well be attributed to the customization of the identified needs and the ability to grow with the end-users. As more and more learning is adapted to the Web 2.0 environment, it becomes imperative that faculty and students have the ability to collaborate through instant file sharing in a secure environment. eDocs has met the pedagogical needs of the University of Illinois Springfield. The eDocs solution has created an environment where sharing learning objects through a series of collaborative learning activities assists in supporting teaching and learning.

In a study of student perceptions regarding blended learning, Greener (2008) found that technologies which incorporated the ability to build self-directed learners and encourage collaboration through group work commitment promoted engagement and learning. Ease of collaboration between faculty members, faculty with students, and students with students may be provided through a basic content system as one solution for engagement in teaching and learning. eDocs provided that collaborative environment to assist with the building of an environment in the online classroom where this type of engagement could occur. Almala (2005) posited that when instructional design and materials were well planned and implemented, a quality e-learning experience was the end result for the student. eDocs is one of the technology tools that provides ease of use of the instructional materials in the online course for faculty and students at UIS.

To encourage this collaboration, all UIS users, including faculty, staff, and students were allocated 1 GB of storage space without charge to the individual or unit. Users include faculty, staff, and students in any capacity – online, full-time, part-time, adjunct, etc. Each department was allocated 5 GB of space. Additional storage space could be requested, up to 1GB for individuals and 5GB for departments, upon approval. Additional storage required units or individuals to purchase space. Furthermore, a 1 GB/hour band-width limitation was implemented to prevent users from making mass uploads that might overload the server. If needed, exceptions could be made by advance request.

One of the many features of this particular system is its propensity to be fault-tolerant. eDocs data is hosted between four redundant servers residing in three different buildings, thus providing consistent access in the case of a hardware or electrical failure in any single building or server. This safety measure provides those engaged in collaborative activities freedom from concern regarding their documents becoming lost or irretrievable. Nightly backups are performed and retained for 2 weeks for disaster recovery purposes. Disaster recovery is necessary for not only the unlikely natural disaster occurrence, but also for power outages, human error in mistakenly eliminating large files, and other errors that might occur in the maintenance of software and hardware. Anyone who has ever lost a file on a shared network drive will be able to identify with the need for a strong backup system. Virus scan software runs on the server and deletes infected files. This provides integrity to the file sharing and collaboration process. Data is encrypted on uploading and downloading, while the files are in use for additional security. eDocs is hosted on a recognized secure port (443).

eDocs uses Web-based Distributed Authoring and Versioning (WebDAV), which allows users to mount eDocs storage as a virtual drive on their computers' desktops. This means that eDocs looks like any other network drive, with a "drag-and-drop" interface. This allows users to edit documents on the remote eDocs server, no matter where the users are physically located. It is almost as easy to use files on a WebDAV server as those stored in local directories. In a 2004 EDUCAUSE presentation, the technical aspects and limitations of the WebDAV components and their impact on twenty-five universities regarding interoperability, sharing and scalability of the WebDAV systems in a teaching and learning environment were reported (Kapor, McCredie, & Sreebny, 2004).

As noted in the Kapor, McCredie, and Screebny (2004) presentation, it was important to identify what the limitations of eDocs were. eDocs is not a streaming server, so while audio and video files can be hosted on eDocs, they must be downloaded to play – they will not stream. In addition, server-side scripting does not work on files hosted on eDocs. Files or websites requiring server-side scripting and/or streaming are hosted on a separate server. As faculty became aware of these limitations, they were able to appropriately encourage the use of eDocs as collaborative tool for group projects.

Even noting the limitations above, eDocs supports online collaboration by allowing file sharing between the faculty and students. Faculty are able to make documents readily available to students through e-mail or through a Learning Management System (LMS). Working groups of students are able to share files, edit, and create versions of learning materials and assignments through use of their individual and shared eDocs accounts. A group of students in an online research class participate in an assignment where they review research articles and prepare a presentation about the research that was conducted with peer learners. Through accessing professor-initiated eDocs files, students are able to review a variety of guided research. Students then create their own shared files, store them in eDocs, and share their work with their professor through eDocs links. This simple storage system supports a seamless learning environment that supports the pedagogy of the online classroom.

Future uses of eDocs in teaching and learning are limited only by the imagination and creativity of students and faculty. Faculty will undoubtedly find innovative ways to implement file storage and sharing to improve time management, efficiency, and effective feedback delivery. One of those innovations may be to determine new integrations into the Blackboard LMS at UIS.

A "Blackboard for all" effort was implemented at UIS in 2008. Every course, regardless of the instructional delivery system, face-to-face or online, automatically has a Blackboard area readied for course specific files and assignments. Most faculty on campus utilize the Blackboard LMS for a myriad of instructional purposes. Syllabi are posted, online discussion boards are posted to enhance the learning conversations, course materials are made available, and assessment data is collected. During the fall of 2008, eDocs became an integral part of the Blackboard delivery system. Faculty members upload course documents into their eDocs accounts. A Blackboard building block tool allows the faculty member to create a link that redirects the student to the document stored in eDocs when they click on the link.

Within eDocs, a professor is able to make a single change to the file, create a Blackboard link and the changed files are immediately available to faculty and students. Students who click on the eDocs link in Blackboard automatically have newest version of the file. In the past, a single correction or change required multiple updates within Blackboard for each location where the document might be found. Additionally, there was always the chance that one document would be forgotten in the change process and not be updated until someone found an error. eDocs makes a simple change just that, simple! With one change, students and faculty can be assured of updated course materials.

eDocs also creates a shared learning opportunity for not only the online student and faculty member, but is equally as robust for the face-to-face class whose faculty and students are utilizing shared documents to improve the learning process, to implement collaborative learning activities, and to improve student/faculty research engagement. Research collaborations are supported through the use of sharing documents, literature, and the shared writing of grants and empirical research documentation. Instead of sharing zip files, locating flash drives, or e-mailing large files that may or may not be received due to space restrictions on e-mail servers, eDocs creates an environment where faculty can upload their files and share them via a link to their research collaborators whether on or off campus. Collaborations are strengthened through the versioning and subscription features that allow collaborators the option to track the work of their co-authors. This aspect of versioning supports the constructs of best practices in e-learning that provide for learning to be built through shared learning experiences (Almala, 2005).

Communication of policies regarding appropriate usage, storage quotas, and backup maintenance procedures has been implemented through use of printed media, web pages, e-mails, faculty and staff development, and administrative committee structures continued regularly through the decision-making process to complete implementation. Marketing, internal promotion, and training were components that created the successful implementation of eDocs as a collaboration tool to promote learning. Posters across campus, weekly brown bag trainings, e-mail updates, and other internal communications successfully provided the needed leadership to connect the branded eDocs system to the needs of the UIS teaching and learning community.

## Challenges and the Need for Future Research

As with the implementation of any new system, there have been and continue to be challenges. Through a sustained team effort between the original ITS working group, the early adopting faculty and staff, and students, these obstacles have been successfully overcome. Software implementation was one of the first obstacles that created the need for different groups on campus to understand what this new eDocs system would mean for them individually. Additionally, there were some initial concerns that old favorite technologies would cease to be supported by ITS. In particular was the concern that the server housing individual faculty webpages would not be supported. Through communication regarding both current and future needs, faculty, as well as administrators and staff, saw the benefits of using the eDocs system for file sharing and storage which included the ability to create and maintain personal webpages in a relatively easy manner.

The integration of eDocs with Blackboard files has been discussed previously. The initial inability to copy courses in Blackboard to a new semester and retain the eDocs links caused concern among faculty. Staff at UIS worked on this issue and created a bridge that allowed a seamless interface from course copy to use in the new semester with new students. Faculty and students, as end-users, were able to easily utilize files that were uploaded into eDocs in previous semesters.

While platform compatibility was recognized early on as a necessity for any file storage and sharing system, challenges for Mac and PC users remained a concern. IT staff maintained solid records of platform issues and continue to work on solution implementations to decrease platform discrepancies. Browser (Internet Explorer, Safari, Opera, Firefox) compatibility was also identified as an issue that IT staff remain committed to working on to decrease inefficiencies. Some challenges remain as more faculty and staff anticipate integration of eDocs with new technologies.

This anticipation has not been realized in the compatibility of eDocs with Smartphone technologies. eDocs, which is dependent upon the use of a double-click system, is not currently supported when used with Smartphones that are dependent on scrolling and single-click technologies. This challenge is somewhat limiting for students who utilize their Smartphones to access online learning and podcasts. This area of supporting users of m-learning (mobile learning) environments is one that needs future research. Winn (1990) contended that stronger learning occurs through the use of environmental interaction. Winn purported the use of cognitive instructional strategies and design that promotes a holistic environmental approach to learning. M-learning promotes that type of holistic environmental learning and eDocs or other storage and file sharing solutions will need to adapt to that learning environment. Research that substantially indicates both usage patterns, future trends and predication for uses in higher education, as well as the ability for users to access learning objects in a Smartphone environment is necessary to develop an understanding of how a system such as eDocs will continue to be salient to learning.

## CONCLUSION

There is a critical need for universities to manage the explosion of e-mail, documents, discussion, and other content in an effective and efficient manner. Document creation to the ultimate disposition and storage of documents in both the function of teaching and learning, as well as the administrative functions within a university setting, requires process management to provide a secure solution. The ABACUS (central guiding priorities) solution was used to determine the best possible solution. In this chapter, we discussed how The University of Illinois Springfield chose to purchase a Xythos product branded as eDocs to the university community to provide a basic content management solution. This choice has allowed faculty, staff and students to effectively and efficiently collaborate in a Web 2.0 environment that supports the teaching and learning within the university.

## REFERENCES

Almala, A. H. (2005). A constructivist conceptual framework for a quality e-learning environment. *Distance Learning*, *2*(5), 9–12.

Bourque, M., Franklin, S. D., & Updegrove, D. A. (2005, October 19). *Delivering shared, authenticated, ubiquitously accessible storage via WebDAV.* Paper presented at the EDUCAUSE 2005, Transforming the Academy: Dreams and Reality, Orlando.

Cook, V. S., & McElwrath, K. (2009). Interview with Tulio Llosa, Director of Educational Technology, Information Technology Services University of Illinois Springfield and Thomas Sidener, Academic Technology Service Specialist, Information Technology Services (unpublished interviewed.). Springfield, IL.

Greener, S. L. (2008). Self-aware and self-directed: Student conceptions of blended learning. *MERLOT Journal of Online Learning and Teaching, 4*(2).

Information Technology Services. (2009). *eDocs.* Retrieved July 10, 2009, from http://www.uis.edu/its/iss/edocs.html

Information Technology Services Centre. (2004). *Security issues and solutions in internet e-mail.* Retrieved July 8, 2009, from http://www.cuhk.edu.hk/itsc/network/app/email/security/securityissues.html

Kapor, M., McCredie, J. W., & Sreebny, O. (2004). *Chandler/Westwood: Progress in open-source collaboration.* Paper presented at the EDUCAUSE 2003 Annual Conference, Texas.

Nelson, M. R. (2007). U.S. advisory committee on student financial assistance textbook study field hearing III. Portland, OR.

PricewaterhouseCoopers. (2008). *Information security breaches survey 2008.* Retrieved from http://www.pwc.co.uk/eng/publications/berr_information_security_breaches_survey_2008.html

ScanSoft Productivity Applications. (2004). Case study: Scan to PC desktop – The University of Michigan. Peabody, MA: ScanSoft division of Xerox Corporation.

University of Illinois Springfield. (2009). *The office of technology-enhanced learning at UIS.* Retrieved July 10, 2009, from http://otel.uis.edu/Portal/index.asp

U.S. Department of Education Institute of Education Sciences. (2009). *Web-related legal issues and policies, weaving a secure web around education: A guide.* Retrieved July 3, 2009, from http://nces.ed.gov/pubs2003/secureweb/ch_3.asp

Wellman, J. V. (2006). *Costs, prices and affordability: A background paper for the secretary's commission on the future of higher education (Report).* Washington, DC: U.S. Department of Education.

Winn, W. (1990). Some implications of cognitive theory for instructional design. *Instructional Science, 19*, 53–69. doi:10.1007/BF00377985

Xythos Software, Inc. (n.d.) *Document management 2.0 – Web-based collaboration and the road to compliance.* Retrieved April 24, 2009, from http://www.xythose.com/_media/pdf/xythose_wp_docmgmt20.pdf

# Chapter 13
# Integrating New Open Source Assessment Tools into dotLearn LMS

**Paloma Moreno-Clari**
*University of Valencia, Spain*

**Esteban Sanchis-Kilders**
*University of Valencia, Spain*

## ABSTRACT

*The convergence process initiated by the European Higher Education Space (EHES) has changed the approach to the teaching and learning process focusing largely on monitoring the progress of the student. This task can be simplified with the proper use of tools integrated in the universities' Learning Management Systems (LMS). Most of these platforms have modules for managing the evaluation and to create online examinations. However, they do not tend to contemplate ways of evaluation depending on the subjects taught. This chapter presents a tool for continuous evaluation of a variety of subjects and their integration into a LMS based on open source. In the University of Valencia's particular case, the LMS includes facilities to create online examinations, but it does not provide a user-friendly interface. To improve the usability of this module a computer program has been developed to simplify the evaluation process.*

## INTRODUCTION

The development of information and communication technologies (ICT) has generated new needs for the students (Lee & Owens, 2000), (Wong, 2003). In particular, students need current knowledge and specific skills readily applicable to their potential professional area (Marqués, 2008). Also the learning support materials have evolved through the development and use of ICTs, which has opened a vast field for training. All these changes have led to a redefinition of the teaching and learning process (Pallof & Pratt, 2003). This process is essentially collaborative in order to achieve a more active participation of the student.

DOI: 10.4018/978-1-61520-853-1.ch013

Currently, the incorporation of ICT's in higher education has been completed. Most European universities use either commercial or an open source Learning Management System (LMS) to support their programs (Barajas & Gannaway, 2007).

Copyright © 2010, IGI Global. Copying or distributing in print or electronic forms without written permission of IGI Global is prohibited.

However, proper integration means to achieve a practical system of education that incorporates both the best technology and the most appropriate teaching techniques (Zurita & Ryberg, 2005). Content and methodology must be adapted to the new e-learning tools (MacDonald & Thompson, 2005). In this sense, the European Higher Education Space (EHES) recommends a higher participation student within a collaborative teaching and learning process with real quality (Brooks, 2003).

Therefore, the adaptation and growth of an open source LMS cannot be made based on a simple development of additional modules, but requires a thorough analysis of its use. This analysis will serve to detect the most used and useful functionalities to improve the learning process and the educational quality. In addition, this study shows which tools should be modified to facilitate interaction between users (Colla, MacDonald & Thompson, 2005), (Thompson & Randal, 2001).

In the case of the University of Valencia (UV), the study was performed after two courses of use of its open source LMS (This period was considered the implantation phase) (Moreno-Clari & Cerverón-Lleó, 2007). The analysis included a comparison of the main functionalities with a university of similar size and characteristics, which had introduced its platform a year earlier. The main objective was to identify the modules that needed improvement. Based on the comparative results two modules had to be improved for adapting them to new needs of the student's monitorization and evaluation. These modules were: the "Activities module", which was used to manage and evaluate homework, workgroup, practical classes and therefore achieve a continuous evaluation and the "Assessment module", used to create online examinations (Cerverón, Moreno, Cubero, Roig & Roca, 2007).

This chapter presents the implementation of a tool for continuous evaluation and grades management based of an improvement of an existing tools. It also contains the description of a tool to create online examinations and their integration into the existing LMS. In both cases, this new tool can be reused on any platform based on the dotLRN open source software.

The background, the starting point and the events that led to their developments are outlined first. Thereafter, the two tools implementation is described, including the system analysis, the research design and the results of the integration process, always keeping in mind that the results can be generalized to other university's open source systems. Finally, new ideas for further research are outlined and the conclusions of this work are exposed.

## BACKGROUND AND STARTING POINT

The University of Valencia was one of the first Spanish universities that started different strategies of change and modernization. Not only its structures but also the teaching method follows the recommendations brought by the European Space for Higher Education (ESHE) (Moreno & Cerverón, 2007). A new plan was developed to improve the use of ICTs and to enhance the students learning processes by involving them more directly. On the other hand, in order to standardize the available technical tools and to promote their use by the whole academic community, all the tools where unified on a single platform. This platform should provide a unified management of the learning process and communication between groups and communities (Cerverón & Moreno, 2006).

### Selection Process, Implementation, Integration and Functionalities' Improvement of the LMS

The University of Valencia (www.uv.es) is one of the largest universities in Spain. It currently offers on-campus education to about 50,000 students in 18 schools, delivering approximately 1,500 different modules. Although by 2003 the university already counted with an important ICT infrastructure, this was not integrated under a single platform. Instead, every specific centre, department or faculty (student enrolment, data files, accountancy, etc.) used its own set of applications. In 2003, it was decided to integrate all these applications under a single LMS. After a careful analysis of commercial and open source products, the university joined the dotLRN project and personalized the product for the specific needs of the institution (academic year 2004-2005) (Essa, Cerverón, & Blessius, 2005). dotLRN is based on Open ACS and is already used by other large universities in the world supported by a large active community working on its development (Santos, Gonzalez-Boticario, & Barrera, 2005). As the tool is based on open source (Raymond, 1999), (Perens, 1999), it is easier to adapt to particular situations, not only from an organizational point of view but also from a pedagogical one. Even new modules can be developed and added to the platform. As a result, this platform was called Aula Virtual and it is currently used to provide a number of functions for each module taught in the University (e.g. a document repository, an event calendar, a news section, the possibility of creating forums, e-mail and chat services, support for notifications, submission of homework, a repository for learning objects, an application to create Web presentations, weblogs, a photo album, and a FAQs section) (Cerverón et al., 2007).

Three different authentication authorities validate users: LDAP (which is verified in the UV LDAP server and allows the use of the accounts that users have created for all e-services in the university: this fact contributes to integration), LOCAL and EXTERNAL.

Once the system was implemented, the users demanded new functionalities and tools, and this fact motivated their development. Among them, we can emphasize the following ones: the chat package integration in the dotLRN courses; the possibility of mathematical formulation insertion, introducing symbols, in LaTeX or in ASCIIMath (based in MathML); the development of a space within the dotLRN courses where each student has a personal file, replacing classical student cards which professors traditionally asked for (this functionality was developed from the Education Equipment Portlet); the improvement and installation of Communication tools (personal Blogs); the inclusion of contents distribution tools (MMplayer module) or a link to Wikipedia.

From the beginning, the University of Valencia has actively taken part in developing the community of the dotLRN platform and the OpenACS architecture (Moreno, Cerverón, López, & Roig, 2007) by improving functions and correcting bugs like characteristics in the Chat module. The user interface was also translated and new modules were created, based on the demand of the academic community. For example, a module called "Fichas" or cards (Soler-Lahuerta, Cubero, López, Roig & Roca, 2005) was developed. This new function provides the lecturer with all the necessary information about the students, including pictures and information such as grades and personal notes, which can be added. Both, the student and the lecturer can see the notes. This function was integrated with the "Activities" module used to manage the submission of homework. This integration improved the "Activities" function. An example of this "Ficha" or student card is shown in Figure 1.

The use of this LMS also provided a new method to plan and evaluate the use of virtual learning in the daily lecturing process, following the criteria set up of the ESHE (CCU, 2006). Technological cooperation was also promoted by working together with the developing community. In addition, pedagogical cooperation was enhanced due to the reuse of lecturing material for using methods based on standards.

## LMS Open Architecture

The LMS personalized at the University of Valencia was, as mentioned in the previous section,

*Figure 1. Personal ID card*

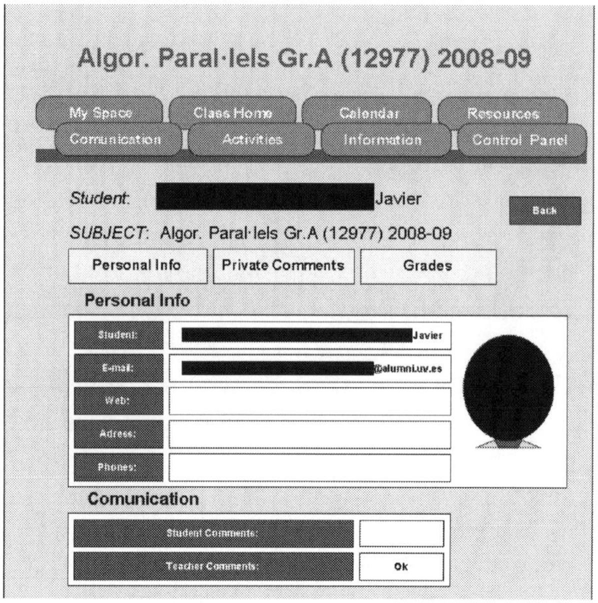


dotLRN. This is a platform implemented upon an open architecture: OpenACS. In this section we describe such kind of architecture and how it was used at the university after the selection process.

In the middle of the 90's, Philip Greenspun proposed a modular set of tools to give a generic answer to virtual communities' necessities (Greespun, 1999). The set of these modular tools was grouped under the name of Arsdigita Community System (ACS) and was released with GNU General Public License (GPL). The first database, which was chosen, was Oracle and the programming language was Tool Command Language (TCL). The present architecture does not depend on the relational database used: it can operate with Oracle or PostgreSQL and has the possibility of operating with other databases, if necessary. The last evolution of OpenACS has been denominated dotLRN, and constitutes an architecture reconstruction to improve the set of applications and the framework infrastructure.

The OpenACS architectures (OACS) is known as an application server and it is integrated by a set of advanced tools that allow programmers to develop Web applications which are oriented to define users' communities. Figure 2 shows the infrastructure and services of the architecture of the OACS and the functionalities and basic component distribution.

There is a common layer, named basic services or infrastructure layer, which is supported by the operating system and the relational database where the system information is stored. Linux is used as operating system. Persistency is obtained using PostgreSQL database, which uses Multi-Version Concurrency Control (MVCC) to manage the database accesses.

Applications use the Infrastructure Layer services and the different modules are available through the Web interface. These modules personalize the users' necessities. The modules are programmed using TCL, Tool Command Language. This programming language is multiplatform interpreted and oriented towards generating Web applications. Some available applications are forums, calendar, news, evaluation, storage area, FAQs, Wimpy Point, etc. Figure 3 shows the available applications in the ACS core.

*Figure 2. Services in OACS architecture*

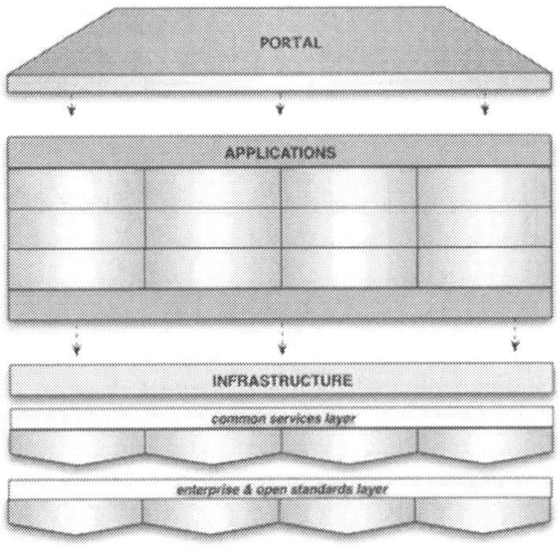


The portal component is located at the final layer. The mission of Web server is to provide the operation interface and interact with the application layer. The Web server technology is AOLserver.

## Motivation

In Parallel to the implementation and improvements once the platform was completely set up (academic year 2005-2006) the statistical data of usage was analyzed. The target of this analysis was to figure out the evolution and use of the platform as well as the detection of possible errors. This analysis would provide useful information to correct errors and improve the overall learning process. The analysis (Stevens, 1999) included a comparative study in the usage of the entire platform and its different tools, based on two different universities, which use the same LMS: dotLRN. The studied tools were those applications located at the portal component (Final layer in the framework architecture, see Figure 3). The universities were: The Galileo University of Guatemala and the University of Valencia (Spain). Both have a similar number of students and lecturers. The Galileo University has one more year of experience on the dotLRN platform, which could lead to better conclusions about the future evolution and use of the platform in the University of Valencia.

The comparative study of the use of different tools shown in Figure 4 demonstrated that although

*Figure 3. Available applications in the ACS core*

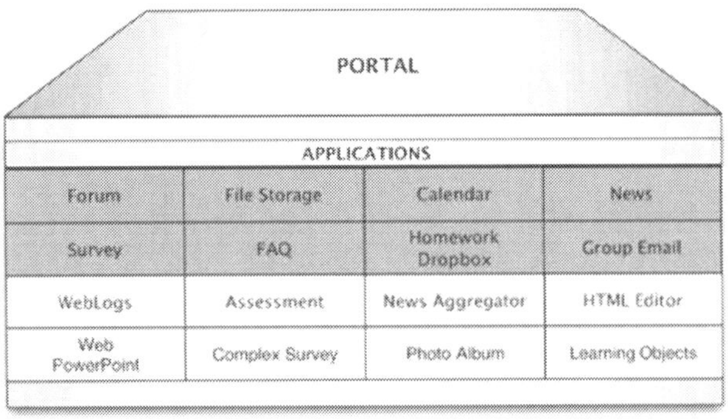

the most used tool on both LMS is the documents repository, the largest differences could be found in tools like Activities and Assessments. In the University of Valencia, the Activities function is used to manage the submission of homework, grades and notes of the students and the Assessment to create online examinations. At Galileo University the use is 36%, but at the University of Valencia it is only 5% (Hernández, Morales, De la Roca & Guerra, 2006). The reason for this difference is not only because of the one year delay in the implantation of dotLRN at the University of Valencia, but also due to the fact that the Galileo University relies more on an online learning method. Therefore, it needs help from these technological tools. This analysis led to a new project, which developed two new tools to evaluate the student's work continuously: on the one hand a project with a main objective to improve the Activities module, named as "Homework Dropbox" in Figure 3 and on the other hand a project with the main objective of improving the Assessments module, named as "Assessment" in Figure 3. These new tools should be integrated on the existing ones, "Homework Dropbox" and "Assessment", at the ACS Architecture's Component Portal (see Figure 3).

## NEW TOOL FOR EVALUATION MANAGEMENT

The functionality to manage continuous evaluations was deployed to fit with the existing modules of Activities and Cards, also developed at the University of Valencia. This new tool allows the organization of the whole evaluation process. It defines the tasks to be fulfilled and their relative weight in the overall grade. Also, it makes sure that the tasks, homework, class expositions, practical classes, etc., are delivered on time, and it gives the chance to set up a complex equation to calculate the final grade of the student (Cubero, López, Roig & Roca, 2006). The improvements with respect to the old tools are two new options: setting up complex evaluation processes and taking care of different evaluation chances available.

The advantages provided by the open source system is used to enhance existing tools, such as the Activities (which were used to manage the submission of homework) and Cards (which were used to save the information, notes and grades of the students as in Fig. 1) modules, based on the results of the analysis done. This implementation has been made to define a new algorithm, which allows setting up the new complex evaluation

*Figure 4. Comparative study of the use of tools In the University of Valencia and the Galileo University of Guatemala*

| USE OF TOOLS | UNIVERSITY OF VALENCIA | GALILEO UNIVERSITY |
|---|---|---|
| DOCUMENTS | 39% | 46% |
| FORUM | 5% | 8% |
| MAIL | 25% | 9% |
| NEWS | 25% | 1% |
| ACTIVITIES ASSESSMENT | 5% | 36% |

processes. Additional improvements, such as exporting the grades in a digital format and the delivery of personalized files to the students have also been added.

It is clear that the evaluation processes can be very different from one subject to another, not only depending on the subject but also on the lecturer (I JORNADAS AV, 2006). Therefore, the first step was to collect flow charts of as many subjects as possible, including technical and non-technical studies. After reviewing all the evaluation possibilities, the new algorithm was set up and defined. Then, the new management of evaluations tool was developed in the Open ACS standard, which maintained and took advantage of the existing functions (Activities and Cards). Based on the lecturers demand, new possibilities such as the file delivery were also included.

The project of implementing this tool in the University of Valencia LMS was divided into two phases. During the first phase the algorithm design was made based on the flow chart of the different evaluation methods. Then the algorithm was programmed and integrated into the LMS.

## Algorithm Design

The evaluation methods are usually based on grades, which were applied to different blocks (like theory or practical part), which can have different weights in the final grade. Each block can be set up by diverse tasks and there can also exist conditions, which have to be fulfilled by the student to pass from one task to the next. These evaluation methods are used in most of the University of Valencia courses. Not only in technical subjects, but also in the humanistic ones there is a practical part, named laboratory. In this practical block students apply to practical cases the concepts they have studied in their theoretical blocks. Nowadays the ESHE challenges base the learning and teaching process on skills that students must reach. These skills are not only on knowledge, but also on abilities and attitudes. Abilities and attitudes can be achieved in the practical or laboratory part of the subject.

To illustrate an instance of the Activities functionality improvement, this paper exposes a practical implementation of the evaluation of an example subject. This subject shows how the algorithm can reflect and evaluate any subject in Higher Education. Figure 5 shows the example subject called TCAD, which has two blocks: theory and laboratory (or practical part to achieve skills on abilities and attitudes), both with the same weight (50%) at the final grade.

After the theory has been explained the students have to pass an exam (Exam Theory February) which if not passed can be repeated at the end of the course (Exam Theory June). When passed, the student keeps the grade of the first or second exam, passed at that time.

At the same time the student has to fulfill three practical tasks in the Lab and to attend a minimum number of Lab sessions. The weight of each task is different and increases as the practical assignments become more difficult. The final grade is the mean value of the Theory and the Lab grades, taking into account that the grade of both parts must be greater than 5 to calculate this mean. The Lab grades would be calculated from the weighted mean of the grades for each of the tasks. If the student is unable to attend the Lab he/she can choose to make an exam on the contents of the Lab (Exam Lab June), which is equivalent to the three different tasks. If a grade is greater than 5, it means that the pupil has passed the task, exam or subject considered.

Attendance at the lab sessions is compulsory for a given percentage. If this part is not fulfilled then the student has to pass the lab part through the exam instead of the three tasks. But Attendance at the lab session is only a condition, which does not affect the final grade.

It seems clear that the student has at least two different paths to pass the subject. These paths are depicted in the flow chart.

Therefore the concept of "path" defines the evaluation process that can be different in each case. This translates itself into the algorithm.

*Figure 5. Flow chart for the evaluation of the subject TCAD*

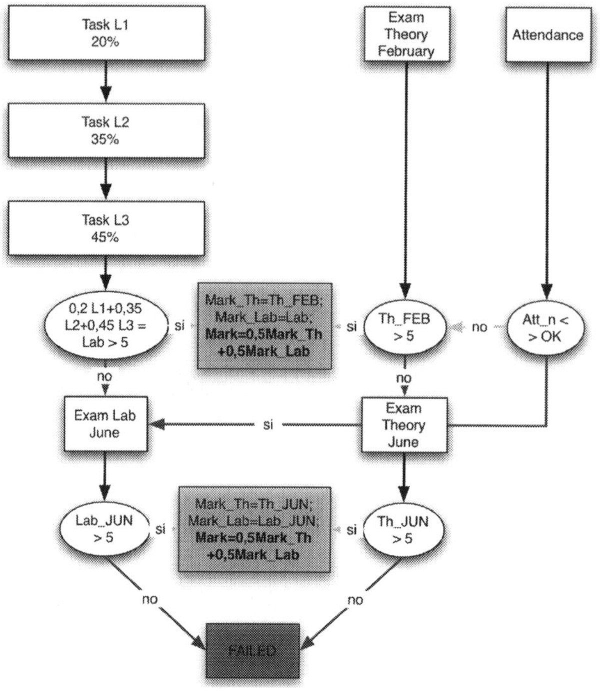


This evaluation process is also used in other subjects, specially the definition of two parts of a subject usually called lab and theory, which are very common in technical and scientifical studies.

The new algorithm applied to the Management of Evaluations (or improved Activities) module of the dotLRN platform translates as follows:

A new type of block has to be defined, which will be called weighted and selectable. Then, as part of the final and total grade, the lecturer can set the weight of the tasks which form the block and he/she can select whether this grade contributes to the final grade or not, depending on which way the student chooses to pass the subject. This selection also depends on the grades that the student has obtained for the different tasks or exams. A screenshot of this set up, once programmed into the application is shown in Figure 6.

The algorithm set up not only includes the new type of block called weighted and selectable, but also a new tool which automatically selects the different grades of the tasks to compose the final grade based on several conditions. These conditions set up the path of passing the whole subject. It is based on the different tasks the student has passed and it reproduces the flow chart (see Figure 7).

## Programming the Algorithm and New Options

The second phase of the implementation of the Management of Evaluations tool, which will be integrated in the preexisting Activities module, is to program the algorithm. The code was based on TCL (Tool Command Language) (Fuente, 2006), (TCL, 2008), JavaScript, and HTML. The LMS database is based on PostgreSQL (Hillar, 2006).

In this section, once the new Management of Evaluations or Improved Activities module has been finished, the way to define the evaluation process is established in the flow chart of Figure 5 as explained.

*Figure 6. Tasks (Attendance 1-n, Exam L June, L1, etc.) defined in the different blocks*

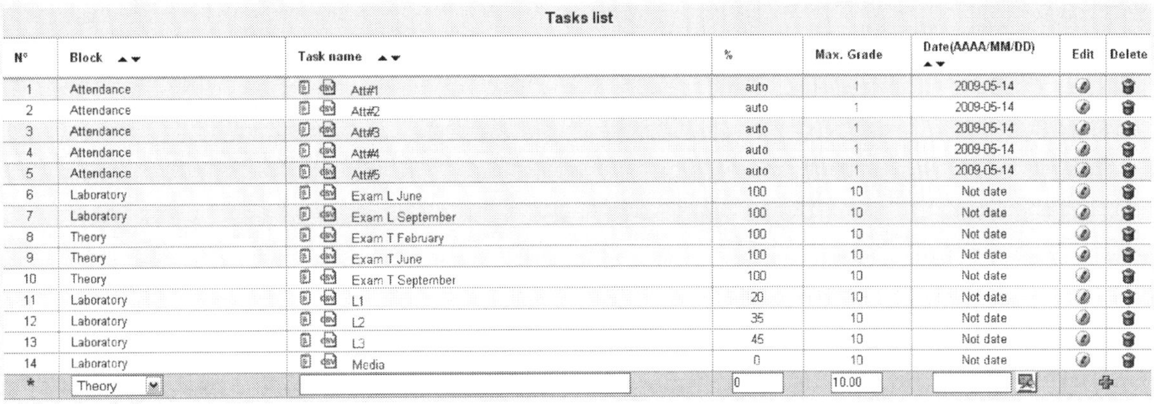

Tasks list

| N° | Block | Task name | % | Max. Grade | Date(AAAA/MM/DD) | Edit | Delete |
|---|---|---|---|---|---|---|---|
| 1 | Attendance | Att#1 | auto | 1 | 2009-05-14 | | |
| 2 | Attendance | Att#2 | auto | 1 | 2009-05-14 | | |
| 3 | Attendance | Att#3 | auto | 1 | 2009-05-14 | | |
| 4 | Attendance | Att#4 | auto | 1 | 2009-05-14 | | |
| 5 | Attendance | Att#5 | auto | 1 | 2009-05-14 | | |
| 6 | Laboratory | Exam L June | 100 | 10 | Not date | | |
| 7 | Laboratory | Exam L September | 100 | 10 | Not date | | |
| 8 | Theory | Exam T February | 100 | 10 | Not date | | |
| 9 | Theory | Exam T June | 100 | 10 | Not date | | |
| 10 | Theory | Exam T September | 100 | 10 | Not date | | |
| 11 | Laboratory | L1 | 20 | 10 | Not date | | |
| 12 | Laboratory | L2 | 35 | 10 | Not date | | |
| 13 | Laboratory | L3 | 45 | 10 | Not date | | |
| 14 | Laboratory | Media | 0 | 10 | Not date | | |
| * | Theory | | 0 | 10.00 | | | |

*Figure 7. Set up of the different conditions (paths) to compose the final grade of the student*

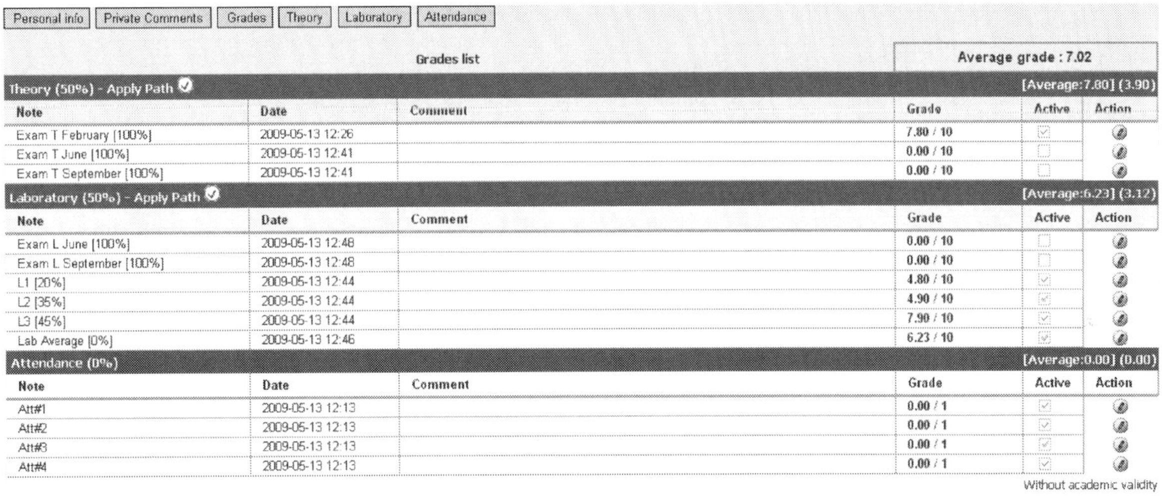

Personal info | Private Comments | Grades | Theory | Laboratory | Attendance

Grades list — Average grade : 7.02

Theory (50%) - Apply Path — [Average:7.80] (3.90)

| Note | Date | Comment | Grade | Active | Action |
|---|---|---|---|---|---|
| Exam T February [100%] | 2009-05-13 12:26 | | 7.80 / 10 | ✓ | |
| Exam T June [100%] | 2009-05-13 12:41 | | 0.00 / 10 | | |
| Exam T September [100%] | 2009-05-13 12:41 | | 0.00 / 10 | | |

Laboratory (50%) - Apply Path — [Average:6.23] (3.12)

| Note | Date | Comment | Grade | Active | Action |
|---|---|---|---|---|---|
| Exam L June [100%] | 2009-05-13 12:48 | | 0.00 / 10 | | |
| Exam L September [100%] | 2009-05-13 12:48 | | 0.00 / 10 | | |
| L1 [20%] | 2009-05-13 12:44 | | 4.80 / 10 | ✓ | |
| L2 [35%] | 2009-05-13 12:44 | | 4.90 / 10 | ✓ | |
| L3 [45%] | 2009-05-13 12:44 | | 7.90 / 10 | ✓ | |
| Lab Average [0%] | 2009-05-13 12:46 | | 6.23 / 10 | ✓ | |

Attendance (0%) — [Average:0.00] (0.00)

| Note | Date | Comment | Grade | Active | Action |
|---|---|---|---|---|---|
| Att#1 | 2009-05-13 12:13 | | 0.00 / 1 | ✓ | |
| Att#2 | 2009-05-13 12:13 | | 0.00 / 1 | ✓ | |
| Att#3 | 2009-05-13 12:13 | | 0.00 / 1 | ✓ | |
| Att#4 | 2009-05-13 12:13 | | 0.00 / 1 | ✓ | |

Without academic validity

First, the lecturer defines the different blocks, which are going to be evaluated by a given grade. All these grades together will compose the final grade of the subject. Usually the type of block will be the most general one, which is weighted and selectable. For example of the chapter, three different blocks are defined: theory, lab and attendance.

The next step is to configure all the tasks that form each block together with their different attributes. For example, the student has to pass the three lab tasks (L1, L2 and L3) all together giving them a different weight to each other (20%, 35% and 45%) and finally, the theoretical exam (Th_FEB of Th_JUN) must be passed by the student (see Figure 6).

After that, the paths have to be set up. All the different conditions, which define the final grade, have to be defined in the right order. The weighted mean of the tasks L1, L2 and L3 is calculated to compose the lab grade (Lab). If the result is smaller than 5, the student has to take the lab exam (Exam Lab June). In theory, the student has to pass the February exam and if he/she fails with a grade less than 5 then the exam will have to be repeated in June by the student (see Figure 7).

It is possible to set up an evaluation path for each block. Finally, the grade of the subject (Mark) is calculated from the mean value of the grades in each block.

When the student evaluates a block, the lecturer applies the conditional calculation and the program selects (activates) the right grades to calculate the resulting grade of the student depending on if he has fail a part or not and so setting up the right evaluation path (see Figure 8).

## NEW TOOL TO CREATE ONLINE EXAMINATIONS

The other function which had been decided to improve after the in depth study was the Assessment one. This module, as it has been explained, allows lecturers to create examinations or assessments online. With this tool the lecturer can also select the type of answer (multiple choice, short answer, long answer, uploaded from an external file); fix some presentation related parameters (order, number of questions per page); establish the duration of the exam; reuse components (questions, sections and answers); and import and export questions and/or complete exams. These examinations or assessments are mainly tests and can be active in different periods of time previously determined. In addition, the online examinations are automatically corrected. The students will therefore be aware of how much they know in regards to the contents of a theme at the moment they do the exam. This module can be very useful in the new concept of education. Nevertheless, as it has been said, this tool is one of the less used in the University of Valencia LMS. The main reason is due to the unfriendly interface in dotLRN and the fact that every new question must be created separately, as shown in Figure 9.

As a result of the in depth analysis made, it was concluded that, although this was an excellent tool for student progress monitoring, its use was scarce. One of the main reasons identified was that the use of the tool was far from intuitive. Consequently (and in order to encourage the use of this module), it was decided that it was necessary to simplify the creation of online examinations or assessments. In this respect, the easiest and most effective approach was to

*Figure 8. Evaluation state after the ways algorithm is applied. On the right in the column "Active" it can be seen that not all grades are selected*

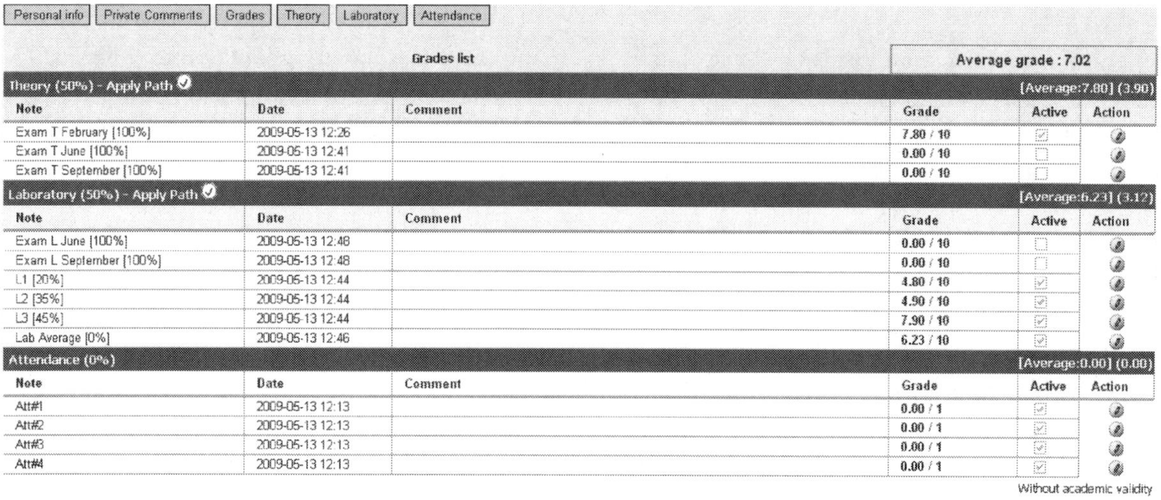

*Figure 9. Setting of a question in the dotLRN Assessment module*

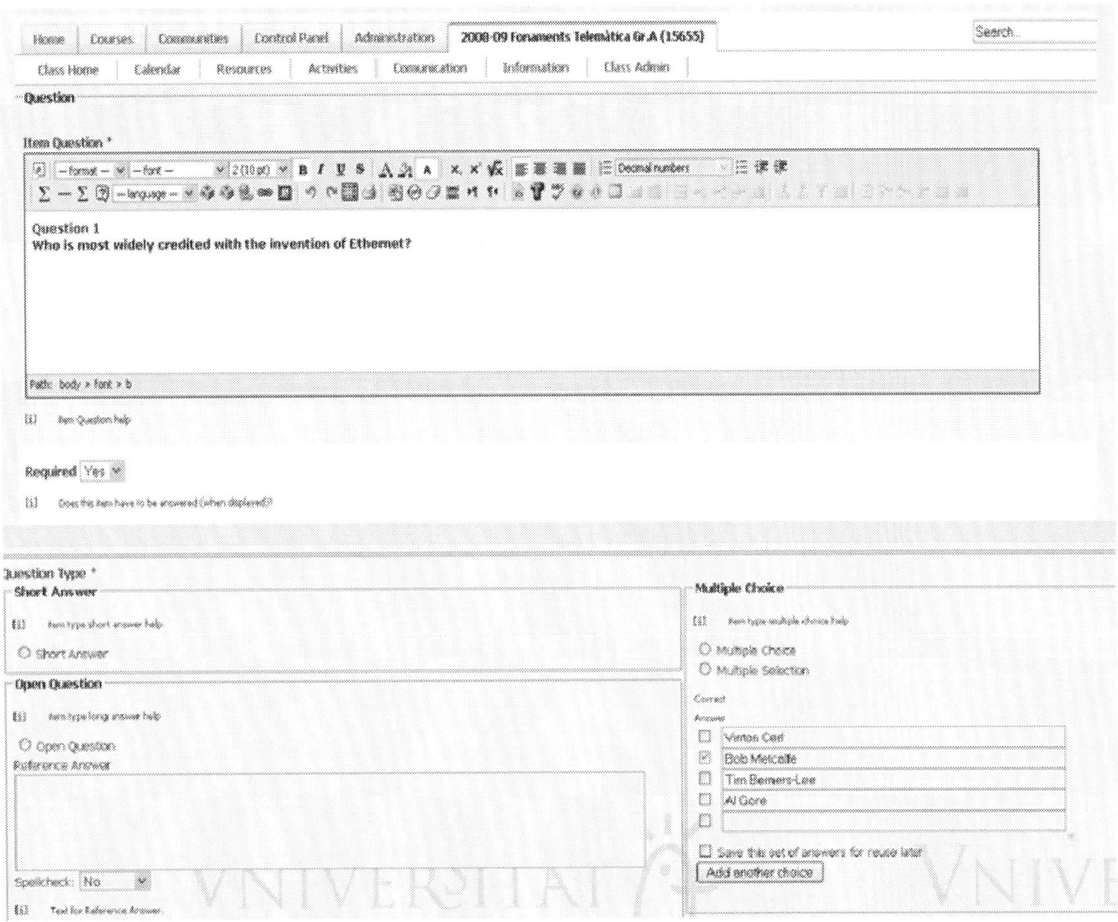


use an external tool, which allowed lectures to create online examinations or assessments, and export them into Aula Virtual once they had been created. The University of Valencia was using Respondus (del Ramo & López-Soler, 2008), commercial software that also allowed the user to edit, import and export examinations created in other formats. The main inconvenience of using this product is the monetary cost of the license associated with it.

For this reason, the decision was to replace this software with an open source tool to provide the same functionality which can be integrated easily on other open source LMS and not only into the University of Valencia's one. The improvement of the Assessment Module was the inclusion and integration in the system of a new tool to create online examinations or assessments easier.

As a result of an investigation of existing tools, it was concluded that the University Pompeu Fabra had already produced a java open source application called QAED (Martinez, 2005). This program allows the user to create multiple choice examinations or assessments and import and export from/to version 2 of the ims-qti specification (by using Java library named qti-lite) (Sayago, 2005). Although this tool does not implement a large part of the functions required by the University of Valencia, it was considered both, a good candidate for extension and a better alternative than starting a new application from scratch.

## Analysis

Once the target of the project had been established, an extensive study was carried out in which the two tools involved were analyzed (dotLRN's Assessment module and QAED).

### Requirements

The analysis of the dotLRN's Assessment module produces the requirements that the extended version of QAED (QAED LINK, 2005) should comply with to fulfill the particular needs of the University of Valencia. One difference in functionality is that the Assessment module offers various types of questions, unlike QAED that only offers multiple choices. Moreover, examinations created in dotLRN contain many options that can be activated or modified by lectures, in order to adjust the examinations to the type of evaluation process that best suits their needs.

Another important difference is in the export function. dotLRN generates an XML file which complies with version 1 of the ims-qti specification (compressed in ZIP format). Therefore, our extension of QAED must also allow the user to export the examinations in this format (QAED supported version 2 instead).

The last issue to consider is that dotLRN's examinations contain three major components. The basic unit of an examination is the question. Questions are grouped into sections and an examination is composed of multiple sections. A question cannot be found inside an examination unless it belongs to a section. Each section contains a series of parameters, which indicate how they should be presented, their score, and the time needed for their completion. In QAED, examinations are composed into multiple-choice questions only. This makes it necessary to adapt our extension of QAED so that examinations have the same structure as in dotLRN.

### QAED

Once the requirements have been determined, the first stage is to analyze QAED to fully understand its internal structure and to evaluate the effort required to implement the extensions.

QAED was developed in Java. The fact that QAED is a desktop application implies that it uses javax to implement its graphic interface. The GUI (Graphical User Interface) appears to be divided into three sections. The menu bar at the top can be used to edit the examinations. Below this bar there are two panels. The left panel contains a chart that visualizes the structure of the examinations that are currently open. The right panel shows the possible actions and content of the components that are selected in the chart.

The original version of QAED uses three libraries, namely JDOM, Xerces, Xalan and jQTI-Lite. These libraries are used to manage XML documents and will be the ones used to provide import and export functionalities. JDOM, Xerces and Xalan facilitate the generation of XML documents (to export files) and the extraction of information from existing ones (to import files). JQTI-Lite is a library that is used to import and export documents in a format compatible with version 2 of the ims-qti specification.

## Design

### Data Structure

A first design decision is based on the data structure required to support the new components that dotLRN offers. The most logical and intuitive structure has to consider a Java class per type of component: Assessment, Section and Question. However questions can be of different kinds, the class Question is declared as abstract and a subclass is created for each type of question supported: MultipleChoiceQuestion, ShortAnswerQuestion,

LongAnswerQuestion, UploadedAnswerQuestion (for questions which require the upload of an external file). This class structure (illustrated in Figure 10) is used to store information concerning the different components. This information is obtained as user input (using data forms) or from XML files (by importing data), and used to execute the assessments or to export them to other formats.

Collections (represented by a black diamond in the diagram of Fig. 10) have been implemented using the class java.util.List. The root component for the structure is the class Repository. This class was already implemented in the previous version of QAED, but has been modified to support the extensions. A repository may contain folders (a recursive structure which may also contain other folders, apart from assessments and sections). Both Assessment and Section objects can be directly linked to a Repository or Folder. A Repository can also be stored on file with a PC.

## Graphical User Interface

### Menu Bar

The next step was to decide how the new functions should be incorporated into the existing application. For this purpose, a new menu group was added into the menu bar (called "UV"). This menu contains three second-level menus: "UV Assessment", "UV Section" and "UV Question", each of them providing a set of actions, which are active only when a component of that type is selected (except for the case of creating a component, which is always active, as long as a repository has already been created).

New items have also been incorporated into the toolbar that appear just below the menu. These allow us to create new assessments as well as sections and questions by pressing the button with the appropriate icon. The button which contains an 'x' symbol can be used to eliminate the component that is selected at the time. Additionally, the two buttons with arrows can be used to export and import examinations (the export button will only be available when a "UV assessment" is selected).

Both the new menu and the buttons use event listeners to carry out the actions. These methods can be found in the main class of the GUI (QTILiteAppMainFrame), along with all the codes needed to create the main application window.

### Forms and Wizards

Once the menu had been created, the next step was to create the forms that would allow us to create and edit the University of Valencia's components. To create the components, it was decided to use simple wizards (a series of JDialogs that will successively appear one after the other, until all the necessary information has been filled in). In

*Figure 10. UML static class diagram that represents some of the major application classes required storing an assessment or online examination*

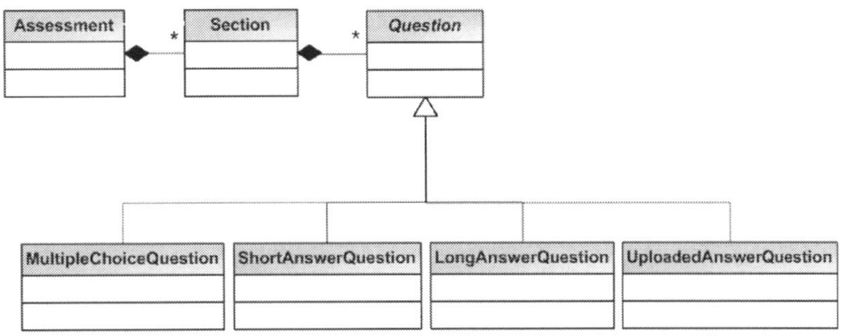

all the cases, the first panel contains a flow chart representing the structure of the entire repository and the user is asked to select the new component's location. Since components can only be placed in certain places, depending on their type, (questions inside sections, sections inside assessments, and assessments inside folders or repositories), error messages are generated if an invalid location is selected.

**Assessment Creation Wizard**

The wizard to construct a new examinations or assessment is composed of the following dialogs (which are displayed sequentially): Title and Description, Instructions, Answer Options, Pages, Timing Options and Other Options. The sequence of forms is explained below.

- First, the title and Description dialog is displayed. This is the only dialog that contains three buttons ("Cancel", "Complete", and "Accept"). The reason for this is that there are two types of examination, namely "Simple" and "Expert". By clicking on the "Accept" button, a simple examination is created. Likewise, clicking on the "Complete" button creates an "Expert" examination. In both cases the button's listener is triggered and checks that the title field is not empty.
- The Instructions dialog is opened. This contains a text area to introduce the information needed by the student to therefore perform the examination.
- The Answer's Option dialog takes control. This dialog contains three numeric fields and therefore, the Accept button's listener will verify that they contain valid content (an Integer).
- The Pages dialog opens up, providing a set of tabs (JTabbedPane objects) that are browsed to fill in the different pages. At the Start and in the End Pages there is a rich text editor that will introduce the HTML code that generates into the expected attribute of the corresponding class. To provide this rich text editor, an external open source java library called Ekit has been used.
- The Timing Options dialog is used to specify the examination time. Another external java library has been used for this purpose (JCalendar). This is used to insert two calendars inside the dialog, which will allow us to select the examination's start and end date. Besides, a set of drop down lists is provided to select the start and end time (only one of the checked boxes associated with then have been selected). By using a list, the introduction of invalid values is avoided. If both the start and end checked boxes are selected, the Accept button listener will verify that the end time and date are after the start time and date.
- Finally, the Other Options dialog allows that an IP mask is specified. In this case, the Accept button listener checks that the IP mask has a valid value.

**Section Creation Form**

To create a new UV section on a selected location, a single dialog is required. This dialog contains a set of tabs (JTabbedPane) that the user needs to browse in order to specify all different parameters. The main one is the Visualization Mode tab. In QAED, visualization modes determine how sections are presented to students. Each repository has a list of visualization modes, and these can be applied to sections. The Visualization Mode tab provides a drop down list, which contains all visualization modes available for the current repository to allow the user to select one of them for the section. This drop down list also contains an option called "New". If this is selected a new dialog that allows the user to create a new visualization mode is displayed. This dialog is composed of two tabs. The first one contains mandatory fields, whereas the second one contains those that are optional.

### Question Creation Dialog

To create a new Question within a section, another simple form is used. Again, a set of tabs (JTabbed-Pane) allows the user to specify all the parameters required. The form contains two mandatory fields: the description and the type of question (which are selected from a drop down list). Once a question type has been selected from the drop down list, a new panel appears inside the tab. This panel shows the type of question selected and requests the values for all the parameters associated with this type. For example, if a multiple-choice question was selected, it would be necessary to specify if the answer is multiple or unique, the text for each proposed answer and which ones are correct. In this case, the "Accept" button listener verifies that all mandatory fields have correct values and then stores the values in an appropriate structure for each question type.

### Edition Dialog

A dialog has also been implemented to edit the components that have been created using the forms and wizard described above. These dialogs are very similar to those used for creation, but present a few exceptions.

First, all dialogs used for editing purposes are internal frames (subclasses of JInternalFrame). This makes it possible to place them inside the panel located at the right hand side of the main frame. Second, the edition of an examination is not performed using a wizard. Instead, an equivalent functionality has been provided using a set of tabs in a single form. Finally, a new tab has been introduced in the cases of assessments and sections. In the assessments, the new tab includes information on the sections it contains, and allows the user to modify their contents. Likewise, the new tab in the section's dialog includes information on the questions it contains, and therefore allows the user to edit them. In both cases, the new tab is composed of a drop down list and five buttons namely "Edit", "Remove", "Add", "Move", and "New". The "Edit" button opens an edition dialog for the selected question or section and it allows us to edit all its attributes. The "Remove" button eliminates the selected question or section from the list of components that are being edited. The "Add" button opens up a dialog in which the repository's chart will be find and will allow the user to add a question or section to the component which is being editing. The "Move" button moves or copies the selected component to a location in the repository tree (erasing it or not from its previous location). Finally, the "New" button opens the creation dialog for the selected question or section.

### Import and Export Functions

In this section the implementation of the classes required to deliver the import and export functions are described. These classes use an external java library called JDOM. This library provides a series of java classes useful to read, write and manipulate XML files. Particularly, it allows the programmer to convert an XML file into java objects and vice versa (create valid XML files from a set of java objects).

To export an examination the corresponding option is selected from the menu (or from the toolbar located just below). Then, a typical file browser dialog requesting the location and file name is displayed. Once these are specified, the XML file is generated and then compressed into a ZIP file. The XML file generation process involves creating each element (one by one) using JDOM, and storing them into the correct place in the XML file, according to specification of version 1 in the ims-qti.

In the case of an import operation, the XML document is first extracted from the ZIP file selected. This is then used to generate the tree structure of a QAED assessment, using JDOM. Once the import operation has been completed, the Assessment java object created is stored in the root of the repository.

## FUTURE RESEARCH DIRECTIONS

Comparative analysis made at The University of Valencia also showed that Documents was the most used tool. Nevertheless, dotLRN does not have the possibility of dynamic content creation, which could be visualized and edited at the platform. For that reason, an active collaboration between two dotLRN Community members, University of Galileo of Guatemala and University of Valencia is maintained, in order to create a content creation tool for Learning Objects related to the dotLRN LORS package. The use of this package will allow creating adaptive and collaborative learning models. On the other hand, it will allow creating re-usable contents not only in different courses and groups of one institution, but also in other LMS. This interoperability will be guaranteed by using standards in contents creation.

Some short term objectives can be mentioned in this collaboration such as: providing a simple interface to create web pages, to easily include and manipulate web assets such as flash, videos, images, etc., to provide a set of web templates, easy to manage, like PPT templates, to provide collaborative content creation tools while being able to set up roles easily, which also has a folder, subfolders and pages for ordering free of "standards" and approaches, so professors with basic text edition knowledge can use it. If the created tool already uses existing content creation tools, it is important to try and keep it as customization, rather than a fork. Mid and Long Term objectives have also been fixed. Some of them are: to automatically export to SCORM or IMS-CP, to import and publish to LORS with one click feature, to define workflow, and to integrate third parties API or resources such as Flickr, Google, Amazon, YouTube and others.

The way to reach the objectives is to start making a full review of the tools available, possible approaches, etc, and to build a prototype that comes close to the short term objectives. The starting point has been an in depth study of such type of tools: XOWiki, LORSM, LORS Central built-in editor, etc…

Keeping in mind the main objective of improving the University of Valencia's LMS, other future directions of research is to integrate the Multimedia Server of the university with the platform. This would allow using and reusing the objects of this server as Learning Objects, to integrate Synchronous Learning Software based on open source, as DimDim, to improve the academic community experience and the learning and teaching process is also a new idea.

## CONCLUSION

The convergence process towards the ESHE has changed the universities subjects' evaluation process. Now, less weight is given to the examinations. This is due to the fact that the new learning techniques introduced evaluate the students' daily work, instead of relaying on a single test mark. This leads to the definition of a set of tasks to be fulfilled by the students and evaluated by the lecturer. These different tasks can belong to different blocks and can have different weights in the final mark. The student can also choose between different evaluations paths to reach his/her final mark. Therefore, he/she needs to know what the value of his/her mark is at any given moment during the semester. These issues must be considered when developing the new software tool of Management of Evaluations which has been integrated into the Activities and Cards functionalities of the University of Valencia's LMS and simplifies all this work.

After the new algorithm had been set up, the new program of Management of Evaluations was developed and tested with different flow charts deployed for all different evaluation processes, which were successfully reproduced with the new software tool. Therefore almost all the evaluation processes in the University of Valencia can be set up with this new tool. With the old Activities tool all the possible evaluation processes could not be set up, and therefore, lecturers, who were interested in its use were unable to manage it and did not stick to the dotLRN original module.

Right now, the new Management of Evaluations tool has already been installed on the LMS development platform and the Computer Services Office is testing it to deliver a bug free version to the academic community. At the beginning of the academic year 2009-2010, it will have been provided as an option that everybody will be able to use. Of course, the development will continue to further improve this tool.

As for the Tool to create online examinations or assessments, integrated into the Assessment functionality of the University of Valencia's LMS, it has been said that the original Assessment module provided by dotLRN allowed lecturers to setup online examinations (Moreno & Cerverón, 2006). Despite the potential of this tool, it presented a low usage rate (Moreno, Cerverón & Arevalillo, 2008). This was in part due to an uncomfortable GUI and the need to prepare the examinations online. To solve these limitations an existing open source product (called QAED) was extended to provide easy of use. An open source desktop application was implemented. It fulfilled the objective of creating, exporting and importing examinations, which complied with those in the dotLRN assessment module.

This new online examination created software offers with the following advantages: a) it allows lecturers to prepare examinations offline; b) simplifies the creation process; c) reduces the time required to prepare an examination considerably and d) it can be used at no cost, avoiding expensive licenses required by other alternatives.

Finally these two new tools are implemented on an open source LMS and therefore can be used in any other higher education institution which uses dotLRN. This fact also improves the cooperation with the open source community (dotLRN and OpenACS). Any kind of large problem can be split into smaller problems and solved by small working groups. This allows the improvement of tools like dotLRN (Moreno et al. 2007). In the future, it is intended to continue the cooperation with OpenACS (Hernández, 2005), dotLRN (dotLRN, 2009), the E-LANE project (ELANE, 2009) and the INNOVA group of UNED (INNOVA, 2009) to share experience and reach new objectives.

With these two developments, it is expected that usage rates for these recently improved modules of Activities and Assessment is increased and lecturers are encouraged to make use of these tools.

## ACKNOWLEDGMENT

The authors would like to thank the Computer Services Office (SIUV) and the European Convergence Office (OCE) of the University of Valencia for their help and support.

## REFERENCES

Barajas, M., & Gannaway, G. (2007). Implementing e-Learning in the Traditional Higher Education Institution. *Higher Education in Europe*, *32*(2-3), 111–119. doi:10.1080/03797720701840609

Brooks, L. (2003). How the Attitudes of Instructors, Students, Course Administrators, and Course Designers Affects the Quality of an Online Learning Environment. *Online Journal of Distance Learning Administration*, *6*(4).

Cerverón, V., & Moreno, P. (2006). Aula Virtual: an e-Learning Management Platform of Universitat de València Development based on Open Code and Collaborative Software. In *Proc. M-ICTE, IV Conferencia Internacional sobre Multimedia y Tecnologías de la Información y Comunicación en Educación* (Vol. 2). Badajoz, Spain: FORMATEX.

Cerverón, V., Moreno, P., Cubero, S., Roig, D., & Roca, S. (2007). Universitat de València's Aula Virtual: a Single Integrated LMS for a University. In *Proc. IADIS e-Learning 2007 Conference Proceedings*, IADIS Digital Library, Lisbon, Portugal.

Colla, J., MacDonald, C., & Thompson, T. L. (2005). Structure, Content, Delivery, Service, and Outcomes: Quality e-Learning in higher education. *International Review of Research in Open and Distance Learning*, *6*(2).

Consejo de Coordinación, U. en colaboración con la Comisión para la Renovación de Metodologías Educativas en la Universidad. (2006). *Propuestas para la renovación de las metodologías educativas en la universidad.* [Online]. Retrieved April 25, 2009, from http://www.micinn.es/univ/ccuniv/html/metodologias/docu/PROPUESTA_RENOVACION.pdf

Cubero, S., López, A., Roig, D., & Roca, S. (2006). Implantación de.LRN en la Universitat de València Estudi general. Integración de la plataforma y desarrollos propios. In Proc. Foro e-learning integral: Estratregias y experiencias, Universidad Galileo, Guatemala.

Del Ramo Romero, J., & López-Soler, N. (2008). Respondus: un programa para la creación y publicación de cuestionarios. *@tic. Revista d'innovació Educativa. dotLRN Consortium.* [Online]. Retrieved April 25, 2009, from http://www.dotlrn.org

*E-LANE Project.* [Online]. Retrieved April 25, 2009, from http://www.e-lane.org

Essa, A., Cerverón, V., & Blessius, C. (2005). An Enterprise Open-Source Learning Management System. In *Proc. Educause*. Orlando: LRN.

Fuente, L. (2006). Máquina virtual utilizada para el desarrollo del proyecto. *Proyecto e-lane*. [Online]. Universidad Carlos III de Madrid. Retrieved April 25, 2009, from http://e-lane.org/vmware/

Greespun, P. (1999). *Philip and Alex's Guide to Web Publishing*. San Francisco: Morgan Kaufmann.

Hernández, R. (2005). OpenACS: robust web development framework. In *Proc. Tcl/Tk 2005 Conference Proceedings*, Portland, Oregon.

Hernández, R., Morales, M., De la Roca, M., & Guerra, V. (2006). Proyecto E-LANE en la Universidad Galileo. *Resumen resultados E-LANE/UG v. 1.0.* [Online]. Retrieved April 25, 2009, from http://ges.galileo.edu/fs/view/articulos/resultados-UG.pdf

Hillar, G. (2006). PostgreSQL 8.1.4: robusto y fácil de administrar. *Mundo Linux: Sólo programadores Linux*, 86.

*I Jornadas de experiencias docentes con el Aula Virtual de la Universitat de València.* (2006). [Online]. Valencia, Spain: Universitat de València.

INNOVA. (n.d.). *Innovation Section of UNED.* Universidad Nacional de Educación a Distancia. [Online]. Retrieved April 25, 2009, from http://innova.uned.es/

Lee, W., & Owens, D. L. (2000). *Multimedia-Based Instructional Design: Computer- Based Training in Web-Based Training, and Distance Learning.* New York: Wiley.

MacDonald, C., & Thompson, T. (2005). Structure, Content, Delivery, Service, and Outcomes: Quality e-Learning in higher education. *International Review of Research in Open and Distance Learning*, *6*(2).

Marqués, P. (2008). Impacto de las TIC en la enseñanza universitaria. *Didáctica, Innovación y Multimedia, 4*(11).

Martinez, J. (2005). *JQTI-Lite Library: Developer's Guide (1)*. [Online]. Retrieved April 25, 2009, from http://www.tecn.upf.es/~ssayag/leteos/jQTILite/docs/jQTILite10_Devel_Guide.pdf

Moreno, P., & Cerverón, V. (2006). Plataforma tecnológica para potenciar los procesos de enseñanza-aprendizaje: desarrollo en la Universitat de València basado en software libre y colaborativo. In *Proc. SIIE 06. VIII Simposio Internacional de informática aplicada a la enseñanza. 8th International Symposium on Computers in Education* (Vol. 2).

Moreno, P., & Cerverón, V. (2007). Enhanced Learning Through Blended Learning: Educative Innovation at the Universitat de València. In *Proc. iNEER, International Conference on Engineering Education*. Retrieved April 25, 2009, from http://icee2007.dei.uc.pt/proceedings/index.html

Moreno, P., Cerverón, V., & Arevalillo, M. (2008). Análisis del uso universitario de plataformas de gestión del aprendizaje y su relación con la innovación educativa y la calidad docente en el marco del EEES. Resultados en la Universitat de València y procedimientos para la generalización. In Proc. Conferencia sobre Software Libre en Educación Superior, Valencia.

Moreno, P., Cerverón, V., López, A., & Roig, D. (2007). LRN Consortium: International Collaboration for Developing a Learning Management System. Experience from the Universitat de València. In *Proc. iNEER, International Conference on Engineering Education*. Retrieved from http://icee2007.dei.uc.pt/proceedings/index.html

Moreno-Clari, P., & Cerverón-Lleó, V. (2007). Encouraging Blended Learning and ICT Use at Universitat de València to Improve the Learning Process with the.LRN Platform: Best Practices and Tools. In *Proc. EATIS 07, Euro American Conference on Telematics and Information System, EATIS'07*, Algarve.

Pallof, R. M., & Pratt, K. (2003). *Collaborating Online: Learning Together in Community*. San Francisco: Jossey-Bass.

Perens, B. (1999). *The Open Source Definition. Open Sources: Voices from the Open Source Revolution*. Sebastopol, CA: O'Reilly & Associates.

*QAED LINK*. (2005). [Online]. Retrieved April 25, 2009, from: http://gti.upf.edu/leteos/newnavs/qaed.html

Raymond, E. (1999). Shut Up and Show Them the Code. *Linux Today* [Online]. Retrieved April 25, 2009, from http://linuxtoday.com/stories/7196.html

Santos, O. C., Gonzalez-Boticario, J., & Barrera, C. (2005). aLFanet: An adaptive and standard-based learning environment built upon dotLRN and other open source developments. In Proc. Foro hispano de LRN. Madrid: Congreso de usuarios y desarrolladores de LRN.

Sayago, S. (2005). *User Manual v1.0 QAED: Assessments and Question editor (1.0)*. [Online]. Retrieved April 25, 2009, from http://www.tecn.upf.es/~ssayag/QAed10_User_Manual.pdf

Soler-Lahuerta, F., Cubero, S., López, A., Roig, D., & Roca, S. (2005). Desarrollo del módulo fichas en la estructura de la UVEG para la herramienta groupware LRN. In Proc. Foro hispano de.LRN. Madrid: Congreso de usuarios y desarrolladores de LRN.

Stevens, J. A. (1999). *A Modern Approach to Intermediate Statistic* (2nd ed.). Mahwah, NJ: LEA.

*TCL Manual of Work*. (2008). Retrieved April 25, 2009, from http://tmml.sourceforge.net/doc/tcl/index.html

Thompson, P., & Randal, B. (2001). Can E-Learning Spur Creativity, Innovation and Entrepreneurship? *Educational Media International*, *38*(4), 289–292. doi:10.1080/09523980110105105

Wong, H. K. (2003). *How to Improve Student Achievement*. Indianapolis, IN: Harry K. Wong Publications.

Zurita, L., & Ryberg, T. (2005). Towards a Collaborative Approach of Introducing e-Learning in Higher Education Institutions. How Do University Teachers Conceive and React to Transitions to e-Learning? In *Proc. 8th IFIP World Conference on Computers in Education*.

Section 4

# Learning Management Systems and Best Practices in Online Education

# Chapter 14
# Improving Hybrid and Online Course Delivery Emerging Technologies

**Nory B. Jones**
*University of Maine Business School, USA*

**Christian Graham**
*University of Maine Business School, USA*

## ABSTRACT

*As educational budgets continue to shrink, colleges and universities have turned to online course delivery as a means of increasing enrollments. In addition, with the proliferation of Internet-based course management and other software that facilitate the learning experience, many traditional courses are adding an online component, creating hybrid courses in different formats. In this chapter, the authors explore different strategies and technology solutions to help instructors develop rich, dynamic courses, whether they are completely online or hybrid courses that use online tools and technologies to augment the traditional class. This chapter covers the advantages and disadvantages of hybrid courses, technologies and practices available for them, emerging technologies such as Second Life™, social networks, dense wavelength division multiplexing, telepresence, satellite networks, and the use of texting in the classroom.*

## INTRODUCTION

Online education is often defined as involving the Internet and web-based technologies to deliver distance education. It can be delivered asynchronously, where the students and instructor do not communicate in real time, using web-based technologies such as asynchronous discussion boards, electronic repositories, and e-mail. It can also be synchronous, where the students and instructor communicate in real time using web-based technologies such as chat rooms or video conferencing over the Internet (Martinez, 2004).

While online education has become routine with 65% of graduate programs across the country using the Internet to deliver classes (Norton and Hathaway, 2008), many colleges and universities are still struggling to discover how to provide a quality educational experience. Menchala and Bekele (2008) found that a variety of technologies and learning styles as well as instructor and administrative support are required to achieve an

DOI: 10.4018/978-1-61520-853-1.ch014

Copyright © 2010, IGI Global. Copying or distributing in print or electronic forms without written permission of IGI Global is prohibited.

engaged, productive, and collaborative classroom experience.

This chapter offers a variety of strategies and technological solutions that educational institutions, academic administrators, and faculty could employ to improve hybrid course delivery in distance education with emerging technologies. Readers could consider this chapter as a resource guide for delivering hybrid education. To this end, the chapter specifically addresses; what is distance education today and hybrid course delivery, advantages and disadvantages of hybrid courses versus "pure" online and traditional courses, emerging software and information systems that support teaching and learning online, and strategies for overcoming existing challenges of the classroom today and tomorrow.

## BACKGROUND

### Distance Education Today

Distance education is often defined as instructional delivery where the student is not in the same physical space as the instructor and other students. Most distance education today is delivered via Internet technologies. Distance courses can be completely synchronous where the instructor and students meet together in virtual environments such as live chat, video, or audio streaming at the same time. They can also be completely asynchronous where the students and instructors use web-based technologies such as discussion forums, blogs, wikis or social networking tools to communicate at different times and on their own schedules. Courses can use a blend of synchronous and asynchronous communications and technologies to enhance the total experience, which we refer to as hybrid classes. In addition, hybrid classes can mix a traditional face-to-face class with different online technologies. For example, many university courses use information systems like Blackboard/WebCT for course management. It is also not uncommon for a live class to be broadcast via streaming Internet video and also archived for later Internet viewing.

The value of distance education continues to grow in importance. In 2007, it was estimated that at least 2.3 million students were enrolled in online classes in the United States (Smathers, 2007). According to the GAO (2007), enrollments in higher education distance classes have almost quadrupled since 1995. The Sloan Consortium reported that over 3.9 million students nationally took at least one online course in the fall 2007 semester. This represented a 12% increase over 2006 statistics (Heck, 2009). Distance courses and programs in higher education represent a real avenue for growth, expanding opportunities for traditional and non-traditional students. This represents an opportunity as well as a reality for most institutions of higher education. With the cost of tuition rising and economic downturns across the country, providing an alternative means of educational delivery benefits everyone. Less time and cost in traveling, more potential course availability, and greater access and convenience all contribute to the value of distance classes (Jackson & Helms, 2008).

## TECHNOLOGIES AND PRACTICES FOR ONLINE EDUCATION

### Hybrid Course Delivery in a Mixed Educational Environment

Traditional classes are those that meet in a physical space in real time with instructors and students present. Pure online classes, which can be synchronous (in real time) or asynchronous (the meetings are not at the same time) are held virtually, usually via the Internet. Finally, hybrid courses use a variety of techniques to combine traditional and online courses. For example, a traditional course may have online asynchronous discussions. Many variations can exist for hybrid courses.

Prior research has revealed some patterns relating to the quality of both online and traditional classes. A consensus is that the course delivery methods, student participation and involvement, the quality of instruction as well as the class culture for learning, the course and program administration, and support all represent vital factors contributing to the quality of the learning experience (Miller & Husmann, 1996). While these factors are important in all forms of class delivery: traditional, online or hybrid, researchers have found that the traditional face-to-face class usually is preferred by students. However, with new, emerging technologies and software products, online or hybrid classes may be able to simulate the rich, contextual face-to-face classroom experience if developed and implemented properly. In an interesting study, Jackson and Helms (2008) found that students offered positive and negative perspectives about hybrid classes. For example, while the distance students had more time to digest and reflect on the material, the in-class students had richer contextual cues and explanations about the content from the professor. Similarly, distance students found that the professors who provided a variety of delivery systems (online content, YouTube videos, discussion forums, etc.) enhanced the depth of the content and class experience. In contrast, the live students still benefited from in-class explanations and the ability to explore a topic as questions arose. Therefore, both systems have their advantages and disadvantages and hybrid classes still benefit and suffer from those factors. The challenge is how to enhance the benefits while reducing or mitigating the negative aspects. New, emerging technologies and software solutions may help.

With the exponential growth of Internet use, the first collaborative technologies were termed Web 1.0. These included e-mail, video-conferencing, chat and discussion forums. One significant form of the discussion forum was called the "Community of Practice" or CoP. (Wenger and Snyder, 2000). Social learning theory suggests that people who form personal networks or communities develop relationships that enhance the learning experience. Therefore, an important component in a distance class is to actively nurture and promote these communities of practice where members regularly engage in sharing and learning, based on common classroom goals and projects. Another benefit to these Web 1.0 communities (often called discussion forums that are divided into specific interest or project groups) is the transmission and spread of tacit knowledge among participants to increase the overall learning experience.

Research in distance learning has shown that a strong sense of community is crucial in the success of distance classes. Another important element in addition to trust and rapport is a shared sense of purpose in the learning community (Conrad, 2008). For example, students who work collaboratively on projects develop camaraderie and a sense of purpose in the achievement of their class goals.

Participants in online classes also succeed when they have a high comfort level with the technologies required for use in the class. For example, using software like Blackboard/WebCT or searching the library's online databases for research must be user friendly and seamless in the course.

New, emerging Internet-based technologies, called Web 2.0 should also be considered for distance classes. These include online social networks, blogs and wikis. As these have become more widespread across many different age cohorts, their use in distance classes should be considered among the many tools to incorporate.

Social networks, such as Facebook, have become the de-facto communication tool among traditional college students. However, these networks are also being adopted by people in all generations and many socio-economics groups. The ability to easily communicate via instant messaging, images and the incorporation of video such as You-Tube holds great promise in the virtual classroom. These new technologies and tools also have features that greatly facilitate work among students, which include shared calendars to coordinate work and scheduling, document editing, white board collaboration, and task management capabilities. Social desktops are new tools that offer a desktop area where many different applications (calendars, forums, messaging, etc.) can be accessed by collaborators in a unified manner. Even widely accepted distance educational technologies like Blackboard (WebCT) have now incorporated Web 2.0 features. For example, Blackboard 9.0 has added blogs and journals that students and professors can use to easily capture and share their reflections and knowledge (Redden, 2009). These updated distance technologies are also using the digital dashboard approach commonly found in the business world to make different applications easily available to students and professors.

Another value of incorporating social networking technologies involves the ability to engage the students and increase their comfort level with the technologies and thus, with the course. According to the Technology Acceptance Model (TAM), user attitudes influence their acceptance and adoption of technologies (Lee et al., 2003). Therefore, by incorporating widely used and familiar technologies like Facebook into an online course, student can "hit the ground running" when they begin the course, feeling confident in their use of technology and the ability to communicate effectively with fellow students and the professor. Another interesting example of a social networking tool is Twitter. This performs similar function as Facebook, and is also a widely used and accepted communication platform. According to Ritter (2008), "At Penn State, twitter has changed the culture on campus and has given us ways to connect across our university that we couldn't have imagined. We've used twitter to ask for help, work on projects, discuss topics during conferences, schedule impromptu lunches, and offer things for sale. We've planned meetings, found opportunities to collaborate and have become a much more connected, intelligent, communicative group that now includes people from several Penn State campuses, departments and academic colleges. We are IT professionals, professors, advisers, learning designers, and students. We have used twitter to build a community that now thrives at Penn State." Her statement clearly shows the value of a technology like this in an online class as well as across the campus.

Similarly, widespread use of video technologies such as You-Tube greatly enriches the educational experience, particularly in an online class, where students appreciate the real examples and audio-visual experience.

Beldarrain (2006) encourages the use of new, emerging technologies like Blogs (Weblogs), wikis, and podcasts. Blogs encourage writing and reflection while wikis (central knowledge repositories like Wikipedia) encourage contributions to collective knowledge building. According to Wikipedia http://en.wikipedia.org/wiki/Podcasting, "A podcast is a media file that is distributed by subscription (paid or unpaid) over the Internet using syndication feeds, for playback on mobile devices and personal computers." They can be used to download audio or video files on a computer, creating new representations of ideas and communication. Bonk and Zhang (2006) extend this model by suggesting that these emerging technologies support different learning styles. For example, podcasting would facilitate an audio-learning style while blogs or communities of practice / discussion forums could support an innovative, reflective learning style. Instant messaging, chat, online simulations, and portals represent other potential technologies to create active learning in online classes.

## Advantages and Disadvantages of Hybrid Courses vs. "Pure" Online and Traditional Courses

As discussed earlier, there are significant differences between traditional face-to-face classes vs. pure online classes as well as hybrid classes. While there are challenges in each type of course delivery, there are also commonalities in learning theories that can be used in any of these classes. Table 1 presents a summary of these factors.

Overview of the Advantages & Disadvantages of the different classes:

- Traditional face-to-face courses:
    - Advantages: Personal attention, ability to interact in a real-time environment, adapt to changing conditions or nature of the class and content, and immediacy of responses and interactions among professors and students. The professors and students also have the opportunity to develop trust, rapport and relationships over the semester. Abstract and complex issues are easier to convey and explain in a face-to-face environment.
    - Disadvantages: Inconvenience of time, travel, short, limited duration of the class, personality conflicts, not enough time to digest and reflect on class presentations and discussions.
- Pure online courses:
    - Advantages: Convenience, accessibility, ability to spend time on course content and class discussions before responding.
    - Disadvantages: Lack of rich, contextual cues from face-to-face interactions, potential feeling of separation from class, instructor, classmates, difficulties in using technologies.
- Hybrid courses: The positive aspects of both traditional and online courses can be incorporated to enhance the overall experience such as use of virtual networking tools to bring everyone together. On the other hand, the negative aspects such as a dichotomy between the traditional and online students can also be a disadvantage to a hybrid class. These topics will be explored in greater depth in the next section. This also depends on the tools and technologies used on the hybrid class. For example, traditional classes are experimenting with social networking technologies to communicate, collaborate and simulate virtual worlds with systems such as Second Life™. Therefore, the ease of use of the technologies, the fit to the goals of the course and the student learning styles will all contribute to the advantages or disadvantages of delivering a hybrid course.

If we examine theories for successful learning irrespective of the delivery method, we can find commonalties to promote success in both environments as well as the hybrid model.

**Active Learning**

According to research in learning theory, active learning, otherwise known as problem based learning has been shown to be the most effective method for students to learn, apply, integrate and retain material (Burch, 2000). While it is relatively easy to present students with projects and problems in a traditional classroom, another

*Table 1. Advantages and disadvantages of different course delivery methods*

| | Advantages | Disadvantages |
|---|---|---|
| Traditional face-to-face classes | Richness of the interactions, development of relationships & trust | Inconvenience of time and location. Limitations on time. |
| Pure Online Classes | Convenience, 24/7 access, time to digest and reflect on content | Lack of rich, personal interactions |
| Hybrid Classes | Can present additional online forums for reflections and discussions. May provide additional access to overcome time and geographical constraints. | May present problems if technologies are not used correctly. |

related approach, called distributed problem-based learning (dPBL) has been used effectively in online courses (Wheeler, 2006). In this method, students collaborate on projects using virtual teams. The advantage of using problem-based learning in both types of course delivery is that it motivates and engages students in a tangible problem and creates a sense of shared purpose among the students to solve a significant problem.

In a hybrid course, this becomes more challenging. Over the past six years, teaching hybrid graduate courses, we have attempted to use this approach. In some classes, when allowing students to self select their teams, we have observed that the in-class students select each other while the virtual students tend to select each other by default. What we discovered is that this creates a "divided society" within the class with an "us vs. them" mentality, leading to sub-optimal class cohesion, discussions and knowledge sharing of the class materials and topics.

In response to this, we require all teams to

- include both "live" and "virtual" students. Again, the live students naturally tend to develop richer relationships. However, what we discovered is that by integrating live, video-conferencing into the class and the teams, this divided structure is greatly reduced.
- use a Diverse Mix of Technologies and require ALL students to use them.

Therefore, creating a vibrant hybrid class involves delivery of content and promoting collaboration and communication using a variety of technologies and learning methods.

**Using Web 1.0 and Web 2.0 Technologies**

As discussed in a prior section, experimenting with both Web1.0 technologies (e-mail, discussion forums, web pages, video-conferencing) as well as Web 2.0 technologies (social networks, blogs, wikis, podcasts) also serves to enhance the overall learning experience. People respond to different communication methods and technologies in different ways. In addition, different technologies serve to facilitate learning and communication in different ways. For example, discussion forums can be very effective for thoughtful, reflective knowledge sharing on projects and class topics. In contrast, live, two-way video-conferencing is necessary to introduce people and allow them to develop trust and relationships. Social networking sites promote more spontaneous communications while technologies like YouTube create more relevant meaning by demonstrating concepts and models.

**Creating Trust and Rapport in the Class**

A crucial component to a successful distance class is to establish a culture of participation and interaction among people who have had little or no personal contact. Communities of practice (CoP) which encourage respect and tolerance while clearly communicating shared goals, help to create a vibrant online class. In addition, we believe that by challenging students and creating an intellectually stimulating, creative environment, where students explore areas of interest to them, a culture can be created to improve communication and motivation in the class., thus adding meaning and value to the CoP. As mentioned above, in a hybrid class, it is imperative to require both the in-class students as well as the distance students to communicate and collaborate using the same technologies to create a sense of community and build trust and rapport among everyone in the class. Social constructivism based on constructivist learning theory that suggests construction of new knowledge is enhanced by collaboratively bringing new knowledge and perspectives among members of a learning community, thus greatly contributing to the effectiveness of the class (Cragg et al, 2008). These researchers also found that in-class students tended to rely more on interactions with the professor. In contrast, distance students relied more on the readings and each other. Therefore, in a hybrid class, more student-student interactions between the in-class and distance students as well as with the instructor would potentially enhance the learning experience for both groups.

**Learner Motivation**

Regardless of whether the class is face-to-face, online or hybrid, incorporating some basic elements regarding learner motivations will also facilitate the overall learning experience.

According to Keller (2008), there are four basic prerequisites to motivating students.

1. You must awaken and sustain their curiosity. Therefore, the content must be interesting and relevant to their goals. In a hybrid and online class, this is more challenging. However, using new technologies such as YouTube to clearly animate and illustrate relevant points helps in this construct.
2. The instruction must similarly be relevant to their values and goals. Thus, as mentioned earlier, using problem-based learning with activities and projects perceived to be relevant to their career goals is helpful.
3. They must feel assured that they will be able to succeed in the class. If students feel confident that they can use the technologies effectively and be able to master the class content, they will be more motivated to learn.
4. The consequences of this learning experience must be consistent with the personal incentive of the student; e.g. they must find the class rewarding and be satisfied with it.

## Technologies and Practices Available for Hybrid Courses Including Emerging Software and Information Systems

### Emerging Technologies: Social Networks and Second Life

As discussed in section 2.2.1, social networks such as Facebook, MySpace, and Twitter among others as well as software systems like Second Life have become widely used tools, especially among younger generations. However, the question becomes, do these new, emerging technologies have a valid role in distance education?

We previously argued that developing communities of practice where class colleagues develop trust and rapport for meaningful communications was a vital part of the online class experience. However, are asynchronous Web 1.0 technologies sufficient to tap into this social networking need or should newer Web 2.0 technologies also be incorporated?

According to Moore (2007), "Writing in the *Handbook of Distance Education*, Dede et al. (2007) suggest the emerging technologies are already having a significant impact on the way young people learn. Learning that uses this technology requires new—or at least more highly developed—skills. They are skills in searching the wide variety of information sources; skills in sorting and sieving the infinite volume of information available; skills in synthesizing from multiple sources of information—all a long way from traditional skills needed to assimilate "validated" sources of knowledge like that found in textbooks or a professor's lectures. This searching requires multitasking—using more than one medium at the same time; the sieving requires "Napsterism"—the ability to make good selections; the synthesis means being creative as one recombines other people's content in forms appropriate to one's own purposes, which is, of course, not the same thing as merely copying it. Working in virtual reality also requires a variety of new skills, as one moves in an environment in which one no longer merely observes a phenomenon as in a book or traditional video program but becomes an agent in its creation, or at least collaborating with others in shaping it."

Moore (2007) also suggests, "Social networking technologies should make constructivist, collaborative knowledge-making more natural and popular among learners and eventually one assumes with their teachers; intelligent search engines offer far more than merely acting as speeded-up encyclopedias for students' research, their greater promise—though still some way off—being as tools for determining learning prescriptions for individual students (a Holy Grail that personality researchers have discussed for at least thirty years); risk-taking being one of the first requirements for learning, virtual reality environments offer opportunities for students to experiment in social as well as physical domains that were previously unthinkable. Readers will, I am sure, be able to add their own suggestions to this short list of potential benefits."

On the other hand, he warns that these new technologies can also prove to be distractions if not structured properly within the context of the course. "The communication of tacit knowledge is more likely to occur when there is frequent and intense communication across functions and organizational levels and functions (Hage and Hollingsworth, 2000)."

**Second Life™**

While hybrid classes provide both students and educators with many advantages noted previously in this chapter, it also has some disadvantages. De Lucia, Francese, Passero, and Tortora (2009) stated that while learning management systems like WebCT™ provide a range of interactive and collaborative features, these functions often are not fully used and the course management system is used simply as a supplement to the "live course". The multimedia content uploaded to course management systems is often "static", and rarely used by students unless the instructor makes it a requirements.

My own personal experiences of teaching hybrid classes reinforce this point. I found myself using web-based course management systems such as WebCT™ /Blackboard™ as a repository for assignment problems and postings for students to go and get information rather than a forum for engaged interaction. Additionally, I found that my students often demonstrated little interest in taking full advantage of its many collaboration tools like student homepages, discussion forums, and shared workspaces and often only cared to know how to retrieve assignments and upload completed assignments on time.

The challenge then became, "how I use the Internet as a medium for simulating the classroom experience when I and students do not meet in the same physical space?" One would need an electronic platform to exchange information and share ideas synchronously while at the same time keeping students interested and participating. Second Life™ provides hybrid classes with the functionality that can both take advantage of student's interest in Web 2.0 technologies such as instant messaging, texting, and collaboration tools. At the same time, Second Life™ more accurately simulates the physical classroom experience, allows for greater interaction amongst each other, and establishes a truer sense of presence for all participating.

What is Second Life™? According to its website, Second Life™ is "a free online virtual world created by its residents. A fast growing digital world filled with people, entertainment, experiences, and opportunities." Johnson and Middleton (2008) stated that it could be thought of as a marriage between online video gaming and social networking technologies with many e-commerce opportunities. At Second Life's core is a virtual three dimensional environment where many remote participants can converge and engage in activities that electronically simulate real life experiences (Bourke, 2008). Examples of simulated real life activities that can be done in Second Life™ include: training, socializing with peers, project collaboration, and participating in distance education. These enhance the distance educational experience by providing a simulated environment that students can relate to.

Several universities and colleges have created virtual campuses in Second Life™. Harvard University, San Diego State University, and the University of Maine are all examples of educational institutions that have virtual islands populated with lecture halls, classrooms, and meeting places. Faculty, researchers, and information technology specialists use the virtual space to build content and enhance experiential learning for students by providing them with simulated real-world problems to overcome and simulated virtual goals to achieve.

In one example of how Second Life™ can be used for hybrid course delivery, Gainesville University of Florida runs a course called "Aesthetic Computing" using Second Life's virtual environment. Caronia (2007) stated that 30 students in this class meet "face to face" in a real world classroom then use their Second Life space to create "digital objects" to make their abstract ideas from the course "real". The course professor states that the Second Life™ virtual environment provides several new forms of interaction that includes opportunities for "collaboration, immersion, shared 3-D object construction, complex and general scripting capabilities, and in-world communication.

Another example of Second Life™ for hybrid course delivery is the case of SecondDMI (De Lucia et.al, 2009). In this example an experiment conducted at the University of Salerno analyzed user impressions concerning "presence, awareness, communication, and comfort with the virtual environment." The results of this experiment demonstrated several important factors regarding user acceptance of the virtual environment as a learning space. A few noted results include;

1. Second Life™ as a virtual environment supports synchronous communication and interaction
2. Users who were less familiar with the environment asked for information from others within the environment or looked at them to gain understanding, and
3. Perhaps one of the most important findings, "tutors and the teacher observed that students were really excited" and "many students expressed high interest in the environment to meet class mates".

At the University of Maine's Second Life™ virtual campus Black Bear Island, classes have been taught in a variety of subjects including a graduate level course on "Teaching and Learning in Second Life." This course was a hybrid course that met both online in Second Life™ and in person throughout the length of the semester. Students explored the use of virtual worlds (using Second Life™) for teaching and learning. Students designed, facilitated, and even evaluated quality of their classmate's instruction in Second Life™ by participating in virtual field trips and gained experience teaching in virtual environments by conducting and attending virtual workshops.

We have begun using Second Life™ as medium to facilitate student advising. Each semester faculty are assigned a number of student advisees. Faculty are responsible for providing academic advising about course registration and for providing guidance about which classes students should take to meet their academic goals. However, time and distance is an obstacle for both students and faculty when it comes to scheduling time to meet. Virtual advising in my office located on Black Bear Island removes the barrier of time and place and provides an environment in which our students and we can meet outside of "normal business hours".

Second Life could additionally provide schools and colleges with solutions to the costs associated with developing and maintaining labs. Budget limitations can keep schools and universities from fully committing to developing and keeping labs up to date with relevant technologies. For example, we experienced the problems associated with inadequate technology labs. We needed to "teach out" a small group of information technology students so that they could graduate and so that we incidentally, could finally terminate the IT program that had suffered from dramatically decreasing enrollment. One of the final classes we taught was a networking course. The college simply threw together a small ad hoc network made up of old computers and related periphery equipment. Students complained to the instructor many times about the inadequacy of this network and how they felt like they were being treated like "second class students" by the college that was terminating their program. This treatment was evidenced according to the students by their woefully inadequate network lab. A virtual networking lab could have solved this problem.

Virtual laboratories are seemingly a good low cost alternative for training and teaching when budgetary and physical space is constrained. According to Weekes (2007) developing labs in Second Life is an alternative that organizations are looking at using to train employees and students. Weekes noted that companies like Cisco Systems are using Second Life as a platform to bring together employees to update them on new technologies. Manpower, a job placement organization, also uses Second Life to explore and exchange ideas about work in virtual environments. Manpower stated that the Second Life environment will be used to "train for the real world," meaning that employees in training could find themselves learning how to run machines, perform job functions, and understand businesses processes all within the virtual environment and using those skills learned virtually in the real world!

Organizations, schools, and universities looking to keep costs down associated with training are looking at virtual worlds more closely (anonymous, 2009). In doing this though, employees have to feel like they are "really apart of the world" which can be difficult to do in other distance learning environments. Virtual worlds like Second Life then provide a training environment in which a truer sense of presence is established between the instructor or trainer to students and between student to student interaction. Michelin IT professionals needed to find a way to train IT professionals how to understand their enterprise architecture. They found that using the virtual world of Second Life was a highly effective and successful medium (Driver, 2008). Second Life provided the organization with a "highly interactive environment that combines presentations and hands on workshops."

Second Life could therefore represent a useful tool for schools and universities to train students how to use, setup, and install networks for example. Virtual labs in Second Life also bring the added benefit of keeping trainees and students interested acknowledged by both Weekes (2008) and Driver (2008). Providing a virtual environment that allows students to learn the skills colleges and universities promise to deliver in a manner that is cost effective, scalable, and even keeps students interested creates a win-win learning environment for all participants involved.

## DWDM Technology and Initiatives

Dense Wavelength Division Multiplexing (DWDM) is an optical transmission technology that enables existing fiber optic networks to increase its bandwidth. DWDM allows multiple optical signals on different wavelengths to be transmitted on a single optical fiber strand (Littman, 2002). As a result, DWDM technology increases the capacity of fiber links by multiple orders of magnitude which translate into significant improvements in bandwidth capacity and network speed.

To illustrate the significance of DWDM's potential to increase network data transfer rates, its bandwidth per fiber increased from 10 Gbps in the year 2000 to 400 Gbps in the year 2007. This increase in data transfer speed is relevant to educational institutions that have seen their own networks needs grow over the last few years. Schools that need to transport video, images, and audio transmissions for a variety of educational and training needs have had to look for new ways to manage network traffic and meet the network demands of supporting Web 2.0 Internet collaboration tools. DWDMs growth in bandwidth capacity means that academic institutions can increase their schools network bandwidth without having to incur significant cost related to installing newer and higher capacity network cables (Maxemchuk, Ouveysi, and Zuckerman, 2007).

**Benefits of DWDM Technology**

A DWDM network is made up of several optical devices such as add-and-drop multiplexers, optical gateways, optical switches and optical routers (Littman, 2002). A benefit of DWDM is that it can interoperate easily with Internet protocols (IP), asynchronous transfer mode (ATM), synchronous optical networks (SONET), and Ethernet transport networks meaning that it can be coordinated to work with most existing school networks. A second benefit of the DWDM technology is related to its cost effectiveness. Generally, the largest expense related to upgrading an existing fiber optic network is the labor. Digging up cables and replacing them with fiber optics with higher capacity bandwidth cables is so time consuming and labor intensive, organizations can significantly reduce their costs of upgrading by utilizing technologies that can take advantage of existing network infrastructure. DWDM provides this solution because it improves the efficiency of each fiber contained within an already existing fiber optic network (Maxemchuk, et.al, 2007).

Dense Wavelength Division Multiplexing provides another benefit to its users. Increased bandwidth reduces network congestion. As more and more users access the network and their data becomes more and more complex and large, a packet-switched network performs poorly, becomes unstable, and can experience oscillations in synchronizations (Chen, Fawaz, Martignon, and Pujolle, 2006). Routing protocol packets can also be lost or delayed as a result of network congestion. As a result traffic loops, black holes, and disconnected regions within the network can occur which only adds increased frustration for users transmitting data over a congested network (Chen, et. al, 2006). DWDMs ability to significantly increase the optical networks bandwidth and resulting speed will drastically reduce packet-switching errors. The implementation of DWDM technologies then can play a vital role in ensuring that distance education programs at schools and universities are successful and that students and faculty have a more enjoyable less frustrating experience.

## Telepresence Technologies for Collaboration in e-Learning Environments

According to Leonard and Staman (2003) telecommunication technologies available today combined with related hardware and software technology makes innovative IP video and data transmissions possible. Distance education programs that would like to create a truer sense of community for their courses can now use videoconferencing technologies to create an engaged near-"in person" experience to remote students. Not only will these tools enhance the learning experience for distance education programs but can also add value to a traditional classroom setting.

Telepresence solutions are now established by combining the technologies of the Internet, videoconferencing, and an application layer such as Cisco's WebEx™ to establish a virtual presence. According to Chen, Fuchs, Nyland, Towles, and Welch (2000), telepresence solutions provide a truer sense of community by allowing participants from different geographic locations to work collaboratively and virtually as if they were in the same room. Telepresence then truly has the ability to totally alter almost every distance learning teaching and training situation.

deHaan and Diamond (2007) describe telepresence as a sense of "being there". Participants in online environments do not "simply transmit and receive information passively via a technology". These students also form unique experiences and relationships with each other. The quality of the application software or user interface provides the user with information and influences the nature of the relationship between the user and the media. Interactivity within the online or virtual environment is also necessary in establishing telepresence. Interactivity describes the extent to which users can modify and form the online environment in real time. deHaan and Diamond, (2007), state three other attributes characterize the level of "presence" in a virtual or online environment and they are:

- **Speed:** the rate at which the users input and contributions can be added to the environment
- **Range:** the number of opportunities for action at any point in time during engagement in the online environment
- **Mapping:** the ability of a system to map its control of changes in the online or virtual environment in a natural and predictable manner

A number of technologies and techniques exist that support distributed learning in synchronous, real-time environments (Leonard, et.al, 2003). Beyond simply including YouTube or Video feeds in a classroom, teachers can use web meeting spaces, video over Internet protocols, and network collaboration tools to expand the reach of their classes to remote sites and / or have guest lecturers from remote locations speak to their classrooms using these technologies.

When incorporating video over Internet protocol (VOIP) in a classroom Leonard, et.al (2003) suggests three possible solutions that add value to the student's experience.

- **Video on demand:** this creates a "one-to-many" experience because lectures, movies, tutorials, and similar material is stored and then downloaded by distance students when the material is needed.
- **Live streaming video:** this also creates a "one-to-many" user experience in that although the students can hear and see in real-time, the speaker is not interacting with the audience.
- **Interactive video:** provides a opportunity for a synchronous "in-person" experience that is "one-to-many" as well as "many-to-many"

The interactive video option provides the greatest chance of true community and e-collaboration of the three options. In this scenario, a teacher teaches form one location while students present in the same classroom and students from remote locations hear, see, and communicate with the teacher in synchronous "real-time."

## Network Technology, Services, and Solutions for Creating Telepresence

Schools and colleges wanting to create a truer sense of presence for remote students may consider taking advantage of videoconferencing technologies available through Cisco™ and POLYCOM®. Cisco's WebEx™ program for example provides a suitable solution for delivering real-time videoconferencing and shared documents to participant's desktops. POLYCOM® provides similar videoconferencing technologies as well. POLYCOM® View Station Exchange (VSX) and Personal Video Exchange (PVX) systems provide high-quality real-time voice and video transmissions to up to 20 remote users anywhere at any time with an Internet connection. Its multi-network capability support Internet protocol (IP) and integrated services digital networks (ISDN) and is session initiation protocol (SIP) compliant. The videoconferencing system also interoperates with Microsoft Live Communication Server and Nortel Networks Multimedia Communication Server (MCS) 5100 and 5200 products. ("POLYCOM® VSX 8000", n.d).

If schools and colleges do not want to make a heavy investment in IT upgrades to achieve telepresence in classrooms interactive video tools such as Windows Netmeeting™ comes preinstalled on Windows 2000, XP and is available for free download for Windows 95/98 and NT. Windows Vista uses Microsoft's newest version called Netmeeting™. Netmeeting is a voice over Internet (VoIP) and multi-point videoconferencing solution. As a preinstalled or free download Windows Netmeeting is an affordable solution for delivering interactive videoconferencing. In order to implement Netmeeting users only need a web camera and a microphone.

## Wireless Technologies Impact on Education

Today's students are becoming more mobile due to the increased access to mobile technologies (Nagel, 2008). As a result, network technologies that are emerging include the increased use of mobile devices combined with wireless networks. Wireless devices use wireless network interface cards in place of NICs and require the IEEE 802.11 wireless protocol to specify hardware requirements and determine how messages can be packaged and processed for transmission over a wireless LAN (Kroenke, 2007).

Satellite networks are a particularly attractive solution for e-learning systems because of the networks global coverage and ability to provide access to remote users through terrestrial wide area networks (Koutsakis, 2006). These wireless networking technologies will have a significant impact in education within the next year because of the use of low-cost grassroots video and collaborative web technology (Nagel, 2008). Essentially grassroots video is user generated videos using readily available mobile devices such as IPod's and video enabled mobile phones. Internet services readily available at little or no cost to the consumer allow students or instructors to share their video's making their content easily available to their intended viewers.

Videos uploaded to an instructor's courseware site using the MP4 format, typically known as a "Podcast" is already in use in many educational institutions. Like E-learning, Podcasting satisfies an increasing need for access to educational content by students. The option of using existing satellite networks and mobile device technology combined with video technology would provide an improvement to E-learning environments synchronously conducted (Guerri, Palau, Pajeres, and Esteve, 2002).

### Example of a Satellite System Used for Videoconferencing

A solution available for real-time videoconferencing in e-learning environments includes the use of a satellite network for multicasting high quality video from the instructor to students using an video camera, a video server, and transmitting the video to remote classrooms via satellite broadcast down-link channel (Guerri, et.al, 2002). Students receiving the video transmissions view the video real-time through an ISDN/Internet connection connected to their workstation pc and return real-time video transmissions of themselves via a web-cam to create a sense of interaction between student and teacher; effectively creating a virtual classroom.

Components required to create the real-time videoconferencing e-learning environment include:

- server and necessary software and hardware on the instructors side, Examples would include a video camera and workstation with an installed video card, and software to process video transmission.
- remote classrooms with a public network that includes Internet access equipped with workstation pc's webcam's, and necessary software to return video images of themselves to the instructor
- multicast link over a HISPASAT satellite.

By developing a videoconferencing system like this, schools and universities should be encouraged by what is possible when considering how to use satellite networks to achieve distance learning solutions. Schools and colleges developing programs to deliver courses to remote locations even in a hybrid environment could consider having live "face time" between both the instructor and students using combinations of telepresence technologies and satellite networks.

## Applications of Satellite Network Solutions in Education

The Abbotsford School District 34 in British Columbia is deploying a new wireless network to improve access for teachers who are increasing using wireless devices in their classrooms. (Nagel, 2007). For example, many teachers in the district are using Tablet PCs to connect to wireless projectors and using laptops to connect to Smart Boards. One of the obstacles that the school district had to overcome was that of continuous and reliable access.

According to Nagel (2007) the school system provided a solution the dependability issue by purchasing 80 MAP-320 and MAP-330 Cloubris MultiService Access Points and a MultiService Controller. The access points cover about a quarter of the schools in the district and the controller allows for centralized management of the network. The school system deployed this wireless network for a number of reasons. One striking reason is that they found a number of their new instructors coming out of the universities had extensive technology-based backgrounds and demanded wireless tools enhancing learning in their classrooms. Kopf (2007) states that Auburn city schools in Alabama have found that by implementing a wireless network for their classrooms has enhanced teacher instruction. Students are benefiting greatly by the wireless network because the instruction has moved from being "static" to more dynamic. More impressive was the fact that instruction has moved from be "teacher-facilitated" to "student-facilitated".

Using Tablet PCs that are wirelessly connected to student's laptops teachers were able to see exactly what students were doing, help individual students that may have had a problem with an assignment, and even redirect students caught not doing their work. (Kopf, 2007). If a teacher saw a student having a problem with a particular question, not only was the teacher able to "take control" of the students desktop and "walk them through completing the problem, teachers could also upload that students problem to a interactive Smart Board and work through the problem with the whole class if it appeared that many students were stuck on the same problem. Due to the proliferation of web enabled devices such as cell phones, PDAs, and laptops, students, teachers, office workers, and so on are demanding real-time access to the information they want when they want it. Wireless technologies have not only made the workforce mobile but it has offered several opportunities for industry's such as healthcare to greatly increase their access to patients and improve quality of service. Lastly, wireless solution are improving cognition in schools and as illustrated in this last section improving instruction by moving from teacher-facilitated to student-facilitated.

## Texting: Using Cell Phones in the Classroom

Today most students have cell phones with them at all times. In many cases instructors have complained that cell phones are a nuisance in class and distract student learning. A primary complaint about cell phones is that the disruptions caused by student's texting each other and receiving or making calls while in class. The argument then against the use of cell phones in class is that these unauthorized uses of cell phones in class negate any possible academic gains that instruction combined with cell phones could provide (Sturgeon, 2007).

This argument however minimizes what cell phones are today. Wains and Mahmood (2008) stated that "today's mobile phones have as much computing power as the computer had a few years back." Mobile phones often come packaged with or can have productivity software packages downloaded to them. MS Office, Acrobat Reader, and other productivity tools are all examples of software that many students can use on their mobile phones. Combine this productivity power in mobile phones with the fact that most students have these sophisticated cell phones and the possibility of using mobile phones for educational purposes seems like a natural next step. The next question is how do we integrate mobile phone usage into the classroom without its use being a distraction? Linquist, Denning, Kelly, Malani, Griswold, and Simon (2007) stated that "mobile phones seem to hold promise for enabling large-scale participatory learning "in the wild": in most any classroom with minimal student cost, without herculean instructor effort to install and maintain the technology."

## SMS Texting

Short messaging service (SMS) is a mobile phone application used to send quick, concise, text based messages. In one example of using mobile phones in education, researchers at the University of Lancaster explored the use of SMS texting as a means for students and instructors to communicate with each other about course work and actually took quizzes via their mobile phones texting services (Wains and Mahmood, 2008). By communicating with each other via mobile phones and participating in actual course work the school had developed a type of hybrid classroom that overcame the issue of time and space which provides an excellent solution to schools that are dealing with the problem of how to provide quality teaching to students in remote locations.

In addition to providing distance education hybrid solutions, texting also offers possible solutions related to student engagement while in the classroom. Each semester professors and instructors with large classrooms observe that many students are not engaged in the classroom discussion and are not asking questions. Much of this could be attributed to several reasons that include but are to limited to; fear of asking a "dumb" question, disinterested in class discussion, and feelings that the lecture is a "one-way" instructor led discussion in which student are expected not to be active participants. Disinterested students will often busy themselves and distract the rest of the class by socializing with friends next to them or texting friends. I actually observed on several occasions students in the same class texting each other!

According to Markett, et al. (2006), developing and incorporating technologies that receive SMS text messages to an instructors PC and then projecting student text messages comments and questions can create a interactive question and answer session during class time that can provide these benefits to the learning process;

1. Students in large classes will demonstrate greater participation in class using the SMS text messaging service
2. Students using SMS text messaging in class will be more engaged
3. Professor / Instructor will be able to better assess student understanding of course material based on SMS texting input and be able to redirect or improve delivery of lecture material as a result

SMS texting is one example of how mobile phones can be integrated into the classroom and distance education hybrid courses to increase student engagement and participation. Texting also provides an easily accessible low cost technological solution to bringing educational opportunities to students in remote locations. Additionally, simply including mobile phones as part of an academic environment schools and colleges can take advantage of several other benfits they bring with them. Today's mobile phones such as the IPhone, Blackberry, and a variety of other multifunctional web-enabled smart phones

*Table 2. Review of existing and emerging technologies*

| | |
|---|---|
| Second Life™ | An online virtual environment that can be used to facilitate "real time" interaction between teacher and student. The virtual environment creates a truer sense of presence with student and instructor avatars |
| Dense Wavelength Division Multiplexing (DWDM) | Significantly increases bandwidth on existing fiber optic networks |
| Telepresence | Create a truer sense of presence with the use of videoconferencing technologies delivered over IP Networks. Can offers a near "in person" experience for students participating remotely |
| Wireless and Satellite Networks | Provides greater access to online classrooms for distance education students spread out over a larger geographic area |
| SMS Texting in the Classroom | Research in this emerging use of cell phones in the classroom demonstrate that SMS texting can be used to increase student engagement and participation |

may be tools that can be used to engage students, reduce discipline problems, and reduce truancy according to Sturgeon (2007). Like laptops and desktop PC's, today's smart phones are a means of putting computing technology in the hands of students.

Mobile phones provide another benefit to schools that are willing to consider adding them to their curriculum. Students equipped with smart phones means that schools do not have to go through the expense of implementing Wi-Fi capability (Sturgeon, 2007). Students would essentially have an always on access to the Internet. Students would be able to write reports, collect data, collaborate with groups, and gather information they need by surfing the Web.

According to Verkasalo (2006), mobile handheld devices such as smart phones provide students with a third benefit. Current online services used by schools such a desktop PCs still cause challenges in the spatial domain. You still need a space to place a computer and network devices such as switches and wireless routers to connect to the Internet. Smart phones with mobile services overcome most cases of spatial and temporal constraints.

To summarize the discussed technologies used to facilitate hybrid distance education, educators and schools should consider several technology solutions that are currently available. As previously stated in this chapter, these technologies offer improvements in network bandwidth, improves the sense of presence for remote students, and can increase student engagement and participation. Table 2 below summarizes these available and emerging technologies.

## CURRENT CHALLENGES: THE CLASSROOM OF TODAY AND TOMORROW

### Creating Equal Quality in Hybrid Classes; Making the Distance Experience "Real"

The lessons learned from prior research indicate that instructors who wish to teach hybrid classes need to follow some basic rules.

First, develop a hybrid system of technologies that will fit the different learning styles of the students. This will include web 1.0 technologies such as e-mail, discussion forums, and videoconferencing. Incorporation of Web 2.0 technologies will integrate the learning styles of younger students, who respond to multi-tasking and more media-rich applications like social networking sites, blogs, wikis, or YouTube type video presentations. However, ALL students must be required to participate in use of the different technologies to facilitate adoption and use of them as well as relationship building and trust.

Instructors in hybrid classes should remember the basic principles in learner motivation and incorporate that into the content, delivery and structure of the classes. This includes developing and maintaining interest and curiosity in the class and content, making the class relevant to their lives and goals, making sure that the technologies are user friendly and the class tasks and goals are attainable and that they students will find the class rewarding and satisfying.

Develop a sense of community and purpose in the hybrid class. In order to avoid an "us vs. them" mentality, make sure that students interact and communicate collaboratively. Develop teams that integrate the in-class with the distance students and make sure that they are all engaged and feel a sense of purpose and engagement in the class.

## FUTURE RESEARCH DIRECTIONS

With regard to the literature and experiences about hybrid distance education in this chapter, several opportunities for future research are possible. Research that evaluates student interactivity and existing videoconferencing technologies to create a truer sense of presence for distance students is one area of continued research while research in the area of skills training in virtual labs is another. Additional research related to virtual labs includes the transferability of skills learned in virtual environments to "real world" environments.

Teaching and learning using Web 2.0 technologies also provide several opportunities for future research. For example, student collaboration and productivity using Wiki technologies offer several opportunities for research in the areas of its efficacy in distance education environments. Lastly, it should be pointed out that even the technologies related to SMS texting in distance education provides seemingly almost endless opportunities for research as cell phones, smart phones, and the like become more and more ubiquitous in modern living.

## CONCLUSION

Many tools and techniques are now available to create telepresence in classrooms and boardrooms today. However, a "build it and they will come" approach will not work for hybrid classes. In other words, colleges and universities cannot simply purchase distance technologies and assume that different delivery systems will be successful. Rather, a successful hybrid class will integrate a variety of technologies and learning methodologies. The savvy instructor will understand the personality and needs of the class and adapt the class content, communications approaches and use of technologies to meet the needs and goals of the students.

The continuing growth of IP networks and their growing reliability supports the development of new innovative videoconferencing techniques and how they are used to improve both traditional classroom learning and online learning. In addition, the evolution of Web 2.0 technologies such as Facebook, Second Life, and blogs among others is contributing to the vibrancy and richness of hybrid classes by bringing different members of the class together into a cohesive community of learners.

Finally, it should be noted that although much of the technology discussed in this paper could potentially improve learning, the true benefits lie in how videoconferencing and other available technologies can bring people together regardless of their geographic location. Therefore, if the correct technologies are selected for the class and if used appropriately, these technologies can facilitate trust, rapport, communication and collaboration to greatly enhance the overall learning experience.

## REFERENCES

Anonymous,. (2009). HOW REAL? *Mechanical Engineering (New York, N.Y.)*, *131*(1), 15–16.

Beldarrain, Y. (2006). Distance Education Trends: Integrating new technologies to foster student interaction and collaboration. *Distance Education*, *27*(2), 139–153. doi:10.1080/01587910600789498

Bonk, C., & Zhang, K. (2006). Introducing the R2D2 Model: Online learning for the diverse learners of this world. *Distance Education*, *27*(2), 249–264. doi:10.1080/01587910600789670

Bourke, P. (2009). Evaluating Second Life for the collaborative exploration of 3D fractals. *Computers & Graphics*, *33*(1), 113–117. doi:10.1016/j.cag.2008.08.004

Burch, K. (2000). A primer on problem-based learning for international relations courses. *International Studies Perspectives*, *1*, 31–44. doi:10.1111/1528-3577.00003

Caronia, N. (2007). Virtual world provides new opportunities for education. *Government Video*, *18*(7), 8–9.

Chen, K., Fawaz, W., Martignon, F., & Pujolle, G. (2006). A novel fault management approach for DWDM optical networks. *International Journal of Network Management*, *16*(5), 337–349. doi:10.1002/nem.601

Chen, W. C., Fuchs, H., Nyland, L., Towles, H., & Welch, G. (2000). Toward a compelling sensation of telepresence: demonstrating a portal to a distant (static) office. In *Proceedings of the Conference on Visualization,* Salt Lake City United.

*Cisco: Welcome to the Human Network.* (n.d.). Retrieved April 30, 2008, from http://www.cisco.com/web/thehumannetwork/index.html

Conrad, D. (2008). From Community to Community of Practice: Exploring the Connection of Online Learners to Informal Learning in the Workplace. *American Journal of Distance Education*, *22*(1), 3–23. doi:10.1080/08923640701713414

Cragg, C., Dunning, J., & Ellis, J. (2008). Teacher and Student Behaviors in Face-to-Face and On-Line Courses: Dealing with Complex Concepts. *Journal of Distance Education*, *22*(3), 115–128.

De Lucia, A., Francese, R., Passero, I., & Tortora, G. (2009). Development and evaluation of a virtual campus on Second Life: The case of SecondDMI. *Computers & Education*, *52*(1), 220–233. doi:10.1016/j.compedu.2008.08.001

deHaan, J., & Diamond, J. (2007). The experience of telepresence with a foreign language video game and video. In *Proceedings of the 2007 ACM SIGGRAPH Symposium on Video Games* (San Diego, California, August 04 - 05, 2007).

(ND). distance education course offerings. *Journal of Distance Education*, *11*(1), 101–110.

Driver, E. (2008). In-World Work:The Immersive Internet Increases Engagement While Reducing Business Costs. *DM Review*, *18*(11), 30.

GAO (United States Government Accountability Office). (2007). *HIGHER EDUCATION: Tuition Continues to Rise, but Patterns Vary by Institution Type, Enrollment, and Educational Expenditures*. GAO-08-245, Report to the Chairman, Committee on Education and Labor, House of Representatives. Retrieved from http://www.gao.gov/new.items/d08245.pdf

Guerri, J. C., Palau, C. E., Pajares, A., & Esteve, M. (2002). A real-time e-learning system via satellite based on JMF and Windows Media. In *Proceedings of the Tenth ACM international Conference on Multimedia,* Juan-les-Pins, France, December 01 - 06, 2002 (pp. 219-222). New York: ACM.

Hage, J., & Hollingsworth, J. (2000). A Strategy for the Analysis of Idea Innovation Networks and Institutions. *Organization Studies*, *21*(5), 971–1004. doi:10.1177/0170840600215006

(ND)... *Harvard Business Review*, *78*(1), 139–145.

Heck, J. (2009). State, national universities see spike in distance-learning enrollment. *Dallas Business Journal.* Retrieved from http://dallas.bizjournals.com/dallas/othercities/wichita/stories/2009/01/12/focus3.html?b=1231736400^1758951

Jackson, M., & Helms, M. (2008). Student Perceptions of Hybrid Courses: Measuring and Interpreting Quality. *Journal of Education for Business*, *84*(1), 7–12. doi:10.3200/JOEB.84.1.7-12

Johnson, R. A., & Middleton, J. M. (2008). Accounting for Second Life. *Journal of Accountancy*, *205*(6), 54–56.

Keller, J. (2008). First principles of motivation to learn and e3-learning. *Distance Education*, *29*(2), 175–185. doi:10.1080/01587910802154970

Kopf, D. (2007). Seattle Public School System Flexing Fiber Muscle. *T.H.E. Journal*. Retrieved on March 10, 2009, from http://www.thejournal.com/articles/20385

Koutsakis, P. (2006). A New Efficient Scheme for the Allocation of Satellite Bandwidth Based on Videoconference Traffic Estimation. *ICWMC*, 16.

Kroenke, D. M. (2007). *Using MIS*. Upper Saddle River, NJ: Pearson, Prentice Hall.

Lee, Y., Kozar, K., & Larsen, K. (2003). The Technology Acceptance Model: Past, Present and future. *Communications of AIS*, (12), 752-780.

Leonard, J., Riley, E., & Staman, E. M. (2003). Classroom and support innovation using IP video and data collaboration techniques. In *Proceedings of the 4th Conference on information Technology Curriculum,* Lafayette, Indiana, USA, October 16 - 18, 2003.

Lindquist, D., Denning, T., Kelly, M., Malani, R., & Griswold, W. G., & Simon, B. (2007). Exploring the potential of mobile phones for active learning in the classroom. In *Proceedings of the 38th SIGCSE technical symposium on Computer science education*. Covington, KY: ACM.

Littman, M. K. (2006). *Building Broadband Networks*. Boca Raton, FL: CRC Press.

Markett, C., & Sánchez, I., A., Weber, S., & Tangney, B. (2006). Using short message service to encourage interactivity in the classroom. *Computers & Education*, *46*(3), 280–293. doi:10.1016/j.compedu.2005.11.014

Martinez, R. (2004). Online Education: Designing for the Future in Appraiser Education. *The Appraisal Journal*, *72*(3), 266–184.

Maxemchuk, N. F., Ouveysi, I., & Zukerman, M. (2005). A quantitative measure for telecommunications networks topology design. *IEEE/ACM Trans. Netw.*, *13*(4), 731–742.

Menchala, M., & Bekele, T. (2008). Learner and instructor identified success factors in distance education. *Distance Education*, *29*(3), 231–252. doi:10.1080/01587910802395771

Miller, J. (2007). The New Education Professionals: The Emerging Specialties of Instructional Designer and Learning Manager. *International Journal of Public Administration*, *30*, 483–498. doi:10.1080/01900690701205970

Miller, M. T., & Husmann, D. E. (1996). A holistic model for primary factors in the ecology of

Moore, M. (2007). Web 2.0: Does it Really Matter? *American Journal of Distance Education*, *21*(4), 177–183. doi:10.1080/08923640701595183

Nagel, D. (2007). Students Cross Cultures Through Videoconferencing. *T.H.E. Journal*. Retrieved November 29, 2007, from http://www.thejournal.com/articles/21660

Nagel, D. (2008). Which Technologies Will Shape Education in 2008? *T.H.E. Journal*. Retrieved February 1, 2008, from http://www.thejournal.com/the/printarticle/?id=21972

Norton, P., & Hathaway, D. (2008). Exploring Two Teacher Education Online Learning Designs: A Classroom of One or Many? *Journal of Research on Technology in Education, 40*(4), 475–495.

Polycom, V. S. X. *8000*. (n.d). Retrieved April 30, 2008, from http://www.ivci.com/videoconferencing_polycom_vsx_8000.html

Redden, E. (2009). Blackboard 9.0. *Inside Higher Ed*. Retrieved from http://insidehighered.com/news/2009/01/27/blackboard

Ritter, S. (2008). *Social Networking in Higher Education*. Penn State World Campus. Retrieved from http://blog.worldcampus.psu.edu/index.php/2008/09/04/social-networking-in-higher-education/

*Second Life*. (n.d.). Retrieved from http://www.secondlife.com

Sharda, R., Romano, N., Lucca, J., Weiser, M., Scheets, G., Chung, J., & Sleezer, C. (2004). Foundation for the Study of Computer-Supported Collaborative Learning Requiring Immersive Presence. *Journal of Management Inquiry, 20*(4), 31–63.

Smathers, D. (2007). *Phi Kappa Phi and the Distance Learner*. Retrieved from http://www.phikappaphi.org/web/Publications/Distance-Learner/index.html

Sturgeon, J. (2007). Talkin' Up a Storm. *T.H.E. Journal*. Retrieved April 6, 2008, from http://www.thejournal.com/articles/21531

*Twitter*. (n.d.). Retrieved from http://twitter.com/

Verkasalo, H. (2006). Empirical observations on the emergence of mobile multimedia services and applications in the U.S., & Europe. In *Proceedings of the 5th international Conference on Mobile and Ubiquitous Multimedia,* Stanford, California, December 04 - 06, 2006.

Wains, S. I., & Mahmood, W. (2008). Integrating m-learning with e-learning. In *Proceedings of the 9th ACM SIGITE conference on Information technology education*, Cincinnati, OH, USA.

Weekes, S. (2007). Get a SECOND LIFE. *Training & Coaching Today*, 29.

Wenger, E., & Snyder, W. (2000). Communities of Practice: The Organizational Frontier.

*What is the MediaTone™ Network?* (n.d.). Retrieved April 22, 2008, from http://www.webex.com/smb/media-tone.html

Wheeler, S. (2006). Learner support needs in online problem-based learning. *The Quarterly Review of Distance Education, 7*(2), 175–184.

# Chapter 15
# Developing Student e-Portfolios for Outcomes-Based Assessment in Personalized Instruction

**Kam Hou Vat**
*University of Macau, Macau*

## ABSTRACT

*This chapter investigates the pedagogical issues of student electronic portfolios (e-portfolios) in the context of personalized instruction for undergraduate education. The discussion elaborates on the educational potential of an e-portfolio system in facilitating an outcomes-based assessment of student achievements. The chapter illustrates practical examples of integrating theory and practice aimed at assisting a meaningful investigation of an e-portfolio system with a focus on inquiry-based student assessment. The objective of such an inquiry is to enhance and encourage student learning, especially learning by doing. Key issues and the necessary institutional support for an outcomes-based and personalized model of education in support of a portfolio learning system are identified. The interrelationship of portfolio assessment to curriculum and pedagogy and required changes to teaching and learning are described. The relevant learning theories that underpin the portfolio form of assessment are deliberated to caution how best to manage the use of e-portfolios for student learning and assessment. Looking beyond, it is expected that the e-portfolio system is an important element to support outcomes-based education involving collaboration from both faculty and students in pursuit of a quality learning experience.*

## INTRODUCTION

Interest in assessment for student learning at colleges and universities has skyrocketed in the late twentieth century and continues to grow. Today there emerges an imminent need on the part of many a university to learn how to do student assessment, and do it the right way to empower student learning. The idea of outcomes-based assessment (OBA) is not new, and it is related to an educational model in which curriculum and pedagogy and assessment are all focused on student learning outcomes. It is an educational process that fosters continuous attention to student learning and promotes institutional

DOI: 10.4018/978-1-61520-853-1.ch015

Copyright © 2010, IGI Global. Copying or distributing in print or electronic forms without written permission of IGI Global is prohibited.

accountability (Driscoll & Wood, 2007). Simply put, the OBA model emphasizes such important practices (Larkin, 1998) as: Faculty publicly articulating assessment information in advance of instruction; students being able to direct their learning efforts to clear expectations; and student progress and acquisition of learning being determined by evidence demonstrated in achieving the learning outcomes. So, the key component in the OBA model of education is outcomes which inform curriculum, teaching and assessment. Maki (2004, p.60) describes a learning outcome as what students should be able to demonstrate, represent, or produce based on their learning histories. Huba and Freed (2000, pp.9-10) describe learning outcomes as teachers' intentions about what students should know, understand, and be able to do with their knowledge when they graduate. For obvious reasons, university faculty is the most appropriate source of student learning outcomes. The issue is how faculty should be empowered in the process of assessment to enhance student learning. One of the most important conclusions about the effect of outcomes on student learning comes from the studies of John Biggs (1999). Biggs found that student achieve deep learning when they have outcomes on which to focus. If students do not know what is important to focus on in their studies, they try to cover all the information, so they skim, they cram, and they stay on the surface. If they have a priority or focus, they are able to dig, to expand, and to achieve depth of understanding. According to Derek Rowntree (1987), if we wish to discover the truth about an educational system, we must look into its assessment procedures. What student qualities and achievements are actively valued and rewarded by the system? How are its purposes and intentions realized? To what extent are the hopes and ideals, aims and objectives professed by the system ever truly perceived, valued, and striven for by those who make their way within it? The answers to such questions are to be found in what the system requires students to do in order to acquire the expected learning outcomes. It is convinced that the electronic transformation of student portfolio assessment, coupled with the context of personalized instruction, with the advent of the Internet technologies should define the de facto curriculum, and promote sustained institutional dialogue about the OBA impact in a learning-centered education.

## BACKGROUND

Today, the use of portfolios have fast become a desired tool for assessing student learning (Zubizarreta, 2009; Johnson, Mims-Cox, Doyle-Nichols, 2006; Banta, 2003) because they are designed to provide authentic evidence of what students know, believe, and are able to do. Assessment for student learning is considered authentic when it focuses on real performance and mastery of a field of knowledge. If instruction is the means by which content, standards, and outcomes are made known to students, then assessment measures the degree to which the standards and outcomes have been achieved. As instructors in higher education, we realize that using portfolios with our students is increasingly transforming the way in which we interact with and engage them in the learning process. This is the kind of appraisal that engages teachers in the process of developing, reviewing, and evaluating portfolios of student work based on explicit criteria and procedures called scoring rubrics. The portfolio is to document what students know and are able to do. Students collect and select pieces of their own work over a period of time as evidence of completing their learning objectives or targets. Usually, students also write a rationale to explain why they think the selected pieces are their best work. Teachers exercise their advising and mentoring roles in the process, recognizing that when instruction is personalized, only authentic forms of assessment can appropriately characterize student performance. Student portfolios may include artwork; essays and other writing samples; logs or journals; notes and reflections;

observation checklist (student and/or teacher); peer evaluations; photographs related to projects; reading inventories and lists; reports (personal or of group work); self-evaluations; solutions to problems; tests and quizzes; video and audio recordings of presentations and performances; and worksheets (Case, 1992; Ryan & Miyasaka, 1995). In a personalized learning environment, the use of electronic student portfolios (Vat, 2009, 2008) to encourage active learning on the part of students is getting more and more popular today. An electronic portfolio (or e-portfolio) could be considered as an extension of the paper-based portfolio, bringing with it the obvious benefit of making a portfolio of evidence portable and shareable anywhere that we have Internet access. In fact, an e-portfolio has a much broader scope as an online collection of reflections and digital artifacts (such as documents, images, blogs, resumés, multimedia, hyperlinks and contact information). Students can use an e-portfolio to demonstrate their learning, skills and development and record their achievements over time to a selected audience.

According to Linda Suskie (2009), authentic assessment for student learning constitutes an ongoing process of learner-centered activities including: a) establishing clear, measurable expected outcomes of student learning; 2) ensuring that students have sufficient opportunities to achieve those outcomes; 3) systematically gathering, analyzing, and interpreting evidence to determine how well student learning matches the expectations; and 4) using the resulting information to understand and improve student learning. Tellingly, the four steps enumerated above do not represent a once-and-done process but a continuous four-step cycle. Namely, in the fourth step, assessment results are used to review and possibly revise approaches to the other three steps, and the cycle begins anew. In fact, portfolio-based assessment is often viewed as part of an integrated, collaborative learning experience. It is convinced that students learn better when their college experiences are not collections of isolated courses and activities but are purposefully designed as coherent, integrated learning experiences in which courses and out-of-class experiences build on and reinforce one another. Understandably, when students can see connections among their learning experiences, their learning is expected to be deeper and more lasting. More importantly, when students are engaged in the consistent production of learning evidence to demonstrate progress completion of different sets of learning outcomes for a course, for a major, or as a comprehensive summative assessment for meeting graduation requirements, the mission of the portfolio is to facilitate the organizing of such evidence to serve as a reflective way of assessing student learning. As mentioned earlier, portfolios are collections of student evidence accompanied by a rationale for the contents and by student reflections on the learning illustrated by the evidence. In her review of literature on portfolios that appeared in Assessment Update, 6:4 (1994), Janet E. Boyle says (Banta, 2003, p.1):

*The portfolio, as an element of authentic assessment, has captured the interest of many instructors who want a more comprehensive way to assess their students' knowledge and skills, to have students actively participate in the evaluation process, and to simultaneously develop students' skills of reflective thinking. These latter features make portfolios an attractive alternative to traditional summative testing.*

Portfolios are considered best when they are planned and purposeful and contain evidence of student efforts, progress, and achievement. When portfolios are used effectively in assessment for learning, students should become active participants in the evaluation process. Indeed, learners must have an intense say in the development of portfolios: selecting evidence, connecting and explaining evidence items, and describing how the evidence illustrates learning. The process of developing a portfolio is full of possibilities for self-assessment and reflection, and the potential for extended and enriched learning. Student portfolios are thereby a good partner with outcomes-based assessment.

## The Context of Authentic Assessment

In many a university around the globe today, faculty are hungry for alternatives to traditional summative testing that will provide more comprehensive ways to assess students' knowledge and skills. They recognize that no single instrument can measure all that student know about a concept or issue, that not every student will be up to giving their best performance on any specific occasion, and that the important element of growth over time cannot be assessed with a single measurement. Thus, assessment as the process of gathering information about student learning, with an attempt to improve the same, must be authentic, regarding the range of students' knowledge and skills. In this light, portfolios can demonstrate a learner's accomplishments, and reveal the range of his or her abilities, talents and learning styles, using a variety of artifacts and media in collections of his or her work completed over a period of time, say, a semester, a year or two, or an entire college career. Another characteristic of authentic assessment that distinguishes from the traditional summative assessment is that too often summative assessment is developed by faculty and administered to students without their involvement – involvement that could deepen and strengthen student understanding. Using portfolios, students are expected to actively participate in the evaluation process, selecting materials to include and combine as evidence of specific learning. Many students take portfolio development very seriously because they plan to use some of the contents to convince potential employers that they have unique skills and talents. Also, since an essential feature of preparing a portfolio is reflecting on the content and explaining how components fit together to illustrate what has been learned, portfolios simultaneously develop students' skills of reflective thinking. Indeed, portfolio assessment as an instance of authentic assessment is adaptable to the needs and intentions of individual learners and the expectations of individual faculty members, programs and degree requirements. It could go beyond the potential of a single piece of evidence of learning in that portfolio provides a holistic picture of the learner's achievements. For example:

- assessment can be structured to include evidence produced early in a course or program, evidence produced midway, and evidence produced at the end.
- assessment can be structured to include drafts of evidence, feedback about the drafts, and final finished evidence.
- assessment can be structured with both required evidence (determined by faculty) and evidence submitted by the learner, with both sets of evidence demonstrating achievement of different sets of learning outcomes.
- assessment can be structured around learning outcomes with evidence generated by varied learning experiences, in varied learning contexts, and in varied forms (e.g., written, video, graphics).

Thus, the versatility of portfolios as a form of authentic assessment, as described above, enables us to see not only what students are learning, but also how they are learning. In fact, besides documenting student learning, the use of portfolios in authentic assessment often reveals the strength and gaps in our curriculum and pedagogy as well as strengths and gaps in student learning.

## The Meaning of Personalized Instruction

Personalized instruction (PI) is often understood as a systemic effort on the part of a school to foster student success (Keefe & Jenkins, 2008; Keefe, 1989). It is an attempt to achieve a balance between the characteristics of the learner and the learning environment. It typically begins with the identification of performance goals and criteria that describe what students should be able to do upon completion of their education. Sequenced instructional materials are then developed to provide a variety of ways for students to accomplish the school learning goals. At its best, the PI context is a management system designed to help students progress in an orderly fashion, taking time and failure out of the learning equation. Yet, PI must begin with learner needs and interests, and fashion the learning environment to meet those needs. It might take place in supervised study, in small groups and in electronic support environments. Students typically work alone, with one other or a few others, in scheduled seminars. Teacher coaching of students occurs on a one-to-one basis, but more often in a cluster of students working in the same area and topic of the curriculum. Since students will necessarily come to any learning experience with different prior experiences – and thus with different starting points for the material to be learned – successful teachers must know how to create experiences that let students access ideas in a variety of ways, yet always press for deeper and more disciplined understanding (Darling-Hammond, 1997, p.12). A personalized approach to education requires, on the part of the instructor, a strong knowledge of subject matter, substantial pedagogical skill, a commitment to helping individual students succeed, and a deep desire to make instruction both thoughtful and interactive. The teacher, acting as adviser and coach (Glasser, 1988; Carroll, 1975), must help students attain a sense of balance in the learning environment between what is challenging and productive and what is beyond the student's present capabilities. Yet, there are two essential aspects for personalized instruction according to Keefe and Jenkins (2008): the culture and the context, that must be put into proper perspective, if a school wishes to develop effective teaching and learning for student success. The cultural aspect establishes the foundation of personalization and ensures that the school prizes a caring and collaborative environment, student diversity, and individual development. The contextual aspect promotes and supports student engagement, thoughtful growth, and proficient performance. The combination of these two aspects should produce a challenging, integrative, learner-centered environment that is interactive and meaningful, but with reasonably structured learning activities, flexible use of time and space, as well as authentic, performance-based assessment of student progress. This conception of personalized instruction is somehow consistent with the blueprint for ongoing improvement in school organization and good practice, rendered by Darling-Hammond (1996): 1) Structures for caring and structures for serious learning, that enable teachers to know students well and to work with them intensely, through such means as smaller school size, student inter-disciplinary clusters, multi-year advisories, and extended time with individual students; 2) Shared exhibitions of student work which makes it clear what the school values and how students are doing. Teachers set standards, create authentic assessments, and display student work in every way possible to provide a basis for what works and what needs to be improved; 3) Structures that support teacher collaboration with a focus on student learning, especially teacher teams for curriculum planning, student advisement, and accountability for student success; and 4) Structures for shared decision making and dialogue about teaching and learning with other teachers, students, and perhaps larger communities. Teachers, often in collaboration with students, should agree on professional development, formulate curriculum, and design evaluation systems. These structures support a workable decentralization of authority and operation.

## WHAT IS ENTAILED IN OUTCOMES-BASED ASSESSMENT?

Based on our discussion in the Background section, there are mainly four iterative stages in outcomes-based assessment (Suskie, 2009): 1) establishing clear, measurable expected outcomes of student learning; 2) ensuring that students have sufficient opportunities to achieve those outcomes; 3) systematically gathering, analyzing, and interpreting evidence to determine how well student learning matches our expectations; and 4) using the resulting information to understand and improve student learning.

### Establish Intended Learning Outcomes

As faculty, the first element in the assessment process is to establish a set of intended leaning outcomes (ILOs), representing our intentions about what students should know, understand, and be able to do with their knowledge when they graduate. In fact, ILOs reflecting the discipline should be developed for each academic program and for each course in the program. ILO statements typically beginning with the phrase, "Students will be able to (SWBAT) ..." are meant to be learner-centered, and developing such statements should reflect a systems approach to teaching in the program. When faculty collectively decide what graduates of an institution or program should know, understand, and be able to do, we are working as a team, rather than as individuals. We are collectively confronting an important question in higher education (Plater, 1998, p.12): What does the degree or certificate that we award mean and how can we prove it? Still, it is worthy to note that learning goals at the institutional level are likely to be more broadly stated than those at the program level, and those at the program level are likely to be more broadly stated than those at the course level. Just as Huba and Freed (200) point out, achieving the more specific learning goals that we develop for a course or even for a particular class period should nonetheless help students make progress toward achieving our program and/or institutional goals.

### Provide Learning Opportunities Leading to ILOs

The second element in the assessment process is to ensure that students have sufficient opportunities both in and outside their courses that help them achieve the intended learning outcomes (ILOs). Namely, if we expect students to achieve our ILOs, we must provide them with learning experiences to acquire what they need to learn. Students' learning is largely affected by the way courses and other required experiences like independent studies, practicum, and internships are organized in the curriculum and the order in which they are taken. Thereby, it is conducive to designing the curriculum as a set of interrelated courses and experiences that will help students achieve the ILOs. Indeed, designing the curriculum by working backward from learning outcomes should help make the curriculum a coherent story of student learning (Plater, 1998, p.11). Consequently, as faculty develop or revise the curriculum, we should scrutinize each of the activities and experiences that we create in our courses and programs and ask ourselves this question: How will this help students achieve the intended learning outcomes of the institution, program, or course?

### Develop Assessment Measures for Student Learning

The third element in the assessment process is to design, or select data gathering measures to assess whether or not our ILOs have been achieved. This element brings to a culmination the previous step of determining learning outcomes because the process of designing assessment measures forces us to come to a thorough understanding of what we really mean by the ILOs (Wiggins & McTighe,

2005). As we develop our assessment measures, we may find ourselves fine-tuning the learning outcomes. Typical assessment measures should include both direct and indirect assessments of student learning (Palomba & Banta, 1999). The former include projects, products, theses, exhibitions, performances, case studies, portfolios, interviews, and examinations. The latter include self-report measures such as surveys distributed to students which can be used both in courses and at the program and institutional levels. In all of these assessments, we ask students to demonstrate what they know or can do with their knowledge, such as to address enduring and emerging issues and problems in their disciplines. Yet, assessment measures (Wiggins, 1989), chosen to provide accurate and useful information for making decisions about learning, are referred to by different names, such as naturalistic assessment because of their intrinsic value; as performance assessments because they require students to demonstrate their learning; and as portfolio assessments because they allow us to evaluate the nature and quality of students' work over time. Whatever they are called, these assessments are effective tools for assessing mastery of factual knowledge, but more importantly, for finding out if students can use their knowledge effectively to reason and solve problems. And any evaluation must be based on subjective judgment using criteria we as faculty collectively develop.

### Use Assessment Results to Improve Learning

The fourth element in the assessment process is to use the assessment results to improve student learning. At the course level, discussions between students and instructors should take place continually with a focus to improve individual student performance using the assessment results as indicators. At the program or institutional level, ongoing review of student achievement should take place among the faculty as a whole. Through discussion of assessment results, faculty should gain insights into the type of learning that is taking place in the program, and be better prepared to make informed decisions about needed program changes. As a result, our school should understand what students can do well and in what areas they have not succeeded. We should also raise questions about the design of the curriculum, about the teaching strategies in use, and even the ILOs. Furthermore, we should develop a better understanding of how better to assess student learning in a way that could build trust for our institution of higher education in the community, especially through sharing summaries of the assessment process with key stakeholders (students, alumni, and advisory groups) to seek additional perspectives. This is indeed an act of public accountability not to be ignored.

## WHAT IS ENTAILED IN PERSONALIZED INSTRUCTION?

Based on Darling-Hammond's (1996) formulations of ongoing improvement in school organization and good practice as described in the Background section, it is believed that any school pursuing the philosophy of personalized instruction should include at least three essential components: teacher dual role as coach and as adviser, student learning characteristics, and collegial relationships.

### Dual Teacher Role

It is believed that the teacher, as the instructional specialist who is closest to the learning situation and best understands the needs and interests of students, must be the indispensable catalyst in the PI environment. Personalized instruction demands that the teacher assumes the dual roles of subject-matter coach, consultant and facilitator on the one hand, and of teacher adviser to mentor selected group of students on the other. As learning coach, the teacher is expected to collaborate with other teachers, student peer tutors, and community resource persons to guide student learning. As teacher adviser, the teacher provides advice, counsel and guidance to students on academic and school adjustment issues.

## Teacher-Coach

The needs of today's students are quite different from those of their counterparts two or three generations ago as our world has experienced several social revolutions and knowledge explosion over the Internet in the past decades. Cognitive and problem-solving skills, also called meta-cognitive skills, are more important today than any particular piece of knowledge. Therefore, the teacher-coach in the school environment must be a facilitator of learning, a learning guide who helps students find appropriate resources and engage in suitable learning activities. Members of the LEC International (Georgiades, et al., 1979; http://www.lecforum.org/) describe such a teacher as not so much educational broadcaster as academic troubleshooter. He devotes fewer hours to lecturing and more to working with students individually and in small groups. He spends a good deal of time preparing basic instructional objectives, analyzing the specific strengths and weaknesses of individual students in relation to those objectives, and investigating and making available a wide range of learning activities and methods that will facilitate student success. He recognizes that each student is a unique human being with his own personal learning needs and style, and he knows that what works well for one may not work at all for another. Indeed, the teacher-coaches must be focused on whether, how, and what students learn, in the environment. They must get outside themselves and inside the minds of students (Lasley, 1998; Perkins, 1992). They must participate in the learning process with their students, in its planning, and its organization, concerned with how students are motivated to learn. Nonetheless, coaching requires time to interact with students, to connect with them, to understand their needs, to provide needed information, advice, and feedback about targeted skills, ideas, or issues. Bransford and Vye (1989) summarized befittingly the role and responsibilities of the teacher-coaches along these lines:

- Coaches monitor and supervise student attempts at problem solving both to give them experience in real problem solving and to keep them from going too far into flawed solutions;
- Coaches help students reflect on their own problem solving, encouraging them to think out loud or even modeling strategies for them;
- Coaches identify what students can already do by letting them solve problems and by providing feedback;
- Coaches help students experience new ways of thinking as guides to their own thinking, to compare and contrast their own ideas with other possibilities;
- Coaches help students comprehend and construct meaning in their experiences using resources related to their needs and interest (not unrelated exercises);
- Coaches use whatever resources are useful to engage students in learning including: presentation, discussion, learning packages, computer-based learning systems, and personal tutoring.

## Teacher-Adviser

In a school geared to foster student success, advisement is an important responsibility of the teacher-adviser who plays a helping role to aid students plan and achieve appropriate career and personal-social goals. In a typical advisement setting, teachers, counselors, and other adults work as a team to promote student adjustment and success in school. Oftentimes, professional counselors serve as advisers to a group of teacher-advisers and help them learn their role and function. School guidance functions that are of concerns to teacher-advisers generally include such areas as: 1) academic program planning, 2) career information, 3) school adjustment issues, and 4) personal-social guidance. To help students personalize their education experience, the teacher-advisers' tasks typically include the following (Keefe, 1983):

- plan student group activities, work with individual students in schedule planning, and counsel students on academic and school adjustment problems.
- collect information about each advisee and provide information as needed on personal and school adjustment, and career planning. Maintain personal folders (portfolios) on each advisee.
- help students recognize their personal aptitudes and interests. Meet with students regularly to discuss their goals, behavior, and academic progress. In particular, serve as a "friend-in-court" for students experiencing adjustment problems.
- function as home-base teachers and chief in-school contact for all persons and agencies concerned with the student. Talk to parents, community persons, prospective employers, and career counselors on behalf of their advisees.

## Student Learning Characteristics

If the goal of personalized instruction is to build a learning environment suited to the aptitudes, needs, and interests of each student, any attempt to provide personalized instruction must begin with knowledge of the learner. Namely, some form of diagnosis is needed to determine what the learning-related characteristics of individual learners are. Indeed, many dedicated teachers spend a lot of time in observing students to find out where they are in the learning process, in checking student progress, and in prescribing learning resources and interventions for more successful performance. This kind of direct feedback and various types of diagnostic assessment are among the principal tools of instruction viewed as coaching, mentoring, facilitating, and advising. Three types of student learning traits that are of interest here include: developmental characteristics, student learning style, and student learning history.

## Developmental Characteristics

Developmental characteristics are those specific stages in individual maturation when certain capacities for learned behavior appear, such as cognitive thinking skill. These characteristics tell us when a student is developmentally ready to learn something. It is understood that if teachers were to personalize student instruction, they must have a good understanding of the learner's developmental traits. Darling-Hammond (1997) called for developmentally attentive schools whose organization and student work must build on student developmental considerations. Learning activities should be based on student needs and legitimate interests rather than, arbitrarily, on generic curriculum guides or the contents of approved textbooks. Diagnosing student developmental characteristics and observing the demands of developmental attentiveness are not to be neglected in today's schools if we are to provide a personalized approach to education. Their importance in program planning and instruction can hardly be overstated.

## Student Learning Style

Learning style encompasses information-processing habits, attitudinal tendencies, and biologically based responses that are typical of the ways a given student learns and prefers to learn. There are three broad categories of learning style characteristics: cognitive, affective, and physiological behaviors that serve as relatively stable indicators of how students perceive, interact with, and respond to the learning environment. They can be measured by a variety of assessment techniques, including the *Learning Style Profile* developed by NASSP (http://www.nassp.org), which assesses 24 independent scales representing four factors: perceptual responses, cognitive styles, study preferences, and instructional preferences. The *Learning Style Profile* and other comprehensive style instruments help teachers identify student style strengths and weaknesses and organize instruction more efficiently and effectively. Learning style diagnosis is now considered as a key element in any attempt to provide a more personalized approach to education.

## Student Learning History

Student learning history (Bloom, 1976, p.69) is a term used to describe the aggregate of personal learning that a student brings to a particular course, class, or school program. A learner's history characterizes his or her instructional readiness which is another broad area of diagnosis. Learning history tells us what a student knows at a given point in his or her learning career – the knowledge and skills the student possesses before beginning a new learning experience. Diagnosis of a learning history involves the determination of what has occurred as a basis for what should occur. Tellingly, existing student knowledge and skills define the expected ground for student success in subsequent learning. Observation, surveys, inventories, and curriculum-referenced tests could best assess these knowledge or skill levels. Indeed, information about student learning history must be made available to teachers in cumulative record folders (student portfolios), in teacher and counselor reports, and from student questionnaires, inventories, and various diagnostic tests.

## Culture of Collegiality

The idea of collegiality is closely related to a school culture of collaboration where teachers and students work together in a cooperative social environment to develop meaningful learning activities. It is considered as an important ingredient of a school advocating personalized instruction. According to John Goodlad (1984, p.242) in his landmark work, *A Place Called Schoo*l, we found students sitting passively in class, listening to lectures, and doing seatwork. The climate of the school was largely nice and pleasant, but teacher-student relationships were perfunctory or cordial but antiseptic. We cannot help but wonder about the flat, neutral emotional ambience of most of the classes observed. Boredom is a disease of epidemic proportions. Of course, students took every opportunity to talk to one another, and were very social, but had little opportunity to actively participate in their schoolwork. Glasser (1986) tells us that if what is being taught does not satisfy the needs about which a student is currently most concerned, it will make little difference how brilliantly the teacher teaches – the student will not work to learn. Wolk (2007) commented that passive schooling creates passive people. If we want people to think, learn, and care about the many dimensions of life, if we want neighbors who accept responsibility of tending to the world and making it a better place, then we need schools and curricula that are actually about life and the world. In personalized environment, students are empowered to pursue work that is meaningful to them; they can satisfy their needs. A constructivist environment and collaborative learning arrangement are found to characterize such a collegial culture:

## Constructivist Environment

The constructivist view of schooling (O'Neil, 1992; Perkins, 1992) holds that people learn by actively constructing knowledge. They weigh new information against their previous understanding, thinking about and working through discrepancies on their own and with others. Finally, they come to a new understanding. This perspective of constructivism borrows from various movements in other disciplines, including social construction of reality in sociology, phenomenology in philosophy, and constructivism in psychology. It is believed that learners can make meaning of what they are learning, and they construct that meaning in light of their prior knowledge and experiences. In a classroom faithful to constructivist views, students must be afforded numerous opportunities to explore phenomena or ideas, to conjecture, to share hypotheses with others, and to revise their original thinking. Time and opportunity for reflective dialogue are critical elements of such a learning environment. Constructivist teachers typically build instruction on student learning styles and skills, and involve students in self-directed learning and in collaborative approaches for a specific topic of study. Students work with their teacher-coaches to improve their cognitive skills and to expand their current experience through reflection, seminars, and other group projects.

## Collaborative Learning Arrangements

Today, numerous evidence exist that students learn better in cooperative groups than when alone (Slavin, 1991, 1995). Cooperative small groups encourage collaboration and afford better socialization than traditional classrooms. Collaborative learning calls for positive inter-dependence among learners, face-to-face interaction, individual responsibility for mastering the target material, and interpersonal skills fostering cooperation and effective working relationships. Collaborative learning arrangements provide an opportunity for students and teachers to work together to verbalize their ideas, to sharpen their thinking, and to capitalize on differences. Students at different levels of school achievement are expected to work together for some good reasons (Glasser, 1986):

- Students gain a sense of belonging by working in small teams;
- Belonging provides the initial motivator for students to do the work. As they achieve some success, they will want to work even harder;
- Stronger students find it need-fulfilling to help weaker students toward a high performing team effort; weaker students find it need-fulfilling to contribute to the team effort. Alone they are able to do little.
- Students do not depend only on the teacher, but also on the team and their own creativity;
- Learning teams provide needed structure to avoid superficiality and support in-depth learning; they have flexibility in the kinds of evidence they present about the knowledge learned or skills achieved;
- Teams can be changed to give all students a chance to work together and to serve on high-scoring teams.

Overall, collaborative learning arrangements are a requisite for a personalized learning environment. These arrangements promote interaction, dialogue, and thoughtful reflection. Together with an enhanced teacher role, and a strong diagnostic component, a culture of collegiality sets the stage for the practice of personalized instruction.

# DEVELOPING STUDENT ELECTRONIC PORTFOLIOS IN EDUCATION

The use of electronic portfolios (e-portfolios) in education for learning and assessment is becoming increasingly popular today (Jafari & Kaufman, 2006; Barrett, 2004; DiBiase, et al, 2002; Cambridge, 2001; Johnson, Mims-Cox,

Doyle-Nichols, 2006). Initially, an e-portfolio may appear as simply a collection of work that has been compiled over a period of time. It is sometimes compared to a scrapbook because it contains artifacts that are selected over time. Yet, the contents of an e-portfolio are often organized to assess competencies in a given standard, goal, or objective and they focus on how well the learner achieves in that area. Through the use of artifacts, which are concrete examples of the student's work, e-portfolios contain evidence of knowledge, dispositions, and skills (Batson, 2002; Brown & Irby, 2000). In particular, an e-portfolio that is used for assessment and evaluation requires the learner to engage in higher levels of thinking through the use of inquiry and reflection (Acosta & Liu, 2006). Inquiry involves a process of collecting, sorting, selecting, describing, analyzing, and evaluating evidence to answer questions on how well the evidence represents the learner's accomplishment of a standard, goal, or objective. The learner is involved in a personal type of action research that entails continual reflection or questioning and resorting of the selected work. The e-portfolio is an ideal tool for meeting the needs of reflective learning which is a form of mental processing applied to gain a better understanding of relatively complicated or unstructured ideas. In the process, the learner is expected to answer how he or she must improve personal practice in order to acquire learning. That way, an e-portfolio is often understood as a user-centered, personalized learning space allowing the user to shape the way an individual presents him or herself to the world. Content and layout can be personalized to create multiple views which meet the specific, differing or changing requirements of the user. This connects well with one of the key tenets of personalized instruction, that students become key partners in the design of learning to suit their needs. Personalized instruction involves thinking about knowledge as an active process. Students get to be informed, active participants in their own learning, they contribute to decisions about what learning can work best for them, and they have a much better understanding of how they are progressing in a specific field of work. Oftentimes, an e-portfolio carries with it the element of being reliably and swiftly updated, as well as easily accessible in terms of the data being tracked. Thereby, an e-portfolio model of education (Vat, 2009, 2008; Platter, 2006; Flanigan & Amirian, 2006; Herbert, 2001) implies a system of empowering the individual to learn and to demonstrate his or her learning acquired over a period of time through an electronic medium of ongoing support. More befittingly, e-portfolios could be considered as personal online spaces for students to access services and store work. They will become ever more useful as learners grow up and start moving between different types of learning and different institutions. They have the potential to provide a central, linking role between the more rigid, institution-led learning management system and the learners' social online spaces.

## The E-Portfolio Context of Student Learning

As online technologies and information resources rise in salience with the advent of the Internet, we are witnessing the emergence of a multi-faceted techno-pedagogic reality in the development of online support for student learning. The e-portfolio model of education could be considered as a result of several important converging forces. Such forces are causing the education community to re-examine where learning takes place and how it could be assessed, how work and knowledge should be managed, who we, education practitioners, really are as we present ourselves to the world, and how we use technology for teaching and learning. This idea of the e-portfolio is said to be a flashpoint "at the converging of imperatives and opportunities in the management of learning for human and social capital development" (Jafari & Kaufman, 2006, p.xxvi). Technically, its context ranges from the simple conceptualization of e-portfolio as a means of capturing student progress through a program of study, involving student work, student reflection, and faculty comments related to activities of teaching and learning (Henry, 2006), to the technological potential (Plater, 2006) which allows faculty and institutions to actually enable each student to have a personally managed, meaningful, coherent, integrated lifelong record of learning that demonstrates competence, transcends educational levels, and is portable across institutions of learning. In fact, e-portfolios are more than storage devices of the learner's best work (O'Brien, 2006). They provide the means for students to set learning goals, monitor and regulate their progress toward these goals (a form of self-directed learning), as well as develop their self-assessment skills. Practically, e-portfolios should serve as the student's pathway from classroom to career.

## The Personalized Aspect of E-Portfolios

As life-long learners, we are always looking for tools to transform our learning experience, to enable learners to become autonomous and enjoy a truly personalized development path. It is believed that the e-portfolio is one of the most significant tools for achieving this goal. It should support the realization of a portfolio-based career, and act as an instrument for social inclusion, allowing us to "tell our story" and celebrate our achievements (Flanigan & Amirian, 2006). In fact, the e-portfolio could facilitate a continuum in the learning space where someone starting an e-portfolio at school, college, university, or work would not have to throw away the investment of years when moving from one episode of life to another. The e-portfolio should be our faithful digital companion, reflecting our digital identity and supporting our learning, and enabling transactions with others in a variety of social networks. For instance, in the professional circles, e-portfolios could become the indispensable tools for reflective practitioners extracting learning from the workplace, and sharing their reflections with their peers to contribute to the development of different professional learning communities.

## The Learning Aspect of E-Portfolios

In a typical learning environment, there are many roles the e-portfolio can play, examples of which include the means of assessing student learning, the means of showcasing outstanding student achievements, and the means of ensuring learner accountability (Acosta & Liu, 2006; Sherman, 2006). Yet, whichever role the e-portfolio might play, there is one aspect that all e-portfolios have in common: namely, the learners must create portfolio elements or artifacts to be presented within the portfolio itself. As instructor or facilitator of e-portfolio learning, the design of sample e-portfolio requirements to document and communicate the learning of skills reflected in the learning process becomes critical. Examples include a learning contract with specific lesson plan detailing what the expected learning should be and the way to demonstrate the acquired learning. Such e-portfolio requirements should delineate the specific artifacts to be created by the learners to complete the process of learning. Indeed, this act of "creation" would necessitate the learning and/or application of a variety of skills related to the learning episode. Importantly, using the e-portfolio requirements as an aid of setting personal learning goals becomes a form of instructional scaffolding. Oftentimes, learners need to articulate clearly the goals of every piece of new learning experience by demonstrating the series of created artifacts to be included in the e-portfolio as evidence of the lessons learned. In this regard, examples of similar works from different learners could be collected into the e-portfolio repository for comparison and evaluation.

## The Design Aspect of E-Portfolios

The advent of Web technology has brought about the currency of e-portfolio, which can not only be considered as an effective way to assess student learning, but also as a vehicle for knowledge development and for career building (Napper & Barrett, 2004). The key behind the e-portfolio movement lies in the empowerment of the learner to take charge of his or her own learning (Ramsdon, 2003; Barrows, 1988). Specifically, the e-portfolio scheme of learning shifts the locus of control from what we faculty teach to what students learn (Acosta & Liu, 2006); namely:

- *Enable students to determine what they need to learn through questioning and goal setting*

It is believed that students should work to identify their knowledge and skill deficits, and to develop strategies in the form of personal learning goals for meeting those deficits. The emphasis is to foster a sense of students' ownership in the learning process. In particular, e-portfolios emphasize analysis and reflection, and the development process, but not merely the product of learning. This process perspective not only raises the cognitive bar, but also shifts the locus of control from not so much what the instructor is doing, to what the student is doing to meet learning objectives. Moreover, the student can reflect on his or her learning and can demonstrate learning to persons outside of the immediate learning environment with the production of relevant electronic artifacts. For example, interested employers could review a student's resume, group project contributions, and other items of interest the student wants to make accessible. Likewise, if teachers, through the e-portfolio support environment, can guide the students in identifying what they already know and what they need to learn, then knowledge gaps and mistakes can be viewed in a positive way such as another opportunity to learn. And students can assume more responsibility in addressing their own learning needs during any instructional episode

- *Enable students to manage their own learning activities*

It is believed that students must be enabled to develop their learning plans, which should describe priorities, instructional tactics, resources, deadlines, roles in collaborative learning situations, and proposed learning outcomes, including presentation and dissemination of new knowledge and skills, if applicable. Traditionally, these instructional events are arranged by teachers to be obeyed by students, in order to accomplish a specified set of pre-determined objectives. Yet, it is not advantageous for students to learn to be self-directed. To manage their own learning activities, students must be guided and supported by the teacher, through the e-portfolio environment, slowly taking on more and more responsibility of their own learning. For example, collaborative learning, inside and outside of the academy, is another feature of the new portfolio model, which should document such efforts as peer-to-peer projects promoting teamwork and communication skills, student-mentor projects (say, internships in the industry) giving students the opportunity to experience the world of work for better understanding of their future profession and workplace culture, student service-learning projects offering students first-hand understanding of societal issues and problems. Whichever type of projects the student is involved, he or she should maintain housekeeping of his or her e-portfolio and allow peers, mentors, and the community to give input, while the instructor at school provides the opportunity for the interactions, and assesses the intended learning outcomes.

## The Curriculum Aspect of E-Portfolios

It is anticipated that the e-portfolio, as a tool to transform teaching and learning, should become a catalyst for curriculum change and a new model of assessment, which should connect the educational mission and objectives with the needs of society. It should also bring students closer to their future profession, and carry learning into students' future careers and possibly into their lifelong devotions. Thereby, the e-portfolio review process should serve as the feedback mechanism to update the academy on the skills required by students as they enter society. Put it simply, if students are immersed in projects that extend into the dynamic workplace and community (rather than the limitations of the campus) then they must demonstrate not only applicability of knowledge, but also flexibility and adaptability. The pedagogical challenge then is to set up connections between academic objectives and societal needs that will update the curriculum by incorporating current global perspectives. It is also expected that faculty members will then be in discussions with interested parties in the community to determine student outcomes. Therefore, the assessment of a course, program of study, and the related discipline will be somehow corroborated with persons outside of the academy. In this regard, the deliberation of an e-portfolio scheme of student learning, including its elements of flexibility should always be an important area of concerns.

## CASE EXPERIENCE

Many of today's users demand a personalized learning experience that extends beyond traditional boundaries to include social networks of peers, evaluators and even external experts. To meet this challenge, the idea of e-portfolios is being adopted at a growing number of universities worldwide to help users analyze patterns in their learning based on intended learning outcomes and performance criteria (Lorenzo & Ittelson, 2005). The Department of Computer and Information Science (DCIS), as a constituent unit of education under the Faculty of Science and Technology at the University of Macau, is installed to offer degree programs in both the undergraduate and graduate levels in Software Engineering. The department has a current population of about 180 undergraduates and 50 graduate students mostly part-time. It has to coordinate per academic year, the enactment of about 20 graduate and 40 undergraduate courses. To help manage course delivery, the university provides course management systems, such as WebCT (since 1998) and MOODLE (since 2008) to teaching staff for their course enactment. Currently, the means of education delivery in our department has largely been didactic; yet, we are quite willing to combine the best of our old values of good teaching through the instructivist approach with the modern-day constructivist way of thinking such as problem-based learning (PBL) (Amador, Miles, & Peters, 2006). We are also interested in the continuing efforts to extend our curriculum and instructional practice over the Internet, blending some continually renewed electronic (mostly Web-based) course support, with our conventional face-to-face interaction between teaching staff and our students. The use of e-portfolios in our DCIS department has not been institutionalized yet. Still, bottom-up e-portfolio efforts from individual academic staff have been encouraged so long as the educational potential for enhancing student learning could be realized in a positive direction. It is believed that the development of e-portfolios could render a program assessment process where teachers and faculty can examine and improve classes and programs based on student achievement of intended learning objectives and standards.

## The PSU Model Adapted

Following a model similar to the Pennsylvania State University's e-portfolio system (http://portfolio.psu.edu), we believe that over the course of a student's college years, the e-portfolio should play a variety of roles. Firstly, it should reinforce the process of student learning by embracing a user-centric approach, and by prompting students to take more responsibility for their own learning. Secondly, it should integrate seamlessly with any Learning Management Systems (LMS) in use, such as our MOODLE environment (http://ummoodle.umac.mo). Thirdly, it should enable students to receive feedback and assessment from peers and others, by showcasing student achievements to multiple audiences. It should also provide a portable demonstration of users' acquisition of knowledge and skills, and place personal learning in a social networking context. Collectively, e-portfolios should enable students to enhance their learning by giving them a better understanding of their skills, as well as where and how they need to improve in order to meet their academic and career goals (Lorenzo & Ittelson, 2005). Besides, e-portfolios should preferably pair social networking and informal learning with traditional classroom education, thus accelerating and expanding student learning. It is believed that harnessing social learning enables institutions to be more responsive and learner-centered across the learning landscape. Students can be linked to an active network of their peers and mentors enabling learning beyond the boundaries of the classroom. More relevantly, the ability to publish in a variety of media within the e-portfolios should give students ultimate control over their learning journey. Thereby, the use of e-portfolios can truly embrace the idea of a personalized learning environment. Consequently, e-portfolios can be utilized by students, faculty and staff, and by the administration of an institution (Lorenzo & Ittelson, 2005; Kahn, 2001) for such functions as: plan educational programs; document knowledge, skills, abilities and learning; track development; define, develop and embark on a career path; evaluate a course, program or institution; and monitor and evaluate personal performance.

## E-Portfolio Designed as a Tool for Assessment

In fact, different portfolios (Stefani, Mason, & Pegler, 2007) have been used by students at traditional universities and colleges where face-to-face teaching is the dominant mode of educational delivery. For example, course portfolios are those assembled by students for individual courses. They document and reflect upon the ways in which the student has met the outcomes for that particular course. Instructor's endorsement is often required to authenticate the course portfolios. Program portfolios are developed by students to document the work they have completed, the skills they have learned, and the outcomes they have met in an academic department or program. The mentor or appraiser could add comments. It could be a requirement for graduation. Students might use a selection from their program portfolio to show to prospective employers. Whatever the primary focus of engagement with students, the use of e-portfolios inevitably adds a strong online element to the activities of teaching and learning. Institutions need to provide electronic support and services; teachers need access and skills to integrate the e-portfolio application into their overall course design, and students need a wide range of electronic abilities in order to develop their e-portfolios. The underlying pedagogy for e-portfolio use is considered the most significant link with online learning support. Our experience has indicated that constructivism (Vat, 2006, 2004; Bangert, 2004) does seem to be the approach worthy of repeated experimentation. The aim of constructivist principles as applied to student learning is to engender independent, self-reliant learners who have the confidence and skill to use a range of strategies to construct their own knowledge (Eklund et al, 2003; Slavin, 1994). Where students are required to develop and maintain their e-portfolios, they are expected to reflect on their learning, consider how to give evidence of their learning and then develop a plan of what they would like to learn. In other words, an e-portfolio implementation of constructivism usually implies a considerable level of learner autonomy and initiative, of learner responsibility for their learning and of opportunities to refine their learning based on feedback from the teacher and their peers. More importantly, the use of e-portfolios can be the basis for several student-centered initiatives (Batson, 2005), including: creating a system to track student work over time, in a single course, with students and faculty reflecting on it; having a more fully informed and constantly updated view of student progress in a program, which is very helpful in formative assessment; aggregating other students' work in a particular course to see how the students as a whole are progressing toward learning goals; and assessing other courses in similar ways that are all part of one major and thus assessing the entire program of study.

## E-Portfolio Positioned as a Tool for Learning

The ease with which the digital form can be adapted, linked and transported is essential to the emergent means of enhancing the use of e-portfolios. One example is an electronic showcase to present student work to prospective employers, or to obtain a place on a post-graduate course. It is a showcase of the student's versatility and an indicator of his or her potential. Besides providing a means of presenting evidence of learning and achievement, the e-portfolio must serve as a reflective document spanning the student's intellectual development and helping learners to become critical thinkers. This idea is often linked to the use of a portfolio as a personal development plan (PDP) (Lorenzo & Ittelson, 2005). As a specific tool of learning, it is not difficult to perceive the development of the e-portfolio over time as an important aspect of learning. The emphasis is on the development process and what this offers the student, rather than merely on a polished end product, no matter how versatile it is. In this light, we identify with DiBiase et al. (2002) concerning the development of a portfolio from simple collection of materials, through selection, reflection and projection to final presentation. These stages could be briefly summarized as follows: a) Collection of materials requires the students, with support from teachers, to save learning artifacts such as assignments, project reports, and presentations that represent achievements, and successes in their day-to-day study; b) Selection of materials requires students to review and evaluate potential portfolio materials to identify those that demonstrate the development of particular skills or achievement of specific standards; c) *Reflection of work done*, requires students to evaluate or assess their own learning through reflective commentary. Students reflect on their own growth and development over time, recognizing achievement of goals and standards, identifying gaps in development or understanding and acknowledging skills requiring further work; d) *Projection of work to accomplish*, requires students, with the teachers' assistance, to compare current achievements or outcomes to standards or performance indicators. They then set learning goals or develop action plans for the future. This stage should link portfolio development and personal development planning (PDP) to support lifelong learning; and e) Presentation of achievements, invites students to share their portfolios with teachers and peers, with an attempt to promote collaborative learning, to foster self and peer evaluation and to further encourage commitment to PDP and lifelong learning.

## E-Portfolio Implementation with Learning Management Systems

Over the years of our trials and errors with different e-portfolio toolsets, we have accrued some experiences on how to assemble and reconfigure some e-portfolio systems to enhance student learning. These experiences come from mainly three sources of interest: a) the desire2learn.com, an e-portfolio software developer; b) the Sakai project, an open source collaborative learning environment (CLE) community; and c) some free Internet tools to create online portfolios for work or school.

## The Desire2Learn.com

The Desire2Learn e-Portfolio (http://www.desire2learn.com/eportfolio) allows users to map their learning journey throughout their lifetime. The e-Portfolio enables users to control what they get out of their learning experience: their goals, their outcomes, and their choice of information recipients. It is a rich and engaging user-centered environment readily accessible both for those familiar with Web 2.0 applications, and for novices. Students choose what to put in their portfolios such as achievements from previous courses, photos and videos from their extracurricular work. The e-Portfolio can be loaded with different formats of information: files, multimedia, personal reflections, presentations, and websites. Users can also decide which specific e-Portfolio elements they publish, and to whom. Using Desire2Learn's assessment engine, any e-Portfolio item can be commented on or evaluated using a rubric that can be created at and shared with any level of the institution. Whether users are presenting materials for assignments, working on group projects, or creating resumes, they could navigate their learning journey with their personalized e-Portfolios. Work becomes thereby organized, searchable, reusable, transportable, and subsequently, more usable and valuable. With Desire2Learn's built-in tools to manage competencies and learning outcomes, assessments by peers and instructors can be made on any e-Portfolio items, from individual documents and files to learning journals, or presentations, enabling users to have a full view of their entire learning path.

Information within the Desire2Learn e-Portfolio can be structured and made consistent across a course, program or organization. Forms and rubrics created by instructors can be shared across the entire organizational structure—providing users with a familiar and consistent environment as they traverse from course to course. For example, a user can fill out a work history form to provide data to the institution. This data can be aggregated to determine the extent of the correlation between the users' grade performance and their real-world experience. Institutions gain knowledge through data on the users' progress toward learning outcomes and the degree of shared learning and collaboration. Instructors can aggregate individual efforts at the course, departmental, and institutional level. Robust reporting allows useful longitudinal and temporal analysis. Equipped with such insight, organizations can shape the learning experience and improve learning outcomes for all users. As users chart their paths, an institution then gains valuable insight into their learning journeys. Organizations could thus create a customarily branded and optimized learning environment with student e-Portfolios as an extension of their campus.

In addition, the Desire2Learn e-Portfolio stores all the artifacts created by a user. Users group artifacts together into collections that can be managed manually or dynamically populated with items such as videos, presentations, pictures or documents that share tags or keywords. Compiling collections and sharing them with learning networks is simplified so that users could focus on learning and not on the software. Reflections throughout the learning process allow students to internalize and synthesize learning beyond the outcomes of traditional memorization. Users are allowed to choose which reflections are to be shared and commented upon. This personalization of the lifelong learning process is at the heart of Desire2Learn e-Portfolio. Besides, users control their presentations: how they look, who sees them, and what level of permission each reviewer has. A range of professional and appealing presentation templates allows users to add flair and individuality while meeting presentation needs. The organization can incorporate their branding while still allowing the user to personalize the look and feel of the presentation. Both evaluators and peers can comment on presentations and even individual items can be assessed.

Moreover, the Desire2Learn e-Portfolio delivers useful assessment capabilities through integration with Desire2Learn's Competencies and Learning Outcome tools. These key features allow for feedback and assessment at a granular level which can be applied to individual artifacts, collections, reflections, or entire presentations. Evaluators can review all comments and assessments made by peers, gaining a more complete picture into the learning process to make a more authentic assessment. Furthermore, e-Portfolio reporting capabilities are built with the user and the institution in mind. As a user, we can view logs of anyone who accessed our e-Portfolios and any changes made to e-Portfolio items. For the institution, a wealth of rich information from the e-Portfolio is available in the data warehouse and reporting tools. Organizations have access to the very detailed information of competency achievements (when they were completed, where, and how they were assessed). Institutions can also gain access to rubrics assessments and frequencies, information from forms such as artifact templates and data can be further aggregated and analyzed.

## The Sakai CLE Project

The Sakai portfolio (http://www.rsmart.com/portfolios) comprises a suite of web-based tools that allow users to store, to organize and to present digital artifacts representing evidence of their teaching, learning or institutional achievement. Sakai's suite of portfolio tools is designed to facilitate the creation of portfolios for self-presentation, reflection, teaching and learning as well as course, program and institutional assessment. By collecting, selecting and presenting subsets of their work, students can create portfolios that showcase coursework, professional experience, academic competency or simply self-expression. Instructors can guide students in their creation of portfolios by designing educational scaffolds that engage them in reflection upon learning in relation to a set of educational outcomes or professional standards. Administrators can use the system as a decision-making and reporting tool. Configured and customized to align with institutional goals and objectives, portfolio sites collect real evidence of teaching and learning that can be correlated with and assessed against course, program, departmental, and institutional objectives.

To create and work with a portfolio in Sakai, we use both the Resources tool and the Portfolios tool in the Sakai CLE environment. First, we collect the materials we want to present in Resources. Then we use the Portfolios tool to present the information. Namely, the process of creating a portfolio involves selecting items from Resources, giving the portfolio a name and an optional expiration date, deciding whether or not to allow comments, determining which site participants or users outside the site will have access to the portfolio, deciding whether or not to make the portfolio public, and optionally notifying others that a portfolio has been shared with them.

One of the most archetypal uses for portfolios suggested in Sakai is the personal representational portfolio, namely, our student portfolio. Personal representation portfolios have a long history of use in disciplines such as art, music, writing and photography, where a culture of presenting samples of one's work has long been the norm. These types of portfolios are generally created to showcase a selection of one's work in a given area, in order to demonstrate talent, experience, skill or development. Personal representation portfolios may also be created to provide evidence of one's development over time across different areas. An example of this use case might be a resume or curriculum vitae, assembled using artifacts from one's online learning environment and shared with potential employers, educational institutions, mentors, peers or other interested parties. Portfolios created for personal representation tend to have both a developmental and a creative focus. They are most effective when they guide users in collecting information about themselves, assist users in developing their virtual identities and facilitate users' presentation of themselves to designated audiences. Some common examples of personal representation portfolios include: digital resumes, professional portfolios, and personal narrative portfolios.

Another example of the Sakai portfolios in use is the teaching and learning portfolios. Teaching and learning portfolios have an educational focus and are generally used to gain insight into a teaching and learning process. They are multi-faceted, guiding students in collecting learning artifacts, reflecting upon these in relation to a linked set of learning standards, objectives or criteria and presenting their work for feedback and evaluation. Teaching and learning portfolios require advanced planning on the part of educational practitioners in identifying learning outcomes, objectives, or criteria used to represent the goals of the teaching and learning process. Many practitioners find that the process of creating a teaching and learning portfolio is as valuable as the actual product for their students. Asking students to reflect upon their learning and present their work in a way that best speaks to their mastery of a subject, issue or experience is a fundamental experiment in meta-cognition that goes beyond what the average student is traditionally asked to do in a classroom. By giving students the opportunity to reflect upon their learning and share their learning artifacts with external audiences, these portfolios seek to make the processes of teaching and learning more transparent as well as accessible. Some examples of teaching and learning portfolios include: general education portfolios, disciplinary portfolios, and extracurricular transcript portfolios.

The third type of Sakai portfolio most commonly created is the assessment and accreditation portfolio. Assessment and accreditation portfolios are generally derived from teaching and learning portfolios and are used to assess the efficacy of a given instructional program or objective. In an age of accountability measures applied to education, this type of portfolio is steadily growing in use. Assessment and accreditation portfolios tend to include quantitative measures of student performance gauged against a set of learning outcomes that have been identified by an instructor, program, department or institution. By using reports that aggregate and analyze data surrounding student learning in relation to a predefined set of educational outcomes, these types of portfolios provide a rich source of information about the actual results of the teaching and learning process and can therefore help institutions align their institutional practice with their stated institutional mission or goals. Institutions may present this data along with representative artifacts to demonstrate their progress in fostering learning in accordance with their goals. The results of assessment portfolios are thus a valuable resource for the accreditation process. In support of accreditation or program assessment, they are usually combined with portfolios for teaching and learning to aggregate and analyze assessment data and identify representative artifacts of learning. Some examples of assessment and accreditation portfolios include: institutional outcomes assessment portfolios, departmental outcomes assessment portfolios, and institutional accreditation portfolios.

## The Free Internet Tools

Oftentimes, within the course delivery constraint of a semester, it is found that the use of free Internet tools to introduce to students the educational potential of portfolios in their learning and subsequent professional development planning, has been well received. It is also experienced that Michele Martin's free guides from

Google.com (http://www.google.com/notebook/user/17615569108845553326) have been very helpful in showing our students how to create their personal portfolios over the Web. It is our students' feedback that following Martin's six steps to creating their online portfolios has been very intuitive and instrumental to their reflective growth: 1) identify the purpose of our portfolio; 2) identify, create, organize our artifacts; 3) identify the technology tools we will use; 4) set up a portfolio structure and table-of-contents; 5) create the portfolio; and 6) market or share our portfolios.

Identifying the purpose of the portfolio should help our students determine the structure and format for their portfolios, the artifacts to select, and the tools for creating the portfolios. Most importantly, students should identify the audience for their portfolios, including such issues as: What is their level of comfort with technology? What are they expecting to see or find in their portfolios? What kind of story do they want the student to tell about him or herself? What does the student want to highlight for this audience?

Selecting the artifacts to be included in the student portfolios must be a thoughtful process since there are typically numerous artifacts for telling different stories of student learning. Examples of artifacts include: resume or record of work samples which could include documents, presentations, reports, online materials; copies of credentials such as degrees, certificates, licenses; other records of achievements, such as newspaper stories of work accomplished; recommendations and commendations from schools; transcripts of academic records; and project records. Whichever artifacts selected, depends on the purpose of the portfolio.

There are a variety of tools students can use to create their online portfolios. Helen Barrett provides a great list at (http://electronicportfolios.org/categories.html). Yet, in selecting what free tools to use to create the portfolios, students are often reminded of the following:

- Do you already have an online presence, such as a blog? If so, how do you want to connect your blog and your portfolio?
- Do you want people to be able to interact with your portfolio through feedback comments?
- Do you want your portfolio to be public or private?
- Do you want to use your own name as your URL for your portfolio?
- Is your portfolio primarily for employment or is it for documenting your own learning?
- What kinds of artifacts do you want to include? Which tool presents artifacts in the best way possible?

Tellingly, students also need help in identifying the structure of their portfolios. It is convinced that every portfolio should have: a) a welcome page explaining the purpose of the portfolio, and any other information the audience may need to know upfront; b) tips on navigating through the portfolio, like a video introduction to the portfolio; c) contact information, including an e-mail address at a minimum, and the various places one can be found online, such as LinkedIn, Facebook, and Twitter. Additional sections, depending on the purpose of the portfolio, could include the following: a biography, work history (for work portfolio), educational credentials and background, competencies, personal beliefs statements, personal strengths, achievements, volunteer work, conference presentations, project works, and class assignments from schools.

The exact details to create online portfolios depend on which free tool to use. For example, if we would like to use WikiSpace.com, then the following source from Michele Martin is an excellent guide (http://www.scribd.com/doc/2238100/Using-WikiSpaces-for-Your-Eportfolio). If the use of Google Sites is preferred, then the following How-to-guide developed by Helen Barrett is also found to be another favorite site among students (http://sites.helenbarrett.net/portfolio/

how-to). Once the portfolios have been created, it is time to begin marketing them, especially if students are using them to sell their professional skills or search for a job. Some typical tips used by students include:

- Buy our name as a domain (e.g., www.marysmith.com) and map our portfolio to our name. This service is available from WikiSpaces.com and Wordpress.com. That way, we will have an easy to remember URL to post on our resume.
- Put links to our portfolios on our blogs and social networking profiles (LinkedIn, Facebook).
- Put the link in our e-mail signatures.
- Include the link on our resumes.
- Include a link with online applications if the applications allow for such inclusion.

## A Course Enactment Example

To put our discussion into perspective, the following course example serves to illustrate an outcomes-based design of a sophomore-level major course in our undergraduate *Computer and Information Science* program. In this course, one meaningful exercise for all the participating students is to incrementally construct in a course of one semester their individual e-portfolios using the free Internet tool provided by Google Sites (http://sites.google.com) as a summary of their learning experiences. The specifics of our course learning are hereby presented below in the context of ***SFTW 241 Programming Languages Architecture (I)***, offered in the spring semester of every school year.

### Course Description

This is the first of a 2-course sequence (SFTW241 compulsory + SFTW342 optional) introducing the concepts, techniques, and models of computer programming. The concepts are organized in terms of computation models introducing different techniques for programming and reasoning about programs. Example computation models covered in this course include the imperative and object-oriented programming and reasoning techniques. Each computation model is based on a core language introduced in a progressive way, by adding concepts one after the other. Languages of interest include both software and hardware description languages, such as C versus Verilog, C++ versus Java, Ada versus VHDL. Other contemporary languages considered of interest in inquiry-based learning could include: JavaScript versus Ajax, Ruby on Rails versus Groovy on Grails.

### Course Syllabus

There are a number of conceptual items selected to be dealt with in SFTW241. They are listed below alongside a brief description of the topics involved in student learning.

### SFTW 241 - Intended Learning Outcomes (ILOs)

It is expected that at the end of their study of SFTW241, students will be able to perform the following based on the items of interest identified:

**Item 01**: *Overview of Programming Languages*

1. Summarize the evolution of programming languages illustrating how this history has led to the paradigms available today.
2. Identify at least one distinguishing characteristic for each of the programming paradigms covered in the course.
3. Evaluate the tradeoffs between the different paradigms, such as structured programming and object-oriented programming, considering such issues as space efficiency, time efficiency, safety and power of expression.

*Table 1.*

| *Concept* | *Topics Identified* |
|---|---|
| Overview of Programming Languages | History of programming languages; brief survey of programming paradigms (procedural, object-oriented, and concurrent languages); the effects of scale on programming methodology |
| Virtual Machines | The concept of a virtual machine; hierarchy of virtual machines; intermediate languages |
| Introduction to Language Translation | Comparison of interpreters and compilers; language translation phases (lexical analysis, parsing, code generation, optimization) |
| Declarations and Types | The conception of types as a set of values together with a set of operations; declaration models (binding, visibility, scope and lifetime); type-checking overview; garbage collection; |
| Abstraction Mechanisms | Procedures, functions, and iterators as abstraction mechanisms; parameterization mechanisms (reference versus value); activation records and storage management; type parameters and parameterized types; modules in programming languages |
| Object-Oriented Programming | Object-oriented design; encapsulation and information-hiding; separation of behavior and implementation; classes and subclasses; inheritance (overriding, dynamic dispatch); polymorphism (subtype polymorphism versus inheritance); class hierarchies; collection classes and iteration protocols; internal representations of objects and method tables |
| Type Systems | Data types as set of values with set of operations; data types including elementary types, product and co-product types, algebraic types, recursive types, arrow (function) types, and parameterized types; type-checking models; semantic models of user-defined types, including type abbreviations, abstract data types, and type equality |
| Programming Languages Design | General principles of language design; design goals; typing regimes; data structure models; control structure models; abstraction mechanisms |

4. Distinguish between programming-in-the-small and programming-in-the-large.

**Item 02**: *Virtual Machines*

1. Describe the importance and power of abstraction in the context of virtual machines.
2. Explain the benefits of intermediate languages in the compilation process.

**Item 03**: *Introduction to Language Translation*

1. Compare and contrast compiled and interpreted execution models, outlining the relative merits of each.
2. Describe the phases of program translation from source code to executable code and the files produced by these phases.

**Item 04**: *Declarations and Types*

1. Explain the value of declaration models, especially with respect to programming-in-the-large.
2. Identify and describe the properties of a variable such as its associated address, value, scope, persistence, and size.
3. Demonstrate different forms of binding, visibility, scoping, and lifetime management.
4. Defend the importance of types and type-checking in providing abstraction and safety.
5. Evaluate tradeoffs in lifetime management, such as reference counting versus garbage collection.

**Item 05**: *Abstraction Mechanisms*

1. Explain how abstraction mechanisms support the creation of reusable software components.
2. Demonstrate the difference between call-by-value and call-by-reference parameter passing.
3. Defend the importance of abstractions, especially with respect to programming-in-the-large.
4. Describe how the computer system uses activation records to manage program modules and their data.

**Item 06**: *Object-Oriented Programming*

1. Justify the philosophy of object-oriented design and the concepts of encapsulation, abstraction, inheritance, and polymorphism.
2. Design, implement, test, and debug simple programs in an object-oriented programming language.
3. Describe how the class mechanism supports encapsulation and information hiding.
4. Design, implement, and test the implementation of "is-a" relationships among objects using a class hierarchy and inheritance.
5. Compare and contrast the notions of overloading and overriding methods in an object-oriented language.
6. Explain the relationship between the static structure of the class and the dynamic structure of the instances of the class.

**Item 07**: *Type Systems*

1. Formalize the notion of typing.
2. Describe each of the elementary data types.
3. Explain the concept of an abstract data type.
4. Recognize the importance of typing for abstraction and safety.
5. Differentiate between static and dynamic typing
6. Differentiate between type declarations and type inference.
7. Evaluate languages with regard to typing.

**Item 08:***Programming Languages Design*

1. Evaluate the impact of different typing regimes on language design, language usage, and the translation process.
2. Explain the role of different abstraction mechanisms in the creation of user-defined facilities.

## SFTW 241: Learning Opportunities Leading to ILOs

Besides our regular lectures with slide presentations, there are various learning activities designed into student course work throughout the semester, including: readings, individual (plus optional) assignments, pair assignments, and project assignment. With the help of our UMMOODLE environment, we also managed to induce online discussion (mostly asynchronous) to help students achieve our ILOs. It should be noted that in order to provide personalized feedback to our student work, almost all the student assignments are given an assignment Wiki inside our UMMOODLE to stimulate student reflection and collaboration to accomplish their coursework. The following is an excerpt from our Spring-2009 SFTW241 coursework to illustrate our discussion. For details, please refer to the student e-portfolios URLs listed at the end of this section:

There is one project in the Spring-2009 semester of SFTW241, in which students are to form into group-based learning units to stimulate one another in their project work (programming in C++, Java, and Ada), which is designed to occupy one and half month in the semester, spanning mainly throughout the month of May. The portion of project work is designed to take care of the ILOs from ***Item 06 (1, 2, 3, 4, 5, 6)***.

Besides, to prepare students to perform their project work, in the month of April, students are to work in pair with their chosen pair partner, to complete two pair assignments (programming in Java and C++), to get them warmed up for the project assignment, because each project team is to be formed by two pairs of students. This portion of the pair assignment takes care of the following ILOs: first assignment supporting ***Item 03 (1,2), Item 04 (1, 2, 3, 4, 5)***; and second assignment supporting ***Item 05 (1, 2, 3, 4), Item 06 (1, 2, 3, 4),*** and ***Item 07 (1, 2, 3, 4, 5)***.

Moreover, to get students prepared for their pair assignments, each student is given an individual

*Table 2.*

| *Items of Evaluation* | *Semester Scores Allocation* |
|---|---|
| Individual Assignment | 5 |
| Pair Assignments | 5 |
| Team Assignment (Wiki Housekeeping) | 5 |
| Project Assignment | 35 |
| Final Examination | 40 |
| Student e-Portfolio | 10 |
| Bonus <Optional Exercises> | 5 |
| Bonus <Semester Interview> | 5 |
| *Total Accrued* | 100 with 10 points bonus |

assignment (programming in C and C++) in the month of March to get familiar with the basic rubric of evaluation, in a chosen topic related to the context of SFTW241. A similar rubric is used to evaluate student's course work throughout the semester, including the pair and project assignments. This individual assignment supports the ILOs from ***Item 02 (1, 2), Item 04 (1, 2, 3, 4, 5),*** and ***Item 05 (1, 2, 3, 4).***

The coursework of SFTW241-2009 also provides 3 optional exercises in ANSI C to get students up to speed in their programming exercise starting from the Individual Assignment. Students are not obliged to complete these exercises, but those who completed them shall have bonus score allocated towards the end of the semester. These optional exercises are meant to review algorithms as models of computational processes.

For the sake of completeness, reading assignments have also been given to students throughout the semester, to stimulate their online discussion. Such readings largely support the ILOs from ***Item 01 (1, 2, 3, 4), Item 03 (1, 2),*** and ***Item 08 (1, 2)***.

## SFTW 241 – Assessment Measures Devised

In particular, the assessment scheme we used in the Spring-2009 semester of SFTW241 is as follows:

*Table 3.*

| *Total Semester Points* | *Semester Grade Awarded* |
|---|---|
| 93 - 100 | A |
| 88 - 92 | A- |
| 83 - 87 | B+ |
| 78 - 82 | B |
| 73 - 77 | B- |
| 68 - 72 | C+ |
| 63 - 67 | C |
| 58 - 62 | C- |
| 53 - 57 | D+ |
| 50 - 52 | D (Pass) |
| Below 50 | F (Fail) |

And semester grades are awarded according to the following system of score accounting:

The specific rubric conceived to evaluate each of the Optional Exercises, Individual Assignment, Pair Assignments, and Project Assignment, is composed of the following items of interest:

1. Assignment statement of purpose
2. System description
3. System analysis
4. System design
5. Data structures and algorithms used
6. Program design including structure chart and architectural components
7. Concept of operation
8. ANSI C/C++, Java or Ada coding
9. Test-run under Windows XP Eclipse CDT/JDT/GNATBench (Ada)
10. Program source documentation and user guide

This rubric is designed based on the pre-requisite requirements of SFTW241, namely, students are expected to have proficiency in at least the ANSI C programming language, and two semester's learning in data structures and algorithms with ANSI C.

### SFTW 241: Sample Course Alignment Grid

The course alignment grid serves to check the course alignment between the teaching and learning activities of SFTW241 and the student learning outcomes. The grid produced here is not the exact grid we used in the Spring-2009 semester of SFTW241, for it is meant to convey the importance of course alignment. Typically, we used an Excel spreadsheet and filled in the course outcomes on the vertical axis. On the horizontal axis, we listed the class sessions, readings plus other resources, assignments, and assessments. Oftentimes, we also include labs, guest speakers, and other distinctive teaching and learning activities that we planned for our classes. Once both axes were complete, we could plot the outcomes for each of the items in the horizontal axis. The grid serves as a matrix, allowing faculty to chart the relationship between our course activities and course learning outcomes. Such a grid makes this relationship visible and easy to analyze and understand.

### SFTW 241: Student e-Portfolios on Google Sites

Listed below are some students' e-portfolios completed after their taking SFTW241 in the spring semesters of both 2008 and 2009. I would like to express my appreciations to these students for allowing their e-portfolios URLs to be shown here for readers' convenience.

## FUTURE TREND FOR E-PORTFOLIOS

The essence of e-portfolio lies in its support of deep learning (Barrett, 2004; Weigel, 2002; Salomon & Perkins, 1989) by facilitating the connections among different learning experiences, which occur in various contexts and environments (Tosh, Werdmuller, Chen, Light & Haywood, 2006). In fact, the idea of a portfolio has long been used to demonstrate progress over time, to represent samples of best work, and to prepare for job or career searches. Yet, advances in Web technologies as well as the availability of higher capacity memory storage at lower cost, have increased the opportunity and potential of electronic portfolios to support student learning in a variety of courses, environments, and experiences, both inside and outside the classroom. Through e-portfolios, we are witnessing the emergence of intentional learners who are able to adapt to new environments and situations, synthesize knowledge and experiences from a variety of sources, and seek out opportunities for continued learning throughout their lives (Huber & Hutchings, 2004). Research on student

*Table 4.*

| *Activity* | *Outcome 1* | *Outcome 2* | *Outcome 3* | *Outcome 4* | *Outcome 5* |
|---|---|---|---|---|---|
| | | | | | |
| Class 1 | X | | | | |
| Class 2 | X | | | | |
| Class 3 | X | X | | | |
| Class 4 | | X | X | | |
| Class 5 | | X | X | | |
| Class 6 | | X | X | | |
| Class 7 | | | X | | |
| Class 8 | | | | X | |
| Class 9 | | | | X | |
| Class 10 | | | X | | X |
| Class 11 | | | X | | X |
| Class 12 | | X | | X | X |
| | | | | | |
| | | | | | |
| Reading A | X | X | | | |
| Reading B | | X | X | | |
| Reading C | | | X | X | |
| Reading D | | | | X | X |
| | | | | | |
| | | | | | |
| Assignment 1 | X | X | | | |
| Assignment 2 | | X | X | | |
| Assignment 3 | | X | X | X | |
| Assignment 4 | | | X | X | X |
| | | | | | |
| | | | | | |
| Assessment 1 | X | X | | | |
| Assessment 2 | | X | X | | |
| Assessment 3 | | X | X | X | |
| Assessment 4 | | | X | X | X |

engagement with learning (Ramsdon, 2003; LaSere Erickson & Weltner-Strommer, 1991) suggests that when students perceive that they have choices in how to learn subject matter, they are more motivated to move beyond just information acquisition to gaining a deeper understanding of the subject (Entwistle, 1998; Marshall, 1996; Marton & Saljo, 1984). E-portfolio tools could be characterized by a focus on learner control, a customized learning environment, and the ability to digitally represent and share formal and informal learning experiences with others. Such features can be used to enhance both social and intellectual interactions in various learning contexts, including academic, workplace, and community. Likewise, at the core of the emerging landscape of e-portfolio

*Table 5.*

| SFTW241 - 2009 | SFTW241 - 2008 |
|---|---|
| http://sites.google.com/site/fstda727244/Home | http://sites.google.com/site/fstda627169/Home |
| http://sites.google.com/site/fstda727263/Home | http://sites.google.com/site/fstda627212/Home |
| http://sites.google.com/site/fstda727279/Home | http://sites.google.com/site/fstda627398/Home |
| http://sites.google.com/site/fstda727303/Home | http://sites.google.com/site/fstda627463/Home |
| http://sites.google.com/site/fstda727573/Home | http://sites.google.com/site/fstda627600/Home |
| http://sites.google.com/site/fstda728127/Home | http://sites.google.com/site/fstda628120/Home |
| http://sites.google.com/site/fstda728146/Home | http://sites.google.com/site/fstda628591/Home |
| http://sites.google.com/site/fstda728152/Home | http://sites.google.com/site/fstda628706/Home |
| http://sites.google.com/site/fstda728259/Home | http://sites.google.com/site/fstda628572/Home |
| http://sites.google.com/site/fstda728860/Home | http://sites.google.com/site/fstda627861/Home |

is an emphasis on integration and synthesis of learning, irrespective of where that learning occurs. According to Tosh et al. (2006), the learning model of e-portfolio can be characterized by three working elements: reflection, meaning the learner maps out his or her thoughts on a course, a piece of work, or more general experiences; communication, meaning the learner communicates his or her reflections to others (students, staff, tutors, and instructors); and sharing, meaning the learner gives selected others (typically knowledge users) access to his or her material including reflections, artifacts, and other tangible and intangible resources. The mutual interactions among these elements exercised in the overlapping domains of academic, workplace and community, become the dynamic forces to transform students into active participants in their learning rather than the passive recipients of information (Batson, 2002). Yancey (2001, p.83) reiterates that "the engaged learner, one who records and interprets and evaluates his or her own learning, is the best learner." It is expected that tools and practices that comprise the emerging landscape of e-portfolios should support such activities not only on a personal level but also on a social level. The result is naturally a heightened intentionality to learn through an enhanced self-awareness acquired through reflection, communication and sharing in the learner's domains of concerns.

## CONCLUSION

The premise in our analysis in this chapter lies in the assumptions of the meaning of collaboration in higher education. We believe in the following: If schools are to improve, staff must develop the capacity to function as professional learning communities (PLC's). If schools are to function as PLC's, they must develop a collaborative culture. If schools are to develop a collaborative culture, they must overcome a tradition of teacher isolation. If schools are to overcome their tradition of teacher isolation, teachers must learn to work in effective, high-performing teams. If schools are to support effective teamwork to enhance student learning, there must be some technology-enhanced environment to enable learning among teachers and students. And the concept of e-portfolio system fits right in to provide the mechanism of a learner-centered collaborative knowledge environment to stimulate and facilitate a learning-centered knowledge sharing culture to enhance student achievement. The impact of an e-portfolio for students' housekeeping of their own learning histories, aided by an insightful institutional push, should serve as a transformative path to enable students to tap into (or rediscover) their own sense of wonder and excitement about their present life and future possibilities. Of critical concern here is the rationale for developing electronic portfolios to support knowledge and learning activities referring mainly to the decisions that define expectations, enable empowerment, or verify performance of the people or units involved. Nonetheless, infusing reflective practice and designing fundamentally different ways to evaluate student work requires changes in practice. Under the peculiar umbrella of positive change, we must be open to the transformative impact of such an e-portfolio organizational effort on the intellectual and social capital of the school itself. Our discussion in this chapter has hereby been situated around story-telling the issues underlying the design of a collaborative e-portfolio culture and system for the learning enterprise. Hopefully, it has provided a sense-making perspective on the challenge of outcomes-based assessment in personalized instruction to overcome barriers to knowledge construction and sharing through the e-portfolio movement.

## REFERENCES


Acosta, T., & Liu, Y. (2006). ePortfolios: Beyond assessment. In A. Jafari and C. Kaufman (Eds.), Handbook of research on ePortfolios (pp. 15-23). Hershey, PA: Idea Group Reference.

Amador, J. A., Miles, L., & Peters, C. B. (2006). *The practice of problem-based learning: A guide to implementing PBL in the college classroom.* Bolton, MA: Anker Publishing Company, Inc.

Angelo, T. A. (1999, May). Doing assessment as if learning matters most. *AAHE Bulletin,* 3-6.

Bangert, A. (2004). The seven principles of good practice: A framework for evaluating online teaching. *The Internet and Higher Education,* *7*(3), 217–232. doi:10.1016/j.iheduc.2004.06.003

Banta, T. W. (2003). *Portfolio assessment: Uses, cases, scoring, and impact.* San Francisco, CA: Jossey-Bass.

Barrett, H. (2004). *Electronic portfolio as digital stories of deep learning – emerging digital tools to support reflection in learner-centered portfolios.* Retrieved on June 30, 2009, from http://electronicportfolios.org/digistory/epstory.html

Barrows, H. S. (1988). *The tutorial process.* Springfield, IL: Southern Illinois University Press.

Batson, T. (2002). *Electronic portfolio boom: What's it all about?* Retrieved June 30, 2009, from http://www.campustechnology.com/Articles/2002/11/The-Electronic-Portfolio-Boom-Whats-it-All-About.aspx

Biggs, J. B. (1999). *Teaching for quality learning at university.* Buckingham, UK: Open University Press.

Bloom, B. S. (1976). *Human characteristics and school learning.* New York: McGraw-Hill.

Bransford, J. D., & Vye, N. J. (1989). A perspective on cognitive research and its implications for instruction. In L. B. Resnick & L. E. Klopfer [ASCD Yearbook] (Eds.), Towards the thinking curriculum: Current cognitive research. Alexandria, VA: Association for Supervision and Curriculum Development.

Brown, G., & Irby, B. J. (2000). *The career advancement portfolio* (2nd ed.). Thousand Oaks, CA: Corwin.

Cambridge, B. L. (2001). *Electronic portfolios: Emerging practices in student, faculty, and institutional learning.* Sterling, VA: Stylus.

Carroll, A. W. (1975). *Personalizing education in the classroom.* Denver: Love Publishing.

Case, R. (1992). On the need to assess authentically. *Holistic Education Review, 5*(4), 14–23.

Darling-Hammond, L. (1996). The right to learn and the advancement of teaching: Research, policy and practice for democratic education. *Educational Researcher, 25*(6), 5–17.

Darling-Hammond, L. (1997). *The right to learn: A blueprint for creating schools that work.* San Francisco, CA: Jossey-Bass.

DiBiase, D., Zembal-Saul, C., Sabre, J., Howard, D., Rademacher, H., Burlingame, P., et al. (2002). *Using e-Portfolio at Penn State to enhance student learning: Status, prospects and strategies.* E-education Institute, The Pennsylvania State University. Retrieved July 9, 2009, from https://www.e-education.psu.edu/files/e-port_report.pdf

Driscoll, A., & Wood, S. (2007). *Developing outcomes-based assessment for learner-centered education: A faculty introduction.* Sterling, VA: Stylus.

Eklund, J., Kay, M., & Lynch, H. (2003). E-learning: Emerging issues and key trends. *Australian Flexible Learning Framework.* Retrieved May 10, 2008, from http://www.flexiblelearning.net.au

Entwistle, N. (1998). Approaches to learning and forms of understanding. In B. Dart & Boulton-Lewis (Eds.), Teaching and learning in higher education. Melbourne: ACER.

Flanigan, E. J., & Amirian, S. (2006). ePortfolios: Pathway from classroom to career. In A. Jafari & C. Kaufman (Eds.), Handbook of research on ePortfolios (pp. 102-111). Hershey, PA: Idea Group Reference.

Georgiades, W. D., Keefe, J. W., Lowery, R. E., e al. (1979). *Take five: A methodology for the humane school.* Los Angeles, CA: Parker and Son.

Ghosh, R. A. (2006). *CODE: Collaborative ownership and the digital economy.* Boston: MIT Press.

Glasser, W. (1986). *Choice theory in the classroom.* New York: Harper Perennial.

Glasser, W. (1988). *The quality school.* New York: HarperCollins.

Goodlad, J. I. (1984). *A place called school.* New York: McGraw-Hill.

Henry, R. J. (2006). ePortfolio thinking: A provost perspective. In A. Jafari & C. Kaufman (Eds.), Handbook of research on ePortfolios (pp. 54-61). Hershey, PA: Idea Group Reference.

Herbert, E. (2001). *The power of portfolios.* San Francisco: Jossey-Bass.

Huba, M. E., & Freed, J. E. (2000). *Learner-centered assessment on college campuses: Shifting the focus from teaching to learning.* Boston: Allyn and Bacon.

Huber, M. T., & Hutchings, P. (2004). *Integrative learning: Mapping the terrain.* Washington, DC: Association of American Colleges and Universities.

Jafari, A., & Kaufman, C. (Eds.). (2006). *Handbook of research on ePortfolios.* Hershey, PA: Idea Group Reference.

Johnson, R. S., Mims-Cox, J. S., & Doyle-Nichols, A. (2006). *Developing portfolios in education: A guide to reflection, inquiry, and assessment.* Thousand Oaks, CA: SAGE Publications.

Kahn, S. (2001). Linking learning, improvement and accountability: An introduction to electronic institutional portfolios. In Cambridage, B. L. (Eds.), *Electronic portfolios: Emerging practices in student, faculty, and institutional Learning* (pp. 135–158). Washington, DC: AAHE.

Keefe, J. W. (1983). Advisement – A helping role. *The NASSP Practitioner*, *9*(4), 1–16.

Keefe, J. W. (1989). Personalized education. In Walberg, H. J., & Lane, J. J. (Eds.), *Organizing for learning: Toward the 21st century.* Reston, VA: National Association of Secondary School Principals.

Keefe, J. W., & Jenkins, J. M. (2008). *Personalized instruction: The key to student achievement* (2nd ed.). Lanham, MD: Rowman & Littlefield Education.

Larkin, J. (1998). *Workshop materials: CSUMB outcomes-based education*. Seaside, CA: CSUMB.

LaSere Erickson, B., & Weltner-Strommer, D. (1991). Knowing, understanding, and thinking: The goals of freshman instruction. In *Teaching college freshmen*. San Francisco: Jossey-Bass.

Lasley, T. J. II. (1998). Paradigm shifts in the classroom. *Phi Delta Kappan, 80*(1), 84–86.

Lorenzo, G., & Ittelson, J. (2005, October). Demonstrating and assessing student learning with e-portfolios. *Educause Learning Initiative, 3*(19).

Maki, P. (2004). *Assessing for learning: Building a sustainable commitment across the institution*. Sterling, VA: Stylus.

Marshall, H. (1996). Recent and emerging theoretical frameworks for research on classroom learning: Contributions and limitations. *Educational Psychologist, 31*(3/4), 147–244.

Marton, F., & Saljo, R. (1984). Approaches to learning. In Marton, F., Hounsell, D., & Entwistle, N. (Eds.), *The experience of learning. Edinburg: Scottish Academic Press.*

Napper, V. S., & Barrett, H. C. (2004). Information and technology: Assessment and ePortfolio. In *Proceedings of the Society for Information Technology & Teacher Education Conference* (pp. 39-40).

O'Brien, K. (2006). ePortfolios as learning construction zones: Provost's perspective. In A. Jafari & C. Kaufman (Eds.), Handbook of research on ePortfolios (pp. 74-82). Hershey, PA: Idea Group Reference.

O'Neil, J. (1992). Wanted: Deep understanding: Constructivism posits new conceptions of learning. *ASCD Update, 34*(3), 1–8.

Palomba, C. A., & Banta, T. W. (1999). *Assessment essentials*. San Francisco, CA: Jossey-Bass.

Perkins, D. (1992). *Smart schools: Better thinking and learning for every child*. New York: Free Press.

Plater, W. M. (1998, November). So… Why aren't we taking learning seriously? *About Campus*, 9–14.

Plater, W. M. (2006). The promise of the student electronic portfolio: A provost's perspective. In Jafari, A., & Kaufman, C. (Eds.), *Handbook of research on ePortfolios* (pp. 62–73). Hershey, PA: Idea Group Reference.

Ramsdon, P. (2003). *Learning to teach in higher education* (2nd ed.). London: Routledge Falmer.

Rowntree, D. (1987). *Assessing students: How shall we know them?* (2nd ed.). London: Kogan Page.

Ryan, J. M., & Miyasaka, J. R. (1995). Current practice in testing and assessment: What is driving the changes? *NASSP Bulletin, 79*(573), 1–10. doi:10.1177/019263659507957302

Salomon, G., & Perkins, D. N. (1989). Rocky roads to transfer: Rethinking mechanisms of a neglected phenomenon. *Educational Psychologist, 24*(2), 112–132. doi:10.1207/s15326985ep2402_1

Senge, P. M. (1990). *The Fifth Discipline: The Art and Practice of the Learning Organization*. London, U.K.: Currency Doubleday.

Sherman, G. (2006). Instructional roles of electronic portfolios. In Jafari, A., & Kaufman, C. (Eds.), *Handbook of research on ePortfolio* (pp. 1–14). Hershey, PA: Idea Group Reference.

Slavin, R. E. (1991). Synthesis of research of cooperative learning. *Educational Leadership, 48*(5), 71–82.

Slavin, R. E. (1994). Student teams and achievement divisions. In Sharon, S. (Ed.), *Handbook of cooperative learning methods* (pp. 3–19). Westport, CT: Greenwood Press.

Slavin, R. E. (1995). *Cooperative learning: Theory, research and practice* (2nd ed.). Needham Heights, MA: Allyn and Bacon.

Stefani, L., Mason, R., & Pegler, C. (2007). *The educational potential of e-portfolios: Supporting personal development and reflective learning*. New York: Routledge.

Suskie, L. (2009). *Assessing student learning: A common sense guide* (2nd ed.). San Francisco, CA: Jossey-Bass.

Tabaka, J. (2006). *Collaboration explained: Facilitation skills for software project leaders*. Boston: Addison Wesley.

Tapscott, D. (1997). *The digital economy: Promise and peril in the age of networked intelligence*. New York: McGraw Hill.

Tosh, D., Werdmuller, B., Chen, H. L., Light, T. P., & Haywood, J. (2006). The learning landscape: A conceptual framework for ePortfolios. In Jafari, A., & Kaufman, C. (Eds.), *Handbook of research on ePortfolio* (pp. 15–32). Hershey, PA: Idea Group Reference.

Vat, K. H. (2004). Toward a learning organization model for student empowerment: A teacher-designer's experience as a coach by the side. In *Proceedings of the 2004 IADIS International Conference on Cognition and Exploratory Learning in Digital Age* (pp. 131-140), Dec. 15-17, Lisbon, Portugal.

Vat, K. H. (2006). Developing a learning organization model for problem-based learning: The emergent lesson of education from the IT trenches. *Journal of Cases on Information Technology*, *8*(2), 82–109.

Vat, K. H. (2008). E-portfolio and pedagogical change for virtual universities. In Putnik, G. D., & Cunha, M. M. (Eds.), *Encyclopedia of network and virtual organizations* (pp. 508–515). Hershey, PA: Information Science Reference.

Vat, K. H. (2009). An e-portfolio scheme of flexible online learning. In P. Rogers, G. Berg, J. Boettcher, C. Howard, L. Justice, & K. Schenk (Eds.), Encyclopedia of distance learning (2nd Ed.) (pp. 941-949). Hershey, PA: Information Science Reference (IGI Global Inc.).

Weigel, V. B. (2002). *Deep learning for a digital age: Technology's untapped potential to enrich higher education*. San Francisco, CA: Jossey-Bass.

Wiggins, G. (1989, May). A true test: Toward more authentic and equitable assessment. *Phi Delta Kappan*, 703–713.

Wiggins, G., & McTighe, J. (2005). *Understanding by design* (2nd ed.). Alexandria, VA: Association for Supervision and Curriculum Development.

Wolk, S. (2007). Why go to school? *Phi Delta Kappan*, *88*(9), 648–658.

Yancey, K. B. (2001). General patterns and the future. In Cambridge, B. (Eds.), *Electronic portfolios: Emerging practices in student, faculty, and institutional learning* (pp. 83–87). Washington, DC: American Association of Higher Education.

Zubizarreta, J. (2009). *The learning portfolio: Reflective practice for improving student learning* (2nd ed.). San Francisco, CA: Jossey-Bass.

# Chapter 16
# Best Practices for Teaching and Designing a Pure Online Science Classroom

**Ricardo Javier Rademacher Mena**
*Futur-E-Scape, LLC, USA*

## ABSTRACT

*With the modification of the 50/50 rule by the Higher Education Reconciliation Act of 2005, the purely online university has become increasingly popular and thus so has the purely online science class. In this chapter, the author will use over a decade of teaching physics and math at traditional offline and pure online universities to compare the two. In the process, the author will uncover what techniques have successfully carried over from the traditional to the online environment and how physics education research and technology are changing the physics classroom. The main purpose of this chapter is to identify best practices in designing and teaching online science courses and to provide recommendations on improving existing online science classrooms. Throughout the chapter, Moodle™, an open source LMS, will be used to showcase and implement the ideas being presented.*

## INTRODUCTION

In 2005, the Higher Education Reconciliation Act (2005) modified the 50/50 rule which mandated that distance education classes could only constitute 50% of a student's total class load in order for a university to qualify as a Title IV institution. Through the relaxation of this constraint, the purely online university has gained popularity and with it the purely online science class. In creating science classes for this new pure online environment, an old paradigm has been squeezed into a new medium. This has worked in some areas, but failed in most. As well, there is little in the way of experienced online science course designers and teachers. While these design problems may be endemic to the online environment, creating a science class brings with it unique problems such as lab requirements and math prerequisites. These problems trickle down to the individual charged with teaching these courses for they are required to not only adapt to this new pure online environment but also work within the

DOI: 10.4018/978-1-61520-853-1.ch016

Copyright © 2010, IGI Global. Copying or distributing in print or electronic forms without written permission of IGI Global is prohibited.

limitations of the technology and curricula itself.

In this chapter, the author will use over a decade of experience teaching physics at traditional offline and pure online universities to compare these two modes of instruction and derive best practices in designing and teaching a science class purely online. While the focus will be exclusively on physics, the methods and ideas should be applicable to any online science class. In Section one, the different types of physics classrooms currently encountered are presented. This will serve as a review of the current educational space as it regards to physics. Several of the terms and concepts used later will be presented here. In Section two, a traditional offline physics class is compared to a pure online physics class. As the online class is derived from the offline class, one might expect to see more similarities than differences; yet this is not the case. Sections three and four are the bulk of the chapter, where the focus is exclusively on the pure online physics classes and how to best design and teach them. These sections have been subdivided by the types of assignments one will likely encounter in an online class. In the process of reviewing these classrooms, the techniques that have been successfully carried over from the traditional to the online environment will be uncovered as well as how to improve the content of existing pure online science classes.

Throughout the chapter, Moodle™ (http://www.moodle.org) will be used to present examples of a pure online classroom. Moodle™ is an open-source Learning Management System (LMS) which allows the user to deploy an online class easily and inexpensively. With a wealth of add on modules and a strong support network, Moodle™ is gaining popularity as the LMS of choice with over 36,000 active sites in over 200 countries (*Registered Moodle sites,* 2009).

## Section 1: Physics Classroom Review

Less than twenty years ago, the only physics classroom was the traditional offline classroom as encountered in any brick and mortar high school or university. Only recently with the advent of the Internet and dedicated research into physics education has this classroom been transformed. In this section, four different types of post-secondary physics classrooms are reviewed. This review follows a chronological order and shows how technologies have influenced these classrooms. This section will also serve as a foundation for the rest of the chapter as several terms used later will be defined here as well as setting the stage for the comparisons of Section two.

### The Traditional Offline Classroom

The first classroom reviewed has remained relatively unchanged for hundreds of years. The traditional offline classroom refers to the conventional brick and mortar classroom. While the size and content of these classes may be different, their setup is essentially the same in pre-or post secondary education. These classrooms are typically broken up into two distinct rooms: the lecture hall and the lab room. Broadly speaking, the lecture hall will present the theoretical understanding of physics while the lab room will present its experimental side.

As the name implies, the lecture hall is where the professor will stand in front of their students to talk about the day's material. The professor may outline that week's content, solve a few problems related to the material, talk about the subject, or any number of other non-interactive activities. These classes usually last an hour a day, 2 to 4 times a week (*McMicken College, 2008, pg. 153).* The entire class can take anywhere from three months to a full year depending on the scope of the material being covered and a school's particular term system (e.g., semesters or quarters). A typical lecture hall will seat from 10's to 100's of students while the lab may have room for no more than 20 to 30 students. It is therefore not uncommon to have several lab rooms for every one lecture hall.

As a complement to the lecture, the professor or one of their teaching assistants will perform experiments and demos in the lab room. This lab is performed on a different day then the lecture and may last from two to four hours but no more than once a week. During this lab, the students are broken up into teams and become lab partners. The lab partners then coordinate in performing hands-on experiments and recording their results. Either during the lab or after, the lab partners must once again collaborate to write up their results. This write up, known as the lab report, follows a canonical "hypothesis, procedure, data, analysis, conclusion" format. As well as the lecture and lab, the students are also tested on the material via homework they do at home and quizzes they take in the lecture hall.

The curriculum for a traditional physics class varies greatly, from conceptual physics to advanced quantum mechanics and beyond. For the purposes of this chapter, the physics content will be restricted to conceptual algebra-based freshman classical mechanics. This will cover Newtonian motion broken up into kinematics (the study of time, position, velocity, and acceleration) and dynamics (the study of force, work, energy, and momentum). Being conceptual in nature, the math is at most single equation solving algebra, known as "plug n' chug." The emphasis in these classes is less on problem solving and more on a broad understanding of the topics of physics.

In Figure 1, a course description for the first few weeks of a freshman physics mechanics course based on MIT's OpenCourseWare (http://ocw.mit.edu) is shown. This outline is representative of what is likely to be encountered in any first year physics course. This course is presented so as to frame the discussion in terms of the curricula that might be covered by any traditional physics course. While Lectures 16 and onward are at a very advanced level, the previous 15 week's lectures stand as a good example of what might be covered in any traditional physics course.

## The Interactive Offline Classroom

The traditional offline classroom is jokingly referred to as the "teach and preach" classroom based on the idea that the physics professor is at the front of a classroom "preaching" physics to the rest of the class as a form of "teaching". This non-interactive mode of education has come under fire lately and has experienced a revolution as a result. Based on over two decades of work, Physics education research (PER) has become a research field of its own and is starting to influence how physics is taught (McDermott, 2001). Its main objective is to understand how people learn physics through life and then use that knowledge to help teach the subject more effectively. One of the most applied results from this research is the idea that collaboration is critical for education. Thus there is greater emphasis on working in small groups and having the professor act as less of a purveyor of facts and more of a facilitator by helping the student along their journey but not telling them exactly where to go.

In terms of the physical classroom, PER has lead to several new types of interactive classrooms. In all cases, the curriculum is the same as in a traditional physics classroom and thus they cover the same amount of material in the same amount of time. The difference is in how these classes are set up and how this curriculum is implemented. The most drastic of these classrooms is Studio Physics (Wilson, 1994). This approach requires that the lecture hall and the lab room be combined into one physical space, the Studio. Within this new space, students are broken up into groups of no more than four or five. The classroom is

*Figure 1. MIT OpenCourseWare classical mechanics class*

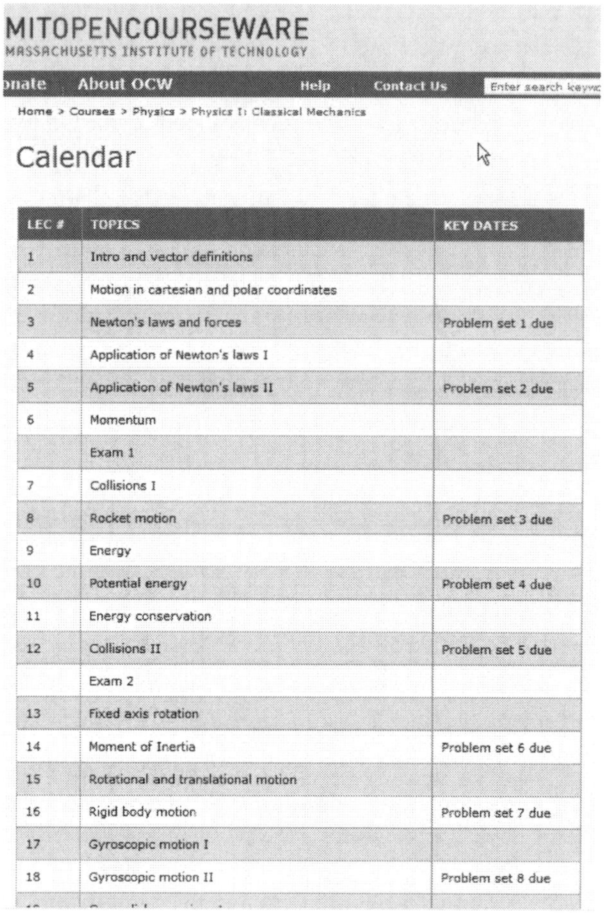

MITOPENCOURSEWARE
MASSACHUSETTS INSTITUTE OF TECHNOLOGY
onate About OCW Help Contact Us Enter search keyw
Home > Courses > Physics > Physics I: Classical Mechanics

Calendar

| LEC # | TOPICS | KEY DATES |
|---|---|---|
| 1 | Intro and vector definitions | |
| 2 | Motion in cartesian and polar coordinates | |
| 3 | Newton's laws and forces | Problem set 1 due |
| 4 | Application of Newton's laws I | |
| 5 | Application of Newton's laws II | Problem set 2 due |
| 6 | Momentum | |
| | Exam 1 | |
| 7 | Collisions I | |
| 8 | Rocket motion | Problem set 3 due |
| 9 | Energy | |
| 10 | Potential energy | Problem set 4 due |
| 11 | Energy conservation | |
| 12 | Collisions II | Problem set 5 due |
| | Exam 2 | |
| 13 | Fixed axis rotation | |
| 14 | Moment of Inertia | Problem set 6 due |
| 15 | Rotational and translational motion | |
| 16 | Rigid body motion | Problem set 7 due |
| 17 | Gyroscopic motion I | |
| 18 | Gyroscopic motion II | Problem set 8 due |

set up as a series of workbenches in which each group will sit through a lecture and work on a lab. The professor will start each class with a small non-interactive lecture but then spend most of the time helping the students do interactive assignments like worksheets, demos, or small experiments. Other examples of PER inspired classrooms are Mazur's Peer Instruction (Mazur, 1997) and the Just-In-Time Teaching project (Novak, 1999). All of these classrooms rely on constant interaction between student and teacher as well as more interaction among the students themselves. For example, several active classrooms make use of "clickers", a classroom responses system (*Clickers*, 2009). The clickers are used in a traditional lecture hall where the professor will ask a question to the class as a whole and then gather input as students use their individual clickers to answer the question. The professor can then modify their lecture or add new material based on the students' responses. This piece of interactivity helps the class progress at a pace that is most beneficial to all students in the class. This new generation of interactive classrooms also strongly relies on the integration of computers with experiments. It is very common for the computer to take over all data recording duties during a lab. This has the effect of freeing the student to observe the phenomenon and learn physics instead of focusing on taking data and performing calculations. The interactive classroom thus represents the pinnacle of offline science classrooms by combining dedicated research with advanced technologies.

## The Hybrid Online Classroom

As the Internet provided a new means of communication, many schools combined its distance education potential with their current ground-based studies. In what is known as a hybrid class, portions of the class are performed at a traditional classroom in an offline physical environment while other portions are completed online in a virtual environment. Using early versions of current LMS systems such as Blackboard™ (http://www.blackboard.com), these classes were some of the first attempts at teaching physics through the internet. Most hybrid classes follow the traditional classroom curriculum and thus will last just as long and cover the same material as a traditional offline class. Typically the lecture and lab component are done offline while the quizzes and homework are submitted online. Hybrid classes are a great option for offline universities since they can be easily integrated into their existing infrastructure. As well, they represent the easiest way to take advantage of the Internet by relegating the admittedly tedious job of grading homework and quizzes to the LMS and allowing the professor to focus on the lectures and labs. However, it also demands that the professor be at two places at once, maintaining both an online and an offline presence. This can be a real challenge to professors who have been teaching offline all their lives or have time intensive research duties.

## The Pure Online Classroom

The pure online classroom is so named because 100% of the class is taught online. These classes are taught entirely through the use of an internet browser and LMS's such as Moodle™ or eCollege™ (http://www.ecollege.com). These classes tend to last between 6 to 8 weeks with a student commitment of 10 hours a week (*2009-2010 Undergraduate Academic Catalog*, 2009; *2009/2010catalog of programs*, 2009). The amount of material covered in a pure online physics class is generally broader and shallower than any other type of physics classroom. This is largely due to the role that the pure online physics classroom plays in a student's career as a requirement for graduation and nothing more.

All classes the author has experienced have followed the format shown in Figure 2, with the course being broken up into weeks. The first week (known as "Week 0" or "Course Home") serves as a central repository for information about the class as well as a place for students to post class-wide questions. Students will gain access to Course Home well before the class starts so they may familiarize themselves with the requirements of the class. Week 1 is the formal start to the course and its layout is replicated in each subsequent week. Each week is broken up into a graded and a non-graded assignment. The non-graded weekly assignments typically consist of that week's lecture as well as a Q&A forum for the student to raise questions about that week's material. In Moodle™ for example, a lecture can take the form of an active lesson (such as a MS PowerPoint™ presentation) or a passive lesson (such as a MS Word™ page). The graded weekly assignments vary widely from week to week and in content. Based on the author's experience, the types of graded weekly assignments in a pure online science class can be broken up into three major types: Discussion, Math, and Lab based assignments.

- *Discussion Based Assignments*: This assignment begins with a question that is asked to the entire class. Each student is then required to post an individual answer to this question in what is known as the "original post". These assignments are meant to be interactive and therefore require the students to post more than just the original post. These extra posts can be a response to something the professor posted or more likely a response to something another student posted. These are

*Figure 2. Teacher view of a hypothetical pure online physics classroom as implemented in Moodle™*

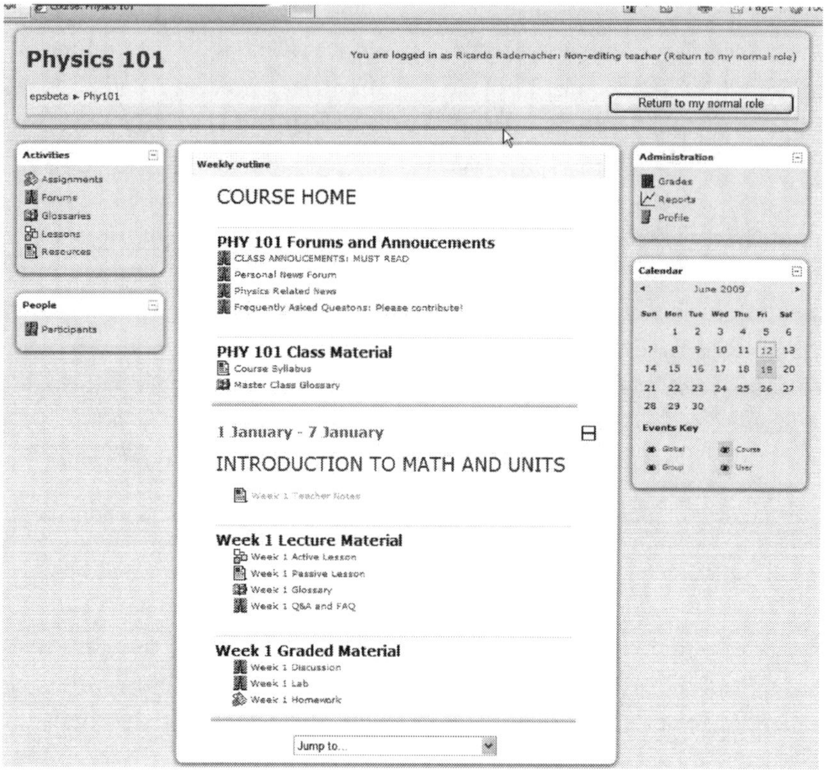


known as the "reply posts" and all the pure online universities the author has experienced have set policies on the minimum number of replies required of students and teachers per assignment. An example of a discussion based assignment is presented in Figure 3. In Moodle™, the name for this type of assignment is the "Forum" and can have several options from every student starting their own discussion thread to an interactive Q&A discussion. The grading criterion for these assignments is generally based on the quality of the original post and the quality and quantity of the reply posts.

- *Math Based Assignments*: A math based assignment consists of problem solving questions with little or no follow up from the class or teacher. Assignments such as worksheets, quizzes, and homework all fall into this category. In the case of a quiz, the LMS will usually handle everything from timing the quiz to grading the assignment to giving the student feedback. In the case of homework, the methods very but generally involve asking a set of math questions and then having the student submit their answers through a post or uploaded document. Figure 4 shows an example of this type of homework. The student is generally not required to interact with the rest of the class on this assignment and the grading criterion is usually based around straightforward right or wrong answers.
- *Lab Based Assignments:* There is often a lab requirement in a physics class. Offline there are special rooms dedicated to these labs. In the online class, the lab based assignments fulfill this requirement. In this assignment, the student is given a set of

*Figure 3. Student view of a discussion based assignment as implemented in Moodle™*

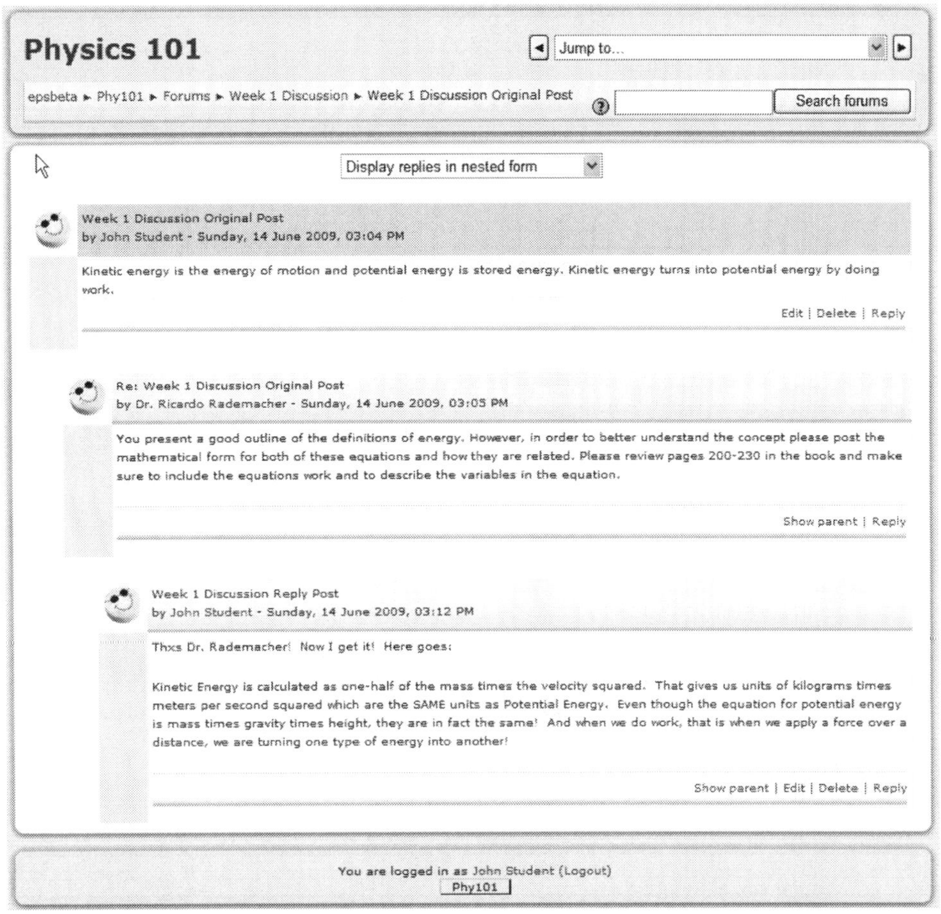


instructions which may involve performing experiments around the house offline or using a physics simulation online. Once they perform the experiment, they are then required to write up their results and submit it for a grade. Students are graded on the thoroughness of their description as well as properly using curriculum materials from the week. In classes with a lab requirement, the pure online classroom has two main options: using online applets such as *Physlets* (Christian & Belloni, 2001) or having the student do the experiments offline in what the author refers to as the "Do It Yourself (DIY)" approach. The canonical lab write up as performed in a traditional physics class is rarely seen as part of the pure online physics class. Rather, performing the lab is followed by answering a set of questions relevant to the experiment or discussing the experience as in a discussion based assignment.

## Section 2: Comparing Traditional Offline to Pure Online Physics Classrooms

Having defined a series of physics classrooms in Section one, this section presents a brief comparison of the traditional offline class to the pure online class. Given that most teachers are only familiar with traditional offline classes, this section

*Figure 4. Teachers view of a math based assignment as implemented in Moodle™*

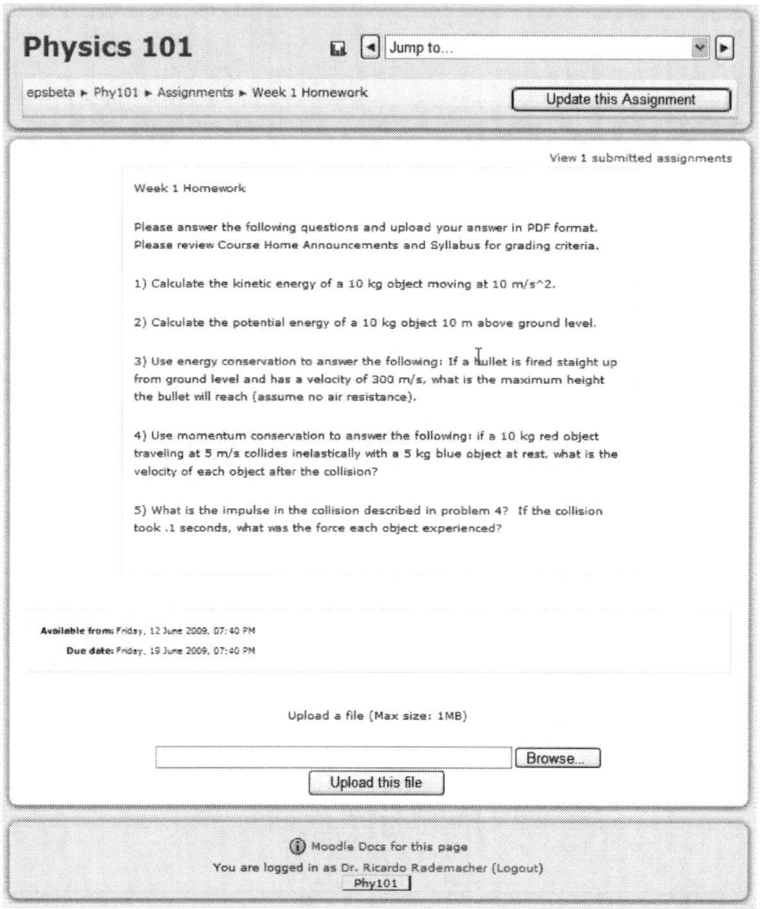


will be a quick primer on the pure online version. As well, the end result of this comparison will be identifying several areas that demand attention when designing or teaching pure online science classes and that will be addressed in subsequent sections.

The most obvious similarity is that both classes have labs, lectures, and math based assignments. The only difference is in how the materials are presented and the lack of a discussion assignment in the offline classroom. The reason for the similarity is understandable given that the offline environment predates the online version. Since most schools and teachers are already used to teaching using this offline paradigm, it is natural that it would be replicated in the online environment. There are also regional and national standards to consider such that a school may be forced to follow the same curriculum online and offline in order to get accredited (*DETC Accreditation Overview*, 2007; ACICS *Accreditation Criteria, Policies, Procedures, and Standards*, 2009). However, while both classes have lab requirements, how these labs are implemented is drastically different. As discussed before, the labs in the pure online class are restricted to experiments the student does on their own or to online simulations. Neither of these methods, however, is an adequate substitute for a real offline lab. How to deal with labs in an online environment will be discussed at length later in the chapter.

Another similarity is the choice of textbooks. Every class the author has taught online has chosen a non-conceptual algebra based textbook like *Fundamentals of Physics* (Halliday, Resnick, & Walker, 2005) or *College Physics* (Serway, Faughn, & Moses, 2003). These books are written for the offline environment and thus tend to align with a semester or quarter system. These books are also very heavy into problem solving as a main component of their educational methodology. There are no physics textbooks known to the author specifically created for a pure online class. This deficiency will be addressed in the next section where several alternatives to these textbooks will be given.

One difference that stands out is the course length. An offline class is bound by the calendar year set forth by the university and thus the majority of classes will fall into either a quarter or semester system. However, pure online universities are not bound by such constraints nor do their students require extended periods of time to achieve deep levels of insight into physics. This is why online classes are about half as long as their offline counterparts. This is understandable considering that most offline universities offer full degrees in some sciences and thus physics would be required as a strong foundation for these degrees. Therefore, an offline curriculum would spend a lot of time on physics as part of the students course load. However, since there aren't many dedicated science degrees in the online environment, the need for physics as an integral foundation of a student's career is minimal or absent. As mentioned already, online classes are not only shorter in length but also in content. Thus the online class is broad and shallow while the offline version is usually narrow and deep. Once again the roots of this discrepancy lie in the requirements that each class fulfills for each university. As seen in a typical syllabus for a university class (Figure 1), there is a linear progression starting with math, continuing through kinematics, and then dynamics. A typical pure online science class will generally also start with some type of math but then abruptly move on to energy, dynamics, radiation, electromagnetism, quantum mechanics, etc. While in the offline class the scientific method is constantly reinforced, in an online class it is rarely mentioned at all. This reinforces the student's preconceptions that physics is nothing but disjoint equations and incomprehensible symbols with no application to their daily lives. Thus until physics becomes a larger part of a student's requirement in the online arena, the short time span and shallow nature of a pure online science class is a fact teachers and designers will have to accept.

One difference that is not readily apparent is how the pure online course is designed and how it is taught. In an offline course or at a traditional university, the instructor is solely responsible for the design and teaching of the class. They are bound by the amount of material they must cover but otherwise can pretty much do it at their pace and their way. In the authors experience however, in a pure online university, one person designs the course and then another completely different person teaches it. The instructor has no control over each week's graded and non-graded materials nor the pacing or presentation of the assignments. This is mostly due to the many oversights that a pure online university imposes on their classes in order to ensure quality courses for all their students and consistency from term to term. This standardization carries a long term advantage whereby the teacher can reuse materials and ideas from previous classes thus making each new class progressively simpler and hopefully more effective. While this may not be the case in every university, the design and teaching best practices will be presented independently so that the reader may focus on what is most relevant to them

## Section 3: Best Practices for Designing an Online Physics Class

In this section, best practices for designing current online science classrooms will be discussed. This section is relevant for course designers and to a lesser extent to teachers. The class will be assumed to be organized around the canonical break up discussed in Section one consisting of a Course Home and then weekly lectures, discussion, lab, and math based assignments (as seen in Figure 2). At the end of this section, Table 2 will summarize these design best practices.

## Designing Non-Graded Assignments

The Course Home should be a repository for all the information that is essential for the students to have during the entire length of the courses. It is critical that there be some form of "webliography," a listing of websites, included so that the student knows of critical sites for information relevant to the class. The syllabus, class expectations, and university policies should also be posted here. Another critical component to the Course Home and every week is the "teacher's eyes only" section. This section will be invisible to the students but accessible to teachers. This is seen in Figure 2 as "Week 1 Teacher Notes", which is gray to indicate that it is hidden to the students. This is a section where the course designers communicate with the teacher about the class. Hints about assignments or tracking changes to the class are good topics to include here. Another important component of the Course Home is the public forum and announcements. The forums provide the class with a place in which the students can communicate about topics related to physics or anything else. In essence it serves a function of the student lounge in an offline university. Announcements should also be regularly posted. These announcements are a section or series of pages in the Course Home that make important public information highly visible to the students. And since Course Home should always visible to the student, the students are guaranteed to see these announcements anytime they enter the class.

When designing an online class, special care should be given to the consistency of the course over the span of the weeks. If there is a discussion based assignment as the first assignment during Week 1,

*Table 1. Offline and pure online physics classroom comparison summary*

| Same | Different |
|---|---|
| Course structure is based around labs, lectures, and math assignments. | A traditional course will typically last at least 3 months.<br>An online course will typically last less than 2 months. |
| Textbooks are chosen from traditional undergraduate offline courses. | A traditional teacher has complete control over their offline classroom.<br>An online teacher cannot make major changes to their online classrooms. |
| | A traditional curriculum tends to be narrow and deep.<br>An online curriculum tends to be broad and shallow. |

*Table 2. Design best practices summary*

| Non-graded | Discussion | Math | Lab |
|---|---|---|---|
| Course home should ideally only have static information | Avoid straightforward math questions that have a set answer | Avoid questions with lengthy or complex problem-solving | Choose from an applet or DIY style lab |
| Each week should have a "Teachers eyes only" section | Use questions based on current events | Base wrong answers on physics misconceptions | Provide meticulous instructions about the lab to students and teachers |
| Each week should be consistent within and without itself | | Upload solutions instead of posting them | Use video clips as a combination of the applet and DIY style lab |
| Use video clips as a way to show physics in action each week | | If available, use the LMS' whiteboard feature | |

then a discussion based assignment should also be the first assignment in the subsequent weeks. This helps build students expectations and guides them by knowing what to expect of each week. It is a very poor practice to vary the number of assignments and their content for each week. If a student has to do three assignments the first week, they will expect to do three assignments of the same type in the second week. If there are more assignments during the second week or they vary in complexity or type, then the student may not be able to properly manage their time to take into account the extra work. This problem is compounded by the fact that future week's content are often inaccessible to the student until the very week in which they have to work on the material so that a student cannot proactively schedule their work until the very week that work is due.

An important practice to be used throughout a course is the use of the videos. A simple search on YouTube™ (http://www.youtube.com) for "physics demos" yields a wealth of pre-recorded demos. These demos showcase several of the concepts that are discussed every week in class and are often accompanied by a spoken explanation of the physics behind the video. They can be used in a lecture to showcase and compliment that week's material. They can be used in a lab based assignments as the object of experimentation and study. They can also be embedded or linked to a discussion assignment to emphasize a point the teacher or student is making. So for example, rather than telling a student that two objects in a vacuum fall at the same rate, a teacher can reference a video which will show it.

The final non-graded best practice is in the area of textbooks. As was previously discussed, pure online universities are using the same textbooks as offline universities. This is despite the fact that the student makeup, curriculum, and course length of the pure online class is radically different than its offline counterpart. In the author's opinion, the following books are much better suited for today's current pure online physics class:

- *Physics for Dummies* (Holzner, 2004) -- While care should be taken to state that reading this book does not make students dummies, the pace and level of presentation of these books is perfect for the online environment. Furthermore, as most students will likely have used other "for dummy" books (or similar books like *The Complete Idiots Guide to Physics* (Dennis, 2003)) and the structure of all these books is the same, it will present physics in terms and at a pace that are familiar to the student. There are books in every subject imaginable with several examples in the sciences (Moore, 2003; Siegfried, 2001; Sterling, 2001).
- *The Cartoon Guide to Physics* (Gonick, Huffman, & Huffman, 1991) -- This is an excellent text especially if the student is in an art or video game design program. While lacking homework problems, it does include the essentials of Newtonian physics. Its cartoon nature and scientific storyline make it an easy and interesting read for any student. It also goes on to cover other topics such as thermodynamics and electricity, making it a suitable text for the broad based nature of the pure online physics class. Aside from physics there are 5 other scientific cartoon guides: chemistry (Gonick & Criddle, 2005), statistics (Gonick & Smith, 1993), the environment (Gonick & Outwater, 1996), genetics (Gonick & Whellis, 1991), and sex (Gonick & Devault, 1999).
- *Conceptual Physics* (Hewitt, 2002) -- This book has been a standard in conceptual physics classes for decades. It follows the traditional textbook structure of presenting the physics material at the beginning of the chapter with homework problems and questions at the end. It also covers a broad range of topics in physics and thus is suitable for any pure online physics class. Since it follows the same layout as

the textbooks currently being used in online physics classes, switching to this book as an alternative to the more advanced texts is highly recommended and easily implemented.

## Designing Discussion Based Assignments

Discussion based assignments are endemic to all pure online classes and may be the only place that many students will actively interact with the instructor. The structure for these assignments is universally the same and was discussed in Section one. The challenge here is to create science questions that foster discussion. Given the dry nature of data taking during labs or problem solving during homework this is not an easy task. It may be tempting to post questions that are too mathematical in nature given that this is a physics class. However, these types of questions will never lead to the type of discussions that will benefit the student. This is an example of a poor discussion question:

**Bad Question:** "What is the kinetic energy of an object of mass 10 kg and velocity 5 m/s? How would your answer change if we double the velocity? …double the mass?"

Notice that this question is purely mathematical in nature and thus has a purely right or wrong answer. It would require the student to solve and post the answers which would then be available for the rest of the class to reference or copy. There is very little discussion possible on this question aside from its answer. Therefore, these are the kinds of questions that are to be avoided in a discussion based assignment. It is much better to relegate these questions to homework or quizzes (i.e.: math based assignments).

Overall, questions should be written that either do not have a straight right or wrong answer or phrased in such a way so as to not lead to a straight mathematical answer. Examples of these types of questions are:

**Good Question:** "How are potential energy and kinetic energy defined? The theory of conservation of energy tells us that energy can turn from one form to another. How does this transformation occur?"

**Good Question:** "The theory of conservation of energy tells us that energy can neither be created nor destroyed. Yet hydroelectric plants are said to "produce" energy. Where does this energy come from? What role does the Sun have in this process?"

Notice that these questions rely on the mathematical definitions of energy yet are phrased in such a way that the student can answer in several different ways of which math is only one. Energy was specifically chosen for this example due to the topical nature in terms of global warming and energy conservation. Examples of how teachers and student interact in these assignments will be given in Section three.

## Designing Math Based Assignments

The math based assignment is an umbrella term for homework and quizzes. This is because their method of presentation and submission are virtually the same in that they do not require the instructor to interact with the student to complete the assignment. The answers to a math based assignment can take the form of an uploaded assignment, a posted answer, or a multiple choice selection the student makes using the LMS.

Special care should be taken when designing these assignments for this may be the only place that the student does any problem solving. Consider that there is very little math and relatively no problem solving in the discussion based assignments. Likewise, labs do not require the student to problem solve insomuch as just observe, record, and analyze. Compounding this problem is the fact that it is a very common practice for the lecture to cover one topic, the discussion another, and the math assignment yet another. This means that by the time the student needs to problem solve in an assignment, they will have very little in-class experience on which to rely. It is therefore important to write these assignments in conjunction with the rest of the weekly materials. As well, make the math based assignment as simple as possible. Avoid lengthy problem-solving or complex mathematics which will discourage the student from attempting the problem at all. Keep all problems so that if the student recognizes the equation then they should be able to "plug" in the numbers and "chug" out an answer. The focus should be to reinforce simple equations and their concepts, not to test a student's math abilities. Good examples of these types of problems are the worked out examples in any textbook chapter. The problems worked out as the chapter develops are simple enough to only require a few steps and it would encourage the student to read through the text if they knew that math problems came from that location. Avoid using the end of chapter questions, the traditional source of homework problems, as a basis for these assignments. They require either too much problem solving or concepts with which the student is not familiar. Another thing to keep in mind is that graphs are not emphasized in the pure online classroom. Due to the inability of common LMSs to draw graphs, there is very little discussion of this topic in the online classroom. For this reason, avoid giving questions that require the student to draw a graph or reference a graph as this is a skill which many students will be lacking.

A good strategy for multiple choice math based assignments learned from PER is to select the wrong answers based on physics misconceptions. Since every LMS system allows for automatic feedback to be given to the student, should a student get a question wrong they can still learn something upon receiving this feedback. For example, the question "What is the force on a ball at the top of its trajectory when it is experiencing freefall?" could have the incorrect option of "Since the ball's velocity at the top of the path is zero, it must be experiencing no force". This incorrect answer is based on the common misconception that if an object is not moving it experiences no forces. Hence if the student chooses this wrong answer, there is a wealth of material on this misconception that the designer can draw upon to help the student better understand the concept. PER inspired texts like *Five Easy Lessons* (Knight, 2002) or *Physics by Inquiry* (Mcdermott, Shaffer, & Rosenquist, 1996) address these misconceptions and provide solid advice on how to phrase the feedback to help disabuse the student of those misconceptions.

Because of the digital nature of these assignments, steps should be taken during the design process to help minimize plagiarism. One way to do this is to provide a list of questions and then have the student choose several from the list which they will turn in for credit. This will minimize plagiarism since it will be obvious when several students choose the same questions to solve. Another way to minimize plagiarism is to not have the solution be posted but rather uploaded. By uploading the assignment to a specific section of the class (such as the "upload a single file" type of assignment in Moodle™ or the "dropbox" in eCollege™), each assignment is only seen by the teacher and not other students.

Finally, if possible design whiteboard functionality into the classroom. For example, in Moodle™ this functionality can be provided by third party modules such as The CovCell Project (http://covcell.org) and DimDim (http://www.dimdim.com). Whiteboards will allow the students and teachers to draw pictures or equations which everybody online at the time can see. This is ineffective as an assignment practice since each student does not have their own personal whiteboard. But it can be effective as a communal place for students and teachers to talk about the assignment or for the professor to show a simple problem to the entire class. Used in this manner it can be part of an office hour style lecture where the teacher can use the whiteboard for a synchronous presentation.

## Designing Lab Based Assignments

In the case of an applet lab, the student is asked to go to a website and use an online simulation in a prescribed manner. They are then asked to run a certain procedure several different times under different conditions and record the results. A popular simulation is the projectile motion lab (e.g., http://phet.colorado.edu/simulations/sims.php?sim=Projectile_Motion). In this applet, a cannon's angle and firepower can be set, as well as the cannonballs mass. The trajectory and variables of the cannonballs flight are then displayed on the screen. The student will then record certain information, such as range or maximum height, as part of the assignment. They can also visualize the trajectory without air resistance (parabola) and with air resistance (skewed parabola). The student will then use their experiences with the simulation to fill out a data table and answer follow-up questions. Students generally find these experiment lots of fun, often comparing them to a video game.

The applets are best implemented when a controlled situation with set results and little discussion is desired. It is easiest to get a consistent set of data from an applet and since the students are online they should have easy access to it. However, it is not the real thing. The applets are always two dimensional while physics is experienced offline in three dimensions. As well, there is no opportunity for the student to go beyond the experiment; they are limited by what the simulation can do as to what area they want to explore. The author therefore suggests this type of lab for more technical classes, where precision and keen observation skills are needed.

In the case of a DIY lab, the student is asked to gather materials from around the house and perform an actual experiment. A typical example is the case of falling objects as shown in Figure 5. In this experiment, a student is asked to gather a piece of paper and a piece of aluminum foil. They are then asked to crumble both into balls of roughly the same size. Upon dropping the objects, they should notice that despite the mass difference, the objects accelerate at the same rate and land at the same time. Due to the subjective nature of the experiment however, much can go wrong though this provides generally provides ample material to discuss in the write up.

The DIY experiments are best for when unpredictability is tolerable and the teacher doesn't mind a lot of off topic discussion. The very nature of the DIY lab means that everyone will be doing the experiment slightly different. It is these differences however that serve as great fodder for discussion. Often, these discussions are more important than the actual write up since they give a personal touch to the material. The DIY approach is also beneficial in that it takes the classroom outside the internet and into students home. This gives them a sense that physics is everywhere which is a very important lesson in all physics courses. However, the unreliability of doing experiments in an uncontrolled environment makes it so that several students give up in frustration or worse, come to the wrong conclusion. This is one reason why it's important to make clear procedures and easy follow-up questions. It is also important for the teacher to keep careful track of discussions; a single incorrect assumption can spread like wildfire through a class if not addressed immediately. In this approach the teacher may or may not ask the students to take data. If they do not or cannot be asked to take data, have the students focus on observing and explaining the phenomena. Instead of asking the students to use a stopwatch to record the time an object falls for example, have them drop several objects of different sizes and compositions and have them write about how each fell the same or different.

In all cases, it is critical that the designer provide clear instructions to the teachers and students regarding a lab. Notes should be placed in the weekly

*Figure 5. Student view of a DIY lab based assignment as implemented in Moodle™*

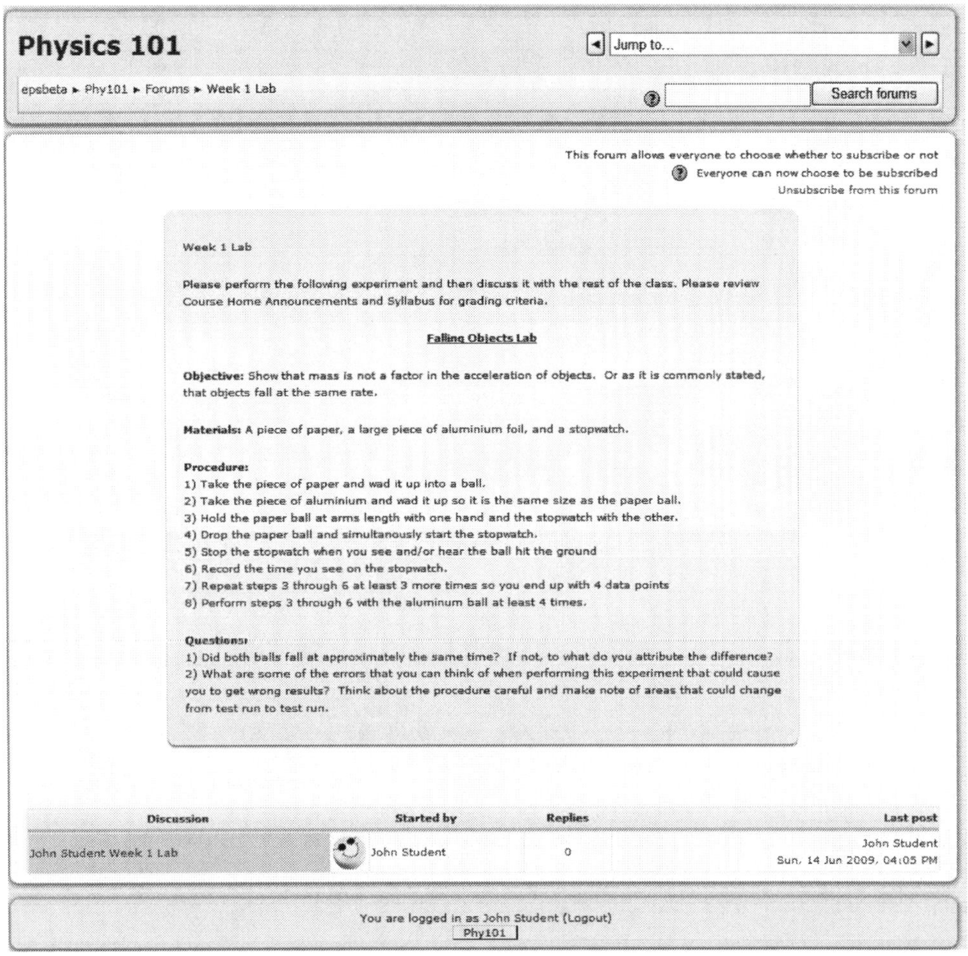


"Teachers Notes" section referencing peculiarities that may be known about the lab or any other comments the designer feels the teacher may need to perform the lab. Student instructions are best presented in an outline and linear format as seen in Figure 5. This way, the student can perform the lab methodically, item list by item list. Paragraph descriptions of procedures will often lead to student doing only what they see in the first sentence and can be unclear. Avoid instructions longer than a sentence and avoid using acronyms or introducing new vocabulary here. Ideally everything the student sees in the lab will have been presented in the other areas of the class. Another idea along the same lines is to provide a lab sheet for download each week. This sheet would repeat the instructions available online as well as provide the student with templates for data taking or other supplementary material. This method has the advantage that the downloaded sheet can then be uploaded as the student's answer, thus ensuring a level of consistency in the completed assignments.

One method that is not seen in the online environment but is seen in the offline environment is the videotaping of an experiment and then analyzing it afterwards. For example, an object can be dropped off the top of a building and video recorded. The professor and students would then

*Figure 6. A projectile motion applet from the University of Colorado at Boulder*

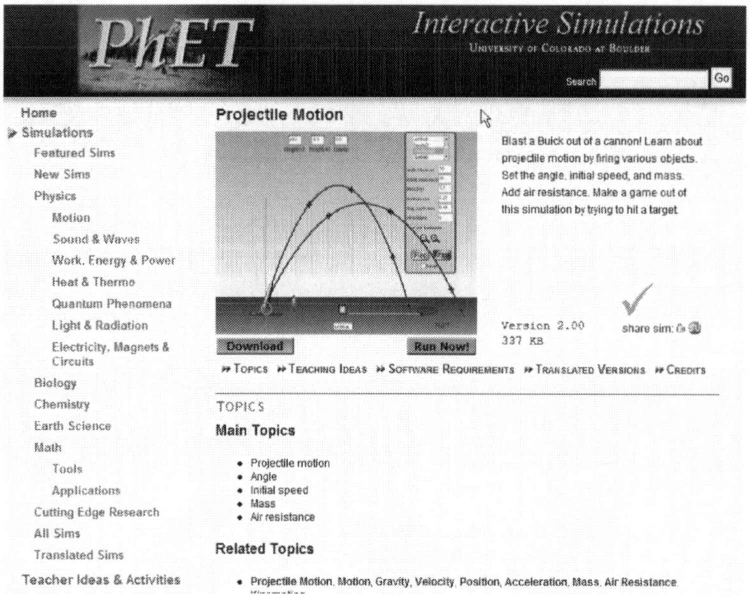


review the tape and analyze it by timing how long it took the object to fall. There is no reason why this could not be done in the online environment as well. YouTube™ has dozens if not hundreds of videos of physics experiments. By posting a link to these videos and then asking the students to analyze it, the strengths of both types of labs are combined: there is the perfect repeatability of an applet lab because the video can be reviewed as much as necessary and the natural realism of a the DIY approach as it is a recording of a live event. This method is very useful in fields that don't have online simulations, such as cellular biology or ecology. In biology for example, videos of cellular processes can be linked to a lab assignment page and used as a reference for students to write up a lab report.

## Section 4: Best Practices for Teaching an Online Physics Class

This section presents some general strategies for teaching the different types of assignments encountered in a contemporary online physics class. Because of the asynchronous nature of the online environment, effective teaching is more about providing consistency and support during the entire week rather than advice and guidance during a few set hours of lecture. At the end of this section, Table 3 will summarize the main teaching best practices.

### Teaching Non-Graded Assignments

These assignments are usually meant to be acted upon by the student alone and don't require a teacher. Examples such as weekly lectures and class announcements are not a dynamic part of the teaching curriculum but should be referenced as much as possible throughout the week. As discussed before, there is often a disconnect between what is presented in the lecture, what is assigned on a homework, and what is tested on a quiz. A best practice to battle this disconnect is to make ample references to non-graded assignments in the graded assignments. For example, in a discussion based assignment, a teacher could make a reference to that weeks lecture as part of

*Table 3. Teaching best practices summary*

| Non-graded | Discussion | Math | Lab |
|---|---|---|---|
| Reference non-graded materials throughout the graded assignments | Be the first to post to all assignments | Require students to upload their answers | Be very visible in the classroom in the case of a DIY lab |
| | Constantly restate and reinforce what constitutes a proper reply | Use "ASCII Math" as an alternative to equation editors | Verify the applet is available online and works before the week starts |
| | Use the Socratic Method for your replies as much as possible | Ask topic specific follow-up questions in cases of suspected plagiarism | Make sure to restate the instructions in your own words |
| | Be mindful of when to use emoticon and acronyms | | |

their response to a student who didn't provide much information in their original post:

*Teacher response: "Good start, but there is more. If you look at this week's lecture you can see what the equations for all these energies look like. Do the following please: research the equations in Chapter 5 of your book and post what the book says about energy that might be relevant to this question."*

Graded assignments from current or past weeks can also be used as response material. This practice is very important for it serves as a motivator for the student to actually do the non-graded assignments and thus be exposed to the material. However as stated in the beginning, non-graded assignments are usually reserved for static material that is meant for the student alone and thus require the least amount of teacher interaction.

## Teaching Discussion Based Assignments

A discussion based assignment will ask a question of the student and then require them to reply one or more times. The instructor is then required to reply to the student's original post and perhaps to one or several of their reply posts. The instructor's job in this assignment is not to provide answers but rather to facilitate understanding; this is why several online universities refer to their instructors as "facilitators" rather than "teachers."

One of the most common practices along the lines of facilitation is the "first post." This is the idea that the teacher be the first person to post in a discussion. This is advantageous for it gives the teacher an opportunity to share their own thoughts on the assignment with the rest of the class. As well, the teacher can use this first post to give hints or guidance to the students based on their past experience with this assignment, such as referencing common mistakes that students might make. Built up in a similar manner to a FAQ section, this first post would grow as the teachers experience with the class grows, making the assignment more enjoyable for teacher and student alike.

It is important for students to gain a conceptual and mathematical understanding of the topic at hand, not just one or the other. Therefore if the student leads with a conceptual answer then the teacher could follow-up with a mathematical one and vice-versa. The following is an example of a student and teacher response:

*Student answer: "Kinetic energy is the energy of motion and potential energy is stored energy. Kinetic energy turns into potential energy by doing work."*

*Teacher response: "You present a good outline of the definitions of energy. Kinetic and potential energy are the two broadest categories for energy. However, in order to better understand the concept please post the mathematical form for both of these equations and how they are related. Make sure to include the equations for work and to describe the variables in these equations"*

Notice that this forces the student to go back and review the equations, how are they are related, and post this research in order to present a more complete understanding of the topic. Had the student instead posted the equations for energy, the facilitator would then ask for a more conceptual answer. As a teacher, it is important to carry several different viewpoints of a topic. That way they can guide a student down a new avenue of investigation without dismissing or minimizing the student's initial contribution.

Be sure to emphasize what constitutes a proper reply and what is unacceptable as a reply. Common requirements are that the reply be from one to two paragraphs in length and that it moves the discussion forward by presenting new information. The most common bad reply encountered is the "Good work" or "I agree" style of reply. These replies are exemplified by the student congratulating another student on their post and nothing else. Variations on this bad post include "I never knew that until today." or "Your post really helped me understand the subject better." As part of a more comprehensive post these comments are welcome but all too often they will be the only thing the student says. By not providing any new information relative to the topic at hand, a "good work" post has no chance of being replied to and thus will not carry the discussion forward. Examples of reply posts that will garner more discussion include "I was looking over your post and I don't understand this part.", "I found the same ideas in my research but here are some new ones I will share with you.", or "I disagree with your conclusion for the following reasons." It is imperative that the teacher post in a meaningful manner themselves so as to reinforce proper posting techniques for the class. The teacher should make numerous references as to what is considered a proper reply during the discussion so as to constantly reinforce this requirement.

A valuable tool in the facilitator's arsenal is the Socratic Method. The Socratic Method relies on not answering question directly, but rather with another question. The follow-up Socratic question should be phrased in such a way so as to encourage the student to do further research on the topic or address a mistake they have made. The question should therefore be a springboard for further discussion and not an endpoint. A Socratic response to the student answer above could be:

*Socratic response: "You present a good outline of the definitions of energy. Yet kinetic energy and potential energy are not the only two types of energies. What are some of the other forms of energy that we see in nature, what are their causes, and how are they related to the energies we have studied this week?"*

This question will lead the student to research other forms of energy, like chemical or nuclear, and then post their relevance online. Due to the amount of press given to global warming and energy conservation, discussions about energy tend to be some of the most fruitful conversations in any physics course. This method is widely used in discussion based assignments since by not directly answering a student, other students who have yet to post an answer are not given any direct information about what is right or wrong. Should a professor divulge the answer to a question directly in a discussion assignment, it is all but guaranteed that the other students will follow suit and post in kind.

In terms of writing style, it is generally preferable to make liberal use of emoticons in posts. Many times the difference between a student's perception of a harsh remark and a gentle reminder boils down to the use of a colon and an end parentheses. The emoticons most relevant for online education are "smiley":) to indicate pleasure in a student's response, "unhappy":(for when they do something wrong or to indicate displeasure, "tongue-out":P for saying something in a silly manner, and "wink" ;) to indicate a hint. Not only do emoticons help set the tone, both good and bad, but they also serve to humanize the discussions. No matter how successful the classroom, science conversations tend to get a little dry. By inserting humor or at least a sunny disposition, students will feel more at ease posting and thus more likely to engage in discussion. The same advice applies to using Internet acronyms, such as LOL (Laughing Out Loud), ATM (At The Moment), or IMO (In My Opinion). However, particular attention should be given to the student makeup. For example, the teacher should not use these techniques in a class with older students who may have never been exposed to these acronyms or emoticons. As well, acronyms can have different meanings such as ATM also standing for an Automated Teller Machine. Thus if a context is not carefully adhered to, an unneeded level of confusion is introduced into the classroom. However, if teaching to a predominately young audience, then the use of emoticons and acronyms may be critical. This is because the Internet generation, with the preponderance of e-mail and other forms of online communication, rely strongly on emoticons and internet shorthand as an essential part of their communication. Using them in a classroom would not only serve to reach out to students but also make it easier for the teacher to convey their ideas.

## Teaching Math Based Assignments

The basic problem with online math based assignments is that most LMSs lack effective methods for including math symbols into the class. For example, Moodle™ provides a way to apply super- and sub-script to an assignment posting while eCollege™ does not. However, even Moodle's™ "insert special character" word editing feature offers only the most basic of ASCII symbols. It is impossible to format a fraction much less a time derivative or a square root symbol.

This leads us to the first math assignment best practice, the use of "ASCII math". For example, the symbol "^" is normally used to indicate an exponent. Thus $v^2$ can be written as "v^2" or $10^3$ can be written as "10^3". This would lead us to write $6x10^3$ (six thousand) as 6x10^3 in ASCII math. Subscripts should be discouraged and can be easily added as extensions to the variable's name. Thus velocity initial vs. velocity final can be written as "vi" and "vf" with no need to subscript. Another example of ASCII math would be expressing $\sqrt{2}$ (the square root of two) as "sqrt(2)." Care should be taken when combining ASCII symbols however. For example initial velocity squared would be "vi^2" and the square root of initial velocity squared would be sqrt(vi^2). A great review of ASCII math can be found at http://mathforum.org/typesetting/ascii.guidelines.html.

Given the pace of contemporary pure online classes, students may have only one week in which to process new symbols. Thus while the context of Newton's second law equation is clear to any physicist, it is merely a bunch of unconnected letters to a student the first time they see it. In the paragraph above, another best practice was used and that is to fully spell out variables and math equations. While this can lead to lengthy posts, it has the advantage of being in a more conversational tone. As well, there is no barrier to the material by way of the symbols. Thus while a student may have no way of approaching "F=ma" on first sight, being told repeatedly that "the force is a result of multiplying an objects mass times it's acceleration" will have meaning the first time they hear it.

The last practice is to require that the students upload their assignments. This technique allows the student to work on their assignment offline on a word processor or a piece of paper which they then scan into the computer. They would then submit these documents as a PDF. The PDF format is recommended since it is machine independent, it best preserves the format the student intended, and cannot be easily altered or copied. This method has the advantage that students with science classes in their background will be more likely to be familiar with working out math problems on paper. They may also have offline tools like MS Office's™ "equation editor" which facilitate putting science symbols into the text. Students can therefore work out the problems in the environment that is most convenient to them and then submit their work in the format required for the class.

## Teaching Lab Based Assignments

How an online lab is taught depends strongly on the form that the lab takes. In the case of an applet based lab, there is generally very little for a teacher to do until the students turn in the final report. Since the strength of these types of labs is that they are very predictable and repeatable, this is usually not a problem. However, several times the site hosting these applets will be down as pure online universities generally do not host these applications themselves. This can lead to a flurry of e-mails as students ask the teacher to solve a problem they have no control over.

Unlike the applet, the DIY lab requires the instructor to do more work up front. The students will be doing the experiment in a very subjective environment and there is often a discussion component to the lab. Hence it is not uncommon to have an ongoing dialogue with one or more students as they do the lab over the course of the week. As a result of this level of interaction, the students tend to talk amongst themselves about what they saw, how they performed the experiment, what their conclusions were, and more. In other words, in a DIY lab the teacher is working with the student interactively on their assignment while in an applet lab they are not. For example, in the object dropping experiment, the student may post that the heavier object fell first. This is a very common expectation and thus a common response to this experiment. When a student says this, it can be followed up by asking them to do other small follow-up experiments that directly challenge the notion that heavier is faster. As well, the teacher should encourage the student to be verbose in how to describe what they did in the lab. The more they talk about what they did, the easier it is to understand where their errors were when things go wrong or what they did right when there are successes.

While a properly designed lab assignment should have a clear set of instructions beforehand, all too often the designer has left the teacher with misleading and often contradictory instructions. While the assignment should give instructions on what to do, the instructor would do well to outline the pitfalls or struggles that the student might have in performing this experiment. Even in the best of cases, it is a good practice for the teacher to place an opening post to the assignment or make an announcement outlining problems and issues the student is likely to run into. By identifying these issues, a teacher can minimize problems or prevent them altogether through a properly worded introduction to the assignment.

## Dealing with Plagiarism

As has been noted throughout this chapter, given the digital and online nature of the assignments in an online class, plagiarism is endemic. This is especially true when math based assignments are implemented as public forums and thus every answer is seen by every student. Hopefully the course designer has designed the assignments so as to minimize plagiarism. As an instructor however, there is often very little that can be done to ameliorate this situation. Teachers are constrained by school policy and their inability to modify the course. Another problem is that proving plagiarism is all but impossible in a math based assignment. In a lab or discussion based assignment, there is enough writing so that plagiarism may be obvious by comparing what is being written and what conclusions are being drawn. However, with a math based assignment, a plagiarized answer may look exactly the same as a non-plagiarized answer. In light of these severe constraints, one way to minimize plagiarism is by asking carefully selected follow-up questions. The teacher would ask questions to a student suspected of plagiarism directly related to the subject they are suspected of plagiarizing. If a student has plagiarized, they will have no idea how to repeat the answer they submitted or follow up on their research. Unfortunately this can merely identify plagiarism; it does not in and of itself prevent it. A technique that can be used if a math based assignment is public is to require the students to upload their answers as PDFs instead of posting them to the classroom or uploading them as MS Word™ Documents. Depending on the LMS and the teachers control over the classroom, this requirement would make the assignment private again, a move that usually doesn't violate university policy. However, even if this upload is public (such as in the case of an attachment to a post), students are less likely to plagiarize if they have to download several documents to get a feel for the right answer. As well, students who "cut and paste" will be easier to find since documents students prepare offline are usually distinct enough that a teacher should be able to tell when someone has copied another's work.

## CONCLUSION

In this chapter, we suggested several optimal practices for designing and teaching pure online physics classes. As a foundation for these practices, four educational spaces in which physics is being taught were presented including the traditional, hybrid, and interactive classrooms. The discussion then primarily focused on the pure online classroom. Section one concluded by presenting the three types of assignments present in the pure online environment: discussion, math, and lab based assignments. In section two we elaborated on the strategies for the design of a pure online classroom including ensuring consistency in the quantity of assignments each week, choosing an appropriate textbook for the class, and choosing between an applet and DIY style lab based assignment. Section three presented recommendations for teaching these classes including the use of ASCII math, the Socratic Method, and emoticons for better communication with students, as well as for requiring students to upload their work whenever possible. As a reference, Table 1 summarizes the differences and similarities between a traditional and a pure online class, while Tables 2 and 3 summarize the design and teaching aspects for physics and mathematics courses.

These practices have been developed by the author as a result of many years of teaching in both the online and offline environment. Considering that the pure online physics class is the newest type of classroom, there is still much room for developing and refinement of relevant teaching and design strategies. As technologies progress, the classroom will be transformed into environment we cannot even imagine today. However for the time being, the LMS is the platform of choice for online teaching and thus adhering to presented practices will benefit both teachers and students.

## ACKNOWLEDGMENT

The author wishes to acknowledge the Massachusetts Institute of Technology for the screenshots of their OpenCourseWare as per their Creative Commons agreement and the University of Colorado at Boulder for the screenshots of their physics applets as per their Creative Commons agreement

## REFERENCES


*2009-2010 Undergraduate Academic Catalog.* (2009). DeVry University, Illinois. Retrieved July 23, 2009 from http://www.devry.edu/uscatalog

*2009/2010 catalog of programs.* (2009). The Art Institute of Pittsburgh. Pennsylvania. Retrieved July 23, 2009 from http://www.artinstitutes.edu/pittsburgh/Admissions/catalog.aspx

*ACICS Accreditation criteria, policies, procedures, and standards.* (2009). Accrediting Council for Independent Colleges and Schools. Washington, DC. Retrieved July 23, 2009 from http://www.acics.org/criteria.aspx

Christian, W., & Belloni, M. (2001). *Physlets.* Englewood Cliffs, NJ: Prentice Hall.

*Clickers.* (2009). Ohio State University. Retrieved from http://telr.osu.edu/clickers

Dennis, J. (2003). *The Complete Idiot's Guide to Physics.* Indianapolis, IN: Alpha Books.

*DETC Accreditation Overview.* (2007). Distance Education and Training Council. Washington, DC. Retrieved July 23, 2009 from http://www.detc.org/freeprublications.html

Gonick, L., & Criddle, C. (2005). *The Cartoon Guide to Chemistry.* New York: HarperResource.

Gonick, L., & Devault, C. (1999). *The Cartoon Guide to Sex.* New York: HarperPerennial.

Gonick, L., Huffman, A., & Huffman, A. (1991). *The Cartoon Guide to Physics.* New York: HarperPerennial.

Gonick, L., & Outwater, A. (1996). *The Cartoon Guide to the Environment.* New York: HarperPerennial.

Gonick, L., & Smith, W. (1993). *The Cartoon Guide to Statistics.* New York: HarperPerennial.

Gonick, L., & Wheelis, M. (1991). *The Cartoon Guide to Genetics.* London: HarperCollins.

Halliday, D., Resnick, R., & Walker, J. (2005). *Fundamentals of Physics.* New York: John Wiley & Sons.

Hewitt, P. (2002). *Conceptual Physics.* Harlow, UK: Addison Wesley.

*Higher Education Reconciliation Act of 2005,* S. 1932, 109th Cong. (2005). Washington, DC: U.S. Government Printing Office

Holzner, S. (2006). *Physics for Dummies.* New York: Wiley.

Knight, R. (2002). *Five Easy Lessons.* Harlow: Addison Wesley.

Mazur, E. (1997). *Peer Instruction.* Englewood Cliffs: Prentice Hall.

McDermott, L. (2001). Oersted Medal Lecture 2001: Physics education research – The Key to Student Learning. *American Journal of Physics, 69*(11). doi:10.1119/1.1389280

Mcdermott, L., Shaffer, P., & Rosenquist, M. (1996). *Physics by Inquiry.* London: Wiley.

*McMicken College of Arts and Sciences course catalog.* (2008). University of Cincinnati, Ohio. Retrieved July 23, 2009 from http://www.uc.edu/courses/a_s.pdf

Moore, J. (2003). *Chemistry for Dummies.* Indianapolis, IN: Wiley Pub.

Novak, G. (1999). *Just-in-Time Teaching.* Englewood Cliffs, NJ: Prentice Hall.

*Registered moodle sites.* (2009). Retrieved June 12, 2009, from http://moodle.org/sites.

Serway, R., Faughn, J., & Moses, C. (2003). *College Physics.* Melbourne, Australia: Thomson-Brooks/Cole.

Siegfried, D. (2001). *Biology for Dummies*. New York: Hungry Minds.

Sterling, M. (2001). *Algebra for Dummies*. New York: Hungry Minds.

Wilson, J. M. (1994). The CUPLE Physics Studio. *The Physics Teacher*, *32*, 518–523. doi:10.1119/1.2344100

# Chapter 17
# Hybrid Dialog:
## Dialogic Learning in Large Lecture Classes

**Tobias Zimmermann**
*University of Zurich, Switzerland*

**Karen-Lynn Bucher**
*University of Zurich, Switzerland*

**Daniel Hurtado**
*University of Zurich, Switzerland*

## ABSTRACT

*Attendance at classical lectures usually leads to rather poor learning success. A wide variety of studies show that while lectures are as effective as any other method for transmitting information, they are inferior in many other dimensions. Lectures are not as effective as discussion methods in promoting thought and they are ineffective at teaching behavioral skills and subject-related values as well as at awakening interest in a subject. Still ex-cathedra teaching is a favored way to cope with a high student-to-teacher ratio. To solve this conflict between organizational and pedagogical requirements, a group of researchers at the Institute of Teacher Education at the University of Zurich has developed a hybrid course setting using an online learning platform. Their setting incorporates a dialog among students within a large lecture class. Furthermore a feedback loop enables the lecturer to continuously adjust the content of the lecture to the learning process of the students. In this article, the authors first present the structure of this setting and then illustrate how to implement it by the web-based open source learning management system OLAT (Online Learning and Training). Based on their research, they focus on key components for the success of their hybrid dialog. They show how individual and group learning can be fostered with corresponding assignments, assessments, and assigned roles such as moderators. Thus, the authors will define their position that the challenge of a large lecture class can be met while successfully implementing social learning and process-oriented assessments of academic achievement.*

DOI: 10.4018/978-1-61520-853-1.ch017

Copyright © 2010, IGI Global. Copying or distributing in print or electronic forms without written permission of IGI Global is prohibited.

## INTRODUCTION

Lecturing still is the most common teaching method in colleges and universities globally (Bligh, 2001, p. 3). The vital role of lectures in academic teaching originates in the ancient Greek academy. Throughout the Middle Ages, the lecture remained the most important academic teaching method until today (McLeish, 1976, pp. 252–254), although it has repeatedly been criticized for being just an oral reproduction of written text (Apel, 1999, pp. 22–30).

Since the 1950s, not only the written word can be reproduced easily, but also the spoken one: In less than hundred years, radio, television, video, as well as computerized multimedia technology (on stationary as well as mobile terminals) became widely available. These developments have lead to traditional lecture being criticized as never before—and with good reason. Today, many faculty and educational researchers are experimenting with streaming lectures, with replacing or supplementing lectures by online tutorials, and so on (e.g. c.f. Brecht & Ogilby, 2008; Glass & Sue, 2008; Guertin, Bodek, Zappe, & Heeyoung, 2007; Spickard, Alrajeh, Cordray, & Gigante, 2002). We are observing these experiments and their outcomes with great interest.

Our research, however, focuses primarily on another aspect. We find that academic teaching often relies too much on the transmittal of information. While transmitting information to students is absolutely necessary for their acquiring knowledge, there is also a range of other learning dimensions which are equally important. For example, we would like students to think independently about subject matter, to acquire values associated with it, and to solve subject related problems. In other words, just knowing facts is not enough to be an expert in a specific subject. And while lectures are good for transmitting information, they are not appropriate for aiming at learning dimensions like independent thinking, value acquisition or problem solving (see section "Background").

Hence, we have developed a hybrid didactic scenario, which aims at learning dimensions beyond acquiring knowledge—without renouncing the benefits of lectures in transmitting information. In the following, we are delivering some empirical background to corroborate our approach (section "Background") and explain the didactic scenario, based on a short introduction to the theory of dialogic learning ("Part I: Hybrid Dialog—a didactic setting to implement a feedback loop in large lecture classes"). Then we explain how to implement our didactic scenario using the LMS OLAT ("Part II: Implementing a dialogical setting using the LMS OLAT"). The last part of our chapter focuses on fostering dialogic online learning ("Part III: "Practical Implications").

## BACKGROUND

During the 20th century, a lot of comparisons between different academic teaching methods have been undertaken. While more than a few of them didn't show any significant results, some trends still can be discerned. They can be summed up to the following three basic propositions (cf. the meta-analyses in Bligh, 2001, pp. 3–20):

1. Lectures appear as effective in transmitting information as other methods.
2. Lectures are less appropriate than discussions when aiming at promoting student thought and the acquisition of procedural knowledge.
3. Lectures are not qualified to change student attitudes and value systems.

Concerning 1): In 298 studies, no significant differences showed up between the declarative knowledge students acquired through the following teaching methods: Lectures, discussions, reading and independent study, inquiry (e.g. projects), and others, mostly audio, TV, computer-assisted learning (Bligh, 2001, pp. 4–8). The meta-analysis of Dubin and Taveggia (1968) came to almost the same conclusion.

That all methods seem equally effective doesn't mean that information only has to be delivered by lectures—on the contrary, from a didactic perspective, applying a variety of methods seems more promising (Meyer & Paradies, 1993). But lectures being the most cost-effective method, cost-benefit considerations speak for their use. Considering the high student to teacher ratios many colleges and universities are faced with today, lectures are the method of choice to cover many students by few teachers.

If studying was only about acquiring declarative knowledge, a dominance of lectures would be a bit repetitive, but not a really noteworthy problem. But as already noted in our introduction, there are other important educational objectives like independent thinking, value acquisition or problem solving.

Concerning 2): Bligh (2001, pp. 8–12) analyzed 73 studies comparing the effect of different methods in promoting (independent) student thinking. His meta-analysis shows two clear tendencies: None of the analyzed methods is more ineffective in promoting student thinking than lectures, and the most effective method are discussions. A study by Bloom (1953, p. 166) documents the different potentials of lectures and discussions to activate student thinking. It shows that lectures prompt significantly more thoughts on the level of simple understanding ($p<.01$), but that discussions stimulate significantly more thoughts concerning problem solving or synthesizing ($p <.01$).

These results can easily be explained when taking into account the results of transfer research: Mental operations can only be learned through active thinking. Hence, subject-specific thinking skills can best be acquired by analyzing and solving concrete subject-specific problems and by associated metacognitive activities (Steiner, 2001, pp. 195–203). And as shown above, these types of mental activities are not promoted by listening to a lecture. In contrast, the frequent changes of speakers and perspectives taking place in discussions are more likely to provoke the analyzing and solving of problems—and first of all, they commit participants to think about the subject at hand and to take up their own position.

These arguments also explain the results of Bligh's (2001, pp. 18–19) meta-analysis concerning procedural knowledge: Compared to lectures, most notably the practice of the focused skills is essentially more effective.

Concerning 3): Compared to discussions and the category "other methods" lectures also cause students less to acquire new or different values or to modify their attitudes (Bligh, 2001, pp. 12–18). This can be justified by the same reasons as given for the acquisition of procedural knowledge: Students develop value systems, interests or social behavior by acting themselves and not by following the mental activities of others.

If the educational objectives of a lecture go beyond the mere acquisition of declarative knowledge and involve students developing subject matter related values, social behavior or procedural knowledge, the method of traditional lecturing does not lead to the desired results. But for economic reasons, lectures can't be simply discarded. Rather, they have to be enhanced by additional methods in order to cover the mentioned objectives. And in this regard, following the research presented above, discussions seem to be most promising.

## HYBRID DIALOG: A DIDACTIC SETTING TO IMPLEMENT A FEEDBACK LOOP IN LARGE LECTURE CLASSES

In lectures with high numbers of participants, hundreds of students meet once a week for one or two hours. In their time together, these students are doing nothing but listening to the lecturer and taking notes. The attitude they take is passive and receptive. That is not to say lectures don't cause students to think, but the provoked thinking mode is mostly reactive, as pointed out in the background section. Furthermore, it seemed to us a waste of potential not to involve the students more actively in the lecture. As a consequence, we searched for a way to enable students to discuss the subject matter presented in our lectures.

Therefore, we have developed a dialogic setting using web technology, which allows students to participate in a dialog about the topics of the lectures (Eberle & Keller, 2003; Ruf & Weber, 2005; Zimmermann, Hurtado, Berther, & Winter, 2008). We successfully implemented this hybrid dialog in an academic lecture for teacher training. This setting aims at multiple goals, the most important of which are:

- Tapping the full potential of social learning by engaging all students in a multidirectional dialog with other students and with the lecturer. Thus the lecture should be more effective in promoting thought and building values concerning the subject matter as well as awakening an interest in it.
- Unburdening the lecturer from the responsibility for the individual learning outcomes and distributing the teaching/learning ratio more equally between lecturer and students.
- Creating a feedback loop between the audience and the lecturer, enabling him to continuously adjust the lecture to the learning development of the students.

Using new media to teach, we have to consider that the benefit of a medium depends on its use by people. No pencil writes on its own, and no book is read by itself. Accordingly, a didactic benefit can only be offered by electronic media if they are in the service of consistent didactic arrangements. Therein, the technologies have to fulfill their clearly defined tasks (Ruf, Frei, & Zimmermann, 2003, p. 192). Or to cut a long story short, we followed the maxim "tools follow concept". (Zimmermann & Haab, 2005, p. 17). Earlier media research has tried for a long time to directly compare the efficacy of different media (e.g. film vs. written text). Meta-analyses with large numbers of such studies produced no winner in this competition (cf. Cohen, Ebeling, & Kulik, 1981). This is because, as Clark (1983, pp. 453–454) already pointed out, we do not really analyze the effects of media but of treatments, i.e. learning arrangements. So the results of so-called media comparisons are about arrangements used, not about media.

In this light, we hold it important to give a short theoretical explanation of the dialogic learning (Ruf & Gallin, 2003), which was the main source for our scenario.

In the dialogic learning, teaching and learning are oriented on the basic pattern of human dialog: During the generation of knowledge, the discussion partners alternately take the role of the speaker and the listener. Simultaneously, the partners guard the smooth processing of the dialog and they continuously take stock of it. Most essential is the permanent change of perspectives that everyone involved is forced to take upon each change of speakers. There are three basic perspectives that constitute a dialogic learning process:

- The perspective of the speaker who is teaching and learning, who lays out the things how he sees them by producing and instructing his view: I see it and I do it this way!
- The perspective of the listener, who follows the speaker by adapting and reconstructing his remarks, queries and tries to incorporate the speaker's view into his own intellectual horizon: How do you see it and how do you do it?
- Likewise, speakers and listeners take up a third perspective: the view from outside upon the course of the dialog from which

emerges a common view of things, a sense of cohesiveness: We all see it this way and that's how we do it. Thus, subject related norms can be derived from the course of the dialog and help to establish the declarative and procedural knowledge that are validated by the academic and/or professional community.

Dialogic Learning transfers essential parts of the knowledge generating dialog in the medium of literality. Thus, it allows for a deeply reflected exchange between a potentially large number of persons. A further promising effect of this approach lies in the fact that newly acquired knowledge is better associated to the previous knowledge, be it declarative or procedural (concerning the essential role of linking new with previous knowledge, cf. Steiner, 2001, pp. 172–173).

In the following, we describe in detail the procedure of our hybrid dialog setting by pointing out the chronological sequence: A) through E) constitute one phase (refer to Figure 1 for an overview).

*Figure 1. Overview of the didactic setting*

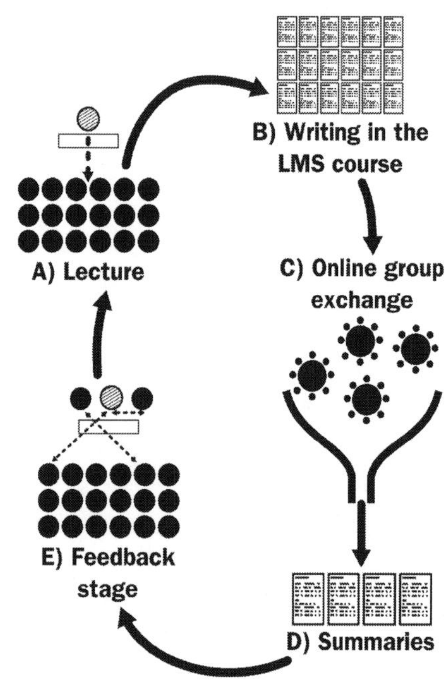


## General Framework

Approximately 200 students attend the type of lecture class in which we successfully implemented our hybrid dialog setting. The students attending the large lecture class need to sign up for a learning group of about 14 members. Each group is facilitated by a moderator. The moderator moderates the online communication but also attends the lecture class as a regular or advanced student. Instructions for the use of the LMS OLAT are given out via email to the students by the assistant of the lecturer. These instructions are necessary in order to make online discussions efficient and to keep the course structure of the LMS as clear as possible. Additionally the students are introduced to three basic rules (netiquette) that guarantee a benevolent online communication: Contributions addressed to another group member ought to be i) task-oriented and specific, ii) constructive and quality-oriented, and iii) personal statements.

A. The workflow starts off with a traditional class held by the lecturer (approx. 90 minutes, the usual length of lectures in the German-speaking area). According to the topic taught an assignment is given at the end of the lecture. It consists of an initial assignment and a feedback assignment: These assignments have to be done by every student in order to meet the requirements for academic achievement (see B and C). The very first assignment is an "introductory assignment", which links the mutual introduction of the students in each learning group with their personal attitudes towards the subject matter.
B. Half of the group members work on the initial assignment. They publish their contributions according to this assignment within 48 hours after the lecture in the file dialog on the LMS. In terms of OLAT the element "file dialog" has already been set for each group by the course manager of the LMS (assistant of the lecturer). Each file dialog

is only accessible to the students assigned to the respective group (see Part II).

C. The other half of the learning group responds to the work done by their fellow students by reflecting and commenting their opinions and arguments according to the feedback assignment given to them. This happens within 72 hours after the lecture and is done in the same file dialog on the LMS as described under B) and creates an exchange in written form between the members of the same learning group. Since the students contribute in writing to the online discussion they are forced to get involved more intensively with the opinions and reasoning of the fellow students. The tasks of writing the contribution and commenting on a contribution alternate within the group from phase to phase with each assignment: Who first published a contribution as an initial assignment will then in the next phase comment on the contributions of the other half of the learning group as a feedback assignment and vice versa. The participation in the online discussion is required in correspondence with the performance record (see also discussion about assessment/ECTS in part III)

D. Within 96 hours following the lecture the moderator summarizes the online discussion by citing the best statements and also considers a new perspective to the discussion. This summary of the contributions is provided to the study group by the moderator as a file in a separate folder in OLAT. In addition the file is passed on to the assistant of the lecturer. The assistant concentrates the summaries he received from every group into a final summary and hands it over to the lecturer. This way the lecturer gets informed about the contents of the discussions which took place online: The lecturer learns which issues arose, what was well understood, and what needs clarification.

E. Last step of D) allows the lecturer to incorporate students' contributions in the following lecture class. At the beginning of each class the lecturer refers to the online discussions on the topics of the past lecture and clarifies issues that have emerged in the discussion. It is also possible to invite students to present their best practice example or controversial contributions in front of class. Due to this feedback stage the attending students are activated in their learning process by associating new contents with foreknowledge, and the lecturer can build upon the actual skill level of the students since he or she gains insight into students' learning processes.

This workflow A) through E) lasts one week and constitutes one phase (it is also possible to customize the duration of the phases; e.g., we have begun to integrate a two week phase twice a semester in order to relieve the time pressure somewhat). With the beginning of a new phase a new initial and feedback assignment is given out to the learning groups. The workflow is repeated continuously and guarantees an effective hybrid dialog.

Our lecture classes are continuously evaluated by two questionnaires, one at the beginning and one at the end of the term. In our case, these are comprehensive surveys which serve the purpose of our research. But also if no educational research is carried out, we recommend conducting a little survey. This enables the lecturers and their assistants to spot potential for improvement of the scenario and to carry out adjustments at relatively short notice.

## IMPLEMENTING A DIALOGICAL SETTING USING THE LMS OLAT

The general requirements of our scenario regarding the technical features of an LMS are quite low. Hybrid dialogs do not depend on specific technical features except for the possibility of asynchronous text-based communication. In smaller classes one can implement a similar scenario to the one described here with a simple discussion forum. However, to implement such a scenario in large lecture classes with several hundreds of students, a powerful, modular LMS allowing customized course environments is needed. Of special importance are the possibility to divide the students into learning groups and a course element allowing the discussion of uploaded files.

The learning environment described in the preceding chapters is realized by using OLAT (Online Learning and Training), an open source, cross-platform LMS solution. Of course, our scenario could also be implemented using other powerful LMSes such as Moodle, Blackboard or Sakai—as long as they provide the necessary features displayed in the following section.

## Features of OLAT

OLAT allows any kind of online learning with very little didactic restrictions. This is achieved by letting the user choose between different elements such as forums, wikis or tests while creating an individual course environment. Thanks to this fully modular approach the course authors can create their tailor-made learning environment with virtually no restrictions, whatever their didactic scenario may be.

The following course elements are currently available in the OLAT course editor (elements marked with * are necessary to implement our course):

- CP learning content* (IMS Content Packages)
- SCORM learning content
- Single page* (with integrated WYSIWYG HTML editor)
- External page
- Wiki
- Forum*
- File dialog* (discussion of uploaded files)
- Folder* (up- and download of files)
- Assessment* (shows results of assessments to learners)
- Task (with drop box, sample solution and scores)
- Test (with scores; IMS QTI standard)
- Self-test (anonymous and no scores; IMS QTI standard)
- Questionnaire* (IMS QTI standard)
- Enrollment* (learners enroll for learning groups)
- Contact form*
- Structure* (grouping of elements)

We will not discuss the capabilities of the elements listed above in depth here. For further information and the features of OLAT in general, see http://www.olat.org. Nevertheless there is a remark to be made regarding some course elements, concerning particularly wikis, questionnaires and tests. They are all typically used to create content by a group of users. Because OLAT is designed in a modular way, course elements one already worked with in an older course can be recycled within a new environment. For example, if one creates a Wiki accompanying a lecture one semester, it can be reused in the following semester in a new course environment without losing the content created by the students a half year ago. Thus user created content can grow over the years. It is even possible to embed one of those elements in multiple course environments at the same time and let users work simultaneously on it.

## Technical Implementation of the Didactic Scenario

The following description of the actual implementation of our didactic scenario is based on OLAT. Nevertheless we try to look at these technical matters from a more generic point of view that provides hints to recreate a course environment similar to our own using any common LMS software. That said, some rather self-explaining details in the configuration of OLAT will intentionally not be discussed here. The following explanations all relate to Figure 2. It shows a screenshot of the course's navigation column as seen by a course manager with all elements folded out (with the exception that it displays only one study group).

The main organizational layer of the course consists of the three areas *public area*, *study group area* and *moderator area*. They determine the visibility of the course elements towards the three different user types: The course managers (lecturer and assistant), the moderators and the students. We created right groups these user types are related to. Right groups allow us to easily adjust the amount of rights (read, write etc.) a group of users has concerning each single element. Thanks to the "enrollment" element **(1)** the users can choose which study group they want to be in—and by doing so they are automatically added to the appropriate right group. As the users just see the enrollment element in the beginning, they are forced to choose a study group before being able to use the course. Thus, the students are divided into learning groups without the need for the administrators to allocate them.

Because the purpose of the public area lies primarily in providing the students with various course-related information, students have read-only access for some parts of this area. For example, the students can download assignments uploaded by the lecturer himself under **(2)** (see "A" in the circle graphic in part I), and documents such as handouts or guides under **(3)**. In these two elements, only the course managers are allowed to upload or change documents. If there is need for technical support, the participants are invited to post their questions publicly in the forum "technical tips" **(6)**, thus fellow students can benefit from

*Figure 2. Navigation column, as seen by a course manager*

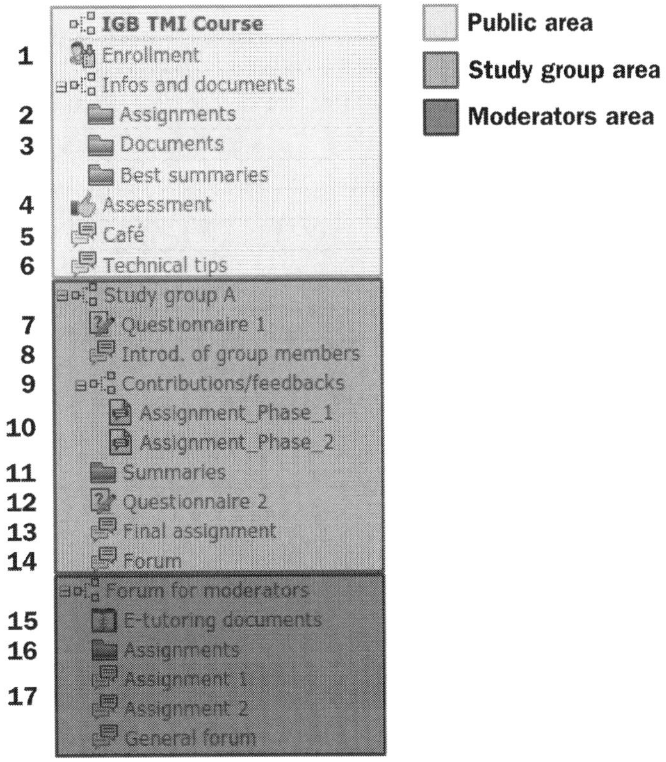

the answers given by the course managers. The forum "café" **(5)** allows a two-way communication as well: On the one hand the course managers can quickly provide news to the students, and on the other hand the students get the opportunity to let their feedback be known by the course managers.

The study group area is the place where most of the learning activities within the course environment happen. It is visible only to the members of the respective study group, again using right groups. All steps between B and D in the circle graphic (Figure 1) in part I happen here. At the start of the semester all students have to write a text following an assignment designed particularly to introduce the participants to each other in the forum "introduction of group members" **(8)**. This allows them to discover the possibilities of OLAT in the beginning within a single forum without being overwhelmed by a host of different elements.

Following this initial contact with the LMS software and their fellow group members, the main part of the learning process for the students happens in the structure "contributions/feedbacks" **(9)**. After receiving an assignment (see letter A in the circle graphic), one half of the learning group proceeds by writing their contributions (letter B). The writing happens offline, the resulting texts are uploaded hereafter in form of rich-text files (RTF) to the corresponding file dialog **(10)** of the current phase. For a screenshot of the actual file dialog element see Figure 3. There, the students are allowed to upload files, but not to delete or change anything. Only one's own feedbacks in the forums may be edited. In general, only moderators and course managers have the right to alter and delete files here. A specialty of the file dialog element is its hybrid nature—one can upload files, just like in the folder element, but each file gets an independent forum attached that has the same possibilities as the regular forum element. This allows the other half of the study group to post their feedbacks during the online group exchange (letter C) directly to the contribution they are referring to. The fact that a fully-fledged forum is attached to each file and not just one discussion thread allows multifaceted discussions with different topics and main feedbacks about one contribution that remain well-arranged nevertheless.

Near the end of the semester the students all have to complete a "final assignment" **(13)**. It is the counterpart of the "introduction of group members"—the students have to analyze their own learning progress fostered by the lecture and the online discussion in a text they post in the forum. Finally the study group area contains all summaries by the group moderators (see letter D), collected in a folder the moderators directly upload their work to **(11)**. Similar to the forum "café" **(5)** where topics of common interest are discussed, each study group area includes a forum **(14)** for the treatment of group-internal matters such as didactic subjects not related with an assignment or organizational questions to the moderator.

The smallest part of the platform is the moderators' area. It is designed to improve the training of the group moderators. It mostly contains elements that we already used in a similar context within the other course areas, such as folders containing assignments to the moderators **(16)** or file dialogs where the moderators post their contributions following our assignments and discuss them by writing feedbacks **(17)**. The "forum for moderators" applies one element only used here. The E-Learning Center of the University of Zurich allowed us to use an extensive, modular self-learning unit dealing with e-moderation they have developed. We integrated the modules relevant to the situation of our e-moderators using the OLAT-element CP learning content (CP=Content Package) **(15)**, since the self-learning unit was provided to us as an IMS content package. The collaboration between different universities or institutes within the same organization is facilitated a lot by the possibility to easily import already existing XML-based CP or SCORM learning content (specifications of the IMS Content Package: http://www.imsglobal.org/content/packaging; Resources concerning

*Figure 3. Feedback to a contribution, as seen by a user*

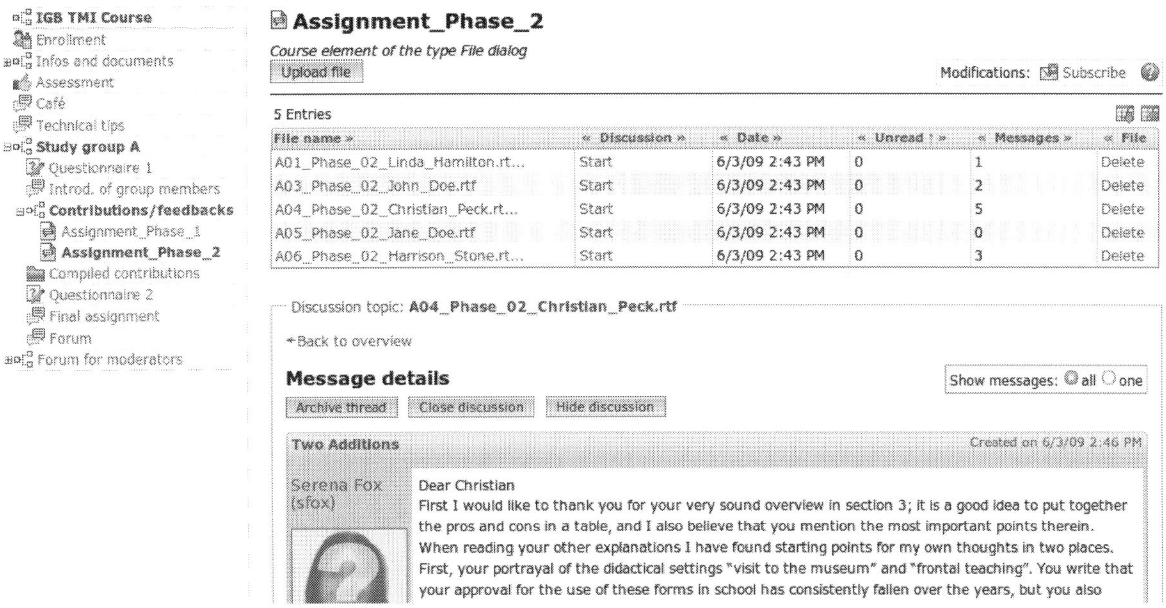


SCORM 2004: http://www.adlnet.gov/Technologies/scorm/SCORMSDocuments/SCORM%20 Resources/Resources.aspx).

## Custom Implementations

There are only a few course elements left we have hardly spoken of yet. One among them is the assessment element placed within the public area **(4)**. It can be used to individually show the students their results in tests or other forms of assessment. The number of points reached and the amount necessary to pass the class are displayed to the user. As the user is identified by his account data, only the information concerning him is shown, this way full privacy can be assured to the students. We use the possibilities of this element to show the participants of our class how many of the 12 necessary textpoints they have already reached (for further discussion see part III). By simply importing a spreadsheet file containing the information about all participants to OLAT we update the database weekly. This file needs to contain at least a column with the usernames of all students whose assessment information should be updated and another column with the current amount of textpoints. It is also possible to add columns just containing pass/fail (y/n) information or comments to the student. This element is a great aid in implementing our textpoint system.

The last element we need to discuss here are the questionnaires. There are two of them: One is shown to the students at the beginning **(7)** and the other at the end of the semester **(12)**. The questionnaire consists of various types of questions, such as single choice, multiple choice or free text. The results are provided to the researcher in a single Excel-file, ready to be processed further. This is more or less the behavior we can expect from tools like this, but we implemented it in a singular way indeed. OLAT allows the use of expert rules, a flexible syntax used to specify the behavior of course elements. We use them to link the first questionnaire **(7)** to the "introduction of group members" **(8)** and the second questionnaire **(12)** to the "final assignment" **(13)**. This means that the student is required to fill out the questionnaire first before gaining access to the related forum ("introduction of group members" or "final assignment"). Thus we virtually get a return rate of 100% (short of technical failures).

A specialty of OLAT is that what the course author sees while creating his course is not what the end-user sees. This is because OLAT provides a course editor view that gives the author full control over virtually every aspect of the course elements. He can freely add, remove and replace elements, but he can also determine under which circumstances a course element is shown or hidden to the end-user. Additionally a visible element can be made inaccessible to users. There are different possibilities to trigger visibility and accessibility of course elements:

- By access right group
- By date
- By expert rule
- By assessment (depending on the result of an assessment)
- By attribute (depending on personal data of the user such as name, field of study etc. This data is provided by AAI (Authentication and Authorization Infrastructure), a system providing one single login to students for different educational online services such as webmail or e-learning. AAI is used in many swiss universities (http://www.switch.ch/aai/about/). The installation of OLAT at the University of Zurich requires users to log in using their personal AAI-account data, hence identification is possible.)

We used three of these possibilities. For example the access to the study group area is controlled by access right groups (called "right groups" in the OLAT nomenclature). Study groups are only shown to members of the respective right group, e.g. study group A is only shown to students that chose group A during enrollment **(1)**. Other elements are activated by date, such as the file dialogs **(10)**. They show up as soon as the respective assignment is given by the lecturer, e.g. the file dialog for discussion phase 2 is visible from the day when the lecture about the topic of assignment no. 2 takes place. And as explained above, by an expert rule the accessibility of the "introduction of group members" **(8)** and the "final assignment" **(13)** depends on a questionnaire which has to be taken first.

We use the features described above to reduce the complexity of the course environment for the students: New course elements only show up when they are used, thus we can start with a quite simple and thus user-friendly environment. This makes up one of the greatest advantages of OLAT: While the course editor is very detailed, only this huge amount of possibilities to influence the behavior of the course and its single elements allows the design of very user-friendly course environments. Through the complexity of the authors' view a maximum of simplicity on the users' part can be achieved.

## PRACTICAL IMPLICATIONS

As it goes for part II we presented usability as a part of the technologies provided for our hybrid dialog. In this section we show how we link these online tools with cooperative work in the large lecture class. We focus on key components which make up our hybrid dialog. On the one hand we are discussing the importance of assignments and their assessment in our hybrid scenario. On the other hand we consider the individuals who fill in different roles to make the scenario work.

### Assignments

Of course, technology is of crucial importance for the realization of our scenario—after all, it couldn't be implemented without using an LMS.

But, as mentioned in Part I, a technical tool needs to be in the service of a pedagogical or didactic concept. Because assignments are a key factor to cooperative learning (Sluijsmans, Prins, & Martens, 2006, pp. 48–50; cf. also Salmon, 2004), they play a major role when using an LMS for cooperative learning.

Without meaningful and authentic assignments, there is little prospect of a fruitful learning dialog as intended by our dialogic scenario. Meaningful in this context means that the assignment has to aim at core problems of the subject matter at hand and that in order to fulfill the assignment, the students are supposed to acquire corresponding knowledge and to discuss their learning progress as well as difficulties encountered with each other. For the same reasons, the assignment should be authentic, that is, as similar as possible to the real tasks a professional in the same field would have to deal with. The authenticity of assignments also has a motivating affect on students and facilitates a transfer of the knowledge acquired into the later professional life (Gibbs & Simpson, 2004, pp. 14–16).

To illustrate the type of assignments we present an example from our scenario: One assignment follows the lecture on the topic of educational theory according to W. Klafki, one of the most influential thinkers in the field of didactic theory in the German-speaking area. The initial assignment instructs the students to take on their future role as a high school teacher and to write a request to their school administration. They are supposed to request an increase of weekly lessons of their subject (e.g. Mathematics or English) and to corroborate their requests with arguments derived from educational theory according to Klafki. As described in Part II, the students upload their contributions as rich-text files to the file dialog element of the current phase. This element automatically attaches a discussion forum to each uploaded file. The feedback assignment instructs the other half of the learning group to respond to these requests by posting a message in this forum. In these replies they take on the role of teachers of the same school who are teaching a subject which is threatened by the requested increase of the other subject. They are to explain to the school administration—also based on the educational theory according to Klafki—why the number of weekly lessons of the other subject should not be increased and why the number of their own subject should be kept up (the detailed technical handling of this interplay of initial contributions and feedbacks is described in Part II, section "Technical Implementation of the Didactic Scenario", (10)).

## Assessment

There are two important issues we want to address in this section. The first is the importance of new instruments to assess cooperative learning as it happens in our dialogic scenario. The second issue discussed here is the difference between grading systems based on punishment and such that are based on positive reinforcement.

In assessing the "whole-task" assignments (Sluijsmans et al., 2006, p. 57) described above, the usual e-assessment tools that mainly focus on item-based testing of declarative knowledge are of no use (Sluijsmans et al., p. 47). Rather, we need a paradigm shift from a test culture to an assessment culture (Birenbaum, 1996; Sluijsmans et al., p. 46; and Zimmermann, in press): "Contrary to more traditional forms of testing, performance assessments in which the students often are confronted with ill-defined problems do not provide clear-cut right or wrong answers. The performance is evaluated in a way that allows informative scoring on multiple criteria" (Sluijsmans et al., p. 50). The instrument widely used for this purpose is the so-called scoring rubric (Sluijsmans et al., p. 50; Allen, 2003, containing a large list with examples).

To assess the quality of the texts written in our scenario, we have also developed a scoring rubric. It aims at the three performance dimensions:

a. quality of the presented subject content,
b. critical thinking, and
c. exchange (concerning only the feedback assignments, aiming at how deeply the thinking of the other student is reflected and discussed).

A further important change needed when shifting towards an assessment culture is an orientation towards potentials and qualities shown in student achievements (each quality displayed in a student work leads to a better grade) instead of the widely established deficit orientation (i.e. everything in a student work that conflicts with established knowledge is marked as false and leads to a lower grade).

In the same direction goes our not punishing students for failing assignments but rewarding them for assignments fulfilled (following the principle of positive reinforcement, cf. Woolfolk, 1993, pp. 221–222). Before this change, students who didn't fulfill each and every assignment were denied the passing of the assessment and had to repeat the lecture in the following year. This caused some dissatisfaction on the side of the students so that we have established a "textpoint system": For each text that has been submitted timely and that is of at least sufficient quality (measured by our scoring rubric), students receive 2 textpoints. If texts are not submitted timely, but within a certain time frame (before the next lecture), and they are of sufficient quality, students receive 1 textpoint. The same is the case when the text is submitted timely but is not qualitatively sufficient—if the student submits a revised version before the next lecture. In the course of the semester, there are 7 to 8 phases (depending on the number of lecture sessions, which can vary). Therefore, students can gather a maximum of 14 to 16 points, if they fulfill every assignment. To release the pressure somewhat and to make the rewarding system work at all (if every point possible would have to be gathered, it would still be a deficit oriented system), we set a threshold of usually 10 to 12 points that have to be achieved in order to pass the assessment. Students who gather less than the required amount of points fail and have to repeat the entire course.

The change from a punishing to the rewarding textpoint system has led to a significantly better judgment of our online scenario: In the last semester under the punishing system, about 10% of the 169 students questioned used the free comment part of our final questionnaire to complain about the "hard regime" and about an exaggerated "school regimentation". In the following semester, under the new textpoint system, only 1 out of 57 students questioned made a remark in that direction. Also in face-to-face and e-mail conversations, we have hardly received any reproaches concerning the textpoint system, in contrast to the often harsh comments concerning the previous punishing system.

The increased acceptance of our new assessment system can also be demonstrated quantitatively. Our final questionnaire includes an item in which the students give a rating about how appropriate they find the online discussions as a course assessment. The scale ranges from 1 (very inappropriate) to 7 (very appropriate). A statistical comparison of these ratings for the fall semester 2007 (n=169) and the fall semester 2008 (n=184) shows that the number of students who rated the online discussions as a very or clearly inappropriate course assessment (values 1 and 2) was reduced to less than half after the change to the textpoint system (37% in 2007, 16% in 2008). In turn, the number of students who rated the online discussions as a clearly or very appropriate course assessment (values 6 and 7) rose from 21% in 2007 to 30% in 2008. Overall, the mean estimated appropriateness rose from 3.63 in 2007 to 4.37 in 2008 (p.000). This data means that the implementation of the textpoint system especially reduced the number of students clearly unhappy with the online discussions as a course assessment from more than a third to less than a sixth of the students. At the same time the number of students clearly happy with the assessment could be increased from about a fifth to almost a third of the students.

## The Role of the Lecturer and Students

The number of large lecture classes at universities strongly indicates that, despite the criticism, the large lecture class still has a role to play in university teaching. The question is how to ensure or improve the quality of the actual standard of large lecture class teaching. The integration of new learning technologies changes the culture of learning putting more weight on learner oriented methods. This includes more communication between learners and teachers as well as among learners and a stronger focus on the development of skills and competences. These claims lead to a new role of the teacher in terms of new forms of learning arrangements, support and assessment (Reinmann, 2005, p. 260).

Within hybrid learning environments, the main focus of effective lecturing shifts to the mediating role the lecturer holds in the learning experience of the students. In particular ensuring cohesiveness in the class requires intervention at critical points to pull together disparate strands of discussion (Bender, 2003, p. 33; Feenberg, 1989, pp. 34–35; Field, 2005, p. 210). Thus the role of the lecturer has to be defined in terms of the students' expectations such as structured lecturing which includes revising the previous lecture. These expectations are met by our hybrid dialog setting.

## The Role of the Assistant of the Lecturer

The tasks of the assistant of the lecturer are very diverse. In our setting the assistant combines different roles in one person. As a course manager the assistant sets all elements in the OLAT course: The numbers of learning groups within a large lecture class as well as the complex course structure of the LMS OLAT with the various numbers of elements demand a course manager. It is possible to appoint a second assistant who is only responsible for the LMS course management and additional technical matters. The assistant in our setting is also the person in charge of supporting the moderators. He or she organizes their instruction at the beginning of the semester and coordinates the assignments for them. He or she also assists the moderators in rating the online contributions of the group members. At last the assistant administrates and makes the final decision over textpoints.

Finally the most important task of the assistant is to bridge between students and the lecturer. By passing the aggregated summaries of contributions to the lecturer he fills the gap of information between them and enables the lecturer to cover the vast content of the online dialog in his lectures.

## The Role of Moderators

The success of cooperative online learning not only depends on well-structured assignments but also on the assistance by moderators allocated to a learning group. Thus our hybrid dialog scenario incorporates moderators moderating the online discussion. Moderators in our setting are students who attend the lecture class on a regular basis but additionally fill the role as a moderator. (It is also possible if not preferable that advanced students fill this role, which is hardly possible in our case, only due to the fact that our course is a post-graduate course with short average duration.) Moderators in our case therefore are not providers of knowledge since they do not possess more knowledge than their fellow students. Instead general tasks of moderators refer to the support of the online activity: fostering the feedback culture among the study group members, initiating questions for controversial discussions, leading attention to certain topics. Moderators are also addressed with organizational questions (e.g. concerning assignment requirements) and smaller technical problems. At the beginning of the semester the moderators are instructed by the assistant of the lecturer and receive a manual for moderators (Zimmermann, Haab, & Schneider-Lastin, 2008). They learn how to administrate the element "file dialog" and other communication tools and how to summarize the contributions at the end of each phase. The latter is the most important aspect of the tasks done by the moderators. Moderators regularly pass on summaries of the contributions published on the LMS OLAT by the study group members to the assistant of the lecturer (see also procedure described in part I). Additionally moderators rate in each phase the two poorest and two best contributions. This rating serves the purpose of a quality control: Poor texts can be refused and very good ones can be passed on to the lecturer by the assistant. In order to do the rating properly moderators exchange their experiences and approaches with their fellow students who also fill the roll as moderators. For these activities they use the moderator's area on the LMS OLAT (see also Part II). The communication among moderators plays a very important role in the development of a common quality agreement and prevents different levels of standards in each learning group.

The time exposure of the tasks done by moderators is compensated with 2 Credit Points (equals 60 hours of study work) according to European Credit Transfer System.

## CONCLUSION

There is a widely spread prejudice that large lecture classes make it impossible to address self-directed learning. This may be true for traditionally held lectures, but there are ways to enhance the unidirectional communication of lectures. As we have shown above, the use of electronic communication technology is a very promising means to establish a multi-directional dialog. However, a promising dialog doesn't emerge just out of applying technology, but has to serve a certain didactic purpose in a pedagogically structured teaching and learning environment. The dialogic learning (Ruf & Gallin, 2003) provides a theoretical and practical basis to design such environments.

Learning Management Systems have to support the implementation of teaching and learning environments. They best can accomplish this by being as shapeable as possible, because sensible and functioning environments are primarily created by a didactical and pedagogical approach. Thus, LMSes have to ensure they do not overrule didactic demands by technical restraints: Tools should be able to follow didactic concepts (Zimmermann & Haab, 2005, p. 17). With OLAT, we have presented a flexible LMS that enables creating customized teaching and learning environments. Above all, this is due to its modular structure and the possibility to customize courses by right groups and expert rules.

Of course, our favoring OLAT for the scenario described in this chapter does not mean that it could not be implemented by other LMSes, because hybrid dialogs do not depend on specific technical features except for the possibility of asynchronous text-based communication. And in smaller classes, one can implement a hybrid dialog with less complex tools such as a simple discussion forum.

Hybrid dialogs offer advantages on different levels. Firstly, the continuous running through the dialogic workflow enables a constant adjustment of the teaching supply by the lecturer and its use by the students. Secondly, such a "just-in-time teaching" (Novak, Gavrin, & Christian, 1999) has proven to be highly motivating, because everybody involved takes responsibility for their own learning as well as for the learning of others. Therefore, it is important to be aware of the specific roles the different people involved (students, moderators, assistants, lecturers) take in the different stages of such a scenario and which demands these roles make on them. Last but not at all least, a hybrid dialog can support a paradigm shift from mere testing to genuine assessing and thus increase intrinsic student motivation by establishing formative feedback processes and positive reinforcement (Zimmermann, in press).

There are some challenges regarding hybrid dialogs that have yet to be addressed. For example, there is a group of about 10-15% of the participating students that, according to our questionnaire, displays very little appreciation of our scenario and an accordingly low motivation. We hypothesize that these are people with a (generally or specifically with regard to educational science) low intrinsic learning motivation. There are also other possible factors that may influence the estimation of hybrid dialogs, such as writing abilities/strategies (cf. Torrance, Thomas & Robinson, 1994) or personality traits (cf. McCrae & Costa, 2003, pp. 37–57). The influences of all these individual factors on the estimation of hybrid dialogs and on the resulting amount of dedication displayed by students have yet to be analyzed (corresponding research is currently conducted by T. Zimmermann in the context of his doctoral thesis).

Another issue is the balance of assessing learning processes and learning products. The assessment applied in our scenario strongly emphasizes learning processes: Initial contributions as well as feedbacks are documentations of students' learning processes and not final products. In contrast, no final learning products such as term papers have to be produced. One may argue that this lopsidedness counterbalances the excessive product orientation in the current academic teaching culture, considering that by far most academic assessments address learning products and hardly account for learning processes. But we hold that this would be a polemic argumentation. Instead, educational science has to search for ways to reconcile process and product orientation, since they address both sides of the same coin. Consequently, possibilities to include learning products and their assessment in our hybrid dialog scenario have to be explored. Therefore, our next step will be to let our students write a brief term paper at the end of our course. In these papers, students will integrate their contributions and feedbacks as well as the feedbacks received into the overall context of the lecture and reflect their learning progress. This adds extra value to the online discussions, as they lead towards the goal of writing a term paper and have therefore to be reconsidered. The term papers could also be reviewed by fellow students to reduce the amount of work on the side of the lecturer and his assistants, who would "merely" have to control the peer reviews. The effects of this measure will have to be analyzed to draw further conclusions about the integration of process and product regarding hybrid dialogic learning.

The described research into the success factors of online learning dialogs on the pedagogical side as well as the development of maximally customizable LMSes on the technological side seem to be the most promising ways to further enhance the potential of hybrid dialogic learning. Overall, we hold that hybrid dialogs have already proven to be very effective means to achieve deeper learning.

## REFERENCES

Allen, M. (2003). *Using Scoring Rubrics.* Long Beach, CA: California State University. Retrieved June 20, 2009, from http://www.calstate.edu/itl/sloa/links/using_rubrics.shtml

Apel, H. J. (1999). *Die Vorlesung: Einführung in eine akademische Lehrform.* Köln: Böhlau.

Bender, T. (2003). *Discussion-based online teaching to enhance student learning: Theory, practice and assessment.* Sterling, VA: Stylus Publishing.

Birenbaum, M. (1996). Assessment 2000: towards A Pluralistic Approach to Assessment. In Birenbaum, M., & Dochy, F. (Eds.), *Alternatives in Assessment of Achievement, Learning Processes and Prior Knowledge* (pp. 3–31). Boston: Kluwer Academic.

Bligh, D. A. (2001). *What's The Use of Lectures?*San Francisco, CA: Jossey-Bass Publishers.

Bloom, B. S. (1953). Thought-Processes in Lectures and Discussions. *The Journal of General Education*, *7*(3), 160–169.

Brecht, H. D., & Ogilby, S. M. (2008). Enabling a Comprehensive Teaching Strategy: Video Lectures. *Journal of Information Technology Education*, *7*, 71–86.

Clark, R. E. (1983). Reconsidering research on learning from media. *Review of Educational Research*, *53*(4), 445–459.

Cohen, P. A., Ebeling, B. J., & Kulik, J. A. (1981). A meta-analysis of outcome studies of visual-based instruction. *Educational Communication and Instructional Journal*, *29*(1), 26–36.

Dubin, R., & Taveggia, T. C. (1968). *The Teaching-Learning Paradox: A comparative analysis of college teaching methods*. Eugene, OR: Center for the Advanced Study of Educational Administration, University of Oregon.

Eberle, F., & Keller, S. (2003). Unterstützung des kooperativen Lernens im Rahmen von Vorlesungen durch ICT. In Schweizerische Zeitschrift für kaufmännisches Bildungswesen, 97(3), 97–117.

Feenberg, A. (1989). The written world: On the theory and practice of computer conferencing. In Mason, R., & Kaye, A. (Eds.), *Mindweave: communication, computers and distance education* (pp. 22–39). Oxford, UK: Pergamon Press.

Field, R. M. (2005). Favourable conditions for effective and efficient learning in a blended face-to-face/online method. In *Ascilite 2005: Balance, Fidelity, Mobility: Maintaining the Momentum,* 4-7 December 2005, Queensland University of Technology. Retrieved June 20, 2009, from http://eprints.qut.edu.au/18100/1/c18100.pdf

Gibbs, G., & Simpson, C. (2004). Conditions under which assessment supports students' learning. *Learning and Teaching in Higher Education*, *1*, 3–31.

Glass, J., & Sue, V. (2008). Student Preferences, Satisfaction, and Perceived Learning in an Online Mathematics Class. *MERLOT Journal of Online Learning and Teaching, 4*(3).

Guertin, L. A., Bodek, M. J., Zappe, S. E., & Heeyoung K. (2007). Questioning the Student Use of and Desire for Lecture Podcasts. *MERLOT Journal of Online Learning and Teaching, 3*(2).

McCrae, R. R., & Costa, P. T. (2003). *Personality in Adulthood: A Five-Factor Theory Perspective* (2nd ed.). New York: Guilford Press.

McLeish, J. (1976). The Lecture Method. In Gage, N. L. (Ed.), *The Psychology of Teaching Methods: The Seventy-fifth Yearbook of The National Society for The Study of Education, Part I* (pp. 252–301). Chicago: The University of Chicago Press.

Meyer, H., & Paradies, L. (1993). *Plädoyer für Methodenvielfalt im Unterricht*. Oldenburg: Oldenburger Vordrucke.

Novak, G., Gavrin, A., & Christian, W. (1999). *Just-in-time teaching: blending active learning with web technology*. Upper Saddle River, NJ: Prentice Hall.

Reinmann, G. (2005). *Blended Learning in der Lehrerbildung. Grundlagen für die Konzeption innovativer Lernumgebungen. Lengerich: Pabst Science Publishers. Richardson, W. (2006). Blogs, Wikis, Podcasts, and Other Powerful Web Tools for Classrooms*. Thousand Oaks, CA: Corwin Press.

Ruf, U., & Gallin, P. (2003). *Dialogisches Lernen in Sprache und Mathematik, Volume 1: Austausch unter Ungleichen. Grundzüge einer interaktiven und fächerübergreifenden Didaktik.* Seelze-Velber: Kallmeyer.

Ruf, U., & Weber, C. (2005). Dialogisches Lernen auf Lernplattformen: Den Dialog in Lehrveranstaltungen an der Hochschule fördern. *Das Hochschulwesen*, *53*(6), 243–247.

Salmon, G. (2000). *E-Moderating: the key to teaching and learning online*. London: Kogan Page.

Salmon, G. (2002). *E-tivities. The Key to Active Online Learning*. London: Kogan Page.

Sluijsmans, D., Prins, F. J., & Martens, R. L. (2006). The Design of Competency-Based Performance Assessment in E-Learning. *Learning Environments Research*, *9*, 45–66.

Spickard, I. A., Alrajeh, N., Cordray, D., & Gigante, J. (2002). Learning about screening using an online or live lecture: Does it matter? *Journal of General Internal Medicine*, *17*(7), 540–545.

Steiner, G. (2001). Lernen und Wissenserwerb. In Krapp, A., & Weidenmann, B. (Eds.), *Pädagogische Psychologie* (4th ed., pp. 137–205). Weinheim: Beltz.

Torrance, M., Thomas, G. V., & Robinson, E. J. (1994). The writing strategies of graduate research students in the social sciences. *Higher Education*, *27*, 379–392.

Woolfolk, A. E. (1993). *Educational Psychology* (5th ed.). Boston: Allyn & Bacon.

Zimmermann, T. (in press). Online-Diskussionen als elektronische Leistungsnachweise Mit Blended Learning vom Test- zum Assessment-Paradigma. In Mandel, S., & Ruedel, C. (Eds.), *E-Assessment*. Münster: Waxmann.

Zimmermann, T., Haab, S., & Schneider-Lastin, W. (2008). *Leitfaden zur E-Moderation.* Zürich: E-Learning-Koordination der Philosophischen Fakultät der Universität Zürich. Retrieved June 20, 2009, from http://www.phil.uzh.ch/institute/elearning/angebot/leitfaeden/ELK-Leitfaden_E-Moderation.pdf

Zimmermann, T., Hurtado, D., Berther, M., & Winter, F. (2008). Dialog mit 200 Studierenden – geht das? Blended Learning in einer Vorlesung mit hoher Teilnehmerzahl. *Das Hochschulwesen* *56*(6), 179–185. Retrieved June 20, 2009, from https://www.zora.uzh.ch/11297/2/Zimmermannua_DialogV.pdf

# Chapter 18
# Software Tools and Virtual Labs in Online Computer-Science Classes

**Vladimir V. Riabov**
*Rivier College, USA*

**Bryan J. Higgs**
*Milford, NH, USA*

## ABSTRACT

*The authors share their experiences in teaching various online computer-science courses (via the Blackboard™ and synchronous web conferencing tools) and using state-of-the-art free-license software tools for conducting online virtual labs and numerous students' projects. The labs were designed to help students explore modern sophisticated techniques of computer-system analysis and design, programming in C/C++ and Java, data communication in networking systems, system simulation and modeling, image processing, multimedia applications, Web development, and database design and management. All the online courses include lab-based projects that provide students with knowledge, instructions, and hands-on experience, and motivate them in selecting topics for technology overviews and research studies.*

## INTRODUCTION

Hands-on technology-exploration experience for students is an integral part of the traditional Computer Science Curriculum. It becomes a challenge for an instructor to transfer such experience to students via online instructions (Higgs & Sabin, 2005; Riabov, 2006a; Sabin & Higgs, 2007). First, a student should have access to state-of-the-art free-license software tools that are used both in academia and the industry. Second, a student should have an opportunity to emulate a computer system (in his/her home or office) by following online tutorials or instructor's lab manuals. Both tasks require the search for adequate software tools and the development of proper tutorials and lab manuals for every computer-science course offered online or as a hybrid one (Higgs & Sabin, 2005; Riabov, 2006a).

This paper contains an overview of the state-of-the-art free-license software tools, tutorials, lab manuals, some lab reports, and related research papers published by Rivier College students who took computer science courses online or in a hybrid format. Examples of lab manuals, students' project

DOI: 10.4018/978-1-61520-853-1.ch018

Copyright © 2010, IGI Global. Copying or distributing in print or electronic forms without written permission of IGI Global is prohibited.

reports and publications are also available on the instructor's website (Riabov, 2009). Advantages of using synchronous web conferencing tools in online course teaching and learning are also discussed.

## TOOLS AND LABS FOR ONLINE COMPUTER-SCIENCE COURSES

### Computer System Analysis and Design

#### Tools for Implementation of the Unified Modeling Language Approach

Visual Paradigm™ for UML (Community Edition) tool (Visual Paradigm, 2009) was selected for exploring and conducting virtual labs in online courses on software engineering and computer/information system analysis and design. This tool allows students to build various diagrams by using the Unified Modeling Language (UML) approach in object-oriented system design (Dennis, Wixom, & Tegarden, 2009). Online tutorials and lab manuals provide students with the instructions and examples of developing Use-Case, Activity, Class, Object, Collaboration, Sequence, and Package diagrams and State-machine charts that have become a part of students' projects. The limited number of diagrams could also be built by using ArgoUML™ (2009) and VIOLET™ (2009) Open Source software tools.

#### Examples of Student's Projects for Object-Oriented System Design

All the software engineering steps in the development and implementation of the individual capstone project by using the object-oriented design approach are described in detail by David Snogles, a Rivier College graduate student, in his article, "Personal Encrypted Talk - Securing Instant Messaging with a Java Application" published recently in the Rivier College Online Academic Journal (Snogles, 2005). The primary goal of the project was to secure Instant Messaging Communication between two parties on the Internet. The Personal Encrypted Talk (PET) system was designed to operate on Microsoft XP and Microsoft 2000 machines with an active connection to the Internet. Microsoft Visio™ with UML Template and Jude™ UML Community Edition (Version 1.4.3) have been used as UML tools for object-oriented system analysis and design. The State Chart has provided details into the interaction of multiple users with the PET application. It illustrates the exchange of secret keys according to the Diffie-Hellman asymmetric key agreement protocol. The sequence diagram was used to solidify the interaction of the main components of the PET system and its end users. UML diagrams were developed for all eleven classes and integrated into a system-level UML class diagram.

The Unified Modeling Language (UML) approach has been used by John Dion, a Rivier College graduate student, in his project, "Musician Web-service Using Ruby-on-Rails, SOAP, FLEX, and AJAX" recently published in the Rivier College Online Academic Journal (John Dion, 2006b). His system provides musicians with a web service to communicate with other musicians and their fans. The large package exchange is implemented. The Use-Case diagrams and User-Interface State charts have been used in the system analysis and design for this project.

### Programming Fundamentals

In programming courses, one issue that commonly arises is whether or not to use an Interactive Development Environment (IDE). It is certainly possible to use only an intelligent editor, combined (usually) with the command line interface for that operating system, programming language compiler and run-time. Suitable intelligent editors typically can be configured (or come preconfigured) for the programming language, providing suitable syntax coloring, syntax completion, and even intelligent reformatting/beautification of code. Examples of such freely available intelligent editors include JEdit (2009), Crimson Editor (2009), TextPad (2009), NoteTab Lite (2009), SourceEdit (2009), and Notepad++ (2009). GNU emacs (2009) is the ultimate configurable editor, but it does have a fairly steep learning curve, and so may not be the best choice for beginners.

The advantages of using an IDE include: 1) An environment that supports more aspects of the language being used for development; 2) Compilation, execution, debugging, and often testing can be performed without leaving the environment; 3) Usually, the easy availability of language-specific help and reference materials. That an IDE typically provides a visual debugging environment, and that it protects the student from exposure to an often user-hostile command line interface, cannot be too strongly emphasized, regardless of the student level.

The disadvantages of using an IDE include: 1) An IDE is typically a more complex environment, which may force students (particularly beginners) to have to learn too many things at once in order to become productive (it is important to note that the BlueJ IDE (2009) is an exception to this, as it is designed for beginners, with a proven pedagogy in mind); 2) The memory and disk footprint of an IDE can become quite significant. However, with today's computers, this is not usually an issue.

## C/C++ IDE

Microsoft Visual Studio™ (2008) is traditionally used for running the face-to-face courses on C/C++ Programming. Visual C++ 2008 Express Edition is available for free download. For teaching these courses online, Visual C++ can be used, or alternatively, some Open Source software tools could be effectively used, e.g., Eclipse (2009) with the CDT (C/C++ Development Tools) Plug-in (2009), & GNU C++ (2009), NetBeans C/C++ support (2009) & GNU C++, the V Portable C++ GUI Framework (2009), and Qt (2009). Higgs and Sabin (2005) have recommended Eclipse with the CDT (C/C++ Development Tools) Plug-in, & GNU C++ (2009) as the leading contender in this area, followed by the Open Source version of Qt (2009). However, since then, NetBeans C/C++ support (2009) (with GNU C++) has improved, and it is now a very viable option.

Perhaps the best way to provide GNU C++ on Windows is to install the Minimalist GNU for Windows (MinGW, 2009). There also exists a MinGW Studio C++ IDE for the MinGW environment (MinGW Developer Studio, 2009). Other C/C++ IDEs include Dev C++ (2009) and Code::Blocks (2009).

Given 1) the rather loose standardization of C++, 2) the tendency of some vendors not to conform to what standards do exist, 3) the very different (and often hard to understand) error messages that C++ compilers produce, 4) the complexities and subtleties of the C++ language, and 5) the common need to develop code to run across different platforms, we have found that it is often beneficial to use more than one IDE and C++ compiler (even different compilers on the same operating system).

The above-mentioned tools have been used for conducting virtual labs in online courses on C++ programming. The lab manuals and numerous exercises that we have used for online instructions are available from the supplements that accompany the textbooks of Brookshear (2009), Dale (2003), and Lippman et al. (2005), or other available relevant textbooks. Students develop the implementations of object-performing operations in basic data structures (e.g., stacks, queues, lists, etc.), and, following lab manuals, explore data storage and data manipulation in computer systems.

## Java IDE

There are many options for Java Interactive Development Environment (IDE) settings that could be used for online instructions. Among others, the good candidates are: Eclipse™ (2009), NetBeans™ (2009), Borland JBuilder™ (2009), and (for beginning programming courses) BlueJ™ (2009).

Higgs and Sabin (2005) have recommended Eclipse (2009), an Open Source project, as the leading contender that is acquiring a great deal of vendor support. NetBeans (2009) is available free for non-commercial use, and is also a very strong contender. Borland™ developed JBuilder, which became one of the most popular Java IDEs. However, in 2006, Borland focused its business strategy towards the Eclipse IDE, and moved JBuilder development to a new subsidiary, CodeGear, which has recently been purchased by Embarcadero Technologies, Inc. In the process, JBuilder's market presence has suffered. CodeGear™ has a freely downloadable version of JBuilder that runs on a number of platforms, including Windows and Linux. Neither NetBeans nor JBuilder are strictly Open Source products.

All of the major Java IDEs (like all of the major C/C++ IDEs) are rather complex, being primarily focused on production programming and productivity for experienced programmers, usually in a team environment. They are less than ideal for the beginning programming students, who can be overwhelmed, unless the instructor augments their documentation with simple instructions on what to do to create a project, populate it, and develop it. However, we have used these IDEs successfully in many programming courses at Rivier College. When helped to understand a very restricted subset of the IDE, students have typically been able to focus on learning the language, and done well.

An exception to this is in the very beginner level programming courses, where we have used the BlueJ (2009) IDE in combination with its associated textbook (Barnes & Kölling, 2004). BlueJ is an academic-inspired product that is focused on the effective teaching of beginning Object-Oriented programming in Java. While the BlueJ IDE could be used without using its associated textbook, the two do really go together very well, and their joint use is recommended for beginning Java programming courses. Our results using the combination have been very encouraging.

Numerous lab manuals and exercises are already available for online instructions. Among others, we recommend lab manuals that accompany the textbooks of Horstmann and Cornell (2007), Flanagan (2005), Barnes and Kölling (2004), Lewis and Loftus (2003), Wu (2006), and Weiss (1999). Examples of implementing recursive algorithms in Java could be found in the book of Pevac (2005).

Many students develop Java sophisticated codes for their individual capstone projects. Implementing the Personal Encrypted Talk, a system for securing instant messaging with a Java application (Snogles, 2005), David Snogles, a Rivier College graduate student, has successfully used Java 1.5 as the development language in the Borland JBuilder X Personal Edition interactive development environment, Java servlets and applets, and Java Frame built using the Swing Library for development of user interfaces.

## Other Programming Languages and Systems

Besides C++ and Java, other programming languages have become very popular of late, both in the real world and for the early teaching of programming techniques. The Scheme (2009) programming language is a popular one for teaching programming principles, and has its own set of development tools, including PLT Scheme (2009). However, we have not used Scheme at Rivier College.

Scripting languages such as Perl, Python, Ruby, and others (including PHP for Web server use) have become very popular, and we taught a Perl Programming course a while ago. Our experience was that Perl is a very eclectic language, with many quirks, and easily abused. We would not use it to teach how to program. All of these languages have active communities and have their own development environments, although it is very common for programmers in such languages to use a minimalist environment (perhaps with syntax highlighting in an slightly intelligent editor). We have plans to start teaching scripting languages in the near future, as a response to the increased relevance of those languages in the real world.

John Dion, a Rivier College graduate student, has successfully used Ruby instructions in development of his project, "Musician Web-service Using Ruby-on-Rails, SOAP, FLEX, and AJAX" recently published in the Rivier College Online Academic Journal (John Dion, 2006b).

Linux systems (notably Ubuntu Linux) have a rich set of tools available, either provided on the initial system install, or by relatively easy separate installation. Ubuntu's system package add/update/remove environment shows many programming environments that one can install, including KDevelop (2009), which supports a number of languages including C/C++ and Ruby. Note that NetBeans is supported on Ubuntu, and other Linux systems, as is Eclipse.

## Data Communication in Networking Systems

The online classes on networking technologies have been designed as virtual hands-on computer labs (Riabov, 2006a), (Wooley, 2002), that help students in understanding study cases and finding the ways of solving them. The OPNET IT Guru™ Academic software package (2005), a unique free-license application that offers all the tools for virtual network model design, simulation, and analysis, was selected for online instruction. OPNET software can simulate a wide variety of different networks, which are linked to each other through routers and switches. Students could work from their home PCs independently to simulate different networks (ATM, Frame Relay, X.25, Fiber Optics, Wireless, etc.) and study visually the impact of various factors (e.g., traffic load, bandwidth, data rate, etc.) on the network (Kumar, 2005). Providing means for analysis and modeling of network performance, OPNET IT Guru™ tool can also be used for studying data message flows, packet losses, link failures, bits errors, etc.

Following the methodology described by K. Brown and L. Christianson (2005), L. Peterson and B. Davie (2005), and R. Panko (2005), twelve online lab assignments have been designed and offered to students for exploring the Ethernet, Token Ring, Switched Local Area Networks (LANs), Frame Relay (FR), Asynchronous Transfer Mode (ATM), Routing Information Protocol (RIP), Open Shortest Path First (OSPF) protocol, Transmission Control Protocol (TCP), Queuing Disciplines, Resource Reservation Protocol (RSVP), Firewalls, and Virtual Private Networks (VPNs). Using this knowledge and skills, students develop their own lab projects (Kumar, 2005) and include virtual lab techniques into their research projects (Sood, 2007a) related to various network application protocols, such as File Transfer Protocol (FTP), Simple Network Management Protocol (SNMP), and Simple Mail Transfer Protocol (SMTP) (Riabov, 2006b).

Hnatyshin and Lobo (2008) have recently published overview of undergraduate data communications and networking projects that use OPNET and Wireshark (2009) software. Confirming our experience, they have also found that these labs help undergraduate students understand fundamental networking concepts through modeling and simulation of computing systems.

Another valuable network simulation tool is OMNeT++™ (Varga, 2004), which is free for academic and non-profit use. OMNeT++ is a public-source, component-based, modular and open-architecture simulation environment with strong GUI support and an embeddable simulation kernel. Its primary application area is the simulation of communication networks and, because of its generic and flexible architecture, it has been successfully used in other areas like the simulation of IT systems, queuing networks, hardware architectures and business processes as well. OMNeT++ is rapidly becoming a popular simulation platform in the scientific community and academia, as well as in industrial settings. Several open source simulation models have been published (Varga, 2004), in the field of internet simulations (IP, IPv6, MPLS, etc), mobility and ad-hoc simulations and other areas.

Martin Milkovich, a Rivier College graduate student in 2004-2005 classes, offered OMNeT++ (Varga, 2004) for modeling and simulating performances of the digital video cluster firm (Milkovich, 2005b). He selected a challenging topic, Digital Video Cluster Simulation, for his CS690 and CS699 research projects, and a few months later he delivered a presentation of his findings at the Winter-2005 International Conference on Simulation Methods held in Florida (Milkovich, 2005a). Nowadays, following his example, other students apply the OMNeT++ tool in their research studies of the clustered Storage Area Networks (Riabov, 2004), wireless sensor networks, and distributed network systems.

## Image Processing and Multimedia Applications

This topic is usually covered in the introductory computer science courses, e.g. in *CS120 Introduction to Computing* course at Rivier College. Following the teaching methodology of Guzdial (2005), a set of virtual (online) labs has been designed to teach students how to manipulate images, sounds, text, and movies. This media computation approach (Guzdial, 2005) is based on basic programming in *Jython* (2009), the specific Java-implemented dialect of the executable *Python* pseudo-code (2009). A special programming environment called *JES* (*Jython Environment for Students*) (2009), (Guzdial, 2005), has been developed and used to make it easier to program in Jython, which allows students to do multimedia across multiple computer platforms.

The labs include Exploring and Modifying Pictures, Creating a Collage, Exploring and Modifying Sounds, Composing Sounds, Modern Music Synthesis, Creating and Modifying Text, Making Text for Web, Creating and Modifying Movies, and Creating Graphical User Interfaces.

## Web Development

Students are usually encouraged to create online Web-based portfolios in computing courses (Higgs & Sabin, 2005), (Sabin, et al., 2005). This educational approach helps students in better learning, course and curriculum management, developing Web design skills, and in promoting their achievements and skills during job interviews or promotions.

Adobe DreamWeaver™ is the gold standard for developing web sites. Unfortunately, it is very expensive for students to purchase, even at an educational discounted price. Adobe's Academic Edition™ is also still too highly priced for many of our students, although a college-wide site license might be the way to go for hybrid courses.

In the past (before 2003), Microsoft FrontPage™ application was frequently used by instructors and students (Sabin, et al., 2005) to create Web-based portfolios. The advantages of using Microsoft FrontPage™ have been described by Higgs and Sabin (2005):

- The ability to easily create web pages/sites without resorting to writing HTML tags directly. Users can construct a web page in a way similar to using Microsoft Word to create a document, using a "what you see is what you get" (*WYSIWYG*) interface.
- The ability to easily publish web pages/sites to a web server, using either FTP or Front Page Extensions.
- The ability to very easily create a web site with a consistent look-and-feel and navigation controls, and to be able to change that look-and-feel for the entire Web site in seconds.

Unfortunately, Microsoft™ Corp. no longer supports Microsoft FrontPage™, and instructors continue searching for alternatives among Open Source platforms. In 2006, Microsoft introduced a suite of Web development products called Microsoft Expression. One of those products, Microsoft Expression Web, was the replacement product for Microsoft FrontPage™, but unfortunately, it was not 100% compatible with its predecessor. The focus of Expression Web is to produce a more powerful Web development product, emphasizing Microsoft's proprietary ASP.Net technology. In the process, they sacrificed, to a considerable extent, the ease of use of FrontPage™. In particular, generating navigation controls in the FrontPage™ approach was replaced with a different mechanism. However, Microsoft Expression Web has reasonable academic pricing.

Unfortunately, there has been a tendency for colleges and their students to use Microsoft Word and other Microsoft Office products to produce Web-based portfolios. This is not a good idea, because those products were not truly designed for the Web, and this practice locks the student into a proprietary (albeit familiar) solution. The HTML code generated by these products is very Microsoft-specific.

Possible replacements for Microsoft FrontPage™ are discussed in the next section.

## Web Authoring

There are basically three aspects to Web authoring: 1) [eXtensible] Hypertext Markup Language [X]HTML editing; 2) Cascading Style Sheet (CSS) editing; and 3) JavaScript editing, which are considered the 'three-legged stool' of Web development. In addition, multiple browser experience should be considered a mandatory component. Browsers are notoriously incompatible, especially Internet Explorer™, and it is important to be able to create web sites that work across the major browsers.

### [X]HTML Editors

Only few Open Source Web-page editors with a *WYSIWYG* interface are available, including Nvu™ (2009), Mozilla Composer™ from Mozilla Suite (2009), and Trellian WebPage™ (2009). According to Higgs and Sabin (2005), Nvu™ seemed to hold the most promise. It is based on the Mozilla Composer™ technology, with some considerable usability improvements. Unfortunately, work on Nvu came to a halt in 2006; Nvu 1.0 is the only version available. A community-driven project, Kompozer™ (2009), maintains the Nvu codebase and fixes bugs until a successor to Nvu is released. The developer of Nvu is currently working on a successor to Nvu, called BlueGriffon™ (2009), which has been rewritten from scratch, based on the rendering engine and development technology used by Mozilla Firefox™. At the time of writing, there is not yet a release of BlueGriffon, but it does seem to look promising, and there are binary versions to download and try out.

One alternative that is available now is SeaMonkey™ (2009), which is also based on the Mozilla Composer technology. Therein lies a little history: The original Netscape Web browser, which included Navigator, Composer and an email program, morphed into the Mozilla Application Suite™. In 2005, Mozilla announced that it would no longer continue to develop its Application Suite, and it was taken over by The SeaMonkey Council, renamed SeaMonkey, and is available from their site (SeaMonkey, 2009). Our experience with it is that it is simple to use for creating a small number of Web pages, but not for Web sites of larger size. It also does not seem to be particularly up to date with the latest Web standards. The latest update apparently fixes some security issues and bugs, but adds no additional functionality.

Trellian™ WebPage (2009) is freeware, and runs only on Microsoft Windows™ (including Vista™). Higgs and Sabin (2005) mention that the product seemed to have become neglected, but it is interesting to note that the latest version has received a considerable facelift, even to an implementation of a menu 'ribbon', reminiscent of the 'ribbon' in Microsoft™ Office 2007 applications.

On the Ubuntu (Linux) platform, Quanta Plus™ (2009) could be an appropriate alternative.

**CSS Editing**

Cascading Style Sheets (CSS) use simple language syntax, but a complex set of rules about what is valid CSS and what is not. One can use a plain editor, or a reasonably intelligent editor, as long as it provides the necessary semantic assistance.

There are a number of intelligent editors that support CSS syntax, but not necessarily with knowledge of the CSS semantics. The reader could be referred to the earlier list of intelligent editors; most of them will support CSS, at least at the syntax coloring level. One very useful little tool, TopStyle Lite™ (2009), is available for free download for Windows systems.

**JavaScript Editing**

JavaScript, while an essential part of Web Authoring, does not fare well with programming language support. Since it typically executes within the confines of a Web browser, the user experience is typically not good. The example of successful implementation of JavaScript could be found in the project, "Personal Encrypted Talk", a system for securing instant messaging with a Java application (Snogles, 2005), developed by David Snogles, a Rivier College graduate student.

**Multiple Browser Experience**

Any Web Authoring course worth its salt needs to address the fact that Web sites must work across multiple Web browser platforms. Web browsers are notoriously incompatible, and things are made worse by the fact that the leading Web browser, Microsoft Internet Explorer™ (currently at version 7) is the least compatible of all the major Web browsers. Students will need to be able to develop Web sites and test them against at least the following modern browsers:

- Microsoft Internet Explorer v.7 (Internet Explorer v.6 is still very common, and even more problematic than v.7; version 8 is close to being released, which will provide its own new issues, despite Microsoft's claims of more standard behavior).
- Mozilla Firefox™ (2009)
- Opera™ Web Browser (2009)
- Apple's Safari™ Browser (also available on Windows) (2009)
- Google's Chrome™ Browser (2009)
- Linux systems also have additional, lesser-known browsers.

Students studying Web authoring should be exposed to as many of these browsers as possible, and encouraged to test their work with as many of them as possible.

**Browser-specific Tools/Environments**

Once JavaScript is introduced into a Web site (and this means most Web sites), the need to be able to debug JavaScript code becomes paramount. Unfortunately, the situation is not good, because different browsers have different JavaScript engines with varying compatibilities.

Opera Web Browser (2009) comes with its own very useful debugging environment. Mozilla Firefox (2009) has the Firebug plug-in (2009), and Web Developer Toolbar (2009). Microsoft Internet Explorer can use the Microsoft Script Editor™ (2009), and the Fiddler Web Debugger™ (2009).

**Web Development IDEs**

The latest crop of production quality IDEs (NetBeans, Eclipse, etc.) typically provide rich support for [X]HTML, CSS, and JavaScript (and more!). However, they tend to be very text/language focused, rather than task-focused (at least for inexperienced students), which can result in a rather complex experience.

In a production environment, such IDEs make a lot of sense – they typically support more sophisticated Web environments such as Java Server

Pages, ASP.Net, PHP, or similar technologies, and provide significant additional 'intellectual leverage' and enhanced productivity. However, this may not be appropriate for an academic environment, except at the more advanced levels of academic achievement.

**Text-Based Web Authoring Tools**

While they are not WYSIWYG tools, quite a number of other text-based Web authoring tools are available, including HTML-Kit™ (2009), which has been very useful in our Web authoring experiences. On Ubuntu Linux, possibilities include Screem™ (2009), Bluefish™ (2009), and Quanta Plus™ (2009).

**Alternatives to [X]HTML/CSS/JavaScript Constructed Web Sites**

An interesting idea to ponder is this: Which students absolutely need to learn how to construct a Web site using the basic languages ([X] HTML, CSS, and JavaScript), and which would do fine with Web-based construction tools, such as Google™ Sites (2009), Blogs (Blogger, 2009; WordPress, 2009), Wikis (PBWiki, 2009) or some other Web-based technology? While it is reasonable to assume that most CS/CIS/IT majors will need to learn the nuts and bolts approach, others probably may never need more than what those Web-based technologies provide.

**Other Utilities**

There are a number of FTP utilities available, so that Web pages may be uploaded to a suitable Web hosting site. Many of the above-mentioned integrated tools provide support for FTP (and other uploading mechanisms). However, sometimes a good standalone FTP client is necessary, or even desirable. One such client on Windows is WinSCP™ (2009). Others are available for Windows, Linux, and other systems.

## Web Server Selection

Apache™ HTTP Server (2009) has been used as the Open Source replacement of Microsoft Internet Information Services (IIS) server. Nowadays, the Apache HTTP Server is the most popular web server, is very well documented, and easily administered (Higgs & Sabin, 2005).

## Web Server Support

One additional advantage that Microsoft FrontPage has is the existence of *FrontPage Server Extensions* that "support features such as hit counters, data collection, e-mail processing, and database processing" (About FrontPage Server Extensions, 2009). The automatic navigation updates and the easy configuration and modification of the overall navigation scheme of the web site are among other important features found by Higgs and Sabin (2005).

The equivalent functionality of these extensions could be implemented through scripting support on the web server. The Open Source options include the Common Gateway Interface (CGI) (2009), Perl (2009), and some other languages, such as PHP (2009), Java Server Pages, and/or Java Servlets (2009).

The CGI is built into the Apache™ Web Server. Other options require that the functionality will be added to a related web server. For example, for PHP, this is the Apache PHP plug-in. In the case of Java Server Pages/Java Servlets, Apache Tomcat™ is the most obvious choice for a Java Enterprise Container™ (Higgs & Sabin, 2005).

In addition, the necessary code should be created to substitute for the Microsoft FrontPage™ Server Extensions, in particular, to provide means for good navigation support.

## Database Development and Management

Microsoft Access™ is usually used as a stand-alone platform in introductory courses on database management. This platform includes two major components: Database Engine and Graphical User Interface (GUI). However, the use of Microsoft Access can be problematic when trying to learn about databases, as opposed to focusing on user-oriented database interfaces. In modern applications, the issues of remotely allocated databases and their management become important, and a proper platform (i.e., Oracle™ database (2009), Microsoft SQLServer™, PostgreSQL™ (2009), etc.) should be selected. The choices among the free-license Open Source databases are limited.

### Database Engine

There are a few high-quality free-license databases, including MySQL™ (2009), PostgreSQL™ (2009), Apache Derby™ (Cloudscape™) (2009), TinySQL™ (2009), Hypersonic™ SQL (HSQLDB, 2009), Firebird™ (2009), and Mckoi™ SQL Database (2009).

MySQL™ is the most popular Open Source database, it runs on many platforms, and it is well documented, with a number of good books describing it. It is also relatively simple to administer (Higgs & Sabin, 2005).

PostgreSQL™ is a well-respected and popular database system that was developed on UNIX systems. Recently, a Windows version has been made available (PostgreSQL, 2009).

Apache™ Derby's history is associated with Informix™, and their acquisition by IBM™. IBM™ has officially open-sourced what was then known as Cloudscape™ and contributed it to the Apache Software Foundation, where it is now an 'incubator' project (Higgs & Sabin, 2005).

TinySQL™, Hypersonic SQL™, Firebird™, and Mckoi™ SQL Database are open source databases with smaller user bases.

### Database GUI

Unlike Microsoft™ database products, most of the Open Source database engines do not include a GUI interface (Higgs & Sabin, 2005). Such interfaces are usually supplied separately, often by third parties. One example of this is that GUI-based interfaces for MySQL are available separately. Java-based GUIs supporting the JDBC interface are particularly prevalent, given the ability of Java to provide platform-independent database and GUI interfaces.

There are a few examples (Higgs & Sabin, 2005) of the Open Source database GUIs: Eclipse™ (2009) with the SQLExplorer™ Plug-in (2009), iSQL-Viewer™ (2009), SQuirreL™ SQL Client (2009), QueryForm™ Database Tool (2009), and FreeQueryBuilder™ (2009). All of the above database GUIs are written in Java. Probably any of these products would be satisfactory. None of them stands out significantly over the others in any obvious way, although platforms such as NetBeans have been making great progress regarding GUI support for Databases – to the point that they may be the preferred GUI-based development interface to any and all of these databases. Higgs and Sabin (2009) have commented that the use of Eclipse™ with SQL Explorer Plug-in would be attractive for students with a reasonable programming background, although NetBeans is definitely in the game.

The other option is to use phpMyAdmin™ (2009), which is a free software tool written in PHP and intended to handle the administration of MySQL (2009) over the World Wide Web. The most frequently used operations are supported by the user interface (managing databases, tables, fields, relations, indexes, users, permissions, etc), while a user still has the ability to directly execute any SQL statement (phpMyAdmin, 2009).

## ONLINE COURSE DELIVERY ISSUES

Blackboard™ has been selected as a management environment tool for running all computer-science online (or hybrid) courses. The course materials are organized on weekly basis by using folders that contain lecture outline slides, links to software downloading sites, software installation instructions, tutorials, lab manuals, and weekly homework assignments. The individual project assignments and final exam questionnaires are allocated in the special "Assignments" section folder. The "Course Materials" folder contains examples on students' projects from the previous course session, URLs to original research papers on course advanced topics, and links to additional resources available on the instructor's web site (Riabov, 2009). A special section on the Blackboard site is designated to students' project presentations and progress reports. Students are also encouraged to submit their final project reports for peer-reviewing and publishing in the *Rivier College Online Academic Journal.*

Unfortunately, Blackboard™ tool (as many other online course management tools) limits live or synchronous communication (typical with face-to-face class communication) to "chat" rooms. One of the new trends in online teaching and learning is "the adoption and integration of web conferencing tools to enable live online classrooms and recreate the ethos of traditional face-to-face sessions" (Sabin & Higgs, 2007). In addition to Blackboard™ course tools (e.g., Discussion Board, Blogs, Collaboration, Self and Peer Assessment, etc.), many online computer science courses at Rivier College utilize state-of-the-art web conferencing software called LearnLinc™ (2009), an iLinc™ Communications, Inc. product, a tool tailored to support "live" online learning through integration of audio and video over the Internet, text messaging, an interactive white board, synchronized browsing, and application sharing. LearnLinc sessions transform online classes into engaging, participatory, and personalized learning environments (Cooper, 2006). Only in the 2006-2007 academic year, Rivier College offered twelve courses, enrolling over 150 students, that used the synchronous online teaching capabilities of LearnLinc. "Students who took courses [offered] in the online or hybrid format experienced a comparable level of interaction, participation, and collaboration as in traditional classes" (Sabin & Higgs, 2007). The students underlined in their course evaluations that the synchronous web conferencing helped them effectively in seeking instructor's assistance, managing time on task, and exercising problem solving skills.

## EXAMPLES OF STUDENTS' PAPERS ON TECHNOLOGY OVERVIEW

In many classes (e.g., CS553 Introduction to Networking Technologies, CS578 Advanced Networks, CS685 Network Management, CS690 System Simulation and Modeling, and CS699 Professional Seminar in Computer Science), the instructor encourages students to conduct research and write project papers on modern networking technologies. The assigned individual projects include all the components of feasibility study, analysis, synthesis, architectural design, application, and evaluation of selected networking systems. The students are motivated to select topics for projects that would be valuable for companies and the community. Usually, they demonstrate their project Web-based portfolios during the job interviews. Such demonstration of their actual professional skills helps students in finding a job immediately after the graduation. Many students' projects are implemented in local companies and the community (Riabov, 2006a).

Seniors and all graduate students have delivered online (via the Blackboard™) their presentations of individual and team projects in the classroom and seminars (Riabov, 2005c; Riabov, 2005d). The instructor helps them in organizing the presentations.

Working on the individual projects in the CS553 Introduction to Networking Technologies and CS578 Advanced Networks classes, students search information on modern networking technologies from the prime sources, such as the Request-For-Comments organization (2009), a repository of the standards in networking industry; IETF Secretariat (2009); WiFi™ Alliance (2009), an association that promotes secure solutions for wireless networking technologies; Fibre Channel Industry Association (2009), an international organization of manufacturers, developers, and end users of the Storage Area Networks, and others.

Many graduate students, who are employed by local networking, computer and IT companies, select topics for individual projects that are based on their experience with the company (Jujjuru, 2008; Milkovits, 2005a; Sood, 2007b; Trull, 2006; Wekhande, 2006). The network computer systems developed by the students under the instructor's supervision are tested in the company's environment and receive a great support of their supervisors and managers (Milkovits, 2005b). This fact helps the students in their job promotion.

## EXAMPLES OF STUDENTS' RESEARCH STUDIES

The author's articles and papers (Riabov, 2000a; 2000b; 2002a; 2002b; 2003; 2004; 2005a; 2005b; 2006b; Sabin, et al., 2005) have a focus on interdisciplinary aspects of teaching the Computer Science classes and designing modern networking systems. The research in these publications includes both disciplinary and interdisciplinary perspectives that challenge the instructors to take a personal responsibility in teaching students according to their different backgrounds in Mathematics, Computer Science, Natural Sciences, and culture.

The special assignments have been designed for the CS690 System Simulation & Modeling course, which encourage the students to grasp understanding of what managers would expect from the results of simulating performances of various network systems and data farms. Working hard on his project, Martin Milkovich, a Rivier College student in 2004-2005 classes, offered a new software tool, OMNeT++ (Varga, 2004), for modeling and simulating performances of the digital video cluster firm. He delivered a presentation, Digital Video Cluster Simulation at the Winter-2005 International Conference on Simulation Methods held in Florida (Milkovich, 2005b). Nowadays, other students apply the OMNeT++ tool in their online research studies of the clustered Storage Area Networks, wireless sensor networks, and distributed network systems.

The authors support graduate students in their efforts of introducing the Computer System Analysis and Design Standards (Wason & McCabe, 1996) to the networking industry, related software tools (Using McCabe QA, 1999), and practices (Riabov, 2005a; Coombs & Coombs, 1998) in the local networking companies. As a result of such efforts, David D. Norman, a Rivier College student in 2001-2003 classes, has been promoted to the position of a Principle Software Engineer at IMPACT Science and Technology, Inc., located in Hollis, NH. He was a guest speaker at the Rivier College Mathematics & Computer Science Lecture Series on November 11, 2003 (Riabov, 2005c), where he shared his experience in software systems engineering and networking technologies with students and faculty.

The students are encouraged to submit summaries of their research projects to professional journals and magazines. Ten graduates, John Dion (2006a; 2006b), David Dwyer (2006), Ajay Kumar (2005; 2006), Jhansi Jujjuru (2008), Martin Milkovich (2005a), David Snogles (2005), Arti Sood (2007a; 2007b), Bruce Trull (2006), Vandana Wekhande (2006), and Robert J. Zupko (2007), have published recently their manuscripts in the *Rivier College Online Academic Journal*. These activities develop a *strong bond* between students and faculty members that will last forever.

## FUTURE RESEARCH DIRECTIONS

In this chapter, we have described our experience of running only few online computer science courses that require intensive technology-exploration experience for students. The development of the Computer Science Curriculum totally offered online is still a difficult task. The large variety of state-of-the-art free-license software tools available for instructors and students is overlapping. This fact requires a thorough selection of the tools for online instructions. Many free-license software tools are still in the development stage and are not accompanied with complete user's manuals and study cases that could be transformed into the virtual lab manuals. Some of the free-license software tools are not used in the industry. This fact could be a significant challenge for colleges that could not afford purchases of professional software editions. These challenges could be addressed in the future studies.

## CONCLUSION

The authors have described the challenges and experience of running several online computer science courses for seniors and graduate students. Their experience of selecting free-license software tools has been in general a very positive one, while at the same time providing useful lessons learned. The authors believe that this virtual-lab and online project-based approach can be effectively applied to future online courses of a similar nature in academia, and believe that the model can be extended to other engineering disciplines beyond the computing sciences. In our practice of running online and hybrid courses, we have found that the using of synchronous web conferencing tools (in addition to traditional Blackboard™ course tools) helped students effectively in seeking instructor's assistance, managing time on task, and exercising problem solving skills.

## REFERENCES

*About FrontPage Server Extensions from Microsoft*. (2009). Retrieved July 12, 2009, from http://office.microsoft.com/enus/assistance/HP011032361033.aspx

*Apache Derby Project*. (2009). Retrieved July 12, 2009, from http://incubator.apache.org/derby/

*Apache HTTP Server Project*. (2009). Retrieved July 12, 2009, from http://httpd.apache.org/

*Apple Safari*. (2009). Retrieved July 12, 2009, from http://www.apple.com/safari/

ArgoUML v.0.26. (2009). *Open Source Software Engineering Tools*. Tigris, Inc. Retrieved July 12, 2009, from http://argouml.tigris.org/

Barnes, D. J., & Kölling, M. (2004). *Objects First with Java: A Practical Introduction using BlueJ* (2nd ed.). Upper Saddle River, NJ: Prentice Hall.

*Blogger*. (2009). Retrieved July 12, 2009, from https://www.blogger.com/start

Blue, J. (2009). *BlueJ – Teaching Java – Learning Java*. Retrieved July 12, 2009, from http://www.bluej.org/

*Bluefish Editor*. (2009). Retrieved July 12, 2009, from http://bluefish.openoffice.nl/index.html

*BlueGriffon*. (2009). Retrieved July 12, 2009, from http://bluegriffon.org/

*Borland JBuilder*. (2009). Retrieved July 12, 2009, from http://www.codegear.com/products/jbuilder

Brookshear, J. G. (2009). *Computer-Science: An Overview* (10th ed.). Reading, MA: Addison-Wesley.

Brown, K., & Christianson, L. (2005). *OPNET Lab Manual to Accompany Data and Computer Communications* (7th ed.). Upper Saddle River, NJ: Pearson Prentice Hall.

*CDT Plugin for Eclipse*. (2009). Retrieved July 12, 2009, from http://www.eclipse.org/cdt/

*Code:Blocks*. (2009). Retrieved July 12, 2009, from http://www.codeblocks.org/

Coombs, C. Jr, & Coombs, C. A. (1998). *Communications Network Test & Measurement Handbook*. New York: McGraw-Hill.

Cooper, K. (2006). Rivier College Launches Online Programs. *Rivier College Online Academic Journal, 2*(2). Retrieved July 12, 2009, from http://www.rivier.edu/journal/RCOAJ-Fall-2006_table.htm

*Crimson Editor*. (2009). Retrieved July 12, 2009, from http://www.crimsoneditor.com/

Dale, N. (2003). *C++ Plus Data Structures* (3rd ed.). Boston, MA: Jones and Bartlett Publishers.

Dennis, A., Wixom, B. H., & Tegarden, D. (2009). *System Analysis & Design, UML Version 2.0: An Object-Oriented Approach* (3rd ed.). Hoboken, NJ: John Wiley & Sons, Inc.

Dev, C. ++. (2009). Retrieved July 12, 2009, from http://www.bloodshed.net/devcpp.html

Dion, J. (2006a). Automated Music Composition: An Expert Systems Approach. *Rivier College Online Academic Journal, 2*(1). Retrieved July 12, 2009, from http://www.rivier.edu/journal/RCOAJ-Spring-2006_table.htm

Dion, J. (2006b). Musician Web-service Using Ruby-on-Rails, SOAP, FLEX, and AJAX. *Rivier College Online Academic Journal, 2*(2). Retrieved July 12, 2009, from http://www.rivier.edu/journal/RCOAJ-Fall-2006_table.htm

Dwyer, D. (2006). Software Reliability Estimations/Projections, Cumulative and Instantaneous. *Rivier College Online Academic Journal, 2*(1). Retrieved July 12, 2009, from http://www.rivier.edu/journal/RCOAJ-Spring-2006_table.htm

*Eclipse*. (2009). Retrieved July 12, 2009, from http://www.eclipse.org/

Fibre Channel Industry Association. (2009). Retrieved July 12, 2009, from http://www.fibrechannel.org/

Fiddler Web Debugger. (2009). Retrieved July 12, 2009, from http://www.fiddlertool.com/fiddler/

*Firebird*. (2009). Retrieved July 12, 2009, from http://firebird.sourceforge.net/

*Firebug*. (2009). Retrieved July 12, 2009, from http://getfirebug.com/

Flanagan, D. (2005). *Java in a Nutshell* (5th ed.). New York: O'Reilly & Associates.

*FreeQueryBuilder*. (2009). Retrieved July 12, 2009, from http://sourceforge.net/projects/querybuilder/

*GNU C++ (GCC)*. (2009). Retrieved July 12, 2009, from http://gcc.gnu.org/

*GNU emacs*. (2009). Retrieved July 12, 2009, from http://www.gnu.org/software/emacs/emacs.html

*Google Chrome*. (2009). Retrieved July 12, 2009, from http://www.google.com/chrome

*Google Sites*. (2009). Retrieved July 12, 2009, from http://sites.google.com/

Guzdial, M. (2005). *Introduction to Computing and Programming in Python: A Multimedia Approach*. Upper Saddle River, NJ: Prentice Hall.

Higgs, B., & Sabin, M. (2005). Towards Using Online Portfolios in Computing Courses. In *Proceedings of the 6th ACM SIG — Information Technology Education Conference (SIGITE 2005)*, Newark, NJ, October 20-22, 2005 (pp. 323-328).

Hnatyshin, V. Y., & Lobo, A. F. (2008). Undergraduate Data Communications and Networking Projects Using OPNET and Wireshark Software. *ACM SIGCSE Bulletin, 40*(1), 241–245. doi:10.1145/1352322.1352222

Horstmann, C. S., & Cornell, G. (2007). Core Java: *Vol. I. Fundamentals* (8th ed.). Upper Saddle River, NJ: Prentice-Hall.

*HSQLDB*. (2009). Retrieved July 12, 2009, from http://hsqldb.org/

*HTML-Kit*. (2009). Retrieved July 12, 2009, from http://www.chami.com/html-kit/download/

iLinc Communications, Inc. (2009). *LearnLinc Tool Panel Overview*. Retrieved July 12, 2009, from http://www.ilinc.com/pdf/trc/ilinc-learnlinc-tool-panel.pdf

*iSQL-Viewer*. (2009). Retrieved July 12, 2009, from http://www.isqlviewer.com/

IT Guru Academic Ed. *OPNET Technologies*. (2009). Retrieved July 12, 2009, from http://www.opnet.com/university_program/itguru_academic_edition/

*JEdit Programmer's Text Editor*. (2009). Retrieved July 12, 2009, from http://www.jedit.org/

Jujjuru, J. (2008). An Overview of Internet Protocol Multimedia Subsystems (IMS) Architecture. *Rivier College Online Academic Journal, 4*(1). Retrieved July 12, 2009, from http://www.rivier.edu/journal/RCOAJ_Spring_2008_table.htm

*Jython*. (2009). Retrieved July 12, 2009, from http://www.jython.org/

*Jython Environment for Students (JES)*. (2009). Retrieved July 12, 2009, from http://coweb.cc.gatech.edu/mediaComp-plan/Mediacomp-Software

KDevelop. *KDE Development Environment*. (2009). Retrieved July 12, 2009, from http://www.kdevelop.org/

*Kompozer: Nvu's Unofficial Bug-Fix Release*. (2009). Retrieved July 12, 2009, from http://kompozer.net/download.php

Kumar, A. (2005). Development of Laboratory Exercises Based on OPNET IT Guru Academic Version. *Rivier College Online Academic Journal, 1*(1). Retrieved July 12, 2009, from http://www.rivier.edu/journal/RCOAJ-Fall-2005_table.htm

Kumar, A. (2006). An Overview of Voice over IP. *Rivier College Online Academic Journal, 2*(1). Retrieved July 12, 2009, from http://www.rivier.edu/journal/RCOAJ-Spring-2006_table.htm

Lewis, J., & Loftus, W. (2003). *Java Software Solutions: Foundations of Program Design* (3rd ed.). Boston, MA: Addison Wesley.

Lippman, S. B., Lajoie, J., & Moo, B. E. (2005). *C++ Primer* (4th ed.). Boston, MA: Addison-Wesley.

*Mckoi Database*. (2009). Retrieved July 12, 2009, from http://www.mckoi.com/database/

Microsoft Inc. (2008). *Microsoft Visual C++ 2008 Express Edition*. Retrieved July 12, 2009, from http://www.microsoft.com/express/vc/

Microsoft Inc. (2009). *Microsoft Script Editor*. Retrieved July 12, 2009, from http://msdn.microsoft.com/en-us/library/aa189846(office.10).aspx

Milkovits, M. N. (2005a). Digital Video Cluster Simulation. In M. E. Kuhl, N. M. Steiger, F. B. Armstrong & J. A. Joines (Eds.), *Proceedings of the 2005 Winter Simulation Conference,* Orlando, Florida, December 2005.

Milkovits, M. N. (2005b). Using OMNeT++ to Simulate a Heterogeneous Digital Video Cluster Fabric. *Rivier College Online Academic Journal, 1*(1). Retrieved July 12, 2009, from http://www.rivier.edu/journal/RCOAJ-Fall-2005_table.htm

Min, G. W. (2009). *Minimalist GNU for Windows*. Retrieved July 12, 2009, from http://www.mingw.org/

*MinGW Developer Studio*. (2009). Retrieved July 12, 2009, from http://programmerstools.org/node/368

*Mozilla Firefox*. (2009). Retrieved July 12, 2009, from http://www.mozilla.com/en-US/firefox/

Mozilla Suite. *The All-in-One Internet Application Suite*. (2009). Retrieved July 12, 2009, from http://www.mozilla.org/products/mozilla1.x/

*MySQL: The World's Most Popular Open Source Database*. (2009). Retrieved July 12, 2009, from http://dev.mysql.com/

*NetBeans*. (2009). Retrieved July 12, 2009, from http://www.netbeans.org/

*NetBeans C and C++ Development*. (2009). Retrieved July 12, 2009, from http://www.netbeans.org/features/cpp/index.html

*Notepad++*. (2009). Retrieved July 12, 2009, from http://notepad-plus.sourceforge.net/uk/site.htm

*NoteTab Lite*. (2009). Retrieved July 12, 2009, from http://www.fookes.com/notetab/index.php

*Nvu: The Complete Web Authoring System for Windows, Mac, and Linux*. (2009). Retrieved July 12, 2009, from http://www.nvu.com/

*Opera Web Browser*. (2009). Retrieved July 12, 2009, from http://www.opera.com/

*Oracle Database Software Downloads*. (2009). Retrieved July 12, 2009, from http://www.oracle.com/technology/software/products/database/index.html

Panko, R. (2005). *Seven OPNET Labs*. Retrieved July 12, 2009, from http://www.opnet.com/services/university/itg_panko.html

*PBWiki*. (2009). Retrieved July 12, 2009, from http://pbwiki.com/

*Perl Directory*. (2009). Retrieved July 12, 2009, from http://www.perl.org/

Peterson, L. L., & Davie, B. S. (2005). *Computer Networks: A System Approach* (3rd ed.). Boston, MA: Morgan Kaufmann.

Pevac, I. (2005). *Recursive Examples in Java*. Ann Arbor, MI: ProQuest Information & Learning.

PHP. *Hypertext Preprocessor*. (2009). Retrieved July 12, 2009, from http://www.php.net/

*phpMyAdmin*. (2009). Retrieved June 30, 2009 from http://www.phpmyadmin.net/home_page/index.php

*PLT Scheme*. (2009). Retrieved July 12, 2009 from http://www.plt-scheme.org/

Postgre, S. Q. L. *The world's most advanced open source database*. (2009). Retrieved July 12, 2009, from http://www.postgresql.org/

*Python*. (2009). Retrieved July 12, 2009, from http://www.python.org/

*Qt Product Overview – single source C++ cross-platform application development for Windows, Linux, Mac*. (2009). Trolltech Corp. Retrieved July 12, 2009, from http://www.trolltech.com/products/qt/index.html

*Quanta Plus*. (2009). Retrieved July 12, 2009, from http://quanta.kdewebdev.org/

*QueryForm Database Tool*. (2009). Retrieved July 12, 2009, from http://sourceforge.net/projects/qform/

Riabov, V. (2000a). *Challenge in the Aerospace University Classroom. AIAA Paper, No. 2000-0802*. Washington, DC: AIAA Inc.

Riabov, V. (2000b). *New Challenges for Aerospace Education Programs in American and Russian Universities. AIAA Paper, No. 2000-0528*. Washington, DC: AIAA Inc.

Riabov, V. (2002a). Simulation Techniques in Hypersonic Aerodynamics. In *Proceedings of the 23rd ICAS Congress*, Paper 181, Toronto, Canada, September 8-13, 2002.

Riabov, V. (2002b). Computational and Analytical Methods in the Aircraft Design Curriculum. In *Proceedings of the 23rd ICAS Congress*, Paper 255p, Toronto, Canada, September 8-13, 2002.

Riabov, V. (2003). Applications of Exponential Box Schemes for Viscous Flows with Combustion and Blowing. In K. J. Bathe (Ed.), *Proceedings of the Second MIT Conference on Computational Fluid and Solid Mechanics,* June 17-20, 2003 (Vol. 1, pp. 1102-1105). Boston, MA: Elsevier.

Riabov, V. (2004). Storage Area Networks. In Bidgoli, H. (Ed.), *The Internet Encyclopedia* (*Vol. 3*, pp. 329–339). Hoboken, NJ: John Wiley & Sons, Inc.

Riabov, V. (2005a). Networking Software Studies with the Structured Testing Methodology. In Petratos, P., & Michalopoulos, D. (Eds.), *Computer Science and Information Systems* (pp. 261–276). Athens, Greece: ATINER.

Riabov, V. (2005b). Three Fast Computational Approximation Methods in Hypersonic Aerothermodynamics. In K. J. Bathe (Ed.), *Proceedings of the Third MIT Conference on Computational Fluid and Solid Mechanics,* June 13-14, 2005 (Vol. 1, pp. 819-823). Boston, MA: Elsevier.

Riabov, V. (2005c). Mathematics and Computer Science Lecture Series (2002 – 2005). In *Rivier College Online Academic Journal, 1*(1). Retrieved July 12, 2009, from http://www.rivier.edu/journal/RCOAJ-Fall-2005_table.htm

Riabov, V. (2005d). Professional Seminar in Computer Science (Spring 2005). In *Rivier College Online Academic Journal, 1*(1). Retrieved July 12, 2009, from http://www.rivier.edu/journal/RCOAJ-Fall-2005_table.htm

Riabov, V. (2006a). Challenging Projects and Virtual Labs in Web-enhanced Networking Technology Classes. *Journal of Computing Sciences in Colleges*, *21*(6), 88–99.

Riabov, V. (2006b). Simple Mail Transfer Protocol (SMTP). In Bidgoli, H. (Ed.), *The Handbook of Information Security: Key Concepts, Infrastructures, Standards and Protocols* (*Vol. 1*, pp. 878–900). Hoboken, NJ: John Wiley & Sons, Inc.

Riabov, V. (2009). *Instructor's Website*. Retrieved July 12, 2009, from http://www.rivier.edu/faculty/vriabov/

Sabin, M., & Higgs, B. (2007). Teaching and Learning in Live Online Classrooms. In *Proceedings of the 8th ACM SIG - Information Technology Education (SIGITE 2007) Conference*, Sandestin, FL, October 18-20, 2007 (pp. 41-48).

Sabin, M., Higgs, B., Riabov, V., & Moreira, A. (2005). Designing and Running a Pre-College Computing Course. *Journal of Computing Sciences in Colleges*, *20*(5), 176–187.

*Scheme*. (2009). Retrieved July 12, 2009, from http://groups.csail.mit.edu/mac/projects/scheme/

*Screem HTML/XML Editor*. (2009). Retrieved July 12, 2009, from http://www.screem.org/

*SeaMonkey*. (2009). Retrieved July 12, 2009 from http://www.seamonkey-project.org/

Secretariat, I. E. T. F. (2009). Retrieved July 12, 2009, from http://www.ietf.org/rfc.html

*Servlet 2.3 and JavaServer Pages 1.2 Specifications*. (2009). Retrieved July 12, 2009, from http://www.jcp.org/aboutJava/communityprocess/first/jsr053/

Snogles, D. (2005). Personal Encrypted Talk - Securing Instant Messaging with a Java Application. *Rivier College Online Academic Journal, 1*(1). Retrieved July 12, 2009, from http://www.rivier.edu/journal/RCOAJ-Fall-2005_table.htm

Sood, A. (2007a). Network Design by Using OPNET™ IT Guru Academic Edition. Software. *Rivier College Online Academic Journal, 3*(1). Retrieved July 12, 2009, from http://www.rivier.edu/journal/RCOAJ-Spring-2007_table.htm

Sood, A. (2007b). Network Access Control. *Rivier College Online Academic Journal, 3*(2). Retrieved July 12, 2009, from http://www.rivier.edu/journal/RCOAJ-Fall-2007_table.htm

*SourceEdit*. (2009). Retrieved July 12, 2009, from http://www.brixoft.net/prodinfo.asp?id=1

*SQLExplorer Plug-in for Eclipse*. (2009). Retrieved July 12, 2009, from http://sourceforge.net/projects/eclipsesql

Squirre, L. *SQL Client*. (2009). Retrieved July 12, 2009, from http://sourceforge.net/projects/squirrelsql

*TextPad, The Text Editor for Windows*. (2009). Retrieved July 12, 2009, from http://www.textpad.com/

*The Common Gateway Interface*. (2009). Retrieved July 12, 2009, from http://hoohoo.ncsa.uiuc.edu/cgi/overview.html

The Request For Comments Editor. (2009). Retrieved July 12, 2009, from http://www.rfc-editor.org/

*TinySQL*. (2009). Retrieved July 12, 2009, from http://sourceforge.net/projects/tinysql/

*TopStyle Lite*. (2009). Retrieved July 12, 2009, from http://www.newsgator.com/individuals/topstyle/default.aspx

*Trellian WebPage*. (2009). Retrieved July 12, 2009, from http://www.trellian.com/webpage/

Trull, B. R. (2006). An Overview of Broadband over Power Line. *Rivier College Online Academic Journal, 2*(1). Retrieved July 12, 2009, from http://www.rivier.edu/journal/RCOAJ-Spring-2006_table.htm

Using McCabe, Q. A. (1999). Columbia, MD: McCabe & Associates.

*V – A Freeware Portable C++ GUI Framework*. (2009). Retrieved July 12, 2009, from http://programming.ccp14.ac.uk/objectcentral/main.htm

Varga, A. (2004). *OMNeT++ Version 3.0 User Manual*. Retrieved July 12, 2009, from http://www.omnetpp.org/

Violet, U. M. L. (Ed.). (2009). Retrieved July 12, 2009, from http://www.horstmann.com/violet/

Visual Paradigm, Inc. (2009). *Visual Paradigm for UML Community Edition (VP-UML CE)*. Retrieved July 12, 2009, from http://www.visual-paradigm.com/product/vpuml/editions/community.jsp

Watson, A. H., & McCabe, T. J. (1996). *Structured Testing: A Testing Methodology Using the Cyclomatic Complexity Metric. NIST Special Publication, No. 500-235*. Gaithersburg, MD: National Institute of Standards and Technology.

*Web Developer*. (2009). Retrieved July 12, 2009, from https://addons.mozilla.org/en-US/firefox/addon/60

Weiss, M. A. (1999). *Data Structures & Algorithm Analysis in Java*. Boston, MA: Addison Wesley.

Wekhande, V. (2006). Wi-Fi Technology: Security Issues. *Rivier College Online Academic Journal, 2*(2). Retrieved July 12, 2009, from http://www.rivier.edu/journal/RCOAJ-Fall-2006_table.htm

WiFi Alliance. (2009). Retrieved July 12, 2009, from http://www.wi-fi.org/OpenSection/index.asp

*WinSCP*. (2009). Retrieved July 12, 2009, from http://winscp.net/eng/index.php

*Wireshark: A Network Protocol Analyzer.* (2009). Retrieved July 12, 2009, from http://www.wireshark.org/

Wooley, B. A. (2002). Utilizing a Computing Lab to Improve Retention and Recruiting of Computer Science and Computer Information Science Students. In *Proceedings of the 18th Annual CCSC Eastern Conference*, Bloomsburg, PA, October 2002.

*WordPress*. (2009). Retrieved July 12, 2009, from http://wordpress.org/

Wu, C. T. (2006). *An Introduction to Object-Oriented Programming with Java* (4th ed.). New York: McGraw-Hill Higher Education.

Zupko, R. J. (2007). Introduction to the IEEE 802.16 Standard: Wireless Broadband Access. *Rivier College Online Academic Journal, 3*(1). Retrieved July 12, 2009, from http://www.rivier.edu/journal/RCOAJ-Spring-2007_table.htm

# Chapter 19
# Managing Case-Based Learning with Interactive Case Study Libraries

**Hao Jiang**
*Pennsylvania State University, USA*

**John M. Carroll**
*Pennsylvania State University, USA*

**Craig Ganoe**
*Pennsylvania State University, USA*

## ABSTRACT

*This chapter discusses a particular pedagogical methodology, case-based learning, and introduces an application that supports case studies. It suggests that authenticity, social interaction, community of practice, and resource accumulation are especially important for design and implementation of case-based learning systems. To make the arguments more vivid, the chapter also introduces a case study library that supports usability engineering education. Some of the suggestions are more related to case libraries or systems alike in particular, and some are valuable for learning management systems in general. The authors hope their study can invoke further research of computer-supported case studies in educational and CSCL communities, and more applications supporting this pedagogical approach will be developed.*

## INTRODUCTION

In a broad sense, distance education is an environment where teaching/learning activities happen when students and teachers are not locally engaged such as in the traditional classroom (Galusha, 1997). Advanced information technologies are an increasingly important component of this educational model. Traditional mass media such as TV has been used to deliver course contents and lectures to learners sitting in front of their television. In this context, Computer-Supported Collaborative Learning (CSCL) is another important an promising area (Stahl, Koschmann & Suthers, 2006).

Taking advantage of information technology, a series of applications have emerged in the edu-

DOI: 10.4018/978-1-61520-853-1.ch019

Copyright © 2010, IGI Global. Copying or distributing in print or electronic forms without written permission of IGI Global is prohibited.

cational domain, addressing various needs and visions. For example, large scale systems such as learning management systems have been used at universities in the United States. With these systems, users (instructors and students) can share learning materials and have various interactions, such as chat, email and synchronous collaboration like BRIDGE (Ganoe, Somervell, Neale, Isenhour, Carroll, Rosson & McCrickard, 2003). There are also applications that have specific focuses. For example, Baker, Quignard, Lund and Séjourné (2003) described the tool they developed to help students mastering argument skills.

Since learning is constituted by many different activities, what should be managed in learning is also multifaceted. Online teaching can take many different forms, and a learning management system can: provide a single user with content management functions (e.g., storing and retrieving); provide simulation of learning subject matter where learners can practice with a simulated object; or build an environment that supports various learning activities and keeps history and context of these activities. However, in this chapter, we will lean toward a collaborative, interactive aspect of an online education tool. Our assumption is that learning outcomes will be improved in an environment that engages learners and teachers in interaction and collaboration.

In this chapter, we will pay special attention to a specific area in teaching and learning, *case studies*, and contribute to this field by elaborating on a type of computer-supported case study applications, case study libraries. Case studies have been employed traditionally as a major and effective pedagogical approach in both formal and informal education. Exploiting contemporary information technology infrastructures, cases have increasingly been put online. We advocate creating interactive case study libraries (Carroll, Ganoe & Jiang 2008) to support learning and education.

This chapter will be organized as following. In the background section, we will briefly discuss the use of case studies in broad educational domains and highlight the pervasiveness as well as importance of case studies, and then we will present challenges and limits of current web-based case collections. Those challenges and limits are, for example, the lack of supporting interactions between users and systems, as well as interaction between users through systems. The resources for case materials are also limited by the nature of case accumulation and the limited means of authoring in current web-based case collections.

To tackle problems faced by computer-supported case studies, we will first analyze this type of activity to identify what is essential to case-based learning/education, and then we will elaborate four key aspects of case studies in the section that follows. In short, we have found that *authenticity, social interaction, communities of practice and resource accumulation and updating* are critical for case studies to succeed. Thus, we take a broader view on the managerial part of learning management system, in which we believe that a system should assist learners and teachers to manage their learning material; and more than that, it should allow users to practice learning activities and wider social interactions.

After analyzing activities in case studies, we will turn our focus to designing technology affordance for those activities and examine how they meet the four requirements enumerated above. To illustrate our analysis, we will describe an interactive case library we are developing, which was designed to support usability engineering education. At the end of this chapter, we return to these four requirements with a more synthetic view and discuss possible directions for future research.

## CASE-BASED LEARNING

The focal issues in education research are how students learn and how educational systems can help to improve learning results. Authentic learning (2000Herrington & colleagues 2000!, 2003; Reeves et al 2004) has been found to be beneficial in attaining learning goals. Authentic learning refers to learning or teaching with authentic contents or materials, such as real-world issues, context and data, and engaging students with authentic activities (Carroll & Rosson, 2005). Authentic activities are activities involved during performing tasks or achieving certain goals that are meaningful and valuable to a certain community. For example, in usability engineering, claim analysis (Rosson & Carroll, 2002), a typical authentic activity, requires designers to base their design arguments and features on theoretical findings as well as practical conditions a particular design situated in and to provide trade-offs of a design.

Herrington and colleagues (2000, 2003) summarized ten features of authentic learning, which for example includes that the tasks students will carry out should have real-world relevance, and students should be able to frame or redefine problems from materials they have in hand. Even in learning elementary skills and concepts for children, playing with physical objects has a significant effect on their mastering concepts of different shapes, etc. (Piaget, 1964). The promise of authentic learning is that it brings learners real-world context, and enables students to learn contextualized knowledge as opposed to decontextualized one.

Case studies, or cases, are descriptions of a specific activity, event, or problem, drawn from the real-world of professional practice. They provide narrative models of practice to students and other novice practitioners. Cases incorporate vivid background information and personal perspectives to elicit empathy and active participation. They include contingencies, complexities, and often dilemmas to evoke integrative analysis and critical thinking. Cases engage the student in the drama of a real situation. They are widely employed in many educational disciplines - in business, medicine, law, and engineering (Williams 1992). For example, the well-known Harvard Business School case collection includes over 7,500 case studies of business decision-making (Garvin 2003). Perhaps coinciding with contemporary recognition that all disciplines incorporate practice (and not merely knowledge), or perhaps just reflecting contemporary pedagogical concern with active learning and critical thinking, cases have become pervasive through the past decade. For example, the NSF-supported National Center for Case Study Teaching in Science includes many case studies in medicine, engineering, environmental science, anthropology, botany, social and cognitive psychology, geology and geography, pharmacy and nutrition, and experimental design (Herreid & Schiller 2005).

This focus on authentic learning activities is based on the hypothesis that learning outcomes will be enhanced if the activities students engage in, and the materials they use, more directly reflect the social and technical contexts of actual scientific and engineering practice. Realistic activities and materials are more intrinsically motivating because they constantly remind learners of the possibilities for meaningfully applying knowledge and skills in the world beyond the classroom (Dewey 1933). Today, many teachers are working to develop and/or acquire realistic instructional activities and materials for their teaching.

Teaching and learning in information technology (IT) related disciplines are demanding with respect to technical knowledge and skill in mathematics, programming, and system architecture, as well as with respect to professional skills in problem-solving, teamwork, project management, professional ethics and values. Recent innovations in IT curricula and educational infrastructures have focused on better integrating these two types of skills through more "authentic" learning activities. For example, team-based projects are pervasive now in undergraduate and professional IT education programs (Dietrich & Urban 1996; Hartfield, Winograd & Bennett 1992; Hayes, Lethbridge & Port 2003; Lamancusa, Jorgensen, Zayas-Castro & de Ramirez 2001). These projects frequently incorporate a range of realistic activities such as requirements interviews, software design, and testing. These are sometimes quite extensive, spanning several weeks, or even an entire semester. Those case collections cover natural sciences to social sciences. Many online case collections are available through the Internet. They are, for example, the National Center for Case Study Teaching in Science; public affairs (i.e., The Electronic Hallway), public policy and women's studies (i.e., Center for Women and public policy) as well as Law study (i.e., Stanford Law School Case Studies Collection).

## MANAGING CASE-BASED LEARNING

### Problems Computer-Supported Case Studies Face

Although Computer supported case-based learning is promising, recent research and practice in CSCL suggest that more work needs to be done (e.g., Carroll & Rosson 2005b; Herrington, Oliver and Reeves 2003; Herrington and Oliver 2000; Kreijns, Kirschner, and Jochems 2003; Kreijns, Kirschner, Jochems, and van Buuren 2007; Reeves, Herrington and Oliver 2004; Stein, Isaacs & Andrews 2004; Xiao, Carroll & Rosson 2007; Grudin, 1988; 1994). Grudin (1988; 1994) highlighted some challenges haunting Computer Supported Collaborative Work (CSCW) systems, such as the difficulty in developing a critical mass of users and resources for the system to achieve a sustainable state; lack of incentive for (potential) users to contribute. Case libraries also suffer from these problems. In current practice, cases are usually created or collected by teachers who adopt this approach, but resources accessible to them are very limited, so case collections are hard to grow. To overcome this, case libraries should ease the way of resource accumulation.

In distance education, a disadvantage, compared with locally organized classroom teaching, is that the instructors and students are distributed in terms of location and also may be distributed in time; in this sense, smooth coordination is under threat. The channel capacity or media richness of communication channels is also impaired to a considerable degree. In locally organized situations, teachers have more control in teaching. Teachers can more closely monitor students' engagement, reactions and attention in the classroom; however, in online situations, this ability to monitor is reduced. These are common problems faced by CSCL systems.

Also, learning actually extends beyond classroom usually. For example, besides in-class activities, students in individual or in teams often have assignments outside of the class. Thus, students and teachers need to access course materials or communicate, when they are not in the class. Information technologies, such as the Internet and database technology, have the potential to connect resource and people distributed and keep information over time. However, in current available online case collections, the power of information technology is not sufficiently exploited. Online case collections are presented to users as static content without supporting further interactions. Users of these sites can only read case materials or download case documents, in forms of word documents, PDFs and HTML, etc., and then reflect on those materials locally. As such, learning largely takes place within individual learners. In longer-term case study activities, weeks or even the whole semester, students not only reflect on existing materials, but also make products they create. The static nature of these case collections makes case libraries hard to support those authentic activities that are crucial to learning.

Communication among users (instructors and learners) is blocked in most current implementations of case collections. Researchers pointed out recently that social interaction is not well supported in CSCL systems. Kreijns et al (2003; 2007) argued that current CSCL systems have failed to support social interaction, compared with functional and educational support, and two pitfalls contribute to this failure. First, people assume that social interaction will always happen or be meaningful, as long as communication happens. However, this is not always true, because communication in real social world is far beyond information exchange. Second, many researchers restrict social interaction to cognitive processes, ignoring other dimensions of social interaction.

To manage case-based learning, we focus on improving learning/educational goals both in individual and group learners and in building a case-based learning platform that serves education in a long run. First, case study libraries should support authentic materials and activities; second, as collective resource and learning platforms, case study libraries should support social interaction among users; third, case study libraries should allow case material accumulation and update through convenient ways; and fourth, case study libraries should develop towards communities of practice. These requirements or concerns are valuable across disciplines. This by no means communicates a context-free point of view. Activities and materials in case studies are domain-specific. However, common requirements on educational systems can be valid and valuable, only designed and implemented in different ways. Importantly, the first requirement that we champion urges authenticity of case study, which targets right on the domain-specific nature of case study.

## Four Requirements for Interactive Case Study Libraries

*Authenticity:* As a form of authentic learning, case studies engage students with real-world problems and context. A digital case library should provide authentic materials. In higher education or professional education, case studies become very complex, since the real world problems are usually complex. For example, as we will describe later on, the case study in usability engineering can last a semester long and consist six phases of software engineering and about 20 types of activities. In this senior undergraduate course, the case studies are mixed with both reflecting on the existing materials and working on real-world usability engineering projects. In these case studies, the students will go out to the users' sites, collect field data by taking field notes, interviewing real users, and evaluate their final products; after they finish the whole usability engineering process, they also create cases. These real-world activities and processes they go through are part authentic learning.

To make case study more useful for gaining problem-solving and other skills in students, authentic activities must be supported in case studies (Carroll & Rosson 2005b). Authentic activities, such as problem-solving and role playing, are not within the materials per se. Practicing those activities will help students gain such ability. To fully exploit the potential of both authentic learning and advanced information technology, case study libraries should focus on authenticity of case materials and activities.

*Social Interaction:* Social interaction, both as a means and an end toward educational goals, is largely ignored in CSCL systems (Kreijns et al 2003; 2007). Social interaction comprises an essential part of authentic activities, especially for collaborative work and human development (Cole 1996; Vygotsky 1978; Wertsch 1985; 1991). One goal of education, especially in professional education, is to help students engage in community of practice.

Social interactions are very critical for higher mental functions development. As many developmental psychologists (Vygotsky 1978; Wertsch 1985; 1991) and anthropologists (Tomasello 1999; 2005) argued, all human higher-order mental functions have their social origins, because all of them develop first on social plane. From this point of view, social interactions are essentially and intrinsically authentic activities at a very basic level. For Vygotsky (1978), higher level mental functions develop at inter-psychological (social) level first, and then by participating in social engagements, individuals internalize those higher mental functions and contents into intra-psychological (individual) plane. The example of child pointing (Vygotsky 1978) shows that the behavior of pointing is symbolized and signified socially, and later becomes internalized as a meaningful symbol of intentional communication in an individual kid. In education research, social interaction is seen as a very important resource and platform for learning (Johnson, Johnson & Stanne 1985). For example, during argumentation, students can exchange ideas and perspectives; these activities can clarify misunderstandings, consolidate knowledge and think things standing in others' shoes.

As educational systems, case study libraries should provide means and encourage learners engaging with each other. Users in a system should be able to exchange ideas, share information, criticize and defense ideas, and collaborate to achieve shared goals. As Kreijins and colleagues (2003; 2007) pointed out, simple communication does not necessarily lead to social interaction. To us, social interaction is more than information exchange; rather, social interaction presumes intersubjectivity (Vygotsky 1978, 1986; Wertsch 1991), construction of common ground (Clark & Brennan 1991; Horton & Keysar, 1996), and joint attention and action (Tomasello 1999; 2005), and so on. During social interactions, such as in argument and counterargument, students are exposed to points of view different from their own. If a student is aware of these different views reflectively, it should help the student think in comprehensive way, and by knowing others' perspectives the student can also consolidate one's own arguments (Baker, Quignard, Lund & Séjourné 2003). We can see this happening in locally organized case study. Students are divided in groups, and asked to provide different solution to an issue, or asked to criticize solution from other groups. A case study library, if it can support these social interactions, will exert more power than letting individual students reflect on case materials alone.

*Community of practice:* The concept of Communities of Practice (COP) has been developed since late 1980s, and applied to many domains: management and organizational study, knowledge management, profession development, etc. When Lave and Wenger (1991; Wenger 1999) conceptualized situated learning and the legitimate peripheral participations, they were talking about a learning process engaging both learning in its narrow sense and doing in practice, both individual and the community at large, and both knowledge of local activities and the social-historical cultural knowledge of the community. Inline with them, we suggest case libraries should develop towards and support communities of practice. Case studies are domain-specific, because they reflect knowledge and activities characterizing those disciplines. During case studies, students are supposed to exercise domain related knowledge and skills valued by the domain they belong, to solve the problems identified by communities.

One of the educational goals in higher education is to help students get to know the communities they are about to enter and prepare them with certain knowledge and skills. One of the core ideas of community of practice is that with practice and learning from others, by explicit instructions from others and/or watching others' doing, learners can move from peripheral participation to more central one. Based on this simple but insightful idea, a good case library should be able to allow and help users to complete this transformation. A learning management system should support this process.

*Resource Accumulation and update:* Systems that support collaboration need a critical mass both in terms of users and resource they carry (Grudin, 1988; 1994). A case study library needs to develop case materials. Although the success of educational digital libraries ultimately depends on users contributing content, current case study libraries or websites do not provide effective incentives and means to support users authoring content via standard schemas so that content can be retrieved effectively (Marlino, Sumner, Fulker,

Manduca and Mogk 2001). This is a version of the "tragedy of the commons" in which resources that benefit most people are given little care. In order to overcome this problem, both the resources and the users of digital libraries should reach critical mass. Furthermore, a case study library should make case contents reflect domain's variety, changes and development in a timely fashion. Case studies facilitate students learning contextualized knowledge, and exercise knowledge *in situ*, so cases representing different situations are of value for learners. All these issues require handy input methods to allow accumulating and updating case materials.

This is an issue of resource management. There is a significant difference between case study libraries and those of other kinds. Libraries like university libraries or other commercial digital libraries have central control over the source of digital contents. Usually, appointed departments and staff, such as acquisition departments, collect those digital contents. However, on the contrary, cases usually come from distributed practice and depend on the very persons who are willing to contribute cases. For example, in many schools, instructors create or collect cases by themselves. In this manner, developing and sharing cases is naturally limited. It also reveals distributed nature of case authoring. This will demand a case study library flexible ways of authoring, and corresponding controls.

## AN EXAMPLE: MANAGING USABILITY CASE STUDY (UCS) WITH USABILITY CASE LIBRARY

In this section, we present a usability engineering case study library we are developing and using in undergraduate usability engineering education at our institution. We had a digital case collection on usability engineering cases starting in 2001. It is a system hosting static hypertext, which consists of six cases. All cases are real-world projects we collected from our collaborators.

The difficulty and complexity of software system development have been realized two decades ago. Software engineering is essentially a social process in which different social groups interact with one another, and different values negotiate and compromise (Curtis, Krasner & Iscoe 1988). Without engaging with real-world problems and activities, students are unable to master knowledge and skills required for their future careers. To address this need, we have used a case-based approach in our usability engineering education (Rosson, Carroll & Rodi 2004). We also developed the UCS library to support this teaching practice.

These case materials are presented and organized as merely read-only, static hyperlinked text. Students can only read cases, and class activities (discussion, claim analysis, etc) can be organized locally. Also, the case contents are not easy to update, so after years the cases are kept the same as the time the system was deployed. Working and teaching in a fast developing discipline like information science, we feel that cases should be updated to capture the development of the field. For instance, using the mouse as an innovative technology is no longer a good example for students who grew up in an era where graphical user interfaces are everywhere.

Realizing these issues above, in 2007, we initiated a UCS redesign and development process, which leads to an interactive digital library that supports online case authoring and other (collaborative) interactions. In the following sections, we will discuss some of the design features and analyze them with the four requirements we proposed in previous section. We will not go into implementation details of the system. The purpose of section is to see how these features address the four requirements we highlighted above. Also, the features are by no means a comprehensive list or must-present features.

## Design Feature and Conceptual Framework

The functional categories in Table 1 summarize the main functions and design concerns in our UCS redesign, which are major affordances for our educational and learning goals. These functional areas are 1) case schema, based on which we design the activities of usability engineering; 2) digital object with metadata, which provide ability for resource accumulation and update, as well as scalability; 3) distributed case authoring, also a major function to enable increasing case materials and collaboration among case authors; 4) (shared) commenting and tagging, which allows individual and collective learning and knowledge sharing. Besides these features, there are several other supportive functions include users management, versioning, authentication and authorization, searching, and logging. These supportive sub-systems make the UCS manageable and help protect resource from corruption. In the following, we will use the UCS as an example to highlight design space and concerns, and we will spend length on how these designs help to manage case-based learning.

## Supporting Authenticity

The first concern in case-based learning is authenticity. This requires analyzing authentic activities within domain carefully. In our situation, we developed a case schema that captures the processes and activities in usability engineering (Rosson & Carroll, 2002) and use it as a protocol to construct cases. This case schema (Figure 1) captures the flow of usability engineering activities and types of documentation corresponding to them. Basically, the system is suggested to support standard phases in system development process: (1) requirements analysis, (2) activity design, (3) information design, (4) interaction design, (5) documentation design, and (6) evaluation. Each of the phases is further decomposed into a set of focal activities: Requirements analysis and usability testing are decomposed into (a) planning, (b) methods and materials, (c) information gathering and (d) synthesis; the four design phases are decomposed into (a) exploration, (b) envisionment and (c) rationale. Totally, there are about 20 categories within a case, including scenarios, documents and artifacts. The system not only collects and provides realistic materials, but also allow and encourage students to perform authentic activities, such as claim analysis that is essential to usability engineering practice (Moran & Carroll 1994). In this activity, students read scenarios in cases or from scenarios they created, saying a problem scenario that captures multifaceted and multilevel issues need to be addressed; and then the students generate design-related claims, as well as trade-offs associated with corresponding claims. By supporting the case schema and idiosyncratic activities in usability engineering, the system moves beyond authentic materials toward more authentic activities.

*Table 1. UCS features supporting four requirements*

| Features | Description | Requirements |
|---|---|---|
| Case schema | Capturing case study activities, procedures and key contents (documents, scenarios, artifacts, etc) | Authenticity |
| Digital object with metadata | Defining data structure and relationship | Resource accumulation and updating |
| Distributed case authoring | Providing means to allow users to contribute remotely | |
| (Shared) Comments and tagging | Providing channels for users to communicate and share information | Community of Practice<br>Social interaction |

Although we have a predefined case structure, it does not mean that the usability engineering processes are fixed or take place in a linear fashion. On the contrary, those processes and activities are highly dynamic and complex. A complete process of software/usability engineering contains multiple stages and different activities. As Carroll and Rosson (2005a, 2005b, 2006) suggested, a case can be narrated linearly, but in real-world situation, events or activities described in a case took place much more dynamically. The activities do not necessarily happen linearly; instead, those activities can take place in parallel, and some activities can repeat over time. For example, usability evaluation can take place throughout the whole usability engineering process, and based on the results of usability evaluation, redesign can happen in other phases.

When users create cases, they will be able to follow the case schema to produce cases with common protocol, so that other users can find materials easily. As we mentioned, the cases hosted in the system come from real-world projects, which keep problems and their context.

Another feature under development is a claims analysis (trade-off analysis) tool, which allows students evaluate design features they have proposed in their usability engineering projects. Claims analysis is a very important and common method used in usability engineering (Moran & Carroll

*Figure 1. Usability engineering process*

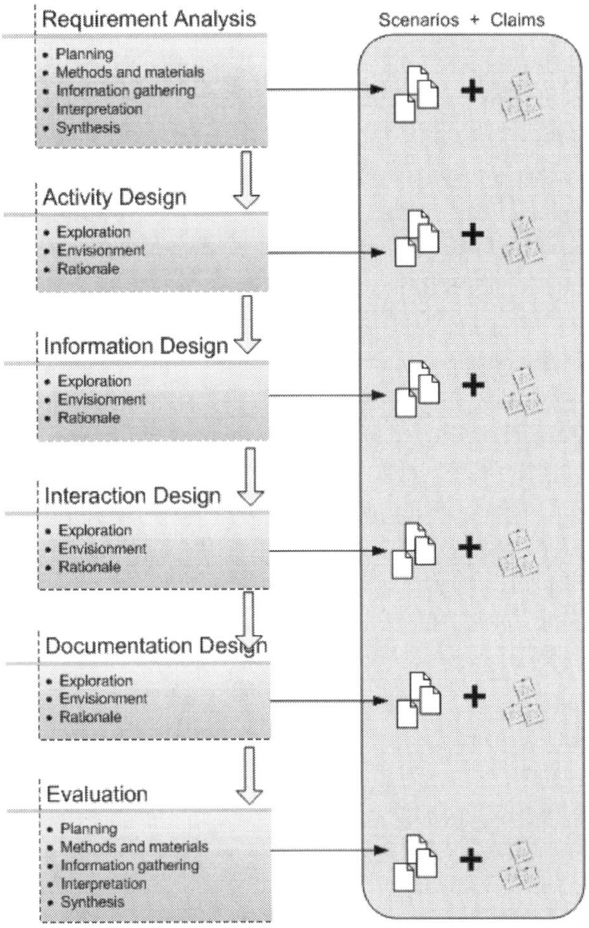

1994). With this tool, students who work on the same project can create claims (design features) and provide pros and cons for each claim. The system also provides statistical information for each claim. For example, the system will calculate how many pros and cons there are for each claim. This information can help student groups evaluate their design more comprehensively. This tool has synchronization mechanisms, so that students can work online at the same time and any new pro or con as well as the statistical information will be shared by the system between users in a synchronous manner.

## Supporting Social Interaction

In the system, we have different ways to support social interactions. For example, the current implementation allows users of the system to post comments on each digital object. Users can contribute and communicate by commenting case contents. Users who have access to a case can make comments on it, providing their thoughts, ideas, suggestions, criticisms, and so on. Other users can review these comments and add their own. Since comments can be shared across users they can promote discussion and convey information during collaboration.

Some other features that engage users socially are under development. For example, we are developing a tagging system that allows users in the system can tag various digital objects and share tags with other users. Tags can help users by providing coded reminders to content or as a way to organize information of interest. In our design, the system will support both shared and private tags. An example of a shared tag might be where an instructor would want to set up tags that will be used by students in her class to find the relevant content for a lesson.

We are also trying to provide group discussion function in the system, so that members of a group can exchange information easily. With embedded group discussion tool, the students do not have to manage their group discussion outside the system; when doing group project, delivering messages in a broadcasting manner will out-perform peer-to-peer chat.

## Managing Communities of Practice

In our system, community of practice is not supported by a specific function or feature; instead, supporting community of practice in the domain of usability engineering becomes an overcharging goal of the system and engaging with system and with other users through various features of the system. The core idea of community of practice is to encourage peripheral participants to move to full participation. To achieve this goal, users of the UCS can observe existing cases, which are complete projects. These cases cover all phases and activities of usability engineering, and contain vivid scenarios. From these cases, students can get to know usability engineering processes.

Other than this, the class is also asked to provide their own analysis on those existing cases. During the whole semester, students are also required to put their semester-long projects on the system. In this sense, learners can learn from reading existing materials, and practice on their own.

Communication and social interactions are also important to build a healthy community of practice. In the system, we provide information about social presence of users. For example, authors of cases will be credited as authors. Everyone who provides claims or trade-off analysis will be credited too. Commenting and shared tagging will help to accumulate knowledge and provide different means for social interactions.

## Supporting Resource Accumulation and Updating

Online case authoring allows users to contribute cases material to the library through the Internet. With this function, users (e.g., instructors) can contribute remotely if they are authorized users with a certain level of access. Many case collections we mentioned earlier do not have convenient authoring method available to wider community members. This will limit the number of people who can contribute case content. With proper access control, remote contribution can aid growth of the case collection. By this mechanism, case authors from different institutions and educational sites can contribute to the library.

Users with proper authorization can create new case materials, edit existing content, and delete objects. All these actions are secured by the authorization subsystem. A related function is version control, which gives users the ability to retrieve previous versions of a digital object from its history. During usability engineering activities (interviewing, on-site note taking, video taping, etc), multimedia data may be collected. In order to fully capture and support authentic materials, the UCS system support digital objects in different media formats, such as video, audio, rich text format, etc. In our current case collection, all cases contain photos taken on-site or pictures of sketches. When authoring cases, authors can upload ready-made files to the case library and refer to those uploaded objects by their URLs.

Allowing online case authoring not only brings more case resource, but also provides a means to increase the sense of community (McMillan 1996; Rovai 2002). Allowing users to create contents is a step toward full participation and engagement within the community, because in our situation, creating a case covers all activities we described in the case schema. In the UCS, we also capture the information such as contributor, which will give case authors visibility and credits.

Figure 2 highlights the data structure we used in the system implementation. A case is package of many documents, scenarios, artifacts and claims and pros and cons. They are relational data. The UCS needs to maintain relationships among data. Also related to this data structure is comment and tag information, which associate with various case materials. For example, users can comment on an

*Figure 2. Data relations of UCS*

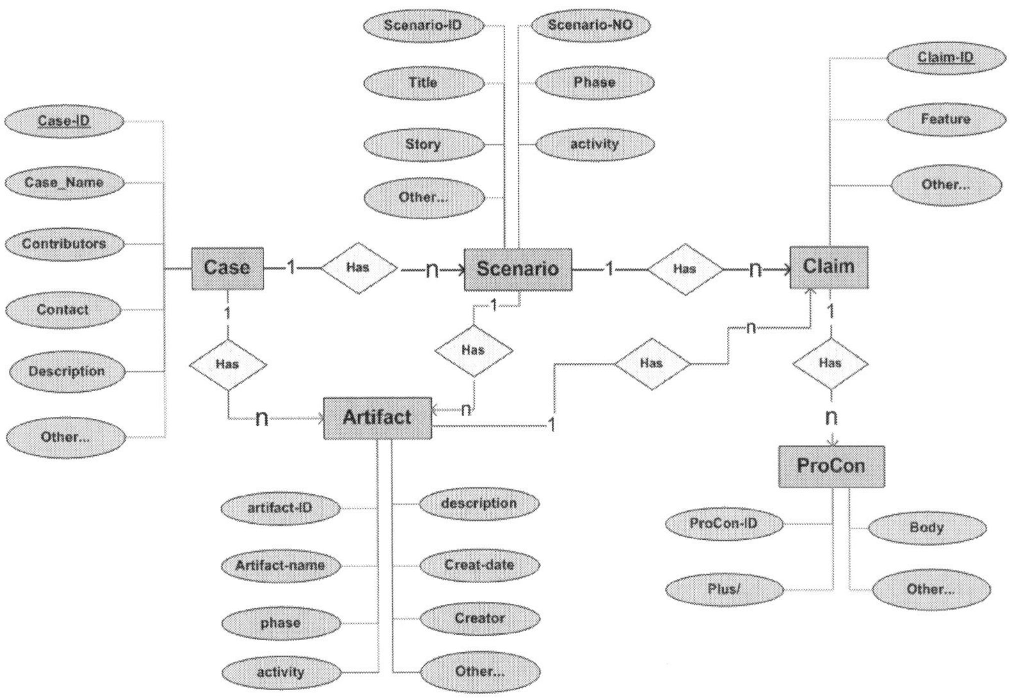

object, saying a scenario, and share that comment with other users. Because of limited space, we will not elaborate on the data structure for the UCS. Different data structures and technologies can be used in different situations, but a common goal is to make the repository extendable and scalable.

## Other Functions

Besides those major functions mentioned, the system has other administrative functions as well, which includes authentication and authorization sub-system, versioning system, and RSS, and user profile management. These sub-systems help the whole system to operate properly or improve usability. Authentication/authorization system secures resources on the UCS library, keeping them from unauthorized access. Versioning system keeps update information of cases, and allow data recovery when data corruption happens. RSS enable uses to subscribe update information of case materials. When students work in team, this awareness feature is very helpful for their collaboration. User profile management allows users to update their profile.

## Implementation and Use in Practice

We want to provide users a light-weight client, so the system follows a server-browser infrastructure. In our implementation, we use FEDORA (http://www.fedora.info) (version 2.2.1) as the digital object repository to store and manage case contents. To organize those contents in data bases, we define meta-information and relational information among digital objects. All services are implemented with java servlets and Java Sever Pages (JSP), and use Ajax and Javascript to program interactive functions on the web.

Figure 3 shows the architecture of the UCS we developed. Users can use standard web browsers to access the system. Figure 4 shows a screen shot of the interactive UCS as currently implemented. The user-interface of the system is mainly written with Java Server Page to layout data. We found that Ajax is very useful in developing interactive

*Figure 3. High level architecture of UCS*

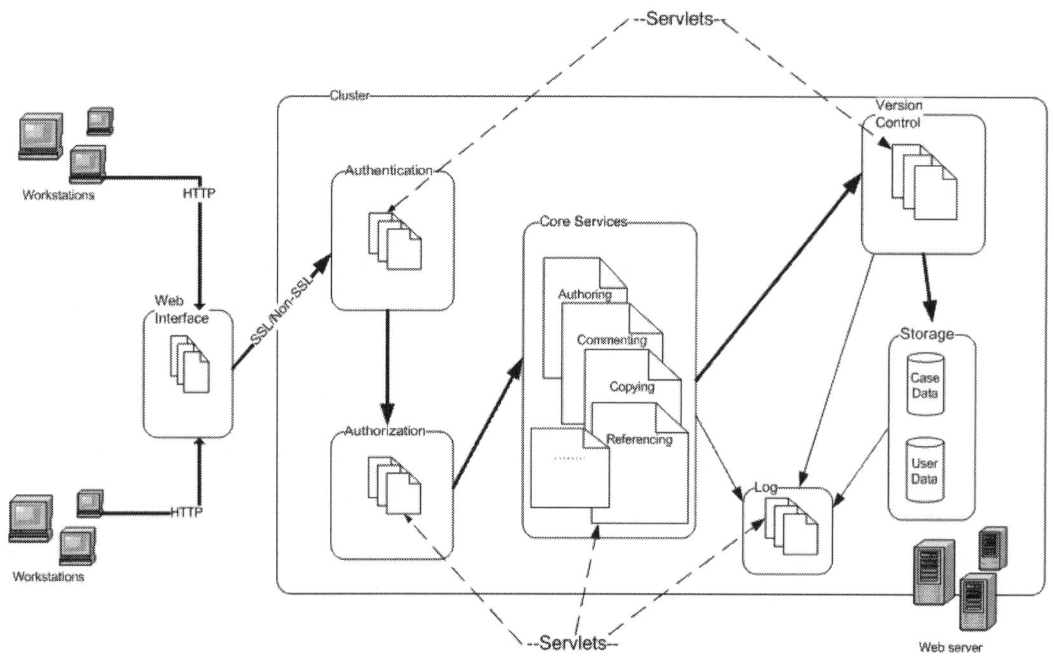

web-systems, since it provides on-demand data fetching ability, without refreshing the whole webpage. For example, when a user click *comments* button, an Ajax call will be made to communicate with Servlet on the server and the webpage will only update comment information.

With the basic functions implemented, we deployed the system on an Internet-accessible server. With the deployed system, we are using it in one of the undergraduate courses in spring 2008 and spring 2009, to support usability engineering education. The size of the class is about 45 to 50. Along with other activities and readings, we offer lectures each week on certain topic of usability engineering. Each week the instructor selects materials corresponding to the topic of that week from the case collection and the students read them and reflect on them. For example, in the week we introduce usability evaluation, the students go through the evaluation section in different cases (e.g., Garden.com, Tapped In, Phone Writer, PAWS, m-Banking, Virtual Science Fair), and they are asked to comment on the approaches used in those cases and come up with ideas and approaches of their own when they are in those situations. The students are working basically in teams and they reflect on the materials and produce course deliverables collectively.

The students also worked on team projects, which are real-world projects. Usually, these projects are applications coming from real-world context: commercial companies, non-profit organization, etc, and they are already put into use or on testing. We asked students to work with those organizations and discover possible usability improvements, redesign and evaluate them. The students post their semester-long projects on the UCS, so those projects become valuable resource for cases. In 2008 spring, the student teams, totally seven, contributed seven cases to

*Figure 4. A screenshot of UCS*

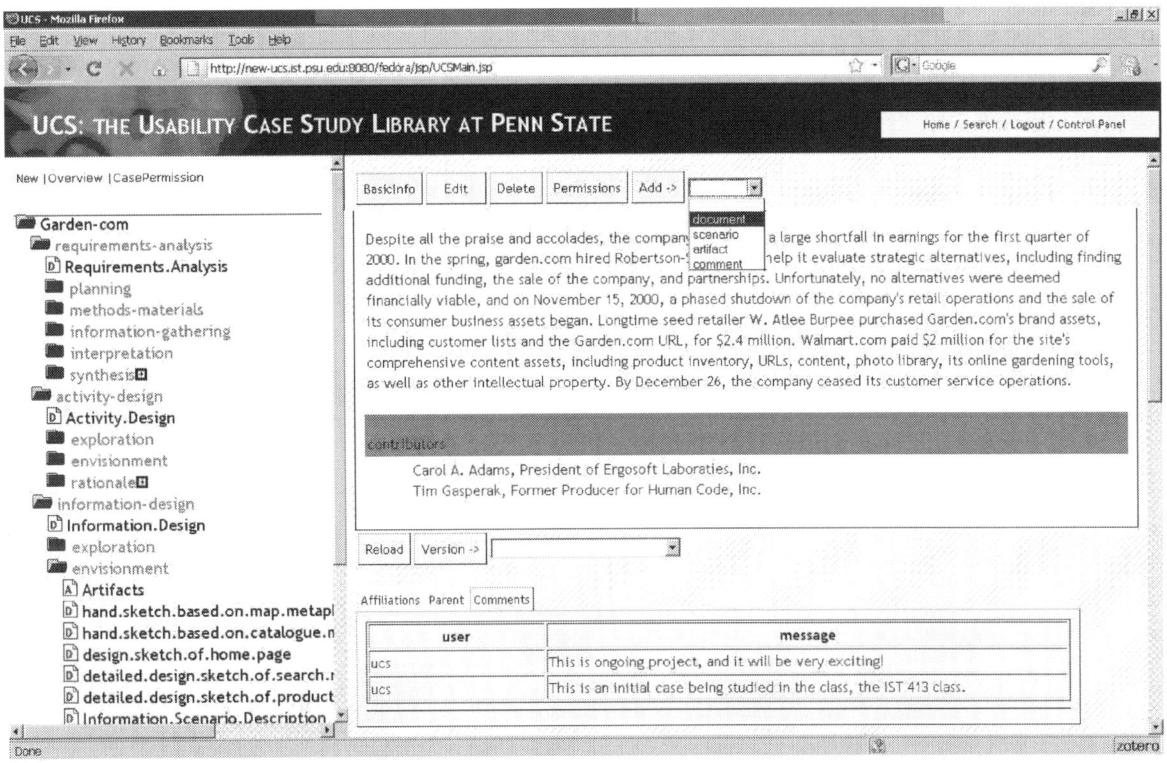

the UCS library. In 2009 spring, the total cases contributed are eight.

## DISCUSSIONS

### A Synthetic View

We have proposed four touchstones for interactive case study libraries at the outset: *authenticity, social interaction, managing community of practice, and resource accumulation*; and introduced our practice with the UCS we have been developing. These four requirements can be employed to guide case study library design as well as evaluate existing systems. They can actually be extended to guide other types of CSCL systems or learning management systems. These requirements do not depend on case studies, but they are very critical for case studies to be effective. For different educational systems, the significance of each of these four requirements may vary, but they are beneficial when taken into account. Although we elaborated them separately, they are interrelated and contribute to one another.

The idea of communities of practice suggests that learning cannot be seen as an individual activity; instead, most learning takes place in social interaction, and knowledge only makes sense in communities of practice. From a Vygotskian perspective, origins of any human higher mental function are social or quasi-social (Vygotsky 1978). Lave and Wenger (1991) elaborated the co-development of individual learners and a community as a whole, when learners move from peripheral to full participation in a community of practice. Looking into the idea of communities of practice more deeply, we found it touches upon a wider range of concepts and thoughts we are moving toward, such as activities, practice, experiencing authentic material, social interactions and rules, etc., so in following discussion, we will revisit the four requirements from the perspective of CoP.

All modern disciplines, both of academia and industry, contain very complex knowledge and activities. Since one of the educational goals of case study library is to help students get familiar with the knowledge and practice of the community they are heading for, a digital case library should be able to present knowledge and processes of that community and support different level of participation, and encourage learners move from peripheral to full participation. That is the core idea of CoP. In the UCS example, the system allows a user to observe others' practice, by reading cases and others' activities; a user can expose oneself to the community by participating in discussion; advanced users can also participate in activities such as creating cases or claim analysis.

Carroll and Rosson (2005a; 2005b; 2006) direct our attention to authentic activities and suggested, for example, that role-playing, decision-making and problem-solving are important authentic activities that should be supported in case-based learning. Since students are supposed to learn knowledge and skills required by specific professional communities through case studies, authenticity becomes a requirement for case studies. The materials and activities a case provides should represent the characteristics of practice in a community. In other words, when we assess case studies and case libraries, we should ask the question that if these systems reflect problems, knowledge and activities that are of value to domain needs. As we argued, a large part of authentic activities is idiosyncratic to communities of practice. For example, building a piece of software and solving a juristic case have very different activities involved. To make case study libraries useful for learners, designers and educational practitioners should explore local activities within domains. In our example, the usability case schema captures activities in usability engineering practice. Design rationale (Moran & Carroll 1994) is one of the core concepts in usability engineering, and giving design rationale in design is a typical activity in this domain. The UCS library encourages and supports students to practice this activity. Claim analysis supported by the UCS provides opportunities for students to exercise practical reasoning with established knowledge and theories and to produce solutions to questions presented in cases. Figure 5 shows an example of claim analysis from one of our cases. In this example, the claim, "formatting product information based on a seed packet metaphor…," is a design element for the website Garden.com. The analysis of this design shows in *Pros and Cons*. Trained with this type of activity, the students will develop their skill of providing design rationale.

With respect to authentic learning, moving from peripheral to center of community of practice requires case libraries to support different level of authentic activities. For example, in many CSCL systems, authentic learning supported is only appropriate for either entry level or advanced level target learners, the large audiences who are in the middle are largely ignored (Means & Olson, 1994). Kollar, Fischer and Slotta (2005) reported that a CSCL system, a system that scaffolds students' argument with scripts, improves students' cognitive ability in general; they also found that the system is useful in varied degree for students with different levels of cognitive ability. The implication is that when designing case study libraries, designers and educators should identify different activities with different level of complexity or difficulties. For example, to an advanced architect, designing a kitchen alone is less authentic than designing a kitchen within a house, because the former situation is not likely to happen in real-world and the latter will impose more constraints and/or opportunities for design. However, designing a solo kitchen is also valuable authentic activity for novice learners. Identifying and designing proper level of activities that fit audiences at different level is one we should concern. An idea we can draw from developmental psychology is the concept of zone of proximal development (Vygotsky 1978), which explains the potential development brought by the joint forces of actual development and help from more competent peers or tutors.

As oppose to local activities bound in communities of practice, another type of authentic activities is fundamental and pervasive. That is social interaction. It is fundamental in that social interaction, as many researchers (Cole 1996; Vygotsky 1962, 1978; Wertsch 1985, 1991) argued, is where human high cognitive ability origin and where collaboration is made possible; it is pervasive in that social interaction takes place in people's everyday life, both professional and personal. Social interaction can be seen as universal authentic activity. CSCL systems must target on social interactions that go beyond just knowledge representation. We see social interactions as complex processes that relate to building common ground, negotiating values, joint attention and action and so forth. Exposed in social interaction, students not only share knowledge with and learn

*Figure 5. An example of claim analysis*

| claims | Pros[+] & Cons [-] | Scenarios |
|---|---|---|
| Formatting product information based on a seed packet metaphor... | ⊞ information is quick and easy to scan<br>⊞ familiar format for experienced gardeners<br>⊟ but may not readily accommodate non-gardening information that is important (like shipping dates, availability, etc.)<br>⊟ but may not be a suitable format for non-plant products | Susan does some spring shopping George Starts A Rose Garden |

from others, but also develop the ability to think in different perspectives, which is essential for cognitive development (Tomasello 1999; 2005). To make high quality social interaction possible, the system must provide channels and support those underlying mechanisms of social interaction.

Resource is another factor that can determine success of a library. To be valuable for learning, cases should be able to reflect different situations and keep up-to-date to capture emerging characters of learning subjects or practice in a community. This demand is especially conspicuous in fast developing disciplines. In the UCS example, online authoring and related functions (e.g., editing, versioning) are designed to meet this need. In online case collections that do not have convenient means of authoring, the chances of increasing case materials are technically blocked to a great degree. Handy online authoring keeps the door open. In schools, it is usually the instructors who gather case materials. However, it is practically hard for any individual teacher to keep cases up-to-date and enriching case contents. With distributed online authoring functions, users (teachers) from different schools can share their cases and also benefit from others' contribution.

## Future Research Directions

The arguments and features of the UCS we presented in this chapter, as stated, are by no means comprehensive. The requirements we advocate come from a community-of-practice perspective. However, other aspects are also important to be considered and supported.

Besides supporting focal activities learners are supposed to master, learning management system could allow learners to manage and reflect on their own learning. For example, *human development* (Vygotsky, 1978) in learners is a core concept and a fundamental goal that we, as educators, are pursuing. Carroll, Rosson, Convertino and Ganoe (2005) have developed an activity awareness framework, of which human development is one of the emphases. They took a social-cultural-historical point of view on human development and suggested a system should be able to attain and provide users with information that shows the progress of individual users' capabilities and role played through time. In this sense, a user portfolio (Chang, 2001) that contains episodes of learning and history of activities will be very constructive.

Another closely related issue is learners' reflection during learning. Schön (1983) argued that learning professional knowledge requires learners to reflect-in-action. Studies on metacognition (Flavell, 1979; Kuhn, 2000) – a set of cognitive ability that "reflects on monitors, or regulates first-order cognition" (Kuhn, 2000, p. 178) – also suggest reflections often bring about better learning outcomes. Applications have been built to support learners reflecting on their reasoning during learning results (e.g., White, Shimoda, & Frederiksen, 1999; Aleven & Koedinger, 2002). These applications provide cognitive scaffolding for the students to help them develop their reasoning skills. A case study library, when it supports learners' practicing tasks at hand, needs to provide means for the learners to reflect on their actions.

Social interaction we mentioned above is still more toward cognitive aspect. However, besides cognitive aspect of social interaction, emotional perspective is worth exploring. For example, when Nardi (2005) discussed on interpersonal communication between people, she highlighted three concepts other than cognitive and information need. Those aspects are affect, commitment, and affiliation. Study of human emotion is an emerging area in HCI and CSCW, and this perspective touches upon another type of human needs. Emotion has been studied from at least 1970's in sociology and even earlier in psychology, but less investigated in HCI and CSCW. Shott (1979) and other researchers made efforts in studying emotion and social life. Emotions convey information of different kind other than information that caters cognitive needs, and emotional intelligence or competency is crucial for learning and adaptation

(Lopes, Brackett, Nezlek, Schutz, Sellin, and Salovey 2004), although may not directly tie to tasks at hand. Emotional aspects of social interaction not only influence individual learners' experience when they interact with other members, but also play a role in maintaining or crumbling a community. Emotional aspects of social interactions, such as trust, subjective well-being, are both important resource and outcome of social capital, a type of capital for community development (Coleman 1988; Resnick 2002).

In terms of motivation of contribution, although technically users can contribute to case libraries with handy means, a further problem is where their motivation comes from. It leads another intriguing topic, *social capital* (Coleman 1988; Portes 1998). Social capital is resources resided in social relations and generated from social interaction among members of a given community. Social capital is resource one, whether as individual or cooperate actor, can potentially get from participating in a social system. To this point, community members here we are referring to are educators who use cases as teaching resource and can potentially contribute. Our assumption is that communities have higher social capital will tend to have better participation and practice. If case study communities have higher social capital, members (e.g., teachers) of those communities will be more willing to contribute cases. Building social capital in communities is not the remedy for all problems, but it will contribute to community building and maintaining in the long run (Resnick 2002).

A learning-object perspective also provides a lens to investigate learning management systems in general and case study libraries in particular. Learning object is considered as any entity, digital or non-digital, that may be used for learning, education or training (IEEE, 2002). Learning objects are supposed to correspond to particular learning objectives (Boyle, 2003). This principle enables created objects to be shared in easy, standard way; and one to one mapping of learning objective to learning object also makes learning objects communicate clear learning goals to the learners. However, a challenging issue is that when education stands at a very high level, such as a level that focuses using knowledge to deal with practical issues, educational/learning goals are too abstract to specify, and therefore it is hard or even impossible to map learning objectives to learning objects. Further research is in need to explore how to create cases as reusable as well as meaningful learning objects.

## CONCLUSION

In this chapter we introduce interactive case study libraries and analyze four aspects to be taken into consideration during design phase. Case study libraries are different from general-purpose learning management systems, since they are domain-specific and provide managerial services (e.g., resource management) as well as content (authentic materials and activities). This type of learning management system uses information technology to assist learners with activities important to local communities, which is usually hard or impossible to achieve with popular general-purpose learning management systems like Blackboard and ANGEL. While a general-purpose learning management system targets pedagogical activities, such as managing learning progress, managing course notes and lecture slides, a case study library aims at expertise and activities that are very domain-specific (such as the UCS) help students to acquire skills and gain knowledge in usability engineering profession.

We take a wider view on what should be managed in educational systems like case study libraries. In this view, not only the materials but also the activities of interest should be managed. In the UCS case, the case schema is used to capture the core activities in usability engineering practice and provide guidance of how a case of usability engineering should be presented and what activities are critical for students to master.

The case-base learning platform should be able to support these activities and, in this way, to bridge the gap between knowledge taught and skills required in real-world practice.

## REFERENCES

Aleven, V., & Koedinger, K. R. (2002). An effective meta-cognitive strategy: Learning by doing and explaining with a computer-based Cognitive Tutor. *Cognitive Science*, *26*, 147–179.

Baker, M. J., Quignard, M., Lund, K., & Séjourné, A. (2003). Computer-Supported Collaborative Learning in the Space of Debate. In Wasson, B., Ludvigsen, S., & Hoppe, U. (Eds.), *Designing for Change in Networked Learning Environments*. Dordrecht, The Netherlands: Kluwer Academic.

Boyle, T. (2003). Design principles for authoring dynamic, reusable learning objects. *Australian Journal of Educational Technology*, *19*(1), 46–58.

Carroll, J. M., Ganoe, C. H., & Jiang, H. (2008). Introducing the digital case library. In Pagani, M. (Ed.), *Encyclopedia of Multimedia Technology and Networking* (2nd ed., pp. 782–788). Hershey, PA: IGI Publishing.

Carroll, J. M., & Rosson, M. B. (2005a). A Case Library for Teaching Usability Engineering: Design Rationale, Development, and Classroom Experience. *ACM Journal on Educational Resources in Computing*, *5*(1), 1–22. doi:10.1145/1101670.1101673

Carroll, J. M., & Rosson, M. B. (2005b). Toward even more authentic case-based learning. *Educational Technology*, *45*(6), 5–11.

Carroll, J. M., & Rosson, M. B. (2006). Case Studies as Minimalist Information. *IEEE Transactions on Professional Communication*, *49*(4), 297–310. doi:10.1109/TPC.2006.885836

*Center for Women and public policy.* (n.d.). Retrieved April 19, 2008, from http://www.hhh.umn.edu/centers/wpp/case_studies.html

Chang, C. (2001). A study on the evaluation and effectiveness analysis of web-based learning portfolio (WBLP). *British Journal of Educational Technology*, *32*(4), 435–458. doi:10.1111/1467-8535.00212

Clark, H. H., & Brennan, S. E. (1991). Grounding in Communication. In Resnick, L., Levine, J., & Teasley, S. (Eds.), *Perspectives on Socially Shared Cognition* (pp. 127–149). Hyattsville, MD: American Psychological Association. doi:10.1037/10096-006

Cole, M. (1996). *Cultural psychology: A once and future discipline*. Cambridge, MA: Belknap/Harvard.

Coleman, J. S. (1988). Social capital in the creation of human capital. *American Journal of Sociology*, *94*, S95–S121. doi:10.1086/228943

Curtis, B., Krasner, H., & Iscoe, N. (1988). A Field Study of the Software Design Process for Large Systems. *Communications of the ACM*, *31*(11). doi:10.1145/50087.50089

Dewey, J. (1933). *How We Think, A Restatement of the Relation of Reflective Thinking to the Educative process*. New York, D. C.: Heath.

Dietrich, S. W., & Urban, S. D. (1996). Database theory in practice: learning from cooperative group projects. In *ACM SIGCSE Bulletin: Proceedings of the twenty-seventh SIGCSE technical symposium on Computer science education, 28*(1), 112-116.

*FEDORA*. (n.d.). Retrieved April 19, 2008, from http://www.fedora.info

Flavell, J. H. (1979). Metacognition and cognitive monitoring: A new area of cognitive-developmental inquiry. *The American Psychologist*, *34*, 906–911. doi:10.1037/0003-066X.34.10.906

Galusha, J. M. (1997). Barriers to Learning in Distance Education. *Interpersonal Computing & Technology*, *5*(3-4), 6–14.

Ganoe, C., Somervell, J., Neale, D., Isenhour, P., Carroll, J. M., Rosson, M. B., & McCrickard, D. S. (2003). Classroom BRIDGE: Using Collaborative Public and Desktop Timelines to Support Activity Awareness. In *Proceedings of the ACM Conference on User Interface Software and Technology* (UIST '03), Vancouver, BC, Canada.

Garvin, D. A. (2003, September). Making the case. *Harvard Magazine*, *106*(1), 56–65.

Grudin, J. (1988). Why CSCW applications fail: problems in the design and evaluation of organization of organizational interfaces. In *Proceedings of the 1988 ACM conference on Computer-supported cooperative work*, Portland, OR, United States. New York: ACM Press.

Grudin, J. (1994). Groupware and social dynamics: eight challenges for developers. *Communications of the ACM*, *37*(1), 92–105. doi:10.1145/175222.175230

Hartfield, B., Winograd, T., & Bennett, J. (1992). Learning HCI design: mentoring project groups in a course on human-computer interaction. In *ACM SIGCSE Bulletin*: *Proceedings of the twenty-third SIGCSE technical symposium on Computer science education, 24*(1), 246-251.

Hayes, J. H., Lethbridge, T. C., & Port, D. (2003). Papers on software engineering education and training: course delivery and evaluation: Evaluating individual contribution toward group software engineering projects. In *IEEE Proceedings of the 25th International Conference on Software Engineering* (pp. 622-627).

Herreid, C. F., & Schiller, N. A. (2005). *The National Center for Case Study Teaching in Science*. State University of New York at Buffalo. Retrieved from http://ublib.buffalo.edu/libraries/projects/cases/ubcase.htm

Herrington, J., & Oliver, R. (2000). An instructional design framework for authentic learning environments. *Educational Technology Research and Development*, *48*(3), 23–48. doi:10.1007/BF02319856

Herrington, J., Oliver, R., & Reeves, T. C. (2003). Patterns of engagement in authentic online learning environments. *Australian Journal of Educational Technology*, *19*(1), 59–71.

Horton, W. S., & Keysar, B. (1996). When do speakers take into account common ground? *Cognition*, *59*(1), 91–117. doi:10.1016/0010-0277(96)81418-1

Johnson, R. T., Johnson, D. W., & Stanne, M. B. (1985). Effects of cooperative, competitive, and individualistic goal structures on computer-assisted instruction. *Journal of Educational Psychology*, *77*(6), 668–677. doi:10.1037/0022-0663.77.6.668

Kollar, I., Fischer, F., & Slotta, J. D. (2005). Internal and external collaboration scripts in web-based science learning at schools. In *Proceedings of CSCL* 2005 (pp. 331-340).

Kreijns, K., Kirschner, P. A., & Jochems, W. (2003). Identifying the pitfalls for social interaction in computer-supported collaborative learning environments: A review of the research. *Computers in Human Behavior*, *19*(3), 335–353. doi:10.1016/S0747-5632(02)00057-2

Kreijns, K., Kirschner, P. A., Jochems, W., & van Buuren, H. (2007). Measuring perceived sociability of computer-supported collaborative learning environments. *Computers & Education*, *49*(2), 176–192. doi:10.1016/j.compedu.2005.05.004

Kuhn, D. (2000). Metacognitive development. *Current Directions in Psychological Science*, *9*(5), 178–181. doi:10.1111/1467-8721.00088

Lamancusa, J. J., Jorgensen, J. E., Zayas-Castro, J. L., & de Ramirez, L. M. (2001). The learning factory - integrating design, manufacturing and business realities into engineering curricula. *International Conference on Engineering Education,* August 6-10, Oslo, Norway (pp. 1-6).

Lave, J., & Wenger, E. (1991). *Situated Learning*. Cambridge, UK: Cambridge University Press.

Lopes, P. N., Brackett, M. A., Nezlek, J. B., Schutz, A., Sellin, I., & Salovey, P. (2004). Emotional intelligence and social interaction. *Personality and Social Psychology Bulletin, 30*, 1018–1034. doi:10.1177/0146167204264762

Marlino, M., Sumner, T., Fulker, D., Manduca, C., & Mogk, D. (2001). The digital library for Earth system education: building community, building the library. *Communications of the ACM, 44*(5), 80–81. doi:10.1145/374308.374356

McMillan, D. (1996). Sense of community. *Journal of Community Psychology, 24*, 315–325. doi:10.1002/(SICI)1520-6629(199610)24:4<315::AID-JCOP2>3.0.CO;2-T

Means, B., & Olson, K. (1994 April). The Link Between Technology and Authentic Learning. *Educational Leadership*.

Moran, T. P., & Carroll, J. M. (Eds.). (1994). *Design Rationale: Concepts, Techniques, and Use*. Hillsdale, NJ: Lawrence Erlbaum Associates.

Nardi, B. (2005). Beyond bandwidth: Dimensions of connection in interpersonal communication. *CSCW, 14*, 91–130.

*National Center for Case Study Teaching in Science*. (n.d.). Retrieved April 19, 2008, from http://ublib.buffalo.edu/libraries/projects/cases/case.html

NSF DUE has supported many projects in the area of computer professionalism and ethics. Examples are the ImpactCS Project (http://www.seas.gwu.edu/~impactcs/) and Dolce (Developing Online Computer Ethics, http://computingcases.org/). Retrieved April 19, 2008.

Piaget, J. (1964). Development and Learning. *Journal of Research in Science Teaching, 2*, 176–186. doi:10.1002/tea.3660020306

Portes, A. (1998). Social Capital: Its Origins and Applications in Modern Sociology. *Annual Review of Sociology*, *22*, 1–24. doi:10.1146/annurev.soc.24.1.1

Reeves, T. C., Herrington, J., & Oliver, R. (2004). A development research agenda for online collaborative learning. *Educational Technology Research and Development*, *52*(4), 53–65. doi:10.1007/BF02504718

Resnick, P. (2002). Beyond Bowling Together: Socio-technical Capital. In Carroll, J. M. (Ed.), *HCI in the New Millennium* (pp. 247–272).

Rosson, M. B., & Carroll, J. M. (2002). *Usability Engineering: Scenario-based Development of Human-Computer Interaction*. San Francisco: Morgan Kaufmann.

Rosson, M. B., Carroll, J. M., & Rodi, C. (2004). Case Studies for Teaching Usability Engineering. *Proceedings of 35th SIGCSE Technical Symposium on Computer Science Education,* Norfolk, VA, March 3-7 (pp. 36-40). New York: ACM Press.

Rovai, A. P. (2002). Building a Sense of Community at a Distance. *International Review of Research in Open and Distance Learning (On-Line), 3*(1). Retrieved April 19, 2008, from http://www.irrodl.org/content/v3.1/rovai.html

Schön, D. A. (1983). *The Reflective Practitioner: How Professionals Think in Action*. New York: Basic Books.

Shott, S. (1979). Emotion and social life: A symbolic interactionist analysis. *American Journal of Sociology*, *84*, 1317–1334. doi:10.1086/226936

Stahl, G., Koschmann, T., & Suthers, D. (2006). Computer-supported collaborative learning: An historical perspective. In Sawyer, R. K. (Ed.), *Cambridge handbook of the learning sciences* (pp. 409–426). Cambridge, UK: Cambridge University Press.

Stanford Law School Case Studies Collection. (n.d.). Retrieved April 19, 2008 from http://www.law.stanford.edu/publications/casestudies/

Stein, S. J., Isaacs, G., & Andrews, T. (2004). Incorporating authentic learning experiences within a university course. *Studies in Higher Education*, *29*(2), 239–258. doi:10.1080/0307507042000190813

The Electronic Hallway. (n.d.). *Public affair case study collection*. Retrieved April 19, 2008, from https://hallway.org

Tomasello, M. (1999). *The Cultural Origins of Human Cognition*. Cambridge, MA: Harvard University Press.

Tomasello, M., Carpenter, M., Call, J., Behne, T., & Moll, H. (2005). Understanding and sharing intentions: The origins of cultural cognition. *The Behavioral and Brain Sciences*, *28*(5), 675–691. doi:10.1017/S0140525X05000129

Vygotsky, L. S. (1978). *Mind in society*. Cambridge, MA: Harvard University Press.

Vygotsky, L. S. (1986). *Thought and Language*. Cambridge, MA: MIT Press.

Wenger, E. (1998). *Communities of Practice: Learning, Meaning, and Identity*. Cambridge, UK: Cambridge University Press.

Wertsch, J. (1991). *Voices of the mind: A sociocultural approach to mediated action*. Cambridge, MA: Harvard University Press.

Wertsch, J. V. (1985). *Vygotsky and the social formation of mind*. Cambridge, MA: Harvard University Press.

White, B. Y., Shimoda, T., & Frederiksen, J. (1999). Enabling students to construct theories of collaborative inquiry and reflective learning: Computer support for metacognitive development. *International Journal of Artificial Intelligence in Education*, *10*, 151–182.

Williams, S. M. (1992). Putting Case-Based Instruction Into Context: Examples From Legal and Medical Education. *Journal of the Learning Sciences*, *2*, 367–427. doi:10.1207/s15327809jls0204_2

Xiao, L., Carroll, J. M., & Rosson, M. B. (2007). Support Case-Based Authentic Learning Activities: A Collaborative Case Commenting Tool and a Collaborative Case Builder. *Human-Computer Interaction* (. *LNCS*, *4553*, 371–380.

# Chapter 20
# Experiences and Opinions of Online Learners:
## What Fosters Successful Learning

**Michael F. Beaudoin**
*University of New England, USA*

## ABSTRACT

*The International Board of Standards for Training, Performance and Instruction (IBSTPI) engaged in a research study in 2007-2009 to survey learners enrolled in online education and training. The project attempts to better understand how students engage in learning in online environments, and to identify the critical factors that affect their satisfaction and success. This data is intended to arrive at a set of competencies critical to online learners' success to be disseminated for use by online learners, as well as providers. This chapter reports findings and analyses of a 58 question survey administered online that yielded 318 responses from American, Israeli, Mexican and Japanese cohorts. The study revealed that the majority of respondents rated their satisfaction and success with online learning as positive, and that these outcomes were largely due to self-motivation and time-management, rather than provider resources. The discussion addresses the impact of instructional resources, to ascertain if characteristics of the online environment are a dominant factor in influencing students' attitudes, behaviors, and learning outcomes. Practical application of findings is considered by articulating a preliminary set of online learner competencies.*

## INTRODUCTION

In 2007, the International Board of Standards for Training, Performance and Instruction (IBSTPI) initiated an ambitious new research project (1) to survey learners enrolled in diverse education and training offerings delivered in an online format. This data is intended to assist the research team in developing a set of universally applicable competencies considered to be critical to success in online and blended learning environments, and that could be widely disseminated for use, not only by online learners, but also by providers designing and delivering online education and training. Knowing

DOI: 10.4018/978-1-61520-853-1.ch020

Copyright © 2010, IGI Global. Copying or distributing in print or electronic forms without written permission of IGI Global is prohibited.

what learners find to be the most effective and least useful features and practices for success in online settings can be instructive to those wishing to enhance the teaching-learning process. Our purpose here is not to arrive at new theories of online teaching and learning, but rather to identify and articulate a set of competencies that are both learner-centered and data-driven.

1. Research team members: Gila Kurtz, Jan Visser, Robin Yap, Katsuaki Suzuki, Peter Cookson

The burgeoning phenomenon of online education has generated several volumes in the past decade presenting instructors` techniques for effective online teaching based on their experiences. These "how to" books (e.g., Pallof and Pratt 2001, Talbot, 2007) are making a useful contribution to the literature and to improving practice, as more faculty and students choose this format for teaching and learning. For the most part, these presentations are largely informed by those authors` experiences, presumably after having taught a number of courses, and hopefully having benefited from at least some student feedback. But it is also quite likely that many faculty novitiates to distance education enter into this new learner landscape with little familiarity with the literature on distance education theory and practice.

Other online educators, however, are quite well grounded in distant teaching theory and practice, and often generate observant and thoughtful advice to their peers via papers and presentations that identify challenges and solutions for online learners, with comments to the effect that: *"The biggest challenge is understanding what it takes to be a successful student."* While these seem like reasonable notions, it is not at all clear what these well-intended statements are based on, if such advice is inspired by valid and reliable data, or if these represent little more than opinions derived from instructors' anecdotal material and personal reflections. And, of course, there is the ultimate question of whether or not such data translates into meaningful and useful information for educational providers and consumers to enhance their teaching and learning.

It cannot safely be asserted, at this point, that our best practices to date have benefited from the input of online learners, obtained through systematic data collection and analysis. De la Teja and Spannaus (2008), among others, have noted that lists of hints for effective online study rarely refer the learner to authoritative sources that might give validity to those recommendations. They also comment that most learner behaviors or characteristics that are advocated are generic and can be applied to learning situations that are not necessarily online. With some institutions worldwide now boasting online enrolments well in excess of 100,000 annually, and if, as likely, it will not be too long before the number of online courses enrollments exceeds face-to-face course enrollments, then distance education managers, course designers and instructors can ill afford to expand their practice without acquiring data directly from the learners and interpreting findings based on such data into reliable and valid conclusions regarding what constitutes successful online learning.

This chapter reports key findings, based on a content analysis of the data obtained through the online learner survey intended to address the need described above. It provides information about what works and does not work, what contributes to and what detracts from successful online learning, and what these learners want and need from their online learning experiences. From these data, the study attempts to identify specific online learner competencies; an initial draft statement of competencies is included in this chapter. A final version of these will be disseminated once the project team has validated its work.

Learning environments (LEs) can be altered to enhance student outcomes, but practitioners and researchers must have means to measure the principal features of those LEs before they make changes that will lead to improved effectiveness in such settings. Qualitative observation, inquiry, interviews, ethnography, case studies, among other forms of assessment, have commonly been used to gather information on LEs. Evaluation of distance education experience is typically concentrated primarily on student outcomes, student/faculty attitudes, and satisfaction. However, to bridge the gap between 3rd party observers' and researchers' views, as well as student/teacher perceptions of what goes on in their LE, a less subjective, qualitative and more economical means of measuring LE features must be created through use of suitable survey instruments, based on validated, efficient, and relevant questionnaires that provide perceptions of LEs from stakeholders' perspectives. (Fraser, 1998)

To this end, a project team of IBSTPI board members spent several months developing a survey with 58 questions, including several open-ended questions, allowing the respondents to tell their "stories" as online learners. A pilot test of the instrument was completed in fall 2007 to identify items needing possible revision, based on initial respondent feedback. After some minor modifications, the final survey tool was administered electronically (in English), via Zoomerang in late 2007- early 2008, yielding 86 responses. The questionnaire was also translated into Japanese and administered in early 2008. A total of 38 completed responses were received by late January 2008. With an approximate 75% response rate, the research team determined that this number constituted an adequate sample to initiate the data analysis, with additional results to be integrated into the report as they became available. The survey tool was translated into Spanish and was administered to online learners in Veracruz, Mexico in March 2008, with a total of 105 completed surveys returned for analysis. Also, the survey was translated in Hebrew, a pilot test was done in March, and after some minor modifications, was administered to online learners in Israel in April 2008, with 44 completed surveys returned.

From the instrument's initial distribution (in English), a total of 83 surveys were completed and returned electronically. Sixty-eight of the respondents are from the United States, 5 from Canada, 5 from various Asian countries, and one each from 5 other Western nations. Because all but five of the 83 respondents are from the US and other Western countries, this cohort is referred to as Western, to distinguish it from other cohorts. Japanese, Mexican and Israeli respondents (38, 105 and 44 respectively) are referred to by their national affiliation. The ratio of Western females to males was nearly 5 to 1; for Japanese, the ratio of females to males was about 2 to 3. The most dominant age group of these two respondent groups was between 31-40: (48 respondents), followed by the age group of 41-50: (35 respondents), over 50: (22 respondents), and 21-30: (20 respondents). None were under 20 years of age. Twelve of the Israeli respondents are male, and 32 are female. The most dominant age group among these respondents was between 21-30 (32 respondents) followed by 7 respondents in the age group of 31-40; 2 respondents were under 20 years of age and 1 respondent was in the 41-50 age group. The gender split among the150 Mexican respondents was nearly equal: 48% female, 52% male. The age range for Mexican students was from 18 to 55. Most (88%) are from the state of Veracruz; the other 12% are from 8 different other Mexican states.

Though not initially planned as an element to investigate in this study, the research team realized, once it was able to identify and obtain responses from four different cohorts representing four distinct countries (Japan, Mexico, Israel, and the United States), that there could be value in adding to its inquiry, the aspect of cultural dynamics in online learning. Rogers (1995) argued that the adoption of technological innovations has often failed because the diffusion process often did

*Table 1. Demographic characteristics of research groups*

| Characteristic | Western respondents | | Japanese respondents | | Israeli respondents | | Mexican respondents | |
|---|---|---|---|---|---|---|---|---|
| | N | % | N | % | N | % | N | % |
| Gender | 73 | 71 | 16 | 42 | 12 | 73 | 72 | 48 |
| Female | 13 | 29 | 22 | 58 | 32 | 27 | 78 | 52 |
| Male | 86 | 100 | 38 | 100 | 44 | 100 | 150* | 100 |
| Total | | | | | | | | |
| Age | - | 16 | - | 16 | 2 | 5 | 7 | 5 |
| Up to 20 | 14 | 39 | 6 | 37 | 32 | 76 | 68 | 46 |
| 21-30 | 34 | 28 | 14 | 29 | 7 | 17 | 52 | 35 |
| 31-40 | 24 | 17 | 11 | 18 | 1 | 2 | 16 | 11 |
| 41-50 | 15 | 100 | 7 | 100 | - | - | 5 | 3 |
| + 51 | 86 | | 38 | | 42 | 100 | 150 | 100 |
| Total | | | | | | | | |

not take into account the cultural beliefs of local communities. This has obvious implications for designing online courses that might be offered across national boundaries to diverse learners who, although they perhaps enrolled in courses in which English is the language of instruction, may possess quite different values, beliefs, and styles of learning (McLoughlin and Oliver, 2000). For example, Gunawardena and LaPointe (2007) identify several aspects of online course environments and activities that could be influenced by cultural variants (e.g., social presence, conflict resolution, group process, participation, help-seeking behavior). With this in mind, the team determined that its data analysis (though not included in this report) should eventually examine the possibility of differing cultural ways of learning and their effect on learner behavior and learning outcomes, as well as relevant competencies that may apply in such settings.

In retrospect, the research team recognizes that perhaps the first administration of the survey should have been limited to respondents only in the US, to more closely match the other cohorts, which represent specific countries. The initial group included only a small number of students (less than 10%) living and studying in some half dozen other countries. Encompassing a wider sample of online learners representing other regions (e.g., Europe) was a desired goal, but not feasible once it was realized how much additional time was required to accommodate just three additional cohorts from Japan, Mexico and Israel. While the research team would have ideally preferred to limit the study to only those learners with experience in three or more online courses, on the assumption that they would have greater insight into the dynamics of the online environment by virtue of exposure to more than just one activity, to qualify the respondents in this manner would be difficult and might also significantly reduce the number of respondents. Thus, some respondents' answers may reflect only a single, short-term online experience.

## Background

As has been noted, much of the recent literature on online education is characterized by its focus on "how to" teach via this medium, and how to optimally utilize the various features available in most instructional platforms, often based on authors' experiences teaching in this milieu. Less evident is if these recommended instructional techniques are based on data derived from systematic analysis of online student behaviors, attitudes, preferences, etc. As this online learner study attempts to arrive at a set of student skills or competencies that can enhance online learning experiences and outcomes, the review of literature focuses on sources that contribute to this effort.

There has been an encouraging increase in the amount of literature published in recent years on various aspects of online teaching and learning and, most recently, on so-called eCompetencies (e.g., Bernath and Sangra's edited volume (2007) of selected papers from the 4th EDEN Research Workshop). This review is limited primarily to those topics relating to key foci of the IBSTPI online learner study: need for learner-based data, online student engagement, student attitudes, behaviors and levels of satisfaction, and competencies applicable to effective online study.

Jones' study (2008) of distance education directors' main concerns revealed their preoccupation with the cost and support of technology, with less attention paid to quality or the level of satisfaction with instruction. While we fully recognize the criticality of technology as the medium by which online instruction is delivered and the milieu in which students process resources, we are alarmed by this continuing preoccupation with technology, seemingly at the expense of adequate attention paid to pedagogy. Accordingly, this study attempts to examine not simply respondents' reaction to the medium, as if they might be observers within it, but more so, to view them as participants in the dynamics which occur in that environment, and how this impacts their learning, for better or worse, in online settings.

Assessments that incorporate stakeholders' perceptions of their learning environment are consistent with predictors of student outcomes (Fraser, 1998). Variables within learning environments can be changed to achieve different affective and cognitive outcomes (Anderson and Walberg 1974). In many studies of environments and outcomes, dimensions of the former have been consistently identified as determinants of the latter (Fraser). Environments which students perceive as positive tend to lead to improved student achievement and attitudes (Chang and Fisher, 2001(a). Research involving LEs has a tradition of considering the association between perceptions of psycho-social characteristics of the environment and of students' cognitive and affective outcomes (Fraser). LE research has demonstrated that across languages, nations, cultures, subject matter, and educational levels, there are consistent associations between environmental perception and student outcomes (Fraser). Personal, institutional and circumstantial variables are critical in affecting student success (Berge and Huang, 2004).

There is need for new instruments to investigate the association between the nature of the distance learning environment and students' enjoyment of studies. Social structure has a strong influence on students` learning and satisfaction, and on the method by which the course is presented (Swan, 2001). Students have need for social connection and a sense of presence in electronically mediated distance education. (Paloff and Pratt, 1999) In these settings, participants can become part of a social milieu with anyone with a computer. Klopfer's (1971) affective categories (e.g., students' attitudes toward subject matter, attitude toward inquiry, enjoyment of learning, interest in subject as a career), are among the variables to be identified and measured to determine if and how certain learning environment characteristics have a positive association with student satisfaction and academic achievement. One instrument devised to investigate these factors is DELES (Walker and Fraser, 2005), which utilized the intuitive-rational strategy in which only items w/ high internal consistency remain. The DELES study revealed that LE characteristics have positive associations with student satisfaction.

Student engagement is, among other factors, defined in important ways by the environments and cultures in which it occurs. The convergence of parallel forces - rapid advancement of instructional technology, the collapse of barriers that has kept much of humanity separated, and the educational opportunities that are now available to anyone-anytime-anyplace - is changing higher education. Expanding instructional delivery to a diverse population is the new reality, and that population is the new majority of learners. As Inoue (2007) claims, this makes it difficult for students enrolled in online courses to visualize, or relate to fellow classmates who now represent cultural diversity related to the various geographical locations to which they pertain.

These advances in distributed learning technologies have made courses accessible to diverse learners worldwide, resulting not only in a diversity of cultures, but also in differences in learning styles. We must be alert to the fact that there are substantial differences in interaction and communication beyond the actual words being said, or in the case of online dialogue, the words being written or read (Januszweski and Molenda 2008).

Accrediting agencies require educational institutions to support goals relating to diversity; this typically refers to gender, race, age, socio-economic status, and the like. It generally does not take into account differences that are culture based, though this latter differentiation may represent the most dramatic distinctions in the way students learn. Technology has facilitated interpersonal communication exchanges, but has it, in fact, truly blurred the distinctions between cultures, nations, and races? If, as Inoue claims, the effectiveness of instructional delivery has improved access among diverse populations to common educational environments, exposing people to the culture and norms of others, we must be sure this experience becomes a valuable element in the learning process, rather than an impediment to it.

It is to be recognized that the online learning environment is itself a unique cultural context. Students come to this setting with preconceptions based on prior experiences in virtual situations as well as off-line contexts. They bring with them differing norms, levels of proficiency, communication styles, comfort levels, expectations, etc. Some of these attributes and behaviors are appropriate to educational contexts; some are not. And Internet skills alone do not determine competency; it is also Internet efficacy that enables them to adapt to effective usage of this medium. Thus, an effective online teacher must be constantly vigilant regarding the interface between teacher, learner, content, and technology that forms that particular educational context (Anderson 2008).

As has been noted, this study is essentially one that attempts to better understand how students engage in online learning and what they themselves view as the most crucial elements that determine the effectiveness of such engagement. Factors included are: interaction with the medium, with the materials, with one another, with the instructor, and ultimately, with the environment in which they work and apply new knowledge and skills acquired through their engagement with these various stimuli. Coates (2006) maintains that despite the emergence of online education, there appears to be a reliance on serendipity to produce patterns of use for teaching and learning. And despite the proliferation of studies into online education in the past decade, and widespread levels of adoption, most research has focused on financial, technical and administrative aspects of these learning systems. In particular, Coates notes there are very few published works on student engagement, and most of these are utilitarian, not conceptual. Furthermore, student engagement studies are, in large part, based on assumptions about campus learning environments that ignore the implications of online learning, and so present constraints in terms of advancing understanding of student engagement in these alternate settings.

Coates (2006) further contends that most publications are more interested in technology than in learning, are descriptive and prescriptive, and few offer an approach that contributes to understanding educational dynamics. Indeed, students are viewed as users rather than learners. Accordingly, such studies lack theoretical contextualization and empirical, thus limiting their generalizability and make them dependent on changing technologies. Consequently, Coates argues, there is much we still do not know about how students engage in online settings with activities and conditions that promote learning. He notes five benchmarks in assessing student engagement: academic challenge, active and collaborative learning, student-faculty interaction, enriching educational experiences, and supportive environment.

Does educational technology improve the performance of students, teachers, designers, and organizations? If so, what means are valid and reliable to verify that certain attributes of teachers and learners are those best suited to optimize performance? Beaudoin (1990) argues that increased attention to process over content should be the dominant concern of instructors to enhance learning in distance education environments. Palloff and Pratt (2001), as have others, define the attributes of successful online learners, listing the usual characteristics that typically appear on many such lists which, by now, have become conventional wisdom in the field, yet are often based on instructor intuition. Successful learners are seen as voluntarily seeking further education, having higher expectations, being more self-disciplined, older, enjoying learning for its own sake, demonstrating good thinking skills, able to work independently with limited structure, and recognizing the value of interacting with other online peers. Ashburn and Foden ((2006) identify attributes of meaningful learning, building on the list developed by Jonassen et al. (1999): Intentionality, content centrality, authentic work, active inquiry, construction of mental models, and collaborative work.

It should be noted that student behavior, students' competence as learners, and the quality of the LEs to which they contribute, is not a mere function of one or the other particular course in which they participate but also, and increasingly so, of their lifelong learning experience (both formal and informal) as well as the broad learning landscape in which they operate (Visser and Visser-Valfrey, 2008). The online learner is obviously a key component of the online learning environment, and determining his/her learner characteristics may not guarantee success, but it can help providers understand what factors and motivators can contribute to successful learning experiences in virtual settings. Dabbagh & Banna-Ritland (2005) report on the work of Dille and Mezack (1991), who studied the profile of distance learners, focusing on the locus of control. Students with an internal focus of control (those who attribute success and failure to personal behaviors and efforts) were more likely to succeed and persevere when faced with challenges. They also found that individual learning styles did not prove to be a significant predictor of success (presumably because distance education is inherently accommodating of various learning styles).

Anderson and Garrison (1998) indicate that the isolation which, fairly or unfairly, characterized an earlier era of distance education, has been largely replaced by collaborative learning, suggesting that this trend demands emerging interactive skills of online learners, along with several attendant skills: social, discursive, group and reflective skills. Similarly, the constructivist approach to learning also demands self-directed learning skills. Of course, the presence of these attributes in online learning environments presumes that those designing and delivering such courses possess the requisite competencies that lead to the implementation and application of those features in their courses.

## Summary Discussion of Findings

What new insights can be derived from the results of this study of the experiences and opinions of more than 300 online learners in more than four countries, to further inform our understanding of the dynamics of teaching and learning in the online learning environment? Also, are these findings helpful in ultimately identifying online learner competencies critical for success in education and training settings and, in turn, aiding providers to hopefully enhance their practice in the expanding distance education/training arena? And, can we discern any useful cross-cultural comparisons and contrasts in the responses of these four distinct cohorts? At this stage of analysis, we are prepared to offer at least some preliminary conclusions. Selected highlights of findings are summarized below.

*Onlinestudents' satisfaction:* The study revealed that the majority of respondents rated their satisfaction with their online learning to be Very Good to Good, as compared to Satisfactory or Poor. The range of responses reflected extremely high satisfaction among Mexicans (91%), perhaps due to support received via their affiliation with the sponsoring consortium, to more modest satisfaction levels among Japanese (53%), probably due, at least in part, to this society's aversion to expressing extreme opinions, and so tended to minimize their selection of responses at either end of the continuum.

*Elements influencingsatisfaction:* Among Westerners and Japanese, the course itself (in terms of organization and content) was a major factor in influencing satisfaction. This was followed by the quality of instruction, interaction, convenience, and flexibility. For Mexicans, it was the instructor and the interaction, followed by flexibility and institutional support. Israelis cited convenience and flexibility as key elements. This suggests that, for those who are studying more or less independently, the quality of the course and the convenience of delivery are paramount in determining satisfaction. While for those enrolled through a specific institutional mechanism (e.g., a consortium), it is the services and support derived from that entity that likely influences their ultimate satisfaction with the experience.

To some extent, this may also depend on the reasons why students are enrolled. Are they seeking a formal degree or diploma that constitutes a crucial step in their career development toward their ultimate goal of becoming competent and knowledgeable in a particular area they wish to dedicate themselves to? Or is it an individual course to be finished as quickly and conveniently as possible? Further, if the student has never experienced deep learning in prior online settings through, for example, peer-to-peer interaction, but rather only some relatively superficial participation, then he or she will not likely identify higher level learning as crucial.

*Challenges:* Almost every respondent encountered one or more challenges in their online learning experiences, with time management the dominant challenge for all but Israeli students. Despite institutional efforts to enhance their student support services as enrollments in online courses increase, it seems that most issues are ultimately resolved by students utilizing their own resources, many by simply persevering.

*Items critical forsuccess:* Excepting the Mexican respondents, the strongest determinant for success among these online learners was self-motivation, followed by time management, then capacity to learn with limited support. It is noteworthy that these choices relate to learner attributes, rather than to course features. This result would indicate that for most of these students, success was primarily dependent on learner traits and behaviors, rather than any factors inherent to the course. This implies that key competencies for online learning success emanate from the learner, rather than from characteristics related to the learning environment. The least critical items were: ability to cope with unstructured settings, familiarity with technology, and (perhaps surprisingly, as it tends to counter somewhat the conventional wisdom), relationships with other online learners.

U.S. and Israeli respondents considered the role of instructors to be important to their success (78% and 95% respectively), but this was minimally important to Japanese and Mexican students. It seems that for some, faculty are viewed largely as facilitators, rather than actually providing instruction. Many commented that faculty are most useful in offering support by answering questions, clarifying requirements, etc., rather than serving as the expert in course content. It appears that many students assume that good facilitation and support is what is to be primarily provided to them by instructors in online environments. This finding raises an interesting question: do students any longer need or want to be exposed to experts? It may be that their dependence on teachers has dwindled, and what they have now come to expect is for instructors to be available primarily as resource persons, should some need arise for course support unrelated to actual instruction.

The fact that only a single respondent alluded to the role of online facilitator as that of managing discussion raises an important question here: Do the majority of online learners somehow devalue the instructional role in that setting, compared to face-to-face faculty? And does the fact that the "self-motivation" is considered the most critical item for online success, while relationships with instructors ranked only 5th overall in importance among the choices, represent somewhat of an indictment of faculty effectiveness in this milieu? Or is it more an indication of students' self-sufficiency, as well as the faculty's 'invisibility' in this medium?

Israeli respondents appear to be the most dependent cohort on interaction with their instructor for a successful learning process. The next items, ranked as Very Critical by more than half of these respondents, represent a more intrinsic focus (e.g., student as independent learner: self motivation, ability to manage one's time and capacity to learn with limited support). Far behind, the Israeli respondents ranked factors associated with technology, non-structured settings, achieving leaning goal, enjoying the learning process and expressing their ideas, and finally, interaction with other learners. While the literature emphasizes the importance of learner- learner interaction for meaningful and deep learning (e.g., Anderson 2003), the Israeli respondents ranked this factor last in importance. A partial explanation can be based on the pedagogical model of their online experiences, which encourages individual learning.

Though relationships with other learners was seen by Japanese and Mexican students to be more important than it was to Western and Israeli students, it garnered less than 50% of any cohort's rating as Very Critical, or Critical, while the majority of all respondent groups saw it as relatively unimportant. In view of the significant interest in the literature regarding online interaction and its presumed importance in this environment, some might be somewhat surprised at these results, which suggest these respondents do not consider this aspect of their experience essential to success. But, it is also possible that some interpreted this question as referring more to personal relationships with peers, rather than online interactions.

*Impact of technology on online learners*: Somewhat surprisingly, familiarity with technology came in last in importance among many respondents. Several remarked that limited technical support requires self-sufficiency in dealing with computer problems, and that online learners must come into this environment already technically adept and prepared to deal with technical issues. While this study was not specifically designed to assess the impact of particular software applications or solutions on respondents' experiences and opinions, some conclusions can be arrived at from the findings, at least preliminarily, regarding this key element in the teaching-learning dynamic. There is little evidence that these learners' level of satisfaction and success was significantly affected by the particular software utilized in the online courses in which they were enrolled and engaged. This raises the question: Does all the attention given to technology applications and software solutions to design and deliver "state of the art" online courses actually make as much difference to users as much as they do to instructional designers and others who assume this is the element that defines quality?

*Suggestions for improving online learning*: Despite many suggested improvements, no single item emerged as dominant in this area. This implies that these learners did not encounter conditions in their respective learning environments so egregious as to elicit repeated recommendations for changes to enhance their learning

Thus, at least on the basis of this study, we can conclude that there are several factors that affect students, positively or negatively, that transcend the learning management system operant in their courses. Specifically, as reported elsewhere in this chapter, it is the learners' own internally-driven attitudes and behaviors (e.g., self-motivation, time-management) that ultimately have greater impact their online studies, rather than externally-driven elements, such as provider services, course software, etc.

A subtle but important distinction can be discerned from these responses regarding student satisfaction and success. When referring to satisfaction in their learning experiences, most of these online learners indicated such course elements as content, instruction, interaction (though not technology) affected this judgment. But, when asked about successful learning, most identified learner-directed factors (e.g., self-motivation) In short, it appears to be primarily learner attributes, rather than course features, that ultimately affects students' experiences and opinions of what works and does not work in online settings, regardless of what technology (e.g., software) is employed in their courses.

## Comments Regarding Competencies

The International Board of Standards for Training, Performance and Instruction (IBSTPI), in the work it has undertaken since 1978 to identify and promulgate competencies in various professional endeavors, such as instructional design, defines a competency as "...a knowledge, skill or attitude that enables one to effectively perform the activities of a given occupation or function to the standards expected in employment" (Richey et al, 2001, p. 31). Such competencies are statements of behavior, rather than traits or beliefs, though they may reflect attitudes. Although initially developed to correlate with performance in certain occupations, it occurred to IBSTPI board members that competencies relating to online student behavior could be useful to academe by identifying those skills or attitudes that promote successful learning in the online environment. Further, it was assumed that such competencies would be instructive to those responsible for the design and delivery of online courses so that these educational products and practices would facilitate optimum teaching and learning at a distance.

De la Teja and Spannaus (2008) point out that what is useful information to enhance learning is to know what differentiates the high-performing learner from the average or low-performing learner. This implies that we arrive at a definition of a competent learner, not simply a successful one. Ideally, both instructors and students should be aware of the competencies they put in practice to teach or learn online, even though only some of those competencies are likely to be manifested in the online classroom. Many competencies probably are undetected by instructors in the virtual classroom, though still practiced by learners. This aspect of online learning invites a number of interesting possibilities for further inquiry, such as whether less experienced learners take part more or less actively than their more experienced counterparts, and if they use a greater or lesser repertoire of competencies. Also, are collaborative e-competencies more likely to occur, or more easily be manifested than self-direction competencies? (Borges 2007)

The phenomenon of globalization and the dramatic proliferation of Internet-based technology in just the past decade has fundamentally transformed how teachers instruct and how students learn, regardless of time and place, that both providers and consumers of online products and processes must benefit from optimum ways to utilize this medium. This has contributed to the increased attention in what is described as eCompetencies, which Sceckenberg (2007) defines as the ability to use ICT for teaching and learning in meaningful ways, and in the development of learning objects such as ePortfolios. Weinert (1999) has articulated the concept of action competence, which combines systematically cognitive, motivational components into a coherent dispositional system of knowledge, skills and attitudes.

The draft competencies presented here are based upon several key assumptions about the nature of online learning and of learners, as well as the role of competencies. These assumptions guided the process of developing and articulating this set of competencies, which represent a particular perspective on both the theory and practice of teaching and learning in online environments.

- *Assumption 1*: Online learners are individuals who bring a certain repertoire of characteristics, concerns and competencies to their respective learning situation.
- *Assumption 2:* Online learner competencies pertain to both the teaching and learning process.
- *Assumption 3*: Online learning is a process that can be informed and enhanced by systematic analysis and formulation of principles to guide student-teacher behaviors.
- *Assumption 4:* Awareness and application of online learner competencies can improve performance of both learners and teachers, as well as the quality of learning objects.
- *Assumption 5*: Online learner competencies span novice, experienced and expert users.
- *Assumption 6:* Relatively few online course designers or instructors are aware of learner competencies; if aware, they are often uncertain how to utilize this in their practice.
- *Assumption 7:* Online learner competencies should be useful and meaningful worldwide, though amenable to customization to accommodate learner styles and cultures.
- *Assumption 8:* Online learner competencies should influence the design of instructional resources utilized in this environment.
- *Assumption 9:* Online learner competencies should reflect social and academic values and ethics.
- *Assumption 10:* Analysis and understanding of online learner behavior and attitudes can generate a set of valid and reliable competencies.

## A Preliminary List of Online Learner Competencies

Following is a draft set of online learner competencies derived from the survey data. It presents a single domain (communication) to illustrate the how such competencies are articulated in a variety of domains (e.g., foundations, technology). In the next phase of this project, these would be augmented by performance statements that provide descriptive examples of how such competencies are actually manifested by learners.

### 1. Practice Effective Online Communication Skills

a. Use online communication tools appropriately
b. Manage online time for timely interaction
c. Express pertinent ideas and post relevant questions that are clear, thoughtful, and civil
d. Post social comments in appropriate designated venues
e. Reflect on the effect of online messages before initiating o r responding communications
f. Show respect for others' opinions
g. Recognize and use participative, social, interactive and cognitive meta-cognition styles
h. Follow effective protocols for various asynchronous and synchronous means f communicating
i. Discard dysfunctional communication behaviors once identified.

### 2. Interact Appropriately with the Instructor

a. Read and adhere to instructions and other online protocols
b. Pose clear and concise questions to increase understanding

c. Utilize instructor feedback to advance knowledge
d. Write clear and elaborated questions and comments
e. Engage in civil discourse
f. Present supported counter-arguments.

### 3. Engage in Productive Interaction with Peers

a. Follow guidelines for interaction provided by instructor
b. Initiate and articulate new ideas
c. Present ideas clearly and concisely appropriate and relevant to the audience
d. Use appropriate discussion strategies to achieve desired learning goals
e. Pose questions and statements appropriately to enrich the dialogue
f. Engage in evidence-based learning by drawing on various expert resources
g. Present counter-arguments that are supported
h. Respect differing ideas and opinions and respond accordingly with civility
i. Make provocative comments with sensitivity for misinterpretation in online environments

### 4. Engage in Collaborative Knowledge Building

a. Develop an understanding of what constitutes collaborative knowledge building
b. Contribute actively / regularly to online learning environment
c. Engage in open exchange, sharing, and comparing of information, ideas, and opinions
d. Consider one's ideas carefully before posting them.
e. Practice collaborative skills that build a solid learning community
f. Summarize agreements
g. Seek others' ideas, experiences, and feedback
h. Clarify points with peers before arriving at conclusions
i. Synthesize new ideas and concepts to arrive at higher meaning
j. Discover consistencies and Identify discrepancies between ideas
k. Apply new knowledge to present and anticipated experiences
l. Develop negotiating skills.
m. Be meta-cognitively aware

## Future Research

This study, though ambitious in its scope and intent, must be seen as somewhat preliminary, and hopefully will prompt others to pursue further avenues of inquiry along these lines, as well as broaden the investigation of these phenomena. Areas of future research are often guided by the limitations discovered at the conclusion of earlier efforts. Thus, several suggestions are presented here to augment what this project has thus far provided. For example, the questionnaire utilized in this study attempted to gain extensive data regarding student characteristics, behaviors, and experiences, but it did not ask respondents to indicate if their learning experiences resulted in completion, or not. Though not the focus of this study, this might be useful to know, particularly in view of Herbert's finding (1994) that 25% of online students did not complete the online course in the semester in which they were enrolled. Nor did the study ask if the experience was their first, second, etc. And the survey did not ask students to identify their preferred learning style, or to indicate if the online environment was compatible with that preferred style. The study did not attempt to explicitly solicit respondents' opinions of the quality of their courses or instructors. And, as noted, the diverse cultural representation of these cohorts invites comparative analyses of responses. While it may have not been an initial aspect of this study, the emergence of four distinct cohorts, representing four different countries and cultures, prompted the research team to consider the possibility that, in addition to identifying commonalities among online learner behavior and attitudes, there might also be value in determining if there are any significant differences in these learners that are driven primarily by cultural orientation. These are areas of inquiry that further studies may wish to incorporate.

It must be emphasized that none of the respondent cohorts included in this study are to be seen as necessarily representative of the larger population of online learners that exist in each of their respective countries. Each sample used is relatively small, and in some cases (e.g., Mexican), the respondents are all students enrolled in studies offered via a particular consortium in the state of Veracruz. There are, of course, numerous other entities offering online instruction, and their students might well respond differently to questions posed in this study. Further research of this nature might survey larger sample sizes, and also include additional respondent group representing other countries, especially if comparative analysis of culturally driven learning outcomes is of interest.

## Preliminary Conclusions

What can we conclude, at least preliminarily, from a content analysis of these data?

- That online learners are willing to candidly share their experiences and opinions;
- That from a systematic and comprehensive survey of online learners, we can identify certain experiences and opinions that inform us as to what works and does not work and why;
- That we can discern patterns of behavior, characteristics, preferences and dislikes among a representative sample of online learners that inform us about what may be expected of others in similar environments;
- That there are many common elements reflected in the responses of online learners, regardless of the particular program, instructional delivery mode, or cultural setting;
- That online learners are willing to accept primary responsibility for the success or failure of their experiences, rather than see outcomes as dependent on the resources of provider institutions;
- That issues of time management and self-motivation are overriding factors in learners' perception of their success in online situations;
- That the technical elements (e.g., instructional platform software) while important, impact learner experiences and outcomes less than the attitudes and behaviors learners themselves bring into the setting;
- That, a strong majority of online learners had positive experiences, and most would enroll in online courses in the future;
- That from a careful content analysis of findings regarding what contributes to students' success in such settings, we can arrive at a set of valid and reliable competencies applicable to most online contexts;
- That these competencies can inform both teachers and learners as to what constitutes the most effective means for ensuring success, through product design and delivery, online interactions, student behaviors, and other elements that impact both content and process;
- That findings from investigations encompassing diverse learners can be especially useful to programs that attract a broad spectrum of learners representing diverse cultures, ages, experiences, and abilities;
- That additional systematic research is warranted to better understand those factors that contribute most to how online students learn most effectively and efficiently in such settings.

## Summary Statement

How to consider findings derived from user-driven data based on their experiences that might contradict provider-driven data from experts whose findings are based on conclusions from their experience? While the student feedback presented in this report is important, it should not be taken to negate expert opinion, particularly if such information is based on research as well as experience.. There is no preferred medium or formulaic method that dictates a single best means of learning in all domains with all learners. Both teachers and students must develop skills whereby both have a repertoire of online activities that are conducive for effective application and adaptation in diverse contexts (Anderson 2008). Indeed, it seems that what can ultimately be concluded is that the effective online instructor possesses the same resilience, flexibility and innovativeness that typifies the successful online learner. Thus, we have taken a critical step in identifying those performance standards that are conducive not only to meaningful learning, but equally as well, to effective teaching.

The online environment is a rich venue for teaching and learning at a distance to occur. Astounding advances in technology have evolved to facilitate this process, to provide an endless array of content, presented in dynamic fashion, in enhanced formats, using synchronous and asynchronous milieu, and making access to educational opportunities more accessible to many in various parts of the globe. The findings reported in this chapter dutifully chronicle many of the redeeming features of this technology that make it feasible for learners to achieve personal and professional aspirations that might otherwise not be attainable. But it must also be recognized that technology is just the medium through which teachers and learners and the resources they share are brought together in a common online space.

And yet, however magical may be this confluence of technological features, whether the latest learning management system, the most appealing software, ever faster hardware, innovative new applications, etc., in the final analysis, it is the human spirit of those who engage in this online community that shapes the value and meaning of those experiences. The inner-directed attitudes, predispositions, and competencies of the participants ultimately transcend the power of whatever instructional tools with which they may be provided.

## REFERENCES

Anderson, G. J., & Walberg, H. J. (1974). Learning environments. In Walberg, H. J. (Ed.), *Evaluating educational performance: A sourcebook of methods, instruments and examples* (pp. 81–98). Berkeley, CA: McCutchan Publishing.

Anderson, T. (2003). Getting the mix right again: An updated and theoretical rationale for interaction. *The International Review of Research in Open and Distance Learning, 4*(2). Retrieved July 13, 2008, from http://www.irrodl.org/index.php/irrodl/article/view/149/230

Anderson, T. (Ed.). (2008). *The Theory and Practice of Online Learning* (2nd ed.). Edmonton, Canada: AU Press.

Anderson, T., & Garrison, D. R. (1998). Learning in a networked world: New roles and responsibilities. In Gibson, C. (Ed.), *Distance Learners in Higher Education* (pp. 97–112). Madison, WI: Atwood Publishing.

Ashburn, E., & Floden, R. (2006). *Meaningful Learning Using Technology-What Educators Need to Know and Do*. New York: Teachers College Press.

Beaudoin, M. (1990). The Instructor's Changing Role in Distance Education. *American Journal of Distance Education, 4*(2). doi:10.1080/08923649009526701

Berge, Z., & Huang, Y. (2004). A Model for Sustainable Student Retention: A Holistic Perspective on the Student Dropout Problem With Special Attention to e-Learning. *Distance Online Symposium, The American Center for the Study of Distance Education, 13*(5).

Bernath, U., & Sangra, A. (Eds.). (2007). *Research on Competence Development in Online Distance Education and E-Learning*. Oldenburg, Germany: IBIS-Vertag der Carl von Ossietzky Universitat.

Borges, F. (2007) e-Competence of Online Students of the Humanities at UOC. In Research on Competence Development in Online Distance Education and E-Learning. Oldenburg, Germany: BIS-Verlag der Carl von Ossietsky Universitatat.

Chang, V., & Fisher, D. (2001, December 2-6). *The validation and application of a new learning environment instrument to evaluate online learning in higher education*. Paper presented at the Australian Association for Research in Education, Fremantle. Retrieved from http://www.aare.edu.au/01pap/cha01098.htm

Coates, H. (2006). *Student Engagement in Campus-based and Online Education*. London: Routledge.

Dabbagh, N., & Bannan-Ritland, B. (2005). *Online Learning- Concepts, Strategies, and Applications*. Upper Saddle River, NJ: Pearson.

De la Teja, I., & Spannaus, T. W. (2008). New online learning technologies: new online learner competencies. Really? In Visser, J., & Visser-Valfrey, M. (Eds.), *Learners in a changing learning landscape: Reflections from a dialogue on new roles and expectations* (pp. 187–211). Dordrecht, The Netherlands: Springer.

DuCharme-Hansen, B., & Dupin-Bryant, P. (2004). *Web-Based Distance education for Adults*. Malabar, FL: Krieger Publishing Company.

Ehlers, U.-D. (2007). Making the Difference in E-Learning: Towards Competence Development and E-Irritation. In Research on Competence Development in Online Distance Education and E-Learning. Oldenburg, Germany: BIS-Verlag der Carl von Ossietsky Universitatat

Fraser, B. J. (1998). *Classroom Environment Instruments: Development, Validity and Applications*. Berlin: Springer.

Gunawardena, C., & LaPointe, D. (2007). Cultural Dynamics of Online Learning. In Moore, M. (Ed.), *Handbook of Distance Education* (2nd ed.). Mahwah, NJ: Lawrence Erlbaum Associates, Publishers.

*IBSTPI Online Learner Project*. (2007). See Appendix A.

Inoue, Y. (2007). *Technology and Diversity in Higher Education*. Hershey, PA: Information Science Publishing.

Januszewski, A., & Molenda, M. (2008). *Educational Technology-A Definition with Commentary*. New York: Lawrence Erlbaum Associates.

Jonassen, D. H., Peck, K. L., & Wilson, B. G. (1999). *Learning with technology: A constructivist perspective*. Upper Saddle River, NJ: Prentice Hall.

Jones, J. (2008). Issues and Concerns of Directors of Postsecondary Distance Learning Programs Regarding Online Methods and Technologies. *American Journal of Distance Education, 22*(1), 46–56. doi:10.1080/08923640701713430

Klopher, L. E. (1971). Evaluation of learning in science. In Bloom, B. S., Hastings, J. T., & Madus, G. F. (Eds.), *Handbook of Formative and Summative Evaluation of Student Learning*. New York: McGraw-Hill.

McLoughlin, C., & Oliver, R. (2000). Designing learning environments for cultural inclusivity: A case study of indigenous online learning at tertiary level. *Australian Journal of Educational Technology*, *16*(1), 58–72.

Palloff, R., & Pratt, K. (2001). *Lessons from the Cyberspace Classroom*. San Francisco: Jossey Bass.

Richey, R., Fields, D., & Foxon, M. (2001). *Instructional Design Competencies: The Standards* (3rd ed.). Syracuse, NY: ERIC.

Rogers, E. (1995). *Diffusion of Iinnovations* (4th ed.). New York: The Free Press.

Swan, K. (2001). Virtual Interactivity: design factors affecting student satisfaction and perceived learning in as ynchronous online courses. *Distance Education*, *22*(2), 306–331. doi:10.1080/0158791010220208

Talbot, C. (2007). *Studying at a Distance- A guide for students*. Berkshire, UK: Open University Press.

Tobin, K., & Fraser, B. J. (1998). A distance education learning environment survey. *Distance Education*, *22*, 306–331.

Visser, J., & Visser-Valfrey, M. (2008). *Learners in a changing learning landscape: Reflections from a dialogue on new roles and expectations*. Dordrecht, The Netherlands: Springer.

Walker, S., & Fraser, B. (2005). Development and Validation of an Instrument for Assessing Distance Education Learning Environments in Higher Education: The Distance Education Learning Environment Survey (DELES). *Learning Environments Research*, *8*(3). doi:10.1007/s10984-005-1568-3

Weinert, F. E. (2001). Concept of competence: A conceptual clarification. In Rychen, D. S., & Salganik, L. H. (Eds.), *Defining and selecting key competencies* (pp. 45–65). Seattle, WA: Hogrefe & Huber Publishers.

## ADDITIONAL READING

Beaudoin, M. (2004). *Reflections on Research, Faculty and Leadership in Distance Education. Oldenburg, Germany: BIS-Verlag der Carl von Ossietsky Universitatat Clyde, W., & Delohery, A. (200). Using technology in Teaching*. New Haven, CT: Yale University Press.

Herbert, M. (1994). Staying the Course: A Study in Online Student satisfaction and retention. *The Online Journal of Distance Learning Administration*, Winter.

Khan, B. (2005). *Managing E-Learning Strategies*. Hershey, PA: Information Science Publishing.

Sanders, W. (2001). *Creating Learning-Centered Courses for the World Wide Web*. Needham Heights, MA: Allyn Bacon.

Schneckenberg, D. (2007). Competence Reconsidered- Conceptual Thoughts on eCompetence and Assessment Models for Academic Stuff. In *Research on Competence Development in Online Distance Education and E-Learning*. Oldenburg, Germany: BIS-Verlag der Carl von Ossietsky Universitatat.

White, K., & Weight, B. (1999). *The Online Teaching Guide*. Boston: Allyn and Bacon.

## APPENDIX A

## IBSTPI Online Learner Project

## Survey

## Consent to Participate in a Survey of Online Learner Competencies

- You are being asked to participate in a research study of online learner opinions, preferences, and competencies relating to your experiences as a student or trainee in this environment.
- You are invited here to complete a survey that will take about 20 minutes of your time. This survey includes questions requiring choices among responses, as well as some open-ended questions.
- By completing this survey, you will be providing information that may assist the International Board of Standards for Training, Performance and Instruction (www.ibstpi.org) in identifying methods for enhancing online instruction by better understanding how learning occurs in such settings. The results are intended to contribute to a better understanding of competencies required for optimal online learning.
- There is no known risk involved.
- Your participation in this research study is entirely voluntary. You may chose not to participate, and there will be no penalties if you chose not to. You may withdraw from this study at any time.
- All information collected during the course of this study will be kept confidential to the extent permitted by law. All data is to be aggregated, and any reports of findings will not identify you as a respondent in any manner. Information from the study may be published, but your identity will be kept confidential in any such publications.
- If you have any questions concerning your participation in this study now or in the future, you may contact the primary principal investigator: Dr. Michael Beaudoin, Professor of Education, University of New England, Biddeford, Maine, USA (mbeaudoin@une.edu)

## Consent to Participate in the Research Study

1. You have read all the above information about this research study. The content and meaning of the information has been explained and is understood. You have no questions before proceeding to the survey. Do you agree to follow the study requirements and take part in the study?

| | |
|---|---|
| Yes | |
| No | |
| Total | |

You may print a copy of this consent form for your records. **Thank you for your participation in this survey.** Click SUBMIT to continue or to end this survey...

The International Board of Standards for Training, Performance and Instruction (IBSTPI) is conducting research into online learning experiences. The purpose of this study is to gather information that will help: -Learners to become better learners -Instructors/trainers to improve their ability to assist the learners using online technology -Training managers to better prepare their staff for online learning -Learning designers to build content that better impacts learning outcomes Your online learning experiences are invaluable to this study. Your participation in the survey strengthens the research into online learning and furthers our collective understanding of how we learn most effectively online. It is often noted that answers to questions give only limited information. We thus encourage you to make use of the opportunities provided in this survey to describe, substantiate and elaborate your answers. Please note your participation is voluntary and will be anonymous on this survey. Information that might identify individuals will not be retained or published. The survey will not involve any risk to you and you will not be paid for responding. IBSTPI will be pleased to email you a copy of the survey results upon request.

**Online Learning Experiences.** Please reflect on up to three (3) significant online learning experiences (individual courses; programs, extensive periods of self-study; virtual classes; etc.) that you have engaged in. For each experience we ask that you respond to the items as fully as possible.

2. To input your experience, please choose from below:

| | | | |
|---|---|---|---|
| First Online Learning Experience | | | |
| Second Online Learning Experience | | | |
| Third Online Learning Experience | | | |
| Skip adding my experience | | | |
| | Total | | |

Click SUBMIT to continue...

3. Experience 1
4. Duration of the experience (indicate whether this number is in hours, days, weeks, months or years).
7. Indicate your level of satisfaction with the experience.

| | | | |
|---|---|---|---|
| Very Poor | | | |
| Poor | | | |
| Satisfactory | | | |
| Good | | | |
| Very Good | | | |
| | Total | | |

Which elements of the experience most influenced your selected level of satisfaction?

8. Element 1:
9. Element 2:
10. Element 3:

What important challenge(s), if any, did you encounter in this learning experience?

11. Challenge 1: Describe how you dealt with this challenge.
12. Challenge 2: Describe how you dealt with this challenge.
13. Challenge 3: Describe how you dealt with this challenge.

What suggestion(s), if any, do you have that could have improved your learning in this experience?

14. Suggestion 1:
15. Suggestion 2:
16. Suggestion 3:
17. Would you like to add your next experience or move on to the next page?

| | | | |
|---|---|---|---|
| Experience 2 | | | |
| Experience 3 | | | |
| Skip adding my experience, go to next page | | | |
| | Total | | |

Click your browser's BACK button to go back and replace a previous response or SUBMIT to continue to the next page...

18. Experience 2
19. Duration of the experience (indicate whether this number is in hours, days, weeks, months or years).
21. Learning support involved (check all that apply)
22. Indicate your level of satisfaction with the experience.

| | | | |
|---|---|---|---|
| Very Poor | | | |
| Poor | | | |
| Satisfactory | | | |
| Good | | | |
| Very Good | | | |

Which elements of the experience most influenced your selected level of satisfaction?

23. Element 1
24. Element 2
25. Element 3

What important challenge(s), if any, did you encounter in this learning experience?

26. Challenge 1 (Describe how you dealt with this challenge)
27. Challenge 2 (Describe how you dealt with this challenge)

28. Challenge 3 (Describe how you dealt with this challenge)

What suggestion(s), if any, do you have that could have improved your learning in this experience?

29. Suggestion 1
30. Suggestion 2
31. Suggestion 3
32. Would you like to add your next experience or move on to the next page?

Click your browser's BACK button to go back and replace a previous response or SUBMIT to continue to the next page...

33. Experience 3
34. Duration of the experience (indicate whether this number is in hours, days, weeks, months or years).
35. Context of the experience
36. Learning support involved (check all that apply)
37. Indicate your level of satisfaction with the experience

Which elements of the experience most influenced your selected level of satisfaction?

38. Element 1
39. Element 2
40. Element 3

What important challenge(s), if any, did you encounter in this learning experience?

41. Challenge 1 (Describe how you dealt with this challenge)
42. Challenge 2 (Describe how you dealt with this challenge)
43. Challenge 3 (Describe how you dealt with this challenge)

What suggestion(s), if any, do you have that could have improved your learning in this experience?

44. Suggestion 1
45. Suggestion 2
46. Suggestion 3

Click your browser's BACK button to go back and replace a previous response or SUBMIT to continue to the next page...

47. So now, tell your learning story, providing examples of what contributed to the success in your online learning experience. It is often difficult to capture experience in a survey or a questionnaire. We ask you to imagine that you are having a conversation with an instructor, a trainer, a colleague or with a fellow learner on the topic of important characteristics of successful online learning.

48. Also, provide examples of what detracted from success in your online learning experience.

Click your browser's BACK button to go back and replace a previous response or SUBMIT to continue to the next page...

Here are some items that we think are critical to being successful online learners. Please reflect on these and indicate how critical you feel each one is to online learners: -Relationships with other online learners -Relationships with online facilitators & trainers -Self-motivation -Familiarity to technology -Ability to express your ideas -Capacity to learn with limited support or in isolation (self directed) -Ability to cope with non-structured learning settings -Confidence that you are able to achieve your learning goals -Ability to manage your time -Enjoying the challenge of learning.

49. Please add your two items
Online Learner Success Items

Click your browser's BACK button to go back and replace a previous response or SUBMIT to continue to the next page...

50. Now rate these 12 items.

| Top number is the count of respondents selecting the option.<br>Bottom % is percent of the total respondents selecting the option. | Very Critical | Critical | Somewhat Critical | Not at all Critical |
|---|---|---|---|---|
| Relationships with other online learners | | | | |
| Relationships with online facilitators & trainers | | | | |
| Self-motivation | | | | |
| Familiarity to technology | | | | |
| Ability to express your ideas | | | | |
| Capacity to learn with limited support or in isolation (self directed) | | | | |
| Ability to cope with non-structured learning settings | | | | |
| Confidence that you are able to achieve your learning goals | | | | |
| Ability to manage your time | | | | |
| Enjoying the challenge of learning | | | | |
| Your first item | | | | |
| Your second item | | | | |
| | | | | |

51. What do you regard as your most rewarding online learning experience(s) and why? Describe.
52. What do you regard as your least rewarding online learning experience(s) and why? Describe.
53. If you could advise those who design and develop online learning, what advice would you give them?
54. Would you consider participating in another online program/course etc.?

| | | | |
|---|---|---|---|
| Yes | | | |
| No | | | |
| Not sure | | | |
| | Total | | |

55. If you have any other observations regarding your online learning experiences, please describe these below.

Please help us to identify the demographics of our survey participants:

56. Gender

| | | | |
|---|---|---|---|
| Male | | | |
| Female | | | |
| | Total | | |

57. Age Group
58. Country where you live

Thank you very much for working with us. Your cooperation is critical for the success of this research project and thus for the success of future online learners. Click your browser's BACK button to go back and replace a previous response or SUBMIT to complete this survey.

# Chapter 21
# Distance Learning Courses:
## A Survey of Activities and Assignments

**Kelley Walters**
*Northcentral University, USA*

**Melanie Shaw**
*Northcentral University, USA*

**David Long**
*University of Florida, USA*

## ABSTRACT

*Drawing on current literature and a survey of online students and instructors from online institutions, the researchers explored the different types of assignments and assessments required for completion of online courses, the Learning Management Systems utilized, and differences between undergraduate and graduate tasks assigned, according to field of study. While there is a significant amount of available research on instructional efficacy in online classes, there are few studies that account for the types of course activities students must complete in distance learning courses. This study details the variety of online assessments and activities assigned to students attending institutions of higher education that are either fully online or utilizing a blended learning format. Recommendations are made for diverse instructional tasks, which can be assigned based on available technology and curricular flexibility. Key terms used in the study are defined at the end of this chapter.*

## INTRODUCTION

Effective learning activities are critical for relevant, scholarly knowledge acquisition. Online instructors are challenged to implement meaningful course activities to help students acquire new skills and demonstrate mastery of course content. Many instructors have flexibility in creating unique course activities for each course, while others adopt a generic syllabus with prescribed activities students must complete to successfully pass the class.

This study details the different types of course activities students must complete in a diverse sampling of online courses. Researchers also investigated the types of assessments students were required to fulfill. Data were tabulated to indicate the different learning management systems (LMS) utilized by survey participants, the format in which

DOI: 10.4018/978-1-61520-853-1.ch021

Copyright © 2010, IGI Global. Copying or distributing in print or electronic forms without written permission of IGI Global is prohibited.

courses were offered, semester length, as well as the indicated frequency and enjoyment level of assigned activities for instructors and students.

There is little research available that describes the types of course activities and assessments desired in distance learning courses. This study fills this void by including the findings from a survey of students enrolled in online classes and instructors of online courses about the types of course activities and assessments assigned. Recommendations are made for instructors and curriculum decision-makers hoping to adopt diverse instructional activities.

## Research Questions

The following research questions were used in this study:

1. What is the typical format and length for an online course? – This question was included to determine if study participants engaged in online or blended course formats; if participants were undergraduate students, graduate students, or instructors, and the duration of the online course.
2. What types of activities and assessments are assigned in online courses?
3. How do participants rate enjoyment of assigned activities?
4. How do assignments and assessment vary between fields of study and level of students?
5. How does the type of Learning Management System (LMS) utilized affect what types of assignments and assessments are used?
6. How does the type of Learning Management System (LMS) utilized affect how students perceive course activities?

## BACKGROUND

In an online course, students may be assigned a variety of learning tasks to assess course content knowledge. For example, papers, multimedia presentations, research, journaling, group activities, course discussions, class readings, and examinations may be required (Oliver, n.d.; Shieh, Gummer, & Niess, 2008). Wild and Quinn (1997) indicated that learning activities in the online environment are fundamentally linked with learning outcomes. Course assignments determine the ways students engage with course content and construct knowledge.

Liao (2006) suggested course assignments be designed to build upon prior knowledge; engage students with real-world issues; and allow student choice in how to approach instructional tasks within the education academic discipline. Online instructors should assign course activities that facilitate deep processing and promote meaningful contextual learning (Ally, 2004). Duffy and Cunningham (1996, as cited in Lombardi, 2007) noted since course activities provide a context for learners to encounter content and information presented in the course readings:

*Instructors [who] provide engaging activities supported by the proper scaffolding can help students develop expertise across all four domains of learning: Cognitive capacity to think, solve problems, and create; Affective capacity to value, appreciate, and care; Psychomotor capacity to move, perceive, and apply physical skills; [and] Conative capacity to act, decide, and commit... Higher education has focused for too long on inculcating and assessing those cognitive skills that are relatively easy to acquire—remembering, understanding, and applying—rather than the arguably more important skills of analyzing, evaluating, and creating. (pp. 8-9)*

Teaching in the online environment requires different pedagogical approaches than the traditional classroom (Shieh, Gummer, & Niess, 2008). Kochtanek and Hein (2000) noted that professors in distance learning courses must act "as facilitators…providing students with opportunities to become more involved in the learning process, in setting their own learning pace, and in contributing to the refinement of the course as a whole as learning becomes more tailored to student needs" (p. 284). Most instructors find teaching in the online course more challenging than the traditional classroom because of increased interaction with students and complications of assessing learning outcomes in online environments (Connolly, Jones, & Jones, 2007). Yet, despite the challenges of online instruction:

*Online higher education should achieve the same high standard of academic rigour [sic] and professional practicability as conventional higher education. With the increase in popularity of online education, cross-continental evidence shows that quality assurance of online education has increasingly become the centre of attention. (Zhao, 2003, p. 220)*

## Growth in the Online Learning Sector

Since the inception of the Internet, online programs in higher education have grown exponentially. In 1990, only 15% of households owned a computer in the United States (U.S. Department of Labor, 1999). By 2005, an estimated 88.9 million U.S. Households owned a computer, and of those, 81.4% had Internet access (Carol, Rivera, Ebel, Zimmerman, & Christakis, 2005). Based on responses of more than 2,500 institutions of higher education, the Sloan Consortium (2008) report noted over the last five years, online enrollments have shown impressive growth across all higher education areas. The number of institutions offering online course options has steadily grown in all disciplines. In addition, almost 4 million students took one or more online courses in the fall of 2007.

With the growth of online access, universities and colleges across the country have risen to meet consumer demand for online programs (Song, Singleton, Hill, & Koh, 2004). Thompson (2009) noted, "information technology has revolutionized the way many people work and interact, but the evidence is that many further education and skill training providers need to pick up the pace of adopting new [online] developments" (p. F4). Fully two thirds of colleges and universities assert that online education is the most significant development in higher education as it offers students flexibility and convenience (Shin & Lee, 2009).

## Assignment Best Practices

Instructors teaching online courses have a multitude of assignment and assessment options. In addition to examinations, essays, research papers, PowerPoint presentations, and discussion entries, instructors have a myriad of possibilities for integrating unique requirements to enhance course assignments. When designing assignments for online courses, it is important that instructors create assignments that are engaging and authentic. Wiggins (1998) suggested the following criteria for creating authentic assignments, particularly in applied courses:

1. *It's realistic.* The task or tasks replicate the ways in which a person's knowledge and abilities are "tested" in real-world situations.
2. *Requires judgment and innovation.* The student has to use knowledge and skills wisely and effectively to solve unstructured problems, such as when a plan must be designed, and the solution involves more than following a set routine or procedure or plugging in knowledge.
3. *Asks the students to "do" the subject.* Instead of reciting, restating, or replicating through demonstration what he or she was taught or what is already known, the student has to carry out exploration and work within the discipline or science, history, or any other subject.
4. *Replicates or simulates the contexts in which adults are "tested" in the workplace, in civic*

*life, and in personal life.* Contexts involve specific situations that have particular constraints, purposes, and audiences. Typical school tests are contextless. Students need to experience what it is like to do tasks in workplace and other real-life contexts, which tend to be messy and murky. In other words, genuine tasks require good judgment. Authentic tasks undo the ultimately harmful secrecy, silence, and absence of resources and feedback that mark excessive school testing.

5. *Assess the student's ability to efficiently and effectively use a repertoire of knowledge and skill to negotiate a complex task.* Most conventional test items are isolated elements of performance – similar to sideline drills in athletics rather than to the integrated use of skills that a game requires. Good judgment is required here, too. Although there is, of course, a place for drill tests, performance is always more than the sum of the drills.
6. *Allows appropriate opportunities to rehearse, practice, consult resources, and get feedback on and refine performances and products.* Although there is a role for the conventional "secure" test that keeps questions secret and keeps resource material from students until during the test, that test must coexist with educative assessment if students are to improve performance; if we are to focus their learning, through cycles of performance-feedback-revision-performance, on the production of known high-quality products and standards; and if we are to help them learn to use information, resources, and notes to effectively perform in context. (pp. 22-24)

When designing assignments for online courses, it is important that instructors consider offering a variety of options.

## Artifacts

Students should be assigned to create artifacts that are professionally or personally useful. Useful course activities are relevant and increase outcomes for all students (Fletcher, n.d.). For instance, education students can create a form to use during classroom evaluations, or business students could create a budget spreadsheet to use for monthly profit and loss statements. It is important for instructors to consider adding assignments that students can/will use in professional situations.

Students should be presented with a variety of instructional task options for each assignment. Talor (1997) noted that students understand and process information in different ways based on individual cognitive strengths. Successful instructors should vary instruction and assessment methods for all students in their classroom. Harris (2005) wrote, "Differentiation…appeals to the talents and learning styles that one finds in classrooms full of lively students. Differentiation also motivates [students]" (p. 18). Moore (2005) indicated that access, one of the five Sloan-C pillars, is critical in the online learning environment. Activities should be tailored to diverse learning activities for at-risk, disabled, and advanced students.

By assigning relevant artifacts with a variety of completion options, online instructors can maximize student achievement and meet the needs of unique learners.

Online instructors only exist because students exist. Without students engaging in distance learning course offerings, there would be no career opportunities in online instruction. Therefore, it is critical that online instructors make learning enjoyable and attainable for all students. The most effective way to ensure that instructional tasks meet the needs of students is to be cognizant of diverse assignment options. Students who enroll in online course are equipped with different types of intelligence, divergent cognitive controls and styles, different learning styles, and varying amounts of prior knowledge. The effective instructor should make a conscious attempt to vary instructional tasks to meet the needs of all the different types of learners in their classes (Santo, 2005). Mupinga, Nora, and Yaw (2006) noted that "each student comes to class with certain learning experiences, expectations, and needs that have to be addressed and to which instructors need to be sensitive, to maximize students' learning experiences (p. 185). Studies show that when instructional strategies are modeled after student' learning styles of choice, achievement increases (Horton & Oakland, 1997, p. 131). Teaching is an intentional act. Each choice must be thought through.

### Innovative Technologies and Higher Order Thinking Tasks

Technology has grown over the last twenty years from being an accessory rarely used as an alternative to the manual typewriter to a vital component of education (Razik & Swanson, 2001). Exposure to new and emerging technology increases student learning, encourages curricular relevance, and provides a contemporary skill set to those who interface with technological advances. Technology is now shaping the world through innovation and ingenuity and students widely embrace available technology through advances including video games, Internet, iPods, and online learning management systems. Educational leaders must undertake the responsibility for integrating technology as systems are impacting the management educational organizations, pedagogy, learning (National School Boards Foundation, 2004).

Students should be provided with an opportunity to use innovative technologies in assignments. There are many options including Boostcast.com which allows users to upload video, audio, pictures, and blogs to create a free video to share; Instructors can create a class blog using a blog site like Livejournal.com, where learners can interact and share course related information; or Screencast-o-matic.com, which allows students to video-capture screen shots with audio and presentation features. Using technology often results in increased student achievement. In a recent study, students who listened to lectures on podcast outperformed peers who attended face to face lectures (Carter, 2009). In addition to these technologies, online students can access chat rooms, e-mails, message boards, and discussion forums.

### Higher Order Thinking Tasks

Students are often required to perform tasks that require basic comprehension of material, however; students can be challenged to deepen their understanding of learned material when required to use higher-order thinking skills. When instructors require higher-order thinking tasks, they enhance students' cognitive skills and abilities. Bloom's Taxonomy (Bissell & Lemons, 2006) is a classification system that orders thinking tasks in a hierarchy from the simplest to the most complex. Bloom's taxonomy provides instructors the possibility of assignments/activities that offer a different way of testing thinking, learning, and being creative (Bogan & Porter, 2005, p. 46). Online instructors should focus instructional tasks on helping students acquire learning in analysis, synthesis, and evaluation of information.

## Online Interaction

Research indicates interactive online courses promote online learner success (Feenberg, 1999). Online faculty should strive to create an interactive, educational environment that incorporates best practices and current technology. Moore (2005) outlined the Sloan Consortium quality framework and five pillars as a helpful tool to ensure high-caliber online interaction. The pillars include learner effectiveness, cost effectiveness and institutional commitment, access, faculty satisfaction, and student satisfaction. In terms of interaction, Moore noted, "Interaction is key: with instructions, classmates, the interface, and via vicarious interaction" (p. 5). Teaching in the traditional, face to face classroom is significantly different than teaching a course online (Zhao, 2003). As Connolly, Jones, and Jones (2007) noted, many online faculty utilize pedagogical practices that were effective in the traditional classroom, but fail to serve the learning needs of online students. In online courses, teaching is most effectively facilitated through class discussions, shared experiences, and activities tailored for the distance learning environment (Kochtanek & Hein, 2000).

Course-based discussions allow students an opportunity to interact and converse about course content. These discussions have gained popularity "as a way to spur intellectual exchange among students outside of regular class time. In fact, a recent survey by the Campus Computing Project finds at least a quarter of college courses [use] discussion areas" (Murray, 2000, para. 1). The discussion area provides students with an area to discuss relevant course topics, a place to get to know one another, and an avenue to share their thoughts (even if the student tends to be more introverted).

Online students can access chat rooms, e-mails, message boards, and discussion forums to interact with colleagues and the instructor. Most online courses require some type of asynchronous collaboration (Song, Singleton, Hill, & Koh, 2004). Feenberg (1999) noted that learning is most effectively facilitated through class discussions and other interactive opportunities. Both the instructor and the student are equally responsible for engaging in a collaborative dialogue throughout the course (Song et al.; Feenberg). Interaction is critical in most learning environments, but it is possibly the most important aspect of a distance learning course (Song, et al.). Online instructors should ensure substantive participation in online courses.

There are a variety of instructional strategies that make online experiences effective. Atkins (2001) advocated the importance of graphics within the course to enhance understanding. He wrote:

*When constructing materials for students who will be learning online, an instructor must remember to be concise, focused and disciplined. As a teacher, it is your responsibility to communicate knowledge in a way that your students can understand. Aim to use words that your pupils will easily understand, and provide concise explanation where you have used a word or phrase that may have baffled them. Above all, make every effort to ensure that your online materials contain interactive content that will engage your learners and encourage them to be proactive. (para. 3)*

Nickerson (2003) asserted that streaming video in online classrooms can provide an experience for students equivalent to traditional lectures. Many learning management systems provide easy uploading of video lectures for students to access throughout the course. Streaming video "opens up another aspect of online education in a meaningful, positive manner" (Nickerson, para. 3).

Broadbent (2002) noted without a workable, effective course design, students will find online courses difficult to navigate. Swan (2001) demonstrated that course design, learner interaction, and active participation were major factors influencing student satisfaction. Berner (2004) noted that the choice of learning management system utilized by the institution, strongly impacts course design.

*The online learning environment is, after all, just another learning environment, in some ways similar to and in some ways different from more traditional environments such as conventional classrooms, seminar rooms, or labs. When we move our class onto the Internet, we should plan for and make the best use of the online environment. The various instructional strategies we use to meet the goals and objectives of our courses are likely to be similar in each environment. However, the ways in which we utilize the strategies will differ as we make the best use of the characteristics and capacities of each environment. (Illinois Online Network, n.d., para. 3)*

Because online courses require content to be presented without traditional lectures, instructional delivery is crucial. Often times, the more visually pleasing aspects of the course receive the most attention from students (Broadbent, 2002), however; it is often the technical aspects of the course that contain important course content. As Kent and McNergney (1999) noted, beautiful graphics and flashy presentation will not overcome fundamental flaws in course design (Kent & McNergney). Christe (2003) also noted the importance of strategic course design in advance of student enrollment, as preparation helps ensure instructors and students feel more comfortable in the online environment. This is particularly important for first-time online faculty and students.

### Online Teaching Skills

Besides having technology proficiency, online instructors must possess the pedagogical skills necessary to teach in online classrooms (Aragon, 2003). In order to facilitate online courses, faculty members need to develop skills in computer literacy and online pedagogical and instructional methods (Aragon). The online instructor must discover ways to encourage autonomy in online learners to instructional intervention, dialogue, and feedback (Shelton, 2005). By incorporating several instructional methods and media, online discussions, chats, icebreakers, and streaming videos, instructors will effectively meet the needs of online learners (Shelton). Online instructors should strive to create an exciting, enticing, and engaging online classroom. The instructional design of the course has a bearing on a learner's motivation, interest, opinions, and attitudes (Maeroff, 2003).

Higher education institutions should provide training and professional development opportunities for potential online instructors to differentiate between traditional and online teaching skills (Aragon). Online professional development is important and provides potential instructors with the tools necessary for successful facilitation in the virtual classroom, especially for elements that replace the traditional lecture and maximize participation (Aragon). There are many ways for an online instructor to facilitate knowledge exchange, other than through traditional lectures and note-taking (Kent & McNergney, 1999). Broadbent (2002) noted, in lieu of the traditional classroom lecture, many online instructors utilize PowerPoint presentations, synchronous chatting, discussion forum activities, group projects, readings, quizzes, papers, and other assignments.

## STUDENT AND INSTRUCTOR PERSPECTIVES

For this study, the researchers created an online survey and requested participation from a wide cross-section of online students and instructors. Access to the desired population was obtained via several online learning forums on Linkedin.com, online college and university websites, and various Yahoo groups. A total of 166 individuals completed the survey. Of those, 11 were undergraduate online students, 58 were online graduate students, 42 were undergraduate online instructors, 16 were graduate online instructors, 26 were instructors for both graduate and undergraduate online courses, 6 were administrators, and 7 listed their affiliation as "other." Two participants noted that they serve in several multiple roles including online graduate student, an online undergraduate instructor, online graduate instructor, and an online administrator.

### Types of Assignments

Survey participants have engaged in a variety of online course assignments. The most frequently assigned online activities were online discussions (58.2%), research papers (48.1%), e-mail correspondences (39.5%), and essays (31.1%). In addition to the options listed on the survey, participants commented that other assignments are also required (See Table 1). Participants also noted activities least required in their online courses. Book reports are rarely required (60.7%), and webquest design (59.6%), and blog participation (41.3%) are infrequently required. By far the most requested course activity was online discussions, as 77.6% of participants ranked it in their top three required course activities.

## Types of Assessments

Participants ranked the types of assessment required in online courses from most to least frequent in this order: Open book multiple choice quiz (24.2%), open book multiple choice exam (21.7%), open book extended response exam (21.1%), portfolios (19.5%), open book extended response (14%), closed book multiple choice exam (10.1%), closed book extended response exam (9.5%), closed book extended response quiz (7.6%), and closed book multiple choice exam (7%). Rankings were similar despite respondent's level of program, discipline, and learning management system used.

## Types of Online Activities Most Preferred

Online students and instructors really enjoy these activities in descending order of preference: Online discussions (37%), PowerPoint presentations (27%), research papers (26%), e-mail correspondence (19%), essays (18%), viewing videos or movies (10%), blogging (9%), viewing webinars (8%) designing webquests (3%), and book reports (2%). When considering the really enjoy and enjoy categories combined; however, the top four preferred activities are: Online discussions (92%), e-mail correspondence (84.5%), essays (79.6%), and research papers (79.1%). Least preferred activities were webquest design (18.6%) and book reports (29.2%).

Another survey question detailed the difficulty level required for assignment completion. Participants believed that e-mail correspondence (85.8%), viewing videos or movies (59.6%), and online discussions (52.5%) were very easy or easy. The most difficult assignments, ranked very difficult and difficult by students and instructors, alike, included research papers (36.8), group projects (30.8), and extended response exams (23.9%).

Using confidence intervals, the researchers assessed which activities survey participants most preferred. The results were statistically significant

*Table 1. Other assignment types reported on survey*

| | | | |
|---|---|---|---|
| Readings | Online Surveys | Creating spreadsheets | Research projects with publication goals |
| Journals | Objective testing | Role playing | Question and answer activities |
| Surveys | Simulations | Annotated bibliographies | Posting to organizational websites |
| Polls | Case studies | Reading summaries | Watching videos on youtube.com |
| Tests | Interviews | Discussion Threads | LinkedIn Group participation |
| Debates | PDF creation | Group presentations | Twitter postings |
| Chats | Visual diagrams | Reflective essays | Project proposals |
| Alcks | Graph Creation | Summarizing reports | Online homework pages |
| Portfolios | Access projects | Chapter reviews | Interactive learning collaborations |
| Wikis | Open Office use | Programming activities | Virtual worlds immersion experiences |
| Math labs | Technical projects | Website creation | Online lecture attendance |

at the 95% confidence level. Data indicate that students prefer online discussions over all other activities (See Table 2). Using regression analysis, controlling for field of study, researchers also determined that students who prefer online discussions rated those discussions as easy.

## Types of Activities that Help Students Retain Course Knowledge

Instructors and students clearly indicated that seven activities best helped students retain knowledge learned in courses: Online discussions (85.5%), essays (81%), research papers (79.1%), PowerPoint presentations (69.1%), e-mail correspondence (62%), group projects (55.3%), and viewing videos or movies (59.1%). Only 5.5% of participants believed that essays did not help students retain knowledge and only 6% did not believe online discussions were of value for student knowledge retention.

## Types of Online Degrees

Participants were enrolled in a variety of degree programs including: Education (36%), Business (35.4%), Technology (13.7%), Psychology (6.2%), Criminal Justice (3.1%), and Health Sciences (1.9%). Other programs listed included Sociology, Statistics, Creative Writing, Liberal Arts, History, Political Science, Arts and Humanities, Spanish, Film and Television Studies, eLearning, Mathematics, Homeland Security, Technical Writing, Rehabilitation Counseling, Organizational Leadership, Chemistry, Cultural Studies, Leadership, Music, Administration, Biology, Philosophy, Professional/Law, IT Management, English as a Second Language, Public Relations, Management, Software Engineering, Building Construction, Paralegal, and Composition.

## Survey Differences According to Fields of Study and LMS

Below are the most frequent course activities assigned by field of study (See Table 3).

Business instructors and students most frequently engage in 9-10 week courses (31.6%). Criminal Justice (60%), Education (46.6%), and Technology (50%) instructors and students typically engage in 15-16 week courses. Psychology instructors and students course lengths were

*Table 2. Preferred course activities one-sample test of preferred course activities*

| | Test Value = 0 | | | | | |
|---|---|---|---|---|---|---|
| | | | | | 95% Confidence Interval of the Difference | |
| | t | df | Sig. (2-tailed) | Mean Difference | Lower | Upper |
| Book Reports | 21.183 | 153 | .000 | 2.325 | 2.11 | 2.54 |
| Research Papers | 42.230 | 162 | .000 | 3.890 | 3.71 | 4.07 |
| Essays | 42.961 | 161 | .000 | 3.864 | 3.69 | 4.04 |
| PPT Presentation | 36.739 | 157 | .000 | 3.816 | 3.61 | 4.02 |
| Online Discussion | 70.834 | 162 | .000 | 4.454 | 4.33 | 4.58 |
| Group Projects | 27.729 | 158 | .000 | 3.208 | 2.98 | 3.44 |
| Videos & Movies | 30.344 | 162 | .000 | 3.466 | 3.24 | 3.69 |
| Webinars | 24.255 | 158 | .000 | 2.975 | 2.73 | 3.22 |
| Blogs | 22.872 | 161 | .000 | 2.864 | 2.62 | 3.11 |
| Webquests | 17.565 | 154 | .000 | 1.897 | 1.68 | 2.11 |
| E-mail | 53.140 | 160 | .000 | 4.050 | 3.90 | 4.20 |

equally distributed (40%) between 7-8 weeks and 11-12 weeks.

Students and instructors in online courses using Angel and WebCT engaged in open book multiple choice exams and quizzes, and extended response exams more than those using Blackboard, eCollege, or Moodle. Blackboard (46%), Angel (61.5%), and WebCT (42.6%) courses most frequently last 15-16 weeks. eCollege (60.9%) courses most frequently last 7-8 weeks, and Moodle courses were evenly distributed (37%) between 5-6 week, 11-12 week, and 15-16 week terms. The following assignments are most frequently assigned per the most common LMS (See Table 4).

## Differences Between Online Undergraduate and Graduate Students

Online undergraduate students most often complete online discussions (63.3%), essays (63.3%), e-mail correspondence (50%), research paper (50%), open book multiple choice exams (45.5%), open book multiple choice quizzes (45.5%), and open book extended response exams (40%). Undergraduates most often enrolled in courses lasting 9-10 weeks (27.3%).

Online graduate students most often complete research papers (72.4%), online discussions (47.4%), e-mail correspondence (32.8%), essays (28.1%), open book extended response exams (19.3%), portfolios (13.8%), and open book multiple choice quizzes (12.1%). Graduate students most often enrolled in courses lasting 11-12 weeks (43.1%).

## Course Length & Format

When responses from both undergraduate and graduate students and instructors were combined, the majority of survey respondents indicated that courses lasted 15-16 weeks (40.4%); however, other courses spanned 1-2 weeks (7.2%), 3-4 weeks (13.9%), 5-6 weeks (21.7%), 7-8 weeks (30.1%), 9-10 weeks (24.7%), 11-12 weeks (28.9%), 13-14 weeks (7.8%), and longer than 16 weeks (4.2%), as noted above.

The respondents participated in both online (83.7%) and blended (44.6%) format courses. Several participants indicated that they teach or participate in blended, online, and face to face

*Table 3. Most frequent course activities by field of study*

| | Business | Criminal Justice | Education | Psychology | Technology |
|---|---|---|---|---|---|
| Research papers | 49.1% | 60% | 50.9% | 70% | 30% |
| Essays | 28.6% | 60% | 34.5% | 60% | 20% |
| Online Discussions | 56.1% | 100% | 59.6% | 60% | 66.7% |
| Group Projects | 29.6% | 40% | 17.5% | 30% | 30% |
| E-mail correspondence | 45.5% | 40% | 31% | 66.7% | 50% |

*Table 4. Most frequently assigned activities by LMS*

| | eCollege | Blackboard | Angel | WebCT | Moodle |
|---|---|---|---|---|---|
| Research papers | 47.8% | 47.5% | 48% | 48.3% | 48% |
| Essays | 26.1% | 28.2% | 33% | 31% | 34.6% |
| Online discussions | 78.3% | 58.5% | 69.2% | 68% | 81.5% |
| e-mail correspondence | 34.8% | 43.3% | 40% | 45.8% | 48% |

programs. Others provided detailed comments including:

- while my program is in a blended format, I cannot attend the residencies due to my employment needs. So, I am only completing the online portion of my program.
- I have learned more on my own than I did in many of my on-site classes at a large university.
- some of my courses are independent study; not online but with a one-on-one tutor.
- my entire master's degree program was held online: even the graduation ceremony, using multimedia and streaming video.
- I attended a university that used blended format before transferring to my current entirely online school.
- the blended format in which scheduled guest lectures can occur and interaction among classmates is encouraged is the best option.
- I love the online format. Although I am having some issues with my current online university, I have completed all of my higher education courses online.
- in my program, the first and final courses require a face to face meeting.
- for entry-level and general education courses, the only option used to be online. Now; however, these courses are offered in a blended format.
- my program requires face to face cohort meetings on alternate weekends.

These responses show that both instructors and students engage in a variety of online learning environments that cannot simply be codified by the terms online and blended format.

## Learning Management Systems

Survey respondents overwhelmingly use Blackboard in their online courses. Of the 166 participants, 107 (64.5%) have used this learning management system. The second most popular system was WebCT with 36.7%. Many participants were unsure of the type of learning management system used (16.9%). There were other LMS reported in use by survey participants including eCollege (13.9%), Angel (15.7%), Compass (13.3%), Moodle (16.3%), and CTools (1.8%). Respondents also listed LMS not included on the survey list. These included both learning management systems and World Wide Web applications. It is unclear if students and instructors are using these Web applications in lieu of or in addition to a traditional LMS. These included University-specific proprietary systems, other proprietary systems, D2L, Google Docs, Yahoo Group, Skype, TRACS, WIMBA, Desire 2 Learn, Newsgroups, Sharepoint, Educator, Centra, WebTyco, SumTotal, Saba, Plateau, Meridian, SkillPort, Wikis, Ning, Sakai, T-Square, Worldclass, iBoard, OLS, and OnCourse.

## CONCLUSION

The researchers surveyed undergraduate and graduate instructors and students to better understand the types of activities assigned in distance learning courses. Drawing on current literature and a survey of online students and instructors from online institutions, the researchers explored the different types of assignments and assessments required for completion of online courses, the Learning Management Systems utilized, and differences between undergraduate and graduate tasks assigned, according to field of study. An analysis of the survey data revealed:

- Online students and instructors utilize a variety of learning formats from blended to exclusively online. Several survey respondents indicated experiences in both types of online formats. Course durations spanned 1-2 week to more than 16 weeks.

When responses from graduate and undergraduate students and instructors were combined, the typical course length was 15-16 weeks.
- Assigned online activities are diverse and include traditional assignments such as research papers and book reports, and innovative, technology-rich assignments like e-mail communication and online discussions. The most favored assignment was online discussions. Survey respondents not only indicated that discussions were enjoyed, but also that discussions helped them retain new knowledge. For these reasons, online course developers should include multiple opportunities throughout a course for online discussions.
- Across all fields, instructors and students are most commonly engaged in the following activities: Research paper writing, essay writing, online discussions, group projects, and e-mail correspondence. There was only a small difference in assignment and assessment variance between fields of study and levels of students as indicated in the research.
- The most commonly utilized LMS is Blackboard. Other LMS included eCollege, Angel, WebCT, and Moodle. Overwhelmingly, learners and instructors engage in online discussions and research paper writing (students) or grading (instructors) regardless of assigned platform. From the research, it does not appear the LMS used had a significant impact on assignment and assessment activity requirements, nor perception of course activities.

It is difficult to distill any best practice applications based on the limited nature of this survey. The most significant trend to emerge from the research is aligned with the Sloan-C pillar on learner effectiveness: Online discussions are a valuable component of distance learning. According to the *Sloan-C Report*, interaction is an essential component of effective learning (Moore, 2005). Instructors and students should interface through online discussions and communication regularly to maximize the benefits of online interaction.

By developing course activities around the creation of relevant artifacts (Fletcher, n.d.), built upon prior knowledge (Liao, 2006) and the use of innovative technologies (Razic & Swanson, 2001), educators can tap into the best of online learning approaches. Employing innovative instructional tasks in an online course positively impacts student achievement and success. Course activities should facilitate deep processing and promote meaningful contextual learning (Ally, 2004) and higher order thinking (Bissell & Lemons, 2006). Teaching in the online environment requires different approaches than the traditional classroom (Shieh, Gummer, & Niess, 2008). This survey provides insight into the rich opportunities available for distance learning course assignments and assessments. In addition to providing access to a wide variety of activities (Kochtanek & Hein, 2000), online instructors should develop strong online teaching skills (Aragon, 2003) to maximize learning. Outstanding educators leave little to happenstance and constantly seek ways to improve teaching practices, increase student achievement, and be more accountable in the online environment.

This research highlighted the diversity of: Online activities, LMS usage, course length, programs of study, and user preferences. In addition, recommendations were made to maximize teaching efficacy based on the literature reviewed. Because online education is a rapidly developing field (Sloan Consortium, 2008), those engaged in online teaching and learning must pioneer new ways of engaging educational practice. These results, coupled with an analysis of scholarly literature on this topic, provide one small step toward deeper understanding of this developing paradigm and better understanding will lead to refined practice.

## REFERENCES


Ally, M. (2004). *Foundations of educational theory for online learning*. Retrieved January 12, 2009, from: http://cde.athabascau.ca/online_book/ch1.html

Aragon, S. (2003). *Facilitating learning in online environments: New directions for adult and continuing education*. San Francisco: Jossey-Bass, Inc.

Bissell, A. N., & Lemons, P. P. (2006, January). A new method for assessing critical thinking in the classroom. *Bioscience*, *56*(1), 66–73. doi:10.1641/0006-3568(2006)056[0066:ANMFAC]2.0.CO;2

Bogan, Y. K., & Porter, R. C. (2005). On the ball with higher-order thinking. *Teaching Pre K – 8, 36*(3), 46-48.

Broadbent, B. (2002). *ABCs of e-Learning: Reaping the benefits and avoiding the pitfalls* (1st ed.). San Francisco: Jossey-Bass.

Carol, A. E., Rivera, F. P., Ebel, B., Zimmerman, F. J., & Christakis, D. A. (2005). *Household and computer Internet access: The digital divide in a pediatric clinical population*. Retrieved May 23, 2009, from http://www.pubmedcentral.nih.gov/articlerender.fcgi?artid=1560660

Carter, D. (2009, March 6). *Top news*. Retrieved May 25, 2009, from http://www.eschoolnews.com/news/top-news/?i=57612

Christe, B. (2003). Designing online courses to discourage dishonesty. *EDUCAUSE Quarterly*, *4*, 54–58.

Connolly, M., Jones, C., & Jones, N. (2007). New approaches, new vision: Capturing teacher experiences in a brave new online world. *Open Learning*, *22*(1), 43–56. doi:10.1080/02680510601100150

Duffy, T., & Cunningham, D. (1996). Constructivism: Implications for the design and delivery of instruction. In Handbook of research for educational telecommunications and technology. New York: MacMillan.

Feenberg, A. (1999). Distance learning: promise or threat. *National Crosstalk*, *7*, 1.

Fletcher, J. M. (n.d.). *Why RTI: Enhancing outcomes for all students*. Retrieved August 2, 2009, from http://ctserc.org/s/index.php?option=com_content&view=article&id=581:why-srbi-enhancing-outcomes-for-all-students-&catid=33:relevant&Itemid=140

Harris, J. J. (2005). Curriculum + differentiation = learning. *Momentum*, *36*(4), 18–21.

Horton, C. B., & Oakland, T. (1997). Temperament-based learning styles as moderators of academic achievement. *Adolescence*, *32*(125), 131–142.

Jonassen, D. H., & Grabowski, B. L. (1993). *Handbook of individual differences, learning, and instruction*. Hillside, NJ: Lawrence Erlbaum Associates.

Kochtanek, T. R., & Hein, K. K. (2000). Creating and nurturing distributed asynchronous learning environments. *Online Information Review*, *24*(4), 280–289. doi:10.1108/14684520010350632

Liao, S. C. (2006). Creating learning scenarios: A planning guide for adult learners. *Convergence*, *39*(1), 101–103.

Lombardi, M. M. (2007, May). *Authentic learning for the 21st century: An overview*. Retrieved January 12, 2009, from http://net.educause.edu/ir/library/pdf/ELI3009.pdf

Maeroff, G. (2003). *A classroom of one: How online learning is changing our schools and colleges*. Houndmills, UK: Palgrave Macmillan.

Moore, J. C. (2005). *The Sloan Consortium quality framework and five pillars*. Retrieved August 2, 2009, from http://www.sloan-c.org/publications/books/qualityframework.pdf

Mupinga, D. M., Nora, R. T., & Yaw, D. C. (2006). The learning styles, expectations, and needs of online students. *College Teaching*, *54*(1), 185–190. doi:10.3200/CTCH.54.1.185-189

Murray, B. (2000, April). Reinventing class discussion online. *Monitor on Psychology, 31*(4). Retrieved May 25, 2009, from http://www.apa.org/monitor/apr00/reinventing.html

Nickerson, M. (2003). *Streaming video in distance education: Giving online humanities instruction a human face*. Retrieved May 31, 2009, from http://www.li.suu.edu/library/courses/humanities1010/

Oliver, R. (n.d.). *Developing online learning environments that support knowledge construction.* Retrieved January 12, 2009, from http://www.edtec.unsw.edu.au/inter/dload/flex_ed/resources/BlendedProgram/EdDesignBits/OliverKnowledgeConstruction.doc

Pogrow, S. (1994). Helping students who just don't understand. *Educational Leadership*, *52*(3), 62–66.

Santo, S. A. (2005, December). *Learning style and Personality Resources*. Retrieved March 1, 2006, from the USD Web Site: http://www.usd.edu/~ssanto/styles.html

Shelton, K., & Saltsman, G. (2005). *An administrator's guide to online education*. Charlotte, NC: Information Age Publishing.

Shieh, R. S., Gummer, E., & Niess, M. (2008, November/December). The quality of a web-based course: Perspectives of the instructor and the students. *TechTrends*, *52*(6), 61–69. doi:10.1007/s11528-008-0220-3

Shin, M., & Lee, Y. (2009, January). Changing the landscape of teacher education via online teaching and learning. *Techniques*, *84*(1), 32–34.

Sloan Consortium. (2008). *Staying the course: Online education in the United States, 2008*. Retrieved August 2, 2009, from http://www.sloan-c.org/publications/survey/pdf/staying_the_course.pdf

Song, L., Singleton, E., Hill, J., & Koh, M. H. (2004). Improving online learning: Student perceptions of useful and challenging characteristics. *The Internet and Higher Education*, *7*, 59–71. doi:10.1016/j.iheduc.2003.11.003

Talor, M. (1997). Learning styles. *Inquiry*, *1*(1), 45–48.

Thompson, M. D. (2004). Faculty self-study research project: Examining the online workload. *Journal of Asynchronous Learning Networks*, *8*(3).

U. S. Department of Labor. (1999, March). *Issues in labor statistics.* Washington, DC: U.S. Government Printing Office.

University of Illinois. (2006). *What Makes a Successful Online Student?* Retrieved May 17, 2009, from http://www.ion.uillinois.edu/resources/tutorials/pedagogy/StudentProfile.asp

Wild, M., & Quinn, C. (1997). Implications of educational theory for the design of instructional multimedia. *British Journal of Educational Technology*, *29*(1), 73–82. doi:10.1111/1467-8535.00047

Zhao, F. (2003). Quality assurance in education. *Bradford, 11*(4), 218-222.

## ADDITIONAL READING


Aldred, L. S., & Reid, B. M. (2003, December). Adopting an innovative multiple media approach to learning for equity groups: Electronically-mediated learning for off-campus students. In G. Crisp, D. Thiele, I. Scholten, S. Barker & J. Baron (Eds.), *Interact, Integrate, Impact: Proceedings of the 20th Annual Conference of the Australasian Society for Computers in Learning in Tertiary Education, Adelaide, 7-10* (pp. 27-36).

Allen, E. L., & Seaman, F. (2007, October). *Online nation: Five years of growth in online learning*. Needham, MA: Sloan Foundation. Retrieved May 27, 2009, from http://www.sloan-c.org/publications/survey/pdf/online_nation.pdf

Bates, A. (2000). *Managing technological change: Strategies for college and university leaders*. San Francisco: Jossey-Bass.

Berner, T. (2004). Less is more: Designing an online course. *Deosnews, 13*, 1-3. Retrieved May 26, 2009, from http://www.ed.psu.edu/ acsde/deos/deosnews.asp

Briggs, D. K. (1982). Adults' recommendations for teaching/learning strategies and andragogy. *Australian Journal of Adult Education, 22*(2), 12–20.

Carnevale, D. (2004, April 30). For online adjuncts, a seller's market. *The Chronicle of Higher Education, 50*(34), A31.

Coldeway, D. O. (1987). Behavior analysis in distance education: A systems perspective. *American Journal of Distance Education, 1*(2), 17–20. doi:10.1080/08923648709526580

Dickenson, G., Agnew, D., & Gorman, R. (1999). Are teacher training and compensation keeping up with institutional demands for distance learning? *Cause/Effect, 22*(3).

Driscoll, M. (2002). *Web-based training: Creating e-learning experiences* (2nd ed.). San Francisco: Pfeiffer.

Garrison, R., & Anderson, T. (2003). *E-Learning in the 21st century: A framework for research and practice*. London: Routledge. doi:10.4324/9780203166093

Gibson, J., & Herrera, J. (1999, January). How to go from classroom based to online delivery in eighteen months or less: A case study in online program development. *T H E Journal, 26*(6), 57. Retrieved May 13, 2009, from http://search.ebscohost.com/login.aspx?direct=true&db=aph&AN=1464348&site=ehost-live

Gilbert, S. D. (2001). *How to be a successful online student*. New York: McGraw Hill.

Hoppe, G., & Breitner, M. H. (2003). *Business models for e-Learning*. Retrieved May 27, 2009, from http://www.wiwi.uni-hannover.de/Forschung/Diskussionspapiere/dp-287.pdf

Kent, T. W., & McNergney, R. F. (1999). *Will technology really change education?*Thousand Oaks, CA: Corwin Press.

Kim, K.-J., & Bonk, C. J. (2006). The future of online teaching and learning in higher education. *EDUCAUSE Quarterly, 29*(4), 22-30. Retrieved May 27, 2009, from http://net.educause.edu/ir/library/pdf/EQM0644.pdf

Knowles, M., Holton, E., & Swanson, R. (2005). *The adult learner: The definitive classic in adult education and human resource development*. Amsterdam: Butterworth-Heinemann.

Milheim, W. (2001). *Faculty and administrative strategies for the effective implementation of distance education*. Retrieved May 17, 2009, from http://www.umsl.edu/technology/frc/pdfs/strategies_for_effective_integration.pdf

National Center for Education Statistics. (2009). *Distance education at degree-granting post-secondary institutions: 2006-2007*. Retrieved March 19, 2009, from http://nces.ed.gov/pubsearch/pubsinfo.asp?pubid=2009044

National School Boards Foundation. (2004). *Technology's impact on education practices*. National School Boards Foundation, Alexandria, VA. Retrieved April 9, 2009, from http://www.nsba.org/sbot/toolkit/tioep.html

O'Lawrence, H. (2006). The influences of distance learning on adult learners. *Techniques*, *81*(5), 47–50.

Pankowski, P. (n.d.). *Faculty training for online teaching*. Retrieved May 17, 2009, from http://www.thejournal.com/articles/16956

Picciano, A. G. (2002). Beyond student perceptions: Issues of interaction, presence, and performance in an online course. *Journal of Asynchronous Learning Networks*, *6*(1), 21–40.

Rovai, A. P., Ponton, M. K., Wighting, M. J., & Baker, J. D. (2007). A comparative analysis of student motivation in traditional classroom and e-learning courses. *International Journal on E-Learning*, *6*(3), 413–432.

Savery, J. R. (2005, Fall). BE VOCAL: Characteristics of successful online instructors. *Journal of Interactive Online Learning*, *4*(2), 141–152.

Simonson, M., & Bauck, T. (2003). Distance education policy issues: Statewide perspectives. Retrieved May 17, 2009, from http://www.questia.com/read/104779442?title=29%3a%20Distance%20Education%20Policy%20Issues%3a%20Statewide%20Perspectives

Smith, P., & Kelly, M. (1987). *Distance education and the mainstream*. London: Croom Helm.

Taylor, M. (2009). *End the university as we know it*. Retrieved May 17, 2009, from http://www.nytimes.com/2009/04/27/opinion/27taylor.html?_r=1&em

The Concord Consortium. (n.d.). *Designing and teaching online courses*. Retrieved May 9, 2008, from http://www.concord.org/courses/designing/

Thornburg, D. (2002). *The new basics: Education and the future of work in the telematic age*. Alexandria, VA: Association for Supervision and Curriculum Development.

Uhlig, G. E. (2002). The present and future of distance learning. *Education*, *122*, 670–673.

Waterhouse, S., & Rogers, R. O. (2004). *The importance of policies in E-learning instruction*. Retrieved May 17, 2009, from http://net.educause.edu/ir/library/pdf/eqm0433.pdf

Whitehead, B. M., Jenson, D. F., & Boschee, F. (2003). *Planning for technology*. Thousand Oaks, CA: Sage.

## KEY TERMS AND DEFINITIONS

**Blended Course:** An online course that requires a face-to-face component for completion.

**Distance Learning:** Any distributed learning approach which allows learners to access course content through methods other than attending a face to face course.

**Extended Response Exam:** A longer assessment that requires essay or short answer responses.

**Extended Response Quiz:** A shorter assessment that requires essay or short answer responses.

**Face-to-Face:** The traditional instructional approach where students physically attend class at a college, university, institution, or off-site location.

**Learning Management System:** Software designed for online learning applications.

**Online Course:** A course that requires no face-to-face interaction for completion.

# Compilation of References


(ND). distance education course offerings. *Journal of Distance Education, 11*(1), 101–110.

(ND)... *Harvard Business Review, 78*(1), 139–145.

*2009/2010 catalog of programs.* (2009). The Art Institute of Pittsburgh. Pennsylvania. Retrieved July 23, 2009 from http://www.artinstitutes.edu/pittsburgh/Admissions/catalog.aspx

*2009-2010 Undergraduate Academic Catalog.* (2009). DeVry University, Illinois. Retrieved July 23, 2009 from http://www.devry.edu/uscatalog

AASCB. (2005). Overview of Assessment. *Assessment Resource Center, AACSB International.* Retrieved January 6, 2006, from http://www.aacsb.edu/resource_centers/assessment/overview-why.asp

*About FrontPage Server Extensions from Microsoft.* (2009). Retrieved July 12, 2009, from http://office.microsoft.com/enus/assistance/HP011032361033.aspx

*ACICS Accreditation criteria, policies, procedures, and standards.* (2009). Accrediting Council for Independent Colleges and Schools. Washington, DC. Retrieved July 23, 2009 from http://www.acics.org/criteria.aspx

Acosta, T., & Liu, Y. (2006). ePortfolios: Beyond assessment. In A. Jafari and C. Kaufman (Eds.), Handbook of research on ePortfolios (pp. 15-23). Hershey, PA: Idea Group Reference.

Ala-Mutka, K., Bacigalupo, M., Kluzer, S., Pascu, C., Punie, Y., & Redecker, C. (2009). *Review of Learning 2.0 Practices.* IPTS technical report. Retrieved April 18, 2009, from http://ipts.jrc.ec.europa.eu/publications/pub.cfm?id=2139

Aleven, V., & Koedinger, K. R. (2002). An effective meta-cognitive strategy: Learning by doing and explaining with a computer-based Cognitive Tutor. *Cognitive Science, 26*, 147–179.

Allen, E., & Seaman, J. (2006). *Making the grade: Online education in the United States, 2006.* Needham, MA: Sloan Center for Online Education. Retrieved September 20, 2008 from http:/./www.sloan-c.org

Allen, I. E., & Seaman, J. (2007). *Online nation: Five years of growth in online learning.* Needham, MA: The Sloan Consortium.

Allen, M. (2003). *Using Scoring Rubrics.* Long Beach, CA: California State University. Retrieved June 20, 2009, from http://www.calstate.edu/itl/sloa/links/using_rubrics.shtml

Ally, M. (2004). *Foundations of educational theory for online learning.* Retrieved January 12, 2009, from: http://cde.athabascau.ca/online_book/ch1.html

Almala, A. H. (2005). A constructivist conceptual framework for a quality e-learning environment. *Distance Learning, 2*(5), 9–12.

Amador, J. A., Miles, L., & Peters, C. B. (2006). *The practice of problem-based learning: A guide to implementing PBL in the college classroom.* Bolton, MA: Anker Publishing Company, Inc.

Anderson, G. J., & Walberg, H. J. (1974). Learning environments. In Walberg, H. J. (Ed.), *Evaluating educational performance: A sourcebook of methods, instruments and examples* (pp. 81–98). Berkeley, CA: McCutchan Publishing.


Copyright © 2010, IGI Global, distributing in print or electronic forms without written permission of IGI Global is prohibited.

Anderson, T. (2003). Getting the mix right again: An updated and theoretical rationale for interaction. *The International Review of Research in Open and Distance Learning, 4*(2). Retrieved July 13, 2008, from http://www.irrodl.org/index.php/irrodl/article/view/149/230

Anderson, T. (Ed.). (2008). *The Theory and Practice of Online Learning* (2nd ed.). Edmonton, Canada: AU Press.

Anderson, T., & Garrison, D. R. (1998). Learning in a networked world: New roles and responsibilities. In Gibson, C. (Ed.), *Distance Learners in Higher Education* (pp. 97–112). Madison, WI: Atwood Publishing.

Angel Learning. (2009). *About angel learning*. Retrieved June 1, 2009, from http://www.angellearning.com

Angelo, T. A. (1999, May). Doing assessment as if learning matters most. *AAHE Bulletin,* 3-6.

Anklam, P. (2001). The Camelot of collaboration: The case of VAX Notes. *Knowledge Management Magazine, 5*(2).

Anonymous,. (2009). HOW REAL? *Mechanical Engineering (New York, N.Y.), 131*(1), 15–16.

Antiplagiat. (2009). *Antiplagiat plagiarism detection service*. Retrieved May 20, 2009, from http://www.antiplagiat.ru

Anwar, M., Greer, J., & Books, C. (2006). Privacy Enhanced Personalization in E-Learning. *Proceedings of the 2006 International Conference on Privacy Security and Trust (PST 2006).* ACM.

*Apache Derby Project*. (2009). Retrieved July 12, 2009, from http://incubator.apache.org/derby/

*Apache HTTP Server Project*. (2009). Retrieved July 12, 2009, from http://httpd.apache.org/

Apel, H. J. (1999). *Die Vorlesung: Einführung in eine akademische Lehrform*. Köln: Böhlau.

Appelt, W. (1999). WWW based collaboration with the BSCW system. In *SOFSEM '99: Proceedings of the 26th Conference on Current Trends in Theory and Practice of Informatics* (LNCS 1725, pp. 66–78). Berlin: Springer.

*Apple Safari*. (2009). Retrieved July 12, 2009, from http://www.apple.com/safari/

Aragon, S. (2003). *Facilitating learning in online environments: New directions for adult and continuing education*. San Francisco: Jossey-Bass, Inc.

ArgoUML v.0.26. (2009). *Open Source Software Engineering Tools*. Tigris, Inc. Retrieved July 12, 2009, from http://argouml.tigris.org/

Armstrong, J. M. A. (1950). *Projects and Their Place in Education*. London: Sir Issac Pitman & Sons, Ltd.

Ashburn, E., & Floden, R. (2006). *Meaningful Learning Using Technology-What Educators Need to Know and Do*. New York: Teachers College Press.

Atan, H., Rahman, Z., & Idrus, R. (2004, June). Characteristics of the Web-Based Learning Environment in Distance Education: Students' Perceptions of Their Learning Needs. *Educational Media International, 41*(2), 103-110. Retrieved May 13, 2009, from doi:10.1080/09523980410001678557

Atkins, D., Seely Brown, J., & Hammond, A. L. *(2007).* A review of the Open Educational Resource movement: achievements, challenges and new opportunities. *Report to the William and Flora Hewlett Foundation. Retrieved-May5, 2009, from*http://www.oerderves.org/wp-content/uploads/2007/03/a-review-of-the-open-educational-resources-oer-movement_final.pdf

Austin, M. J., & Brown, L. D. (1999). Internet Plagiarism: Developing Strategies to Curb Student Academic Dishonesty. *The Internet and Higher Education, 2*(1), 21–33. doi:10.1016/S1096-7516(99)00004-4

Avner, R. A., & Tenczar, P. (1969). *The TUTOR manual*. CERL Report X-4, University of Illinois, Urbana, IL.

Baggaley, J. (2008, May). Where did distance education go wrong? *Distance Education, 29*(1), 39-51. Retrieved July 30, 2009, from doi:10.1080/01587910802004837

Bailey, J. (2006). Web Design Plagiarism. *Plagiarism Today*. Retrieved June 12, 2009, from http://http://www.plagiarismtoday.com/2006/08/02/web-design-plagiarism/

Baker, M. J., Quignard, M., Lund, K., & Séjourné, A. (2003). Computer-Supported Collaborative Learning in the Space of Debate. In Wasson, B., Ludvigsen, S., & Hoppe, U. (Eds.), *Designing for Change in Networked Learning Environments*. Dordrecht, The Netherlands: Kluwer Academic.

Baltes, B. B., Dickson, M. W., Sherman, M. P., Bauer, C. C., & LaGanke, J. S. (2002). Computer-mediated communication and group decision making: A meta-analysis. *Organizational Behavior and Human Decision Processes*, *87*(1), 156–179. doi:10.1006/obhd.2001.2961

Bangert, A. (2004). The seven principles of good practice: A framework for evaluating online teaching. *The Internet and Higher Education*, *7*(3), 217–232. doi:10.1016/j.iheduc.2004.06.003

Banta, T. W. (2003). *Portfolio assessment: Uses, cases, scoring, and impact*. San Francisco, CA: Jossey-Bass.

Barajas, M., & Gannaway, G. (2007). Implementing e-Learning in the Traditional Higher Education Institution. *Higher Education in Europe*, *32*(2-3), 111–119. doi:10.1080/03797720701840609

Barnes, D. J., & Kölling, M. (2004). *Objects First with Java: A Practical Introduction using BlueJ* (2nd ed.). Upper Saddle River, NJ: Prentice Hall.

Barrett, H. (2004). *Electronic portfolio as digital stories of deep learning – emerging digital tools to support reflection in learner-centered portfolios*. Retrieved on June 30, 2009, from http://electronicportfolios.org/digistory/epstory.html

Barrows, H. S. (1988). *The tutorial process*. Springfield, IL: Southern Illinois University Press.

Bassett, E., & Burndt, C. (2006). Beyond course management systems. *EDTECH: Focus on Higher Education October/November 2006*. Retrieved May 9, 2009, from http://www.edtechmag.com/higher/october-november-2006/tech-outlook-2.html

Batson, T. (2002). *Electronic portfolio boom: What's it all about?* Retrieved June 30, 2009, from http://www.campustechnology.com/Articles/2002/11/The-Electronic-Portfolio-Boom-Whats-it-All-About.aspx

Baumgartner, P., Häfele, H., & Maier-Häfele, K. (2002). *E-Learning Praxishandbuch: Auswahl von Lernplattformen. Marktübersicht – Funktionen – Fachbegriffe*. Innsbruck, Austria: StudienVerlag.

Beatty, B., & Ulasewicz, C. (2006). Online teaching and learning in transition: Faculty perspectives on moving from Blackboard to the Moodle learning management system. *TechTrends*, *50*(4), 36–45. doi:10.1007/s11528-006-0036-y

Beaudoin, M. (1990). The Instructor's Changing Role in Distance Education. *American Journal of Distance Education*, *4*(2). doi:10.1080/08923649009526701

Beldarrain, Y. (2006). Distance Education Trends: Integrating new technologies to foster student interaction and collaboration. *Distance Education*, *27*(2), 139–153. doi:10.1080/01587910600789498

Bender, T. (2003). *Discussion-based online teaching to enhance student learning: Theory, practice and assessment*. Sterling, VA: Stylus Publishing.

Bennett, S., & Lockyer, L. (2004). *Becoming an Online Teacher: Adapting to a Changed Environment for Teaching and Learning in Higher Education*. London: Educational Media International, Routledge.

Berge, Z., & Huang, Y. (2004). A Model for Sustainable Student Retention: A Holistic Perspective on the Student Dropout Problem With Special Attention to e-Learning. *Distance Online Symposium, The American Center for the Study of Distance Education, 13*(5).

Bergin, T. (2006). The proliferation and consolidation of word processing software: 1985-1995. *IEEE Annals of the History of Computing, 28*(4).

Bernath, U., & Sangra, A. (Eds.). (2007). *Research on Competence Development in Online Distance Education and E-Learning*. Oldenburg, Germany: IBIS-Vertag der Carl von Ossietzky Universitat.

Berryhill, A. H., & Durrington, V. A. (2006). The online course: The development and implementation of training and support. *Distance Learning*, *3*(2), 51–53.

Bersin, J., Howard, C., O'Leonard, K., & Mallon, D. (2009). *Learning management systems 2009: Facts, practical analysis, trends and provider profiles*. Oakland, CA: Bersin and Associates.

Biggs, J. B. (1999). *Teaching for quality learning at university*. Buckingham, UK: Open University Press.

Birenbaum, M. (1996). Assessment 2000: towards A Pluralistic Approach to Assessment. In Birenbaum, M., & Dochy, F. (Eds.), *Alternatives in Assessment of Achievement, Learning Processes and Prior Knowledge* (pp. 3–31). Boston: Kluwer Academic.

Bissell, A. N., & Lemons, P. P. (2006, January). A new method for assessing critical thinking in the classroom. *Bioscience*, *56*(1), 66–73. doi:10.1641/0006-3568(2006)056[0066:ANMFAC]2.0.CO;2

Bitzer, D. L., Braunfeld, P. G., & Lichtenberger, W. W. (1961). PLATO: An automatic teaching device. *IRE Transactions on Education*, *4*, 157–161. doi:10.1109/TE.1961.4322215

Bitzer, D. L., Slottow, H. G., & Willson, R. H. (1971). *A gaseous discharge display and memory mechanism*. United States Patent 3,559,190. Filing Date: December 22, 1966.

Bizoňová, Z., & Ranc, D. (2007). Model driven LMS platform integration. In *The Third Advanced International Conference on Telecommunications, 2007 (AICT'07)* (pp. 25–29). Washington, DC: IEEE Computer Society.

Blackboard, Inc. (2009). *Blackboard media center press releases*. Retrieved June 10, 2009, from http://www.blackboard.com/Company/Media-Center/Press-Releases.aspx

Bligh, D. A. (2001). *What's The Use of Lectures?* San Francisco, CA: Jossey-Bass Publishers.

*Blogger*. (2009). Retrieved July 12, 2009, from https://www.blogger.com/start

Bloom, B. S. (1953). Thought-Processes in Lectures and Discussions. *The Journal of General Education*, *7*(3), 160–169.

Bloom, B. S. (1976). *Human characteristics and school learning*. New York: McGraw-Hill.

Blue, J. (2009). *BlueJ – Teaching Java – Learning Java*. Retrieved July 12, 2009, from http://www.bluej.org/

*Bluefish Editor*. (2009). Retrieved July 12, 2009, from http://bluefish.openoffice.nl/index.html

*BlueGriffon*. (2009). Retrieved July 12, 2009, from http://bluegriffon.org/

Blumenfeld, P. C., Soloway, E., Marx, R. W., Krajcik, J. S., Guzdial, M., & Palinscar, A. (1991). Motivating project-based learning: Sustaining the doing, supporting the learning. *Educational Psychologist*, *26*(3-4), 369–398. doi:10.1207/s15326985ep2603&4_8

BMBF (Federal Ministry of Education and Research). (2004). *eLearning-Dienste für die Wissenschaft Richtlinien über die Förderung der Entwicklung und Erprobung von Maßnahmen der Strukturentwicklung zur Etablierung von eLearning in der Hochschullehre im Rahmen des Förderschwerpunkts Neue Medien in der Bildung*. Retrieved July 26, 2009, from http://www.bmbf.de/foerderungen/2576.php

Bogan, Y. K., & Porter, R. C. (2005). On the ball with higher-order thinking. *Teaching Pre K – 8*, *36*(3), 46-48.

Boisvert, R. F., & Irwin, M. J. (2006). Plagiarism on the rise. *Communications of the ACM*, *49*(6), 23–24. doi:10.1145/1132469.1132487

Bonk, C., & Zhang, K. (2006). Introducing the R2D2 Model: Online learning for the diverse learners of this world. *Distance Education*, *27*(2), 249–264. doi:10.1080/01587910600789670

Borcea, K., Donker, H., Franz, E., Pfitzmann, A., & Wahrig, H. (2006). Towards Privacy-Aware eLearning. *Proceedings of the 2005 Workshop on Privacy Enhancing Technologies (PET 2005)* (LNCS 3856, pp. 167-178). Berlin: Springer.

Borges, F. (2007) e-Competence of Online Students of the Humanities at UOC. In Research on Competence Development in Online Distance Education and E-Learning. Oldenburg, Germany: BIS-Verlag der Carl von Ossietsky Universitatat.

*Borland JBuilder.* (2009). Retrieved July 12, 2009, from http://www.codegear.com/products/jbuilder

Boss, S., & Krauss, J. (2007). *Reinventing Project-Based Learning: Your Field Guide to Real-World Projects in the Digital Age.* New York: ISTE.

Bourke, P. (2009). Evaluating Second Life for the collaborative exploration of 3D fractals. *Computers & Graphics, 33*(1), 113–117. doi:10.1016/j.cag.2008.08.004

Bourque, M., Franklin, S. D., & Updegrove, D. A. (2005, October 19). *Delivering shared, authenticated, ubiquitously accessible storage via WebDAV.* Paper presented at the EDUCAUSE 2005, Transforming the Academy: Dreams and Reality, Orlando.

Boyle, T. (2003). Design principles for authoring dynamic, reusable learning objects. *Australian Journal of Educational Technology, 19*(1), 46–58.

Bradford, P., Porciello, M., Balkon, N., & Backus, D. (2007). The Blackboard Learning System: the be all and end all in educational instruction? *Journal of Educational Technology Systems, 35*(3), 301–314. doi:10.2190/X137-X73L-5261-5656

Brandon-Hall. (2009). *Brandon-Hall Research LMS KnowedgeBase 2009.* Retrieved June 10, 2009, from http://www.brandon-hall.co/publications/lmskb/lmskb_firms.shtml

Bransford, Brown & Conking. (2000). *Designing Effective Projects: Characteristics of Projects Benefits of Project-Based Learning.* Retrieved June 17, 2009, from http://download.intel.com/education/Common/en/Resources/DEP/projectdesign/DEP_pbl_research.doc

Bransford, J. D., & Vye, N. J. (1989). A perspective on cognitive research and its implications for instruction. In L. B. Resnick & L. E. Klopfer [ASCD Yearbook] (Eds.), Towards the thinking curriculum: Current cognitive research. Alexandria, VA: Association for Supervision and Curriculum Development.

Brecht, H. D., & Ogilby, S. M. (2008). Enabling a Comprehensive Teaching Strategy: Video Lectures. *Journal of Information Technology Education, 7,* 71–86.

Breytenbach, K. (2005, August 12). Poet accused of plagiarism gives back her prize. *Cape Times.* Retrieved June 16, 2009, from http://www.capetimes.co.za/general/print_article.php?fArticleId=2834085

Broadbent, B. (2002). *ABCs of e-Learning: Reaping the benefits and avoiding the pitfalls* (1st ed.). San Francisco: Jossey-Bass.

Brooks, J., & Brooks, M. (1993). *In Search of Understanding: The Case for Constructivist Classrooms.* Upper Saddle River, NJ: Prentice Hall.

Brooks, L. (2003). How the Attitudes of Instructors, Students, Course Administrators, and Course Designers Affects the Quality of an Online Learning Environment. *Online Journal of Distance Learning Administration, 6*(4).

Brookshear, J. G. (2009). *Computer-Science: An Overview* (10th ed.). Reading, MA: Addison-Wesley.

Brown, G., & Irby, B. J. (2000). *The career advancement portfolio* (2nd ed.). Thousand Oaks, CA: Corwin.

Brown, K., & Christianson, L. (2005). *OPNET Lab Manual to Accompany Data and Computer Communications* (7th ed.). Upper Saddle River, NJ: Pearson Prentice Hall.

Bruffee, K. A. (1995). *Collaborative Learning: Higher Education, Interdependence, and Authority of Knowledge.* Baltimore, MD: John Hopkins University Press.

Buckley, W., & Smith, A. (2007). Application of Multimedia Technologies to Enhance Distance learning. *RE:view, 39*(2), 57–65. doi:10.3200/REVU.39.2.57-65

Burch, K. (2000). A primer on problem-based learning for international relations courses. *International Studies Perspectives, 1*, 31–44. doi:10.1111/1528-3577.00003

Burnett, M. N., & Williams, J. M. (2009). Institutional uses of rubrics and e-portfolios: Spelman College and Rose-Hulman institute. *Peer Review, 11*(1).

Cambridge, B. L. (2001). *Electronic portfolios: Emerging practices in student, faculty, and institutional learning.* Sterling, VA: Stylus.

Campus Computing Project. (2002). *The campus computing report.* Encino, CA: Campus Computing.

Card, S., Mackinlay, J., & Shneiderman, B. (1999). *Readings in Information Visualization - using vision to think.* San Francisco, CA: Morgan Kaufmann.

Carnevale, D. (2007). Pearson's purchase of eCollege signals publishers' interest in online education. *The Chronicle of Higher Education, 53*(38).

Carol, A. E., Rivera, F. P., Ebel, B., Zimmerman, F. J., & Christakis, D. A. (2005). *Household and computer Internet access: The digital divide in a pediatric clinical population.* Retrieved May 23, 2009, from http://www.pubmedcentral.nih.gov/articlerender.fcgi?artid=1560660

Caronia, N. (2007). Virtual world provides new opportunities for education. *Government Video, 18*(7), 8–9.

Carroll, A. W. (1975). *Personalizing education in the classroom.* Denver: Love Publishing.

Carroll, J. M., & Rosson, M. B. (2005a). A Case Library for Teaching Usability Engineering: Design Rationale, Development, and Classroom Experience. *ACM Journal on Educational Resources in Computing, 5*(1), 1–22. doi:10.1145/1101670.1101673

Carroll, J. M., & Rosson, M. B. (2005b). Toward even more authentic case-based learning. *Educational Technology, 45*(6), 5–11.

Carroll, J. M., & Rosson, M. B. (2006). Case Studies as Minimalist Information. *IEEE Transactions on Professional Communication, 49*(4), 297–310. doi:10.1109/TPC.2006.885836

Carroll, J. M., Ganoe, C. H., & Jiang, H. (2008). Introducing the digital case library. In Pagani, M. (Ed.), *Encyclopedia of Multimedia Technology and Networking* (2nd ed., pp. 782–788). Hershey, PA: IGI Publishing.

Carter, D. (2009, March 6). *Top news.* Retrieved May 25, 2009, from http://www.eschoolnews.com/news/top-news/?i=57612

Carter, D. (2009, May 7). Blackboard–Angel deal gives pause to some. *eSchool News.* Retrieved July 26, 2009, from http://www.eschoolnews.com/news/top-news/?I=58675

Case, R. (1992). On the need to assess authentically. *Holistic Education Review, 5*(4), 14–23.

Catalyst, I. T. Ltd. (2004). Technical evaluation of selected learning management systems. Technical report, The Open Polytechnic of New Zealand, Wellington, New Zealand.

CDPNet. (2009). *Collaborative plagiarism detection network.* Retrieved June 3, 2009, from http://www.cpdnet.org

*CDT Plugin for Eclipse.* (2009). Retrieved July 12, 2009, from http://www.eclipse.org/cdt/

*Center for Women and public policy.* (n.d.). Retrieved April 19, 2008, from http://www.hhh.umn.edu/centers/wpp/case_studies.html

Ceraulo, S. (2005). Benefits of upgrading to LMS. *Distance Education Report, 9*(9).

Cerverón, V., & Moreno, P. (2006). Aula Virtual: an e-Learning Management Platform of Universitat de València Development based on Open Code and Collaborative Software. In *Proc. M-ICTE, IV Conferencia Internacional sobre Multimedia y Tecnologías de la Información y Comunicación en Educación* (Vol. 2). Badajoz, Spain: FORMATEX.

Cerverón, V., Moreno, P., Cubero, S., Roig, D., & Roca, S. (2007). Universitat de València's Aula Virtual: a Single Integrated LMS for a University. In *Proc. IADIS e-Learning 2007 Conference Proceedings*, IADIS Digital Library, Lisbon, Portugal.

Chaiklin, S. (2003). The Zone of Proximal Development in Vygotsky's Analysis of Learning and Instruction. In Kozulin, A., Gindis, B., Ageyev, V., & Miller, S. (Eds.), *Vygotsky's Educational Theory and Practice in Cultural Context*. Cambridge, UK: Cambridge University Presents.

Chang, C. (2001). A study on the evaluation and effectiveness analysis of web-based learning portfolio (WBLP). *British Journal of Educational Technology*, *32*(4), 435–458. doi:10.1111/1467-8535.00212

Chang, V., & Fisher, D. (2001, December 2-6). *The validation and application of a new learning environment instrument to evaluate online learning in higher education*. Paper presented at the Australian Association for Research in Education, Fremantle. Retrieved from http://www.aare.edu.au/01pap/cha01098.htm

Chao, I. T. (2008). Moving to Moodle: Reflections two years later. *EDUCAUSE Quarterly*, *3*, 46–52.

Chasen, M. (2009). *Future of blackboard and angel*. Keynote presentation at the Angel User Conference, Chicago, IL.

Chen, J., & Lin, T.-F. (2008). Class attendance and exam performance: a randomized experiment. *The Journal of Economic Education*, *39*(3), 213–227. doi:10.3200/JECE.39.3.213-227

Chen, K., Fawaz, W., Martignon, F., & Pujolle, G. (2006). A novel fault management approach for DWDM optical networks. *International Journal of Network Management*, *16*(5), 337–349. doi:10.1002/nem.601

Chen, W. C., Fuchs, H., Nyland, L., Towles, H., & Welch, G. (2000). Toward a compelling sensation of telepresence: demonstrating a portal to a distant (static) office. In *Proceedings of the Conference on Visualization,* Salt Lake City United.

Chism, N. (2004). Using a framework to engage faculty in instructional technologies. *EDUCAUSE Quarterly*, *2*, 39–45.

Christe, B. (2003). Designing online courses to discourage dishonesty. *EDUCAUSE Quarterly*, *4*, 54–58.

Christian, W., & Belloni, M. (2001). *Physlets*. Englewood Cliffs, NJ: Prentice Hall.

*Cisco: Welcome to the Human Network*. (n.d.). Retrieved April 30, 2008, from http://www.cisco.com/web/thehumannetwork/index.html

Clark, H. H., & Brennan, S. E. (1991). Grounding in Communication. In Resnick, L., Levine, J., & Teasley, S. (Eds.), *Perspectives on Socially Shared Cognition* (pp. 127–149). Hyattsville, MD: American Psychological Association. doi:10.1037/10096-006

Clark, R. E. (1983). Reconsidering research on learning from media. *Review of Educational Research*, *53*(4), 445–459.

*Clickers*. (2009). Ohio State University. Retrieved from http://telr.osu.edu/clickers

Clough, P. (2000). *Plagiarism in Natural and Programming Languages: An Overview of Current Tools and Technologies. Research Memoranda, CS-00-05*. UK: Department of Computer Science, University of Sheffield.

Coates, H. (2006). *Student Engagement in Campus-based and Online Education*. London: Routledge.

Cocea, M., & Weibelzahl, S. (2006). Can log files analysis estimate learner's level of motivation? In Schaaf, M., & Althoff, K.-D. (Eds.), *Lernen-Wissensentdeckung* (pp. 32–35). Adaptivität.

*Code:Blocks*. (2009). Retrieved July 12, 2009, from http://www.codeblocks.org/

Cohen, P. A., Ebeling, B. J., & Kulik, J. A. (1981). A meta-analysis of outcome studies of visual-based instruction. *Educational Communication and Instructional Journal*, *29*(1), 26–36.

Cole, J., & Foster, H. (2007). *Using moodle* (2nd ed.). Sebastapol, CA: O'Reilly Media.

Cole, M. (1996). *Cultural psychology: A once and future discipline*. Cambridge, MA: Belknap/Harvard.

Coleman, J. S. (1988). Social capital in the creation of human capital. *American Journal of Sociology*, *94*, S95–S121. doi:10.1086/228943

Colla, J., MacDonald, C., & Thompson, T. L. (2005). Structure, Content, Delivery, Service, and Outcomes: Quality e-Learning in higher education. *International Review of Research in Open and Distance Learning*, *6*(2).

Collis, B., & Moonen, J. (2001). *Flexible Learning in a Digital World: Experiences and Expectations. Open & Distance Learning Series*. London: Kogan Page.

Concannon, F., Flynn, A., & Campbell, M. (2005). What campus-based students think about the quality and benefits of e-learning. *British Journal of Educational Technology*, *36*(3), 501–512. doi:10.1111/j.1467-8535.2005.00482.x

Connolly, M., Jones, C., & Jones, N. (2007). New approaches, new vision: Capturing teacher experiences in a brave new online world. *Open Learning*, *22*(1), 43–56. doi:10.1080/02680510601100150

Conole, G., & Oliver, M. (Eds.). (2007). *Contemporary Perspectives in E-learning Research. Themes, methods and impact on practice*. London, New York: Routledge.

Conole, G., & Warburton, B. (2005). A review of computer-assisted assessment. *ALT-J*, *13*(1), 17–31. doi:10.1080/0968776042000339772

Conole, G., Brasher, A., Cross, S., Weller, M., Clark, P., & Culver, J. (2008). Visualising learning design to foster and support good practice and creativity. *Educational Media International*, *45*(3), 177–194. doi:10.1080/09523980802284168

Conole, G., de Laat, M., Dillon, T., & Darby, J. *(2007, Mar26-28)*. 'Disruptive technologies', 'pedagogical innovation': What's new? Findings from an in-depth study of students' use and perception of technology. *Paper presented at the Biennial Conference on Computer Assisted Learning (CAL 07), Dublin, IRELAND.*

Conrad, D. (2008). From Community to Community of Practice: Exploring the Connection of Online Learners to Informal Learning in the Workplace. *American Journal of Distance Education*, *22*(1), 3–23. doi:10.1080/08923640701713414

Consejo de Coordinación, U. en colaboración con la Comisión para la Renovación de Metodologías Educativas en la Universidad. (2006). *Propuestas para la renovación de las metodologías educativas en la universidad.* [Online]. Retrieved April 25, 2009, from http://www.micinn.es/univ/ccuniv/html/metodologias/docu/PROPUESTA_RENOVACION.pdf

Cook, V. S., & McElwrath, K. (2009). Interview with Tulio Llosa, Director of Educational Technology, Information Technology Services University of Illinois Springfield and Thomas Sidener, Academic Technology Service Specialist, Information Technology Services (unpublished interviewed.). Springfield, IL.

Coombs, C. Jr, & Coombs, C. A. (1998). *Communications Network Test & Measurement Handbook*. New York: McGraw-Hill.

Cooper, K. (2006). Rivier College Launches Online Programs. *Rivier College Online Academic Journal*, *2*(2). Retrieved July 12, 2009, from http://www.rivier.edu/journal/RCOAJ-Fall-2006_table.htm

CopyCatch. (2009). *CopyCatch Software Suite*. Retrieved May 20, 2009, from http://www.copycatchgold.com/

Corich, S. (2005). Let's get ready to Moodle. *Bulletin of Applied Computing and Information Technology*, *3*(3). Retrieved from http://www.naccq.ac.nz/bacit/0303/2005Corich_LMS.htm.

Cotton, J. (1995). *The theory of assessment. An introduction*. London: Kogan Page.

Cragg, C., Dunning, J., & Ellis, J. (2008). Teacher and Student Behaviors in Face-to-Face and On-Line Courses: Dealing with Complex Concepts. *Journal of Distance Education*, *22*(3), 115–128.

*Crimson Editor*. (2009). Retrieved July 12, 2009, from http://www.crimsoneditor.com/

Cuban, L. (1986). *Teachers and Machines: The Classroom Use of Technology Since 1920*. New York: Teachers College Press.

Cuban, L. (2001). *Oversold and Underused: Computers in the Classroom*. Cambridge, MA: Harvard University.

Cubero, S., López, A., Roig, D., & Roca, S. (2006). Implantación de.LRN en la Universitat de València Estudi general. Integración de la plataforma y desarrollos propios. In Proc. Foro e-learning integral: Estratregias y experiencias, Universidad Galileo, Guatemala.

Curtis, B., Krasner, H., & Iscoe, N. (1988). A Field Study of the Software Design Process for Large Systems. *Communications of the ACM, 31*(11). doi:10.1145/50087.50089

Dabbagh, N., & Bannan-Ritland, B. (2005). *Online learning: Concepts, strategies, and applications.* Upper Saddle River, NJ: Pearson Prentice Hall.

Dale, N. (2003). *C++ Plus Data Structures* (3rd ed.). Boston, MA: Jones and Bartlett Publishers.

Darling-Hammond, L. (1996). The right to learn and the advancement of teaching: Research, policy and practice for democratic education. *Educational Researcher, 25*(6), 5–17.

Darling-Hammond, L. (1997). *The right to learn: A blueprint for creating schools that work.* San Francisco, CA: Jossey-Bass.

De Boer, W. (2004). *Flexibility support for a changing university.* PhD thesis, University of Twente, Enschede, The Netherlands.

De la Teja, I., & Spannaus, T. W. (2008). New online learning technologies: new online learner competencies. Really? In Visser, J., & Visser-Valfrey, M. (Eds.), *Learners in a changing learning landscape: Reflections from a dialogue on new roles and expectations* (pp. 187–211). Dordrecht, The Netherlands: Springer.

De Lucia, A., Francese, R., Passero, I., & Tortora, G. (2009). Development and evaluation of a virtual campus on Second Life: The case of SecondDMI. *Computers & Education, 52*(1), 220–233. doi:10.1016/j.compedu.2008.08.001

deHaan, J., & Diamond, J. (2007). The experience of telepresence with a foreign language video game and video. In *Proceedings of the 2007 ACM SIGGRAPH Symposium on Video Games* (San Diego, California, August 04 - 05, 2007).

Del Ramo Romero, J., & López-Soler, N. (2008). Respondus: un programa para la creación y publicación de cuestionarios. *@tic. Revista d'innovació Educativa. dotLRN Consortium.* [Online]. Retrieved April 25, 2009, from http://www.dotlrn.org

*Delfino, M., & Persico, D. (2007). On-line or face-to-face? Experimenting with different techniques in teacher training.* Journal of Computer Assisted Learning.

Deloitte & Touche. LLP (2007). *2006 Canadian technology fast 50 winners.* Retrieved June 11, 2009, from http://www.deloitte.com.

Deming, W. E. (1986). *Out of the crisis.* Cambridge, MA: MIT, Center for Advanced Engineering Study.

Dennis, A., Wixom, B. H., & Tegarden, D. (2009). *System Analysis & Design, UML Version 2.0: An Object-Oriented Approach* (3rd ed.). Hoboken, NJ: John Wiley & Sons, Inc.

Dennis, J. (2003). *The Complete Idiot's Guide to Physics.* Indianapolis, IN: Alpha Books.

*DETC Accreditation Overview.* (2007). Distance Education and Training Council. Washington, DC. Retrieved July 23, 2009 from http://www.detc.org/freeprublications.html

Dev, C. ++. (2009). Retrieved July 12, 2009, from http://www.bloodshed.net/devcpp.html

Devadoss, S., & Foltz, J. (1996). Evaluation of factors influencing student class attendance and performance. *American Journal of Agricultural Economics, 78,* 499–507. doi:10.2307/1243268

Devine, J. *(2005).* Major challenges facing the higher education system in the ICT era. *Retrieved May 26, 2009, from* http://www.elearningeuropa.info/directory/index.php?doc_id=1151&doclng=6&page=doc

Dewey, J. (1933). *How We Think, A Restatement of the Relation of Reflective Thinking to the Educative process.* New York, D. C.: Heath.

Di Domenico, F., Panizzi, E., Sterbini, A., & Temperini, M. (2005). *Analysis of commercial and experimental e-learning systems. Technical report.* Trondheim, Norway: TISIP Research Foundation.

Diaz, V., Garrett, P. B., Kinley, E. R., Moore, J. F., Schwartz, C. M., & Kohrman, P. (2009, May/June). Faculty development for the 21st century. *EDUCAUSE Review*, 47–55.

DiBiase, D., Zembal-Saul, C., Sabre, J., Howard, D., Rademacher, H., Burlingame, P., et al. (2002). *Using e-Portfolio at Penn State to enhance student learning: Status, prospects and strategies*. E-education Institute, The Pennsylvania State University. Retrieved July 9, 2009, from https://www.e-education.psu.edu/files/e-port_report.pdf

Dietrich, S. W., & Urban, S. D. (1996). Database theory in practice: learning from cooperative group projects. In *ACM SIGCSE Bulletin: Proceedings of the twenty-seventh SIGCSE technical symposium on Computer science education, 28*(1), 112-116.

Dion, J. (2006a). Automated Music Composition: An Expert Systems Approach. *Rivier College Online Academic Journal, 2*(1). Retrieved July 12, 2009, from http://www.rivier.edu/journal/RCOAJ-Spring-2006_table.htm

Dion, J. (2006b). Musician Web-service Using Ruby-on-Rails, SOAP, FLEX, and AJAX. *Rivier College Online Academic Journal, 2*(2). Retrieved July 12, 2009, from http://www.rivier.edu/journal/RCOAJ-Fall-2006_table.htm

Dorrian, J., & Wache, D. (2009). Introduction of an on-line approach to flexible learning for on-campus and distance education students: Lessons learned and ways forward. *Nurse Education Today, 29*(2), 157–167. doi:10.1016/j.nedt.2008.08.010

Douglas, I. (2008). Measuring participation in Internet supported courses. In. *Proceedings of the International Conference on Computer Science and Software Engineering, 5*, 714–718.

Douglas, I., & Alemanne, N. D. (2007a). Monitoring participation in online courses. In *Proceedings of the 8th International Conference on Information Technology Based Higher Education and Training* (ITHET 2007), Kumamoto, Japan.

Douglas, I., & Alemanne, N. D. (2007b). Monitoring student participation and effort. In *Proceedings of the IADIS International Conference on Cognition and Exploratory Learning in Digital Age* (CELDA 2007), Algarve, Portugal (pp. 299-302).

Dringus, L., & Ellis, T. (2005). Using data mining as a strategy for assessing asynchronous discussion forums. *Computer & Education Journal, 45*, 141–160. doi:10.1016/j.compedu.2004.05.003

Driscoll, A., & Wood, S. (2007). *Developing outcomes-based assessment for learner-centered education: A faculty introduction*. Sterling, VA: Stylus.

Driscoll, M. P. (2007). Psychological theories of instructional design. In Reiser, R. A., & Dempsey, J. V. (Eds.), *Trends and issues in instructional design and technology* (2nd ed.). Upper Saddle River, NJ: Pearson Merrill/Prentice-Hall.

Driver, E. (2008). In-World Work:The Immersive Internet Increases Engagement While Reducing Business Costs. *DM Review, 18*(11), 30.

Dubin, R., & Taveggia, T. C. (1968). *The Teaching-Learning Paradox: A comparative analysis of college teaching methods*. Eugene, OR: Center for the Advanced Study of Educational Administration, University of Oregon.

DuCharme-Hansen, B., & Dupin-Bryant, P. (2004). *Web-Based Distance education for Adults*. Malabar, FL: Krieger Publishing Company.

Duffy, T., & Cunningham, D. (1996). Constructivism: Implications for the design and delivery of instruction. In Handbook of research for educational telecommunications and technology. New York: MacMillan.

Dwyer, D. (2006). Software Reliability Estimations/Projections, Cumulative and Instantaneous. *Rivier College Online Academic Journal, 2*(1). Retrieved July 12, 2009, from http://www.rivier.edu/journal/RCOAJ-Spring-2006_table.htm

Eberle, F., & Keller, S. (2003). Unterstützung des kooperativen Lernens im Rahmen von Vorlesungen durch ICT. In Schweizerische Zeitschrift für kaufmännisches Bildungswesen, 97(3), 97–117.

*Eclipse*. (2009). Retrieved July 12, 2009, from http://www.eclipse.org/

EduTools. (2009). *CMS: CMS home*. Retrieved June 10, 2009 from http://www.edutools.info/static.jsp?pj=4&page=HOME

Eduventures, Inc. (2008). *Challenges and opportunities in postsecondary outcomes assessment – benchmarking institutional progress*. Boston: Eduventures, Inc., Retrieved March 22, 2009, from http://www.eduventures.com/

Eduventures. (2007). *Postsecondary academic computing: Identifying opportunities beyond LMS*. Retrieved from http://www.eduventures.com

Ehlers, U.-D. (2007). Making the Difference in E-Learning: Towards Competence Development and E-Irritation. In Research on Competence Development in Online Distance Education and E-Learning. Oldenburg, Germany: BIS-Verlag der Carl von Ossietsky Universitatat

Eibl, C., von Solms, B., & Schubert, S. (2006). A Framework for Evaluating the Information Security of E-Learning Systems. In *Proceedings of 2nd Internation Conference on Information in Secondary Schools: Evolution and Perspectives (ISSEP 2006)* (LNCS 4226). Berlin: Springer.

Eissen, S. M., & Stein, D. (2006). Intrinsic Plagiarism Detection. *Lecture Notes in Computer Science*, *3936*, 565–569. doi:10.1007/11735106_66

Eklund, J., Kay, M., & Lynch, H. (2003). E-learning: Emerging issues and key trends. *Australian Flexible Learning Framework*. Retrieved May 10, 2008, from http://www.flexiblelearning.net.au

*E-LANE Project*. [Online]. Retrieved April 25, 2009, from http://www.e-lane.org

Engels, S., Lakshmanan, V., & Craig, M. (2007). Plagiarism detection using feature-based neural networks. In *Proceedings of the 38th SIGCSE technical symposium on computer science education* (pp. 34–38).

Entwistle, N. (1998). Approaches to learning and forms of understanding. In B. Dart & Boulton-Lewis (Eds.), Teaching and learning in higher education. Melbourne: ACER.

Essa, A., Cerverón, V., & Blessius, C. (2005). An Enterprise Open-Source Learning Management System. In *Proc. Educause*. Orlando: LRN.

European Commission. (2007). *The Bologna Process - Towards the European Higher Education Area*. Retrieved April 6, 2009, from http://ec.europa.eu/education/policies/educ/bologna/bologna_en.html

Eynon, R. (2008). *The use of the world wide web in learning and teaching in higher education: reality and rhetoric*. Innovations in Education & Teaching International.

*FEDORA*. (n.d.). Retrieved April 19, 2008, from http://www.fedora.info

Feenberg, A. (1989). The written world: On the theory and practice of computer conferencing. In Mason, R., & Kaye, A. (Eds.), *Mindweave: communication, computers and distance education* (pp. 22–39). Oxford, UK: Pergamon Press.

Feenberg, A. (1999). Distance learning: promise or threat. *National Crosstalk*, *7*, 1.

Fibre Channel Industry Association. (2009). Retrieved July 12, 2009, from http://www.fibrechannel.org/

Fiddler Web Debugger. (2009). Retrieved July 12, 2009, from http://www.fiddlertool.com/fiddler/

Field, R. M. (2005). Favourable conditions for effective and efficient learning in a blended face-to-face/online method. In *Ascilite 2005: Balance, Fidelity, Mobility: Maintaining the Momentum,* 4-7 December 2005, Queensland University of Technology. Retrieved June 20, 2009, from http://eprints.qut.edu.au/18100/1/c18100.pdf

*Firebird.* (2009). Retrieved July 12, 2009, from http://firebird.sourceforge.net/

*Firebug.* (2009). Retrieved July 12, 2009, from http://getfirebug.com/

Fjermestad, J. (2004). An analysis of communication mode in group support systems research. *Decision Support Systems, 37*(2), 239–263.

Flanagan, D. (2005). *Java in a Nutshell* (5th ed.). New York: O'Reilly & Associates.

Flanigan, E. J., & Amirian, S. (2006). ePortfolios: Pathway from classroom to career. In A. Jafari & C. Kaufman (Eds.), Handbook of research on ePortfolios (pp. 102-111). Hershey, PA: Idea Group Reference.

Flavell, J. H. (1979). Metacognition and cognitive monitoring: A new area of cognitive-developmental inquiry. *The American Psychologist, 34*, 906–911. doi:10.1037/0003-066X.34.10.906

Fletcher, J. M. (n.d.). *Why RTI: Enhancing outcomes for all students.* Retrieved August 2, 2009, from http://ctserc.org/s/index.php?option=com_content&view=article&id=581:why-srbi-enhancing-outcomes-for-all-students-&catid=33:relevant&Itemid=140

Foster, A. L. (2002). Plagiarism-detection tool creates legal quandary. *The Chronicle of Higher Education*, , A37.

Fraser, B. J. (1998). *Classroom Environment Instruments: Development, Validity and Applications.* Berlin: Springer.

*FreeQueryBuilder.* (2009). Retrieved July 12, 2009, from http://sourceforge.net/projects/querybuilder/

Fuente, L. (2006). Máquina virtual utilizada para el desarrollo del proyecto. *Proyecto e-lane.* [Online]. Universidad Carlos III de Madrid. Retrieved April 25, 2009, from http://e-lane.org/vmware/

Galusha, J. M. (1997). Barriers to Learning in Distance Education. *Interpersonal Computing & Technology, 5*(3-4), 6–14.

Ganoe, C., Somervell, J., Neale, D., Isenhour, P., Carroll, J. M., Rosson, M. B., & McCrickard, D. S. (2003). Classroom BRIDGE: Using Collaborative Public and Desktop Timelines to Support Activity Awareness. In *Proceedings of the ACM Conference on User Interface Software and Technology* (UIST '03), Vancouver, BC, Canada.

GAO (United States Government Accountability Office). (2007). *HIGHER EDUCATION: Tuition Continues to Rise, but Patterns Vary by Institution Type, Enrollment, and Educational Expenditures.* GAO-08-245, Report to the Chairman, Committee on Education and Labor, House of Representatives. Retrieved from http://www.gao.gov/new.items/d08245.pdf

Garrison, D. R., & Kanuka, H. (2004). Blended learning: Uncovering its transformative potential in higher education. *The Internet and Higher Education, 7*(2), 95–105. doi:10.1016/j.iheduc.2004.02.001

Garvin, D. A. (2003, September). Making the case. *Harvard Magazine, 106*(1), 56–65.

Georgiades, W. D., Keefe, J. W., Lowery, R. E., e al. (1979). *Take five: A methodology for the humane school.* Los Angeles, CA: Parker and Son.

Gerson, S. M. (2000). E-CLASS: Creating a Guide to Online Course Development for Distance Learning Faculty. *Online Journal of Distance Learning Administration, 3*(4). Retrieved June 21, 2009, from http://www.westga.edu/~distance/ojdla/winter34/gerson34

Ghosh, R. A. (2006). *CODE: Collaborative ownership and the digital economy.* Boston: MIT Press.

Gibbons, S. (2005). Library course-management systems: an overview. *ALA Library Technology Reports, 41*(3).

Gibbs, G., & Simpson, C. (2004). Conditions under which assessment supports students' learning. *Learning and Teaching in Higher Education, 1*, 3–31.

Glass, J., & Sue, V. (2008). Student Preferences, Satisfaction, and Perceived Learning in an Online Mathematics Class. *MERLOT Journal of Online Learning and Teaching, 4*(3).

Glasser, W. (1986). *Choice theory in the classroom*. New York: Harper Perennial.

Glasser, W. (1988). *The quality school*. New York: HarperCollins.

Glesne, C. (2006). *Becoming qualitative researchers* (3rd ed.). Boston, MA: Pearson Education.

Glynn, K. A., & Rajendran, K. N. (1993, September). Perceptual-based student outcomes assessment process in the marketing curriculum. *Journal of Education for Business*, *69*, 11–17.

Gnägi, F. (2006). *OLAT–Moodle Comparison*. Retrieved July 31, 2009, from http://www.frentix.com/de/olat/docu/OLAT-MOODLE-en.pdf

*GNU C++ (GCC)*. (2009). Retrieved July 12, 2009, from http://gcc.gnu.org/

*GNU emacs*. (2009). Retrieved July 12, 2009, from http://www.gnu.org/software/emacs/emacs.html

Gonick, L., & Criddle, C. (2005). *The Cartoon Guide to Chemistry*. New York: HarperResource.

Gonick, L., & Devault, C. (1999). *The Cartoon Guide to Sex*. New York: HarperPerennial.

Gonick, L., & Outwater, A. (1996). *The Cartoon Guide to the Environment*. New York: HarperPerennial.

Gonick, L., & Smith, W. (1993). *The Cartoon Guide to Statistics*. New York: HarperPerennial.

Gonick, L., & Wheelis, M. (1991). *The Cartoon Guide to Genetics*. London: HarperCollins.

Gonick, L., Huffman, A., & Huffman, A. (1991). *The Cartoon Guide to Physics*. New York: HarperPerennial.

Gooch, S. J. (1980). *Cybernetic music system*. United States Patent 4,206,675. Filing Date: February 28, 1977.

Goodlad, J. I. (1984). *A place called school*. New York: McGraw-Hill.

*Google Chrome*. (2009). Retrieved July 12, 2009, from http://www.google.com/chrome

*Google Sites*. (2009). Retrieved July 12, 2009, from http://sites.google.com/

Graf, F. (2002). Providing security for eLearning. *Computers & Graphics*, *26*(2), 355–365. doi:10.1016/S0097-8493(02)00062-6

Graziadei, W. D. (1996). VICE in REST. In Harrison, T. M., & Stephen, T. D. (Eds.), *Computer Networking and Scholarly Communication in Twenty-First-Century University* (pp. 257–276). Albany, NY: SUNY Press.

Greener, S. L. (2008). Self-aware and self-directed: Student conceptions of blended learning. *MERLOT Journal of Online Learning and Teaching, 4*(2).

Greespun, P. (1999). *Philip and Alex's Guide to Web Publishing*. San Francisco: Morgan Kaufmann.

Grudin, J. (1988). Why CSCW applications fail: problems in the design and evaluation of organization of organizational interfaces. In *Proceedings of the 1988 ACM conference on Computer-supported cooperative work*, Portland, OR, United States. New York: ACM Press.

Grudin, J. (1994). Groupware and social dynamics: eight challenges for developers. *Communications of the ACM*, *37*(1), 92–105. doi:10.1145/175222.175230

Grund, S., et al. (2002). Concept for interdisciplinary evaluation of learning tools (CIELT). In F. F. C. Lutz, P. Schulz & L. Cantoni (Eds.), 4th International conference on new educational environments (ICNNE), Manno, Lugano, Switzerland.

Guerri, J. C., Palau, C. E., Pajares, A., & Esteve, M. (2002). A real-time e-learning system via satellite based on JMF and Windows Media. In *Proceedings of the Tenth ACM international Conference on Multimedia,* Juan-les-Pins, France, December 01 - 06, 2002 (pp. 219-222). New York: ACM.

Guertin, L. A., Bodek, M. J., Zappe, S. E., & Heeyoung K. (2007). Questioning the Student Use of and Desire for Lecture Podcasts. *MERLOT Journal of Online Learning and Teaching, 3*(2).

Gunawardena, C., & LaPointe, D. (2007). Cultural Dynamics of Online Learning. In Moore, M. (Ed.), *Handbook of Distance Education* (2nd ed.). Mahwah, NJ: Lawrence Erlbaum Associates, Publishers.

Guzdial, M. (2005). *Introduction to Computing and Programming in Python: A Multimedia Approach*. Upper Saddle River, NJ: Prentice Hall.

Hage, J., & Hollingsworth, J. (2000). A Strategy for the Analysis of Idea Innovation Networks and Institutions. *Organization Studies*, *21*(5), 971–1004. doi:10.1177/0170840600215006

Halliday, D., Resnick, R., & Walker, J. (2005). *Fundamentals of Physics*. New York: John Wiley & Sons.

Hampson, K. (2008). Neil Allison: Education, Outcome Assessment, and Technology. *Higher Education Management*. Retrieved May 22, 2009, from http://highereducationmanagement.wordpress.com/2008/12/03/neil-allison-outcome-assessment-and-technology/

Harrington, C. F., Gordon, S. A., & Schibik, T. J. (2004). Course management system utilization and implications for practice: A national survey of department chairpersons. *Online Journal of Distance Learning Administration*, *7*(4).

Harris, J. J. (2005). Curriculum + differentiation = learning. *Momentum*, *36*(4), 18–21.

Hart, M., & Friesner, T. (2004, December). Plagiarism and poor academic practice – A threat to the extension of e-learning in higher education? *The Electronic Journal of e-Learning, 2*(2). Retrieved June 8, 2009, from http://www.ejel.org/volume-2/vol2-issue1/issue1-art25-hart-friesner.pdf

Hart, R. S. (1995). The Illinois PLATO Foreign Languages Project. *CALICO Journal*, *12*(4), 15–37.

Hartfield, B., Winograd, T., & Bennett, J. (1992). Learning HCI design: mentoring project groups in a course on human-computer interaction. In *ACM SIGCSE Bulletin: Proceedings of the twenty-third SIGCSE technical symposium on Computer science education, 24*(1), 246-251.

Hayes, J. H., Lethbridge, T. C., & Port, D. (2003). Papers on software engineering education and training: course delivery and evaluation: Evaluating individual contribution toward group software engineering projects. In *IEEE Proceedings of the 25th International Conference on Software Engineering* (pp. 622-627).

Haymes, T. (2008). The three E-strategy for overcoming resistance to technological change. *EDUCAUSE Quarterly*, *4*, 67–69.

Head, A. J. (2007) Beyond Google: How do students conduct academic research? *First Monday, 12*(8). Retrieved June 16, 2009, from http://firstmonday.org/htbin/cgiwrap/bin/ojs/index.php/fm/article/view/1998/1873

Heck, J. (2009). State, national universities see spike in distance-learning enrollment. *Dallas Business Journal*. Retrieved from http://dallas.bizjournals.com/dallas/othercities/wichita/stories/2009/01/12/focus3.html?b=1231736400^1758951

Helander, M. (1997). *Handbook of Human-Computer Interaction*. New York: Elsevier Science.

Henry, R. J. (2006). ePortfolio thinking: A provost perspective. In A. Jafari & C. Kaufman (Eds.), Handbook of research on ePortfolios (pp. 54-61). Hershey, PA: Idea Group Reference.

Herbert, E. (2001). *The power of portfolios*. San Francisco: Jossey-Bass.

Hernández, R. (2005). OpenACS: robust web development framework. In *Proc. Tcl/Tk 2005 Conference Proceedings*, Portland, Oregon.

Hernández, R., Morales, M., De la Roca, M., & Guerra, V. (2006). Proyecto E-LANE en la Universidad Galileo. *Resumen resultados E-LANE/UG v.1.0.* [Online]. Retrieved April 25, 2009, from http://ges.galileo.edu/fs/view/articulos/resultados-UG.pdf

Herreid, C. F., & Schiller, N. A. (2005). *The National Center for Case Study Teaching in Science*. State University of New York at Buffalo. Retrieved from http://ublib.buffalo.edu/libraries/projects/cases/ubcase.htm

Herrington, J., & Oliver, R. (2000). An instructional design framework for authentic learning environments. *Educational Technology Research and Development, 48*(3), 23–48. doi:10.1007/BF02319856

Herrington, J., Oliver, R., & Reeves, T. C. (2003). Patterns of engagement in authentic online learning environments. *Australian Journal of Educational Technology, 19*(1), 59–71.

Hewitt, J. (2005). Toward an understanding of how threads die in asynchronous computer conferences. *Journal of the Learning Sciences, 14*(4), 567–589. doi:10.1207/s15327809jls1404_4

Hewitt, J., & Teplovs, C. (1999). An analysis of growth patterns in computer conferencing threads. In C. Hoadley & J. Roschelle (Eds.), *Proceedings of the computer support for collaborative learning (CSCL) 1999 Conference,* Dec. 12-15, Stanford University, Palo Alto, California. Mahwah, NJ: Erlbaum.

Hewitt, P. (2002). *Conceptual Physics.* Harlow, UK: Addison Wesley.

Higgs, B., & Sabin, M. (2005). Towards Using Online Portfolios in Computing Courses. In *Proceedings of the 6th ACM SIG — Information Technology Education Conference (SIGITE 2005)*, Newark, NJ, October 20-22, 2005 (pp. 323-328).

Higher Education Quality Council. (1993). *Student Assessment and the Classification of Awards.* London: Higher Education Quality Council.

*Higher Education Reconciliation Act of 2005,* S. 1932, 109th Cong. (2005). Washington, DC: U.S. Government Printing Office

Hill, J. R., Wiley, D., Nelson, L. M., & Han, S. (2004). Exploring research on internet-based learning: From infrastructure to interactions. In Jonassen, D. H. (Ed.), *Handbook of research on educational communications and technology* (2nd ed.). New York: Simon and Schuster/Macmillan.

Hillar, G. (2006). PostgreSQL 8.1.4: robusto y fácil de administrar. *Mundo Linux: Sólo programadores Linux*, 86.

Hines, R., & Pearl, C. (2004, Spring). Increasing Interaction in Web-based Instruction: Using Synchronous Chats and Asynchronous Discussions. [from Academic Search Premier database.]. *Rural Special Education Quarterly, 23*(2), 33–36. Retrieved May 13, 2009.

Hnatyshin, V. Y., & Lobo, A. F. (2008). Undergraduate Data Communications and Networking Projects Using OPNET and Wireshark Software. *ACM SIGCSE Bulletin, 40*(1), 241–245. doi:10.1145/1352322.1352222

Holyfield, S. (2003). *Developing a Shared Understanding of the Managed Learning Environment (MLE) – the role of diagramming and requirements gathering.* Commissioned JISC report. Retrieved July 28, 2009, from http://www.jisc.ac.uk/uploaded_documents/ACF450.pdf

Holzner, S. (2006). *Physics for Dummies.* New York: Wiley.

Horstmann, C. S., & Cornell, G. (2007). Core Java: *Vol. I. Fundamentals* (8th ed.). Upper Saddle River, NJ: Prentice-Hall.

Horton, C. B., & Oakland, T. (1997). Temperament-based learning styles as moderators of academic achievement. *Adolescence, 32*(125), 131–142.

Horton, W. S., & Keysar, B. (1996). When do speakers take into account common ground? *Cognition, 59*(1), 91–117. doi:10.1016/0010-0277(96)81418-1

Hrastinski, S. (2005). *Introducing a synchronous medium in a computer-mediated distance learning course: Towards understanding how student participation is affected* [PDF Document]. Retrieved from http://www.ascilite.org.au/conferences/brisbane05/confblogs/index.php?blog=143

Hrastinski, S. (2006, August). The relationship between adopting a synchronous medium and participation in online group work: An explorative study. *Interactive Learning Environments, 14*(2), 137-152. Retrieved May 13, 2009, from doi:10.1080/10494820600800240

*HSQLDB.* (2009). Retrieved July 12, 2009, from http://hsqldb.org/

*HTML-Kit*. (2009). Retrieved July 12, 2009, from http://www.chami.com/html-kit/download/

Huba, M. E., & Freed, J. E. (2000). *Learner-centered assessment on college campuses: Shifting the focus from teaching to learning*. Boston: Allyn and Bacon.

Huber, M. T., & Hutchings, P. (2004). *Integrative learning: Mapping the terrain*. Washington, DC: Association of American Colleges and Universities.

*IBSTPI Online Learner Project*. (2007). See Appendix A.

*IJornadas de experiencias docentes con el Aula Virtual de la Universitat de València*. (2006). [Online]. Valencia, Spain: Universitat de València.

iLinc Communications, Inc. (2009). *LearnLinc Tool Panel Overview*. Retrieved July 12, 2009, from http://www.ilinc.com/pdf/trc/ilinc-learnlinc-tool-panel.pdf

Information Technology Services Centre. (2004). *Security issues and solutions in internet e-mail*. Retrieved July 8, 2009, from http://www.cuhk.edu.hk/itsc/network/app/email/security/securityissues.html

Information Technology Services. (2009). *eDocs*. Retrieved July 10, 2009, from http://www.uis.edu/its/iss/edocs.html

INNOVA. (n.d.). *Innovation Section of UNED*. Universidad Nacional de Educación a Distancia. [Online]. Retrieved April 25, 2009, from http://innova.uned.es/

Inoue, Y. (2007). *Technology and Diversity in Higher Education*. Hershey, PA: Information Science Publishing.

Ioannou, A., & Hannafin, R. (2008). Deficiencies of course management systems: Do students care? *Quarterly Review of Distance Education*, *9*(4).

Ioannou, A., & Hannafin, R. D. (2008). Course management systems: Time for users to get what they need. *TechTrends*, *52*(1), 46–50. doi:10.1007/s11528-008-0111-7

ISO/IEC. (2005). *ISO/IEC 27001:2005(E): Information technology - Security techniques - Information security management systems - Requirements*. Geneva: ISO Copyright Office.

*iSQL-Viewer*. (2009). Retrieved July 12, 2009, from http://www.isqlviewer.com/

IT Guru Academic Ed. *OPNET Technologies*. (2009). Retrieved July 12, 2009, from http://www.opnet.com/university_program/itguru_academic_edition/

Jackson, M., & Helms, M. (2008). Student Perceptions of Hybrid Courses: Measuring and Interpreting Quality. *Journal of Education for Business*, *84*(1), 7–12. doi:10.3200/JOEB.84.1.7-12

Jafari, A., & Kaufman, C. (Eds.). (2006). *Handbook of research on ePortfolios*. Hershey, PA: Idea Group Reference.

Janssen, J., Erkens, G., Kanselaar, G., & Jaspers, J. (2007). Visualization of participation: Does it contribute to successful computer-supported collaborative learning? *Computers & Education*, *49*(4), 1037–1065. doi:10.1016/j.compedu.2006.01.004

Januszewski, A., & Molenda, M. (2008). *Educational Technology-A Definition with Commentary*. New York: Lawrence Erlbaum Associates.

Jaschik, S. (2009, May 7). Blackboard buys Angel. *Inside Higher Ed*. Retrieved July 26, 2009, from http://www.insidehighered.com/news/2009/05/07/bb

*JEdit Programmer's Text Editor*. (2009). Retrieved July 12, 2009, from http://www.jedit.org/

Jensen, L. A., Arnett, J. J., Feldman, S. S., & Cauffman, E. (2002). It's wrong, but everybody does it: Academic dishonesty among high school students. *Contemporary Educational Psychology*, *27*(2), 209–228. doi:10.1006/ceps.2001.1088

Jeong, A. (2003). Sequential analysis of group interaction and critical thinking in online threaded discussions. *American Journal of Distance Education*, *17*(1), 25–43. doi:10.1207/S15389286AJDE1701_3

Jeong, A. (2004). The combined effects of response time and message content on growth patterns of discussion threads in computer-supported collaborative argumentation. *Journal of Distance Education*, *19*(1), 36–53.

JISC (Joint Information Systems Committee). (2006). *Glossary E-Assessment*. Retrieved July 27, 2009, from http://www.jisc.ac.uk/uploaded_documents/eAssess-Glossary-Extended-v1-01

JISC. (2009). *JISC plagiarism detection service*. Retrieved June 11, 2009, from http://www.submit.ac.uk

Johnson, J. (2007). Technology at the technical college. *University Business, 10*(2).

Johnson, R. A., & Middleton, J. M. (2008). Accounting for Second Life. *Journal of Accountancy, 205*(6), 54–56.

Johnson, R. S., Mims-Cox, J. S., & Doyle-Nichols, A. (2006). *Developing portfolios in education: A guide to reflection, inquiry, and assessment*. Thousand Oaks, CA: SAGE Publications.

Johnson, R. T., Johnson, D. W., & Stanne, M. B. (1985). Effects of cooperative, competitive, and individualistic goal structures on computer-assisted instruction. *Journal of Educational Psychology, 77*(6), 668–677. doi:10.1037/0022-0663.77.6.668

Jonassen, D. H., & Grabowski, B. L. (1993). *Handbook of individual differences, learning, and instruction*. Hillside, NJ: Lawrence Erlbaum Associates.

Jonassen, D. H., Peck, K. L., & Wilson, B. G. (1999). *Learning with technology: A constructivist perspective*. Upper Saddle River, NJ: Prentice Hall.

Jones, E. L. (2001). Plagiarism monitoring and detection - towards an open discussion. In *Proceedings of the Seventh Annual Consortium For Computing in Small Colleges Central Plains Conference on the Journal of Computing in Small Colleges* (pp. 229-236).

Jones, J. (2008). Issues and Concerns of Directors of Postsecondary Distance Learning Programs Regarding Online Methods and Technologies. *American Journal of Distance Education, 22*(1), 46–56. doi:10.1080/08923640701713430

Jujjuru, J. (2008). An Overview of Internet Protocol Multimedia Subsystems (IMS) Architecture. *Rivier College Online Academic Journal, 4*(1). Retrieved July 12, 2009, from http://www.rivier.edu/journal/RCOAJ_Spring_2008_table.htm

Jyothi, S., McAvinia, C., & Keating, J. G. (2007). An interaction visualisation tool for a learning management system. In *Proceedings of the Conference of the Center For Advanced Studies on Collaborative Research,* Richmond Hill, Ontario, Canada, October 22 – 25 (pp. 326-331). New York: ACM.

*Jython Environment for Students (JES)*. (2009). Retrieved July 12, 2009, from http://coweb.cc.gatech.edu/mediaComp-plan/MediacompSoftware

*Jython*. (2009). Retrieved July 12, 2009, from http://www.jython.org/

Kahn, S. (2001). Linking learning, improvement and accountability: An introduction to electronic institutional portfolios. In Cambridage, B. L. (Eds.), *Electronic portfolios: Emerging practices in student, faculty, and institutional Learning* (pp. 135–158). Washington, DC: AAHE.

Kakasevski, G., Mihajlov, M., Arsenovski, S., & Chungurski, S. (2008). Evaluating usability in Learning Management System Moodle. In *Proceedings of the ITI 2008 30th International Conference on Information Technology Interfaces*, June 23-26, Cavtat, Croatia (pp. 612-618).

Kapor, M., McCredie, J. W., & Sreebny, O. (2004). *Chandler/Westwood: Progress in open-source collaboration*. Paper presented at the EDUCAUSE 2003 Annual Conference, Texas.

Katz, L., & Chad, S. (2000). *Engaging Children's Minds: The Project Approach* (2nd ed.). Norwood, NJ: Ablex Publishing Corp.

KDevelop. *KDE Development Environment*. (2009). Retrieved July 12, 2009, from http://www.kdevelop.org/

Keefe, J. W. (1983). Advisement – A helping role. *The NASSP Practitioner*, *9*(4), 1–16.

Keefe, J. W. (1989). Personalized education. In Walberg, H. J., & Lane, J. J. (Eds.), *Organizing for learning: Toward the 21st century*. Reston, VA: National Association of Secondary School Principals.

Keefe, J. W., & Jenkins, J. M. (2008). *Personalized instruction: The key to student achievement* (2nd ed.). Lanham, MD: Rowman & Littlefield Education.

Keller, J. (2008). First principles of motivation to learn and e3-learning. *Distance Education*, *29*(2), 175–185. doi:10.1080/01587910802154970

Keller, J. M. (1983). Motivational design of instruction. In Reigeluth, C. M. (Ed.), *Instructional-design theories and models: an overview of their current status* (pp. 383–434). Hillsdale, NJ: Lawrence Erlbaum Associates.

Kemmis, S., & McTaggart, R. (Eds.). (1988). *The Action Planner* (3rd ed.). Geelong, Australia: Deakin University Press.

Kempfert, T. (2003). UW selects Desire2Learn as its new e-learning system. *Teaching With Technology Today*, *9*(7).

Kienle, A., & Ritterskamp, C. (2007). Facilitating asynchronous discussions in learning communities: the impact of moderation strategies. *Behaviour & Information Technology*, *26*, 73–80. doi:10.1080/01449290600811594

Kingsley, K., & Kingsley, K. (2009). A case study for teaching information literacy skills. *BMC Medical Education*, *9*(1), 7. doi:10.1186/1472-6920-9-7

Kleimann, B., & Wannemacher, K. (2005). E-Learning-Strategien deutscher Universitäten: Fallbeispiele aus der Hochschulpraxis. [Hannover, Germany: Hochschul-Informations-System.]. *HIS-Kurzinformation*, *B4*, 2005.

Klopher, L. E. (1971). Evaluation of learning in science. In Bloom, B. S., Hastings, J. T., & Madus, G. F. (Eds.), *Handbook of Formative and Summative Evaluation of Student Learning*. New York: McGraw-Hill.

Klosgen, W., & Zytkow, J. (2002). *Handbook of data mining and knowledge discovery*. New York: Oxford University Press.

Knight, A., Almeroth, K., & Bimber, B. (2004). An Automated System for Plagiarism Detection Using the Internet. In L. Cantoni & C. McLoughlin (Eds.), *Proceedings of World Conference on Educational Multimedia, Hypermedia and Telecommunications 2004* (pp. 3619-3625). Chesapeake, VA: AACE.

Knight, R. (2002). *Five Easy Lessons*. Harlow: Addison Wesley.

Ko, S., & Rossen, S. (2001). *Teaching online: a practical guide*. Boston: Houghton Mifflin Company.

Kochtanek, T. R., & Hein, K. K. (2000). Creating and nurturing distributed asynchronous learning environments. *Online Information Review*, *24*(4), 280–289. doi:10.1108/14684520010350632

Kollar, I., Fischer, F., & Slotta, J. D. (2005). Internal and external collaboration scripts in web-based science learning at schools. In *Proceedings of CSCL* 2005 (pp. 331-340).

*Kompozer: Nvu's Unofficial Bug-Fix Release*. (2009). Retrieved July 12, 2009, from http://kompozer.net/download.php

Kopf, D. (2007). Seattle Public School System Flexing Fiber Muscle. *T.H.E. Journal*. Retrieved on March 10, 2009, from http://www.thejournal.com/articles/20385

Koutsakis, P. (2006). A New Efficient Scheme for the Allocation of Satellite Bandwidth Based on Videoconference Traffic Estimation. *ICWMC*, 16.

Koutsonikola, V., & Vakali, A. (2004). *LDAP: Framework, Practices, and Trends*. Retrieved July 27, 2009, from http://www2.computer.org/portal/web/csdl/doi/10.1109/MIC.2004

Kreijns, K., Kirschner, P. A., & Jochems, W. (2003). Identifying the pitfalls for social interaction in computer-supported collaborative learning environments: A review of the research. *Computers in Human Behavior*, *19*(3), 335–353. doi:10.1016/S0747-5632(02)00057-2

Kreijns, K., Kirschner, P. A., Jochems, W., & van Buuren, H. (2007). Measuring perceived sociability of computer-supported collaborative learning environments. *Computers & Education*, *49*(2), 176–192. doi:10.1016/j.compedu.2005.05.004

Kroenke, D. M. (2007). *Using MIS*. Upper Saddle River, NJ: Pearson, Prentice Hall.

Kubicek, H. (2004). *Organisatorische Einbettung von E-Learning an deutschen Hochschulen (Technical report)*. Bremen, Germany: Institut für Informationsmanagement.

Kuhn, D. (2000). Metacognitive development. *Current Directions in Psychological Science*, *9*(5), 178–181. doi:10.1111/1467-8721.00088

Kulathuramaiyer, N., Zaka, B., & Helic, D. (2008). Integrated Copy-Paste Checking: Design and Services. In *Proceedings of the 2008 Eighth IEEE international Conference on Advanced Learning Technologies* (pp. 519-523). Washington, DC: IEEE Computer Society. Retrieved from DOI= http://dx.doi.org/10.1109/ICALT.2008.312

Kumar, A. (2005). Development of Laboratory Exercises Based on OPNET IT Guru Academic Version. *Rivier College Online Academic Journal, 1*(1). Retrieved July 12, 2009, from http://www.rivier.edu/journal/RCOAJ-Fall-2005_table.htm

Kumar, A. (2006). An Overview of Voice over IP. *Rivier College Online Academic Journal, 2*(1). Retrieved July 12, 2009, from http://www.rivier.edu/journal/RCOAJ-Spring-2006_table.htm

Kvavnik, R., & Caruso, J. (2005). *ECAR study of students and technology 2005: Convenience, connection, control and learning*. Boulder, CO: EDUCAUSE Center for Applied Research.

Laffey, J., Tupper, T., Musser, D., & Wedman, J. (1998). A computer-mediated support system for project-based learning. *Educational Technology Research and Development*, *46*(1), 73–86. doi:10.1007/BF02299830

Lamancusa, J. J., Jorgensen, J. E., Zayas-Castro, J. L., & de Ramirez, L. M. (2001). The learning factory - integrating design, manufacturing and business realities into engineering curricula. *International Conference on Engineering Education,* August 6-10, Oslo, Norway (pp. 1-6).

Lane, L. M. (2008). Toolbox or trap? Course management systems and pedagogy. *EDUCAUSE Quarterly*, *31*(2).

Lave, J., & Wenger, E. (1991). *Situated Learning*. Cambridge, UK: Cambridge University Press.

Lee, C., Wu, C., & Yang, H. (2008). A Platform Framework for Cross-Lingual Text Relatedness Evaluation and Plagiarism Detection. In *Proceedings of the 2008 3rd international Conference on innovative Computing information and Control*. Washington, DC: IEEE Computer Society. Retrieved from DOI= http://dx.doi.org/10.1109/ICICIC.2008.76

Lee, W., & Owens, D. L. (2000). *Multimedia-Based Instructional Design: Computer- Based Training in Web-Based Training, and Distance Learning*. New York: Wiley.

Lee, Y., Kozar, K., & Larsen, K. (2003). The Technology Acceptance Model: Past, Present and future. *Communications of AIS*, (12), 752-780.

Leonard, J., Riley, E., & Staman, E. M. (2003). Classroom and support innovation using IP video and data collaboration techniques. In *Proceedings of the 4th Conference on information Technology Curriculum,* Lafayette, Indiana, USA, October 16 - 18, 2003.

Lewis, J., & Loftus, W. (2003). *Java Software Solutions: Foundations of Program Design* (3rd ed.). Boston, MA: Addison Wesley.

Liao, S. C. (2006). Creating learning scenarios: A planning guide for adult learners. *Convergence*, *39*(1), 101–103.

Lin, H. (2007). The ethics of instructional technology: Issues and coping strategies experienced by professional technologists in design and training situations in higher education. *Educational Technology Research and Development*, *55*(5), 411–437. doi:10.1007/s11423-006-9029-y

Lindquist, D., Denning, T., Kelly, M., Malani, R., & Griswold, W. G., & Simon, B. (2007). Exploring the potential of mobile phones for active learning in the classroom. In *Proceedings of the 38th SIGCSE technical symposium on Computer science education*. Covington, KY: ACM.

Lippman, S. B., Lajoie, J., & Moo, B. E. (2005). *C++ Primer* (4th ed.). Boston, MA: Addison-Wesley.

Littman, M. K. (2006). *Building Broadband Networks*. Boca Raton, FL: CRC Press.

Lockyer, J., Sargeant, J., Curran, V., & Fleet, L. (2006). The transition from face-to-face to on-line CME facilitation. *Medical Teacher*, *28*, 625–630. doi:10.1080/01421590600922909

Lokken, F. (2009). *2008 distance education survey results: Tracking the impact of eLearning at community colleges*. Washington, DC: Instructional Technology Council.

Lombardi, M. M. (2007, May). *Authentic learning for the 21st century: An overview.* Retrieved January 12, 2009, from http://net.educause.edu/ir/library/pdf/ELI3009.pdf

Lopes, P. N., Brackett, M. A., Nezlek, J. B., Schutz, A., Sellin, I., & Salovey, P. (2004). Emotional intelligence and social interaction. *Personality and Social Psychology Bulletin*, *30*, 1018–1034. doi:10.1177/0146167204264762

Lovett, C., & Robert, T. (2004). We Need an Honest Conversation. *The Chronicle of Higher Education*, 51. Retrieved January 6, 2006, from http://chronicle.com/weekly/v51/i02/02b00601.htm

Lyons, H. (2009). *Case study research methodology for publishing developments in ICT-facilitated learning in higher education - a prescriptive approach*. Innovations in Education & Teaching International.

MacDonald, C., & Thompson, T. (2005). Structure, Content, Delivery, Service, and Outcomes: Quality e-Learning in higher education. *International Review of Research in Open and Distance Learning*, *6*(2).

Maeroff, G. (2003). *A classroom of one: How online learning is changing our schools and colleges*. Houndmills, UK: Palgrave Macmillan.

Maki, P. (2004). *Assessing for Learning: Building a Sustainable Commitment Across the Institution*. Sterling, VA: Stylus Publications.

Malikowski, S. R., Thompson, M. E., & Theis, J. G. (2007). A model for research into course management systems: Bridging technology and learning theory. *Journal of Educational Computing Research*, *36*(2), 149–173. doi:10.2190/1002-1T50-27G2-H3V7

Mangan, K. (2008, February 13). Courtroom showdown begins in blackboard's patent dispute with desire2learn. *Chronicle of Higher Education.*

Markett, C., & Sánchez, I., A., Weber, S., & Tangney, B. (2006). Using short message service to encourage interactivity in the classroom. *Computers & Education*, *46*(3), 280–293. doi:10.1016/j.compedu.2005.11.014

Marlino, M., Sumner, T., Fulker, D., Manduca, C., & Mogk, D. (2001). The digital library for Earth system education: building community, building the library. *Communications of the ACM*, *44*(5), 80–81. doi:10.1145/374308.374356

Marqués, P. (2008). Impacto de las TIC en la enseñanza universitaria. *Didáctica, Innovación y Multimedia, 4*(11).

Martin, B. (2009). *CWE/SANS Top 25 Most Dangerous Programming Errors*. Retrieved April 6, 2009, from http://www.sans.org/top25errors/

Martinez, J. (2005). *JQTI-Lite Library: Developer's Guide (1)*. [Online]. Retrieved April 25, 2009, from http://www.tecn.upf.es/~ssayag/leteos/jQTILite/docs/jQTILite10_Devel_Guide.pdf

Martinez, R. (2004). Online Education: Designing for the Future in Appraiser Education. *The Appraisal Journal*, *72*(3), 266–184.

Mason, R. (1994). *Using communication media in open and flexible learning*. London: Kogan Page.

Mathes, A. (2004). *Folksonomies - Cooperative Classification and Communication Through Shared Metadata Computer Mediated Communication.* Retrieved June 17, 2009, from http://www.adammathes.com/academic/computer-mediated-communication/folksonomies.html

Maurer, H., Kappe, F., & Zaka, B. (2006). Plagiarism – A Survey. *Journal of Universal Computer Sciences, 12*(8), 1050–1084.

Maxemchuk, N. F., Ouveysi, I., & Zukerman, M. (2005). A quantitative measure for telecommunications networks topology design. *IEEE/ACM Trans. Netw., 13*(4), 731–742.

Mazur, E. (1997). *Peer Instruction.* Englewood Cliffs: Prentice Hall.

Mazza, R., & Botturi, L. (2007). Monitoring an online course with the GISMO tool: a Case study. *Journal of Interactive Learning Research, 18*(2), 251–265.

Mazza, R., & Dimitrova, V. (2005). Generation of graphical representations of student tracking data in Course Management Systems. In *9th IEEE International Conference on Information Visualisation*, London (pp. 253-258).

McAndrew, P., & Santos, A. I. *(2009).* Learning from OpenLearn: Research Report 2006-2008. *Retrieved May26, 2009, from*http://kn.open.ac.uk/public/getfile.cfm?documentfileid=15028

McCrae, R. R., & Costa, P. T. (2003). *Personality in Adulthood: A Five-Factor Theory Perspective* (2nd ed.). New York: Guilford Press.

McDermott, L. (2001). Oersted Medal Lecture 2001: Physics education research – The Key to Student Learning. *American Journal of Physics, 69*(11). doi:10.1119/1.1389280

Mcdermott, L., Shaffer, P., & Rosenquist, M. (1996). *Physics by Inquiry.* London: Wiley.

*Mckoi Database.* (2009). Retrieved July 12, 2009, from http://www.mckoi.com/database/

McLeish, J. (1976). The Lecture Method. In Gage, N. L. (Ed.), *The Psychology of Teaching Methods: The Seventy-fifth Yearbook of The National Society for The Study of Education, Part I* (pp. 252–301). Chicago: The University of Chicago Press.

McLoughlin, C., & Oliver, R. (2000). Designing learning environments for cultural inclusivity: A case study of indigenous online learning at tertiary level. *Australian Journal of Educational Technology, 16*(1), 58–72.

*McMicken College of Arts and Sciences course catalog.* (2008). University of Cincinnati, Ohio. Retrieved July 23, 2009 from http://www.uc.edu/courses/a_s.pdf

McMillan, D. (1996). Sense of community. *Journal of Community Psychology, 24*, 315–325. doi:10.1002/(SICI)1520-6629(199610)24:4<315::AID-JCOP2>3.0.CO;2-T

Means, B., & Olson, K. (1994 April). The Link Between Technology and Authentic Learning. *Educational Leadership.*

Menchala, M., & Bekele, T. (2008). Learner and instructor identified success factors in distance education. *Distance Education, 29*(3), 231–252. doi:10.1080/01587910802395771

Meyer, H., & Paradies, L. (1993). *Plädoyer für Methodenvielfalt im Unterricht.* Oldenburg: Oldenburger Vordrucke.

Michinov, N., & Primois, C. (2005). Improving productivity and creativity in on-line groups through social comparison process: new evidence for asynchronous electronic brainstorming. *Computers in Human Behavior, 21*(1), 11–28. doi:10.1016/j.chb.2004.02.004

Microsoft Inc. (2008). *Microsoft Visual C++ 2008 Express Edition.* Retrieved July 12, 2009, from http://www.microsoft.com/express/vc/

Microsoft Inc. (2009). *Microsoft Script Editor.* Retrieved July 12, 2009, from http://msdn.microsoft.com/en-us/library/aa189846(office.10).aspx

Milkovits, M. N. (2005a). Digital Video Cluster Simulation. In M. E. Kuhl, N. M. Steiger, F. B. Armstrong & J. A. Joines (Eds.), *Proceedings of the 2005 Winter Simulation Conference,* Orlando, Florida, December 2005.

Milkovits, M. N. (2005b). Using OMNeT++ to Simulate a Heterogeneous Digital Video Cluster Fabric. *Rivier College Online Academic Journal, 1*(1). Retrieved July 12, 2009, from http://www.rivier.edu/journal/RCOAJ-Fall-2005_table.htm

Miller, F., Seay, R., & Chamberlain, D. (1991). The current status of outcomes assessment in marketing. *Journal of the Academy of Marketing Science, 19*(Fall), 353–362. doi:10.1007/BF02726511

Miller, J. (2007). The New Education Professionals: The Emerging Specialties of Instructional Designer and Learning Manager. *International Journal of Public Administration, 30*, 483–498. doi:10.1080/01900690701205970

Miller, M. T., & Husmann, D. E. (1996). A holistic model for primary factors in the ecology of

Min, G. W. (2009). *Minimalist GNU for Windows.* Retrieved July 12, 2009, from http://www.mingw.org/

*MinGW Developer Studio.* (2009). Retrieved July 12, 2009, from http://programmerstools.org/node/368

Mohr, E. (2007, September/October). Incremental innovations: Changing the culture of teaching and learning. *EDUCAUSE Review*, 6–7.

Monnard, J., & Brugger, R. (2003). Web based course platforms: Evaluation report. Technical report, Edutech mandate of the Swiss Virtual Campus program, Freiburg, Switzerland.

Moore, J. (2003). *Chemistry for Dummies.* Indianapolis, IN: Wiley Pub.

Moore, J. C. (2005). *The Sloan Consortium quality framework and five pillars.* Retrieved August 2, 2009, from http://www.sloan-c.org/publications/books/qualityframework.pdf

Moore, M. (2003, September). Learner Support. [from Academic Search Premier database.]. *American Journal of Distance Education*, 17(3), 141–143. Retrieved May 13, 2009. doi:10.1207/S15389286AJDE1703_1

Moore, M. (2007). Web 2.0: Does it Really Matter? *American Journal of Distance Education, 21*(4), 177–183. doi:10.1080/08923640701595183

Moran, T. P., & Carroll, J. M. (Eds.). (1994). *Design Rationale: Concepts, Techniques, and Use.* Hillsdale, NJ: Lawrence Erlbaum Associates.

Moreno, P., & Cerverón, V. (2006). Plataforma tecnológica para potenciar los procesos de enseñanza-aprendizaje: desarrollo en la Universitat de València basado en software libre y colaborativo. In *Proc. SIIE 06. VIII Simposio Internacional de informática aplicada a la enseñanza. 8th International Symposium on Computers in Education* (Vol. 2).

Moreno, P., & Cerverón, V. (2007). Enhanced Learning Through Blended Learning: Educative Innovation at the Universitat de València. In *Proc. iNEER, International Conference on Engineering Education.* Retrieved April 25, 2009, from http://icee2007.dei.uc.pt/proceedings/index.html

Moreno, P., Cerverón, V., & Arevalillo, M. (2008). Análisis del uso universitario de plataformas de gestión del aprendizaje y su relación con la innovación educativa y la calidad docente en el marco del EEES. Resultados en la Universitat de València y procedimientos para la generalización. In Proc. Conferencia sobre Software Libre en Educación Superior, Valencia.

Moreno, P., Cerverón, V., López, A., & Roig, D. (2007). LRN Consortium: International Collaboration for Developing a Learning Management System. Experience from the Universitat de València. In *Proc. iNEER, International Conference on Engineering Education.* Retrieved from http://icee2007.dei.uc.pt/proceedings/index.html

Moreno-Clari, P., & Cerverón-Lleó, V. (2007). Encouraging Blended Learning and ICT Use at Universitat de València to Improve the Learning Process with the.LRN Platform: Best Practices and Tools. In *Proc. EATIS 07, Euro American Conference on Telematics and Information System, EATIS'07,* Algarve.

Morgan, R. L., Cantor, S., Carmody, S., Hoehn, W., & Klingenstein, K. (2004). Federated security: The Shibboleth approach. *EDUCAUSE Quarterly, 27*(4), 12–17.

Moser, F. Z. (2007). Faculty adoption of educational technology. *EDUCAUSE Quarterly, 1,* 66–69.

Motiwalla, L., & Thompson, J. (2009). *Enterprise Systems for Management.* Upper Saddle River, NJ: Prentice Hall.

*Mozilla Firefox.* (2009). Retrieved July 12, 2009, from http://www.mozilla.com/en-US/firefox/

Mozilla Suite. *The All-in-One Internet Application Suite.* (2009). Retrieved July 12, 2009, from http://www.mozilla.org/products/mozilla1.x/

Mupinga, D. M., Nora, R. T., & Yaw, D. C. (2006). The learning styles, expectations, and needs of online students. *College Teaching, 54*(1), 185–190. doi:10.3200/CTCH.54.1.185-189

Murphy, E., & Manzanares, M. (2008, March). Instant Messaging in a context of virtual schooling: balancing the affordances and challenges. *Educational Media International, 45*(1), 47-58. Retrieved May 29, 2009, from doi:10.1080/09523980701847180

Murphy, R. T., & Appeal, L. R. (1978). Evaluation of the PLATO IV computer-based education system in the community college. *SIGCUE Outlook, 12*(1), 12–28. doi:10.1145/963847.963849

Murray, B. (2000, April). Reinventing class discussion online. *Monitor on Psychology, 31*(4). Retrieved May 25, 2009, from http://www.apa.org/monitor/apr00/reinventing.html

*MySQL: The World's Most Popular Open Source Database.* (2009). Retrieved July 12, 2009, from http://dev.mysql.com/

Nagel, D. (2007). Students Cross Cultures Through Videoconferencing. *T.H.E. Journal.* Retrieved November 29, 2007, from http://www.thejournal.com/articles/21660

Nagel, D. (2008). Which Technologies Will Shape Education in 2008? *T.H.E. Journal.* Retrieved February 1, 2008, from http://www.thejournal.com/the/printarticle/?id=21972

Nardi, B. (2005). Beyond bandwidth: Dimensions of connection in interpersonal communication. *CSCW, 14,* 91–130.

*National Center for Case Study Teaching in Science.* (n.d.). Retrieved April 19, 2008, from http://ublib.buffalo.edu/libraries/projects/cases/case.html

NCATE. (2002). *Professional Standards for the Accreditation of Schools, Colleges, and Departments of Education.* Washington, DC: National Council for the Accreditation of Teacher Education.

Nelson, M. R. (2007). U.S. advisory committee on student financial assistance textbook study field hearing III. Portland, OR.

*NetBeans C and C++ Development.* (2009). Retrieved July 12, 2009, from http://www.netbeans.org/features/cpp/index.html

*NetBeans.* (2009). Retrieved July 12, 2009, from http://www.netbeans.org/

Nicholoson, C., Barnett, S., & Dascher, P. (2005). Curriculum assessment in marketing programs: current status and examination of AACSB core standards at the program level. *Marketing Education Review, 15*(2), 13–26.

Nickerson, M. (2003). *Streaming video in distance education: Giving online humanities instruction a human face.* Retrieved May 31, 2009, from http://www.li.suu.edu/library/courses/humanities1010/

Nielson, J. (1994). *Usability Engineering.* San Francisco: Morgan Kaufmann.

Nielson, J. (1997). *Be succinct! (Writing for the Web).* Retrieved June 21, 2009, from http://www.useit.com/alertbox/9703b.html

Niezgoda, S., & Way, T. P. (2006). SNITCH: A Software Tool for Detecting Cut and Paste Plagiarism. In *Proceedings of SIGCSE'06 conference* (pp. 51 – 55).

Nitterhouse, D. (2003). Plagiarism: Not just an academic problem. *Teaching Business Ethics*, *7*, 215–227. doi:10.1023/A:1025017019246

Norton, P., & Hathaway, D. (2008). Exploring Two Teacher Education Online Learning Designs: A Classroom of One or Many? *Journal of Research on Technology in Education*, *40*(4), 475–495.

*Notepad++*. (2009). Retrieved July 12, 2009, from http://notepad-plus.sourceforge.net/uk/site.htm

*NoteTab Lite*. (2009). Retrieved July 12, 2009, from http://www.fookes.com/notetab/index.php

Novak, G. (1999). *Just-in-Time Teaching*. Englewood Cliffs, NJ: Prentice Hall.

Novak, G., Gavrin, A., & Christian, W. (1999). *Just-in-time teaching: blending active learning with web technology*. Upper Saddle River, NJ: Prentice Hall.

NSF DUE has supported many projects in the area of computer professionalism and ethics. Examples are the ImpactCS Project (http://www.seas.gwu.edu/~impactcs/) and Dolce (Developing Online Computer Ethics, http://computingcases.org/). Retrieved April 19, 2008.

NSF. *(2008)*. Fostering learning in the networked world: learning opportunity and challenge. A 21st Century agenda for the National Science Foundation. *Retrieved May 8, 2009, from* http://www.nsf.gov/publications/pub_summ.jsp?ods_key=nsf08204

*Nvu: The Complete Web Authoring System for Windows, Mac, and Linux*. (2009). Retrieved July 12, 2009, from http://www.nvu.com/

O'Hara, T. (2005, November 26). Blackboard's webct deal spurs antitrust questioning. *The Washington Post*.

O'Leonard, K., & Bersin, J. (2006). *Learning management systems 2006: Facts, practical analysis, trends and vendor profiles*. Oakland, CA: Bersin & Associates, Pike, G. H. (2009). The fences of a patent. *Information Today*, *26*(2).

Oblinger, D. G., & Hawkins, B. L. (2006, January/February). The myth about online course development. *EDUCAUSE Review*, 14–15.

Odinma, A. C., Butakov, S., Grakhov, E., & Bollou, F. (2008). Planning, designing and implementing an enterprise network in a developing nation. *International of Journal of Enterprise Network Management*, *2*(3), 301–317. doi:10.1504/IJENM.2008.018783

Office of Information and Instructional Technology. (2006). *Implementing USG's online learning management system*. Retrieved from http://www.usg.edu/oiit/about/factsheets/webctvista.pdf

Oliver, R. (n.d.). *Developing online learning environments that support knowledge construction*. Retrieved January 12, 2009, from http://www.edtec.unsw.edu.au/inter/dload/flex_ed/resources/BlendedProgram/EdDesignBits/OliverKnowledgeConstruction.doc

Olshansky, S., & Carmody, S. (2009). *Shibboleth Enabled Applications and Services*. Retrieved April 6, 2009, from https://spaces.internet2.edu/display/SHIB2/ShibEnabled

*Opera Web Browser*. (2009). Retrieved July 12, 2009, from http://www.opera.com/

*Oracle Database Software Downloads*. (2009). Retrieved July 12, 2009, from http://www.oracle.com/technology/software/products/database/index.html

Ospelt, T., & Aegerter, D. (2007). *Eine Hochschule—Eine Lernplattform. Eine Anforderungsanalyse*. Diplom Thesis, Zurich University of Applied Sciences, Winterthur, Switzerland.

Otter, S. (1992). *Outcomes assessment in Higher Education*. London: UDACE.

OWASP-Project. (2009). *Open Web Application Security Project*. Retrieved April 6, 2009, from http://www.owasp.org von Solms, B. (2005). Information Security Governance in ICT Based Educational Systems. In *Proceedings of the 4th International Conference on ICT and Higher Education,* Siam University (pp. 109-119).

PAIRwise. (2009). *The PAIRWise system*. Retrieved May 20, 2009, from http://www.pairwise.cits.ucsb.edu/system.htm

Pallof, R. M., & Pratt, K. (2003). *Collaborating Online: Learning Together in Community*. San Francisco: Jossey-Bass.

Palloff, R., & Pratt, K. (2001). *Lessons from the Cyberspace Classroom*. San Francisco: Jossey Bass.

Palomba, C. A., & Banta, T. W. (1999). *Assessment Essentials*. San Francisco, CA: Jossey-Bass.

Panko, R. (2005). *Seven OPNET Labs*. Retrieved July 12, 2009, from http://www.opnet.com/services/university/itg_panko.html

*PBWiki*. (2009). Retrieved July 12, 2009, from http://pbwiki.com/

Pembrook, R. G. (1986). Some implications of students' attitudes toward a computer-based melodic dictation program. *Journal of Research in Music Education, 34*(2), 121–133. doi:10.2307/3344740

Perens, B. (1999). *The Open Source Definition. Open Sources: Voices from the Open Source Revolution*. Sebastopol, CA: O'Reilly & Associates.

*Perl Directory*. (2009). Retrieved July 12, 2009, from http://www.perl.org/

Peterson, L. L., & Davie, B. S. (2005). *Computer Networks: A System Approach* (3rd ed.). Boston, MA: Morgan Kaufmann.

Pevac, I. (2005). *Recursive Examples in Java*. Ann Arbor, MI: ProQuest Information & Learning.

PHP. *Hypertext Preprocessor*. (2009). Retrieved July 12, 2009, from http://www.php.net/

*phpMyAdmin*. (2009). Retrieved June 30, 2009 from http://www.phpmyadmin.net/home_page/index.php

Piaget, J. (1964). Development and Learning. *Journal of Research in Science Teaching, 2*, 176–186. doi:10.1002/tea.3660020306

Picapica. (2009). *Picapica plagiarism detection service*. Retrieved from http://www.picapica.net

Piña, A. A. (2005). Course management systems: Introduction and implications for libraries. *Library Hi Tech News, 24*(5).

Piña, A. A. (2008a). How Institutionalized is Distance Learning? A Study of Institutional Role, Locale and Academic Level. *Online Journal of Distance Learning Administration, 11*(1).

Piña, A. A. (2008b). *Distance Learning and the Institution: Foundations, Importance and Implementation*. Saarbrücken, Germany: VDM Publishing.

Piña, A. A., & Eggers, M. R. (2006). *Teaching, administering and supporting blackboard, webct and desire2learn*. Paper presented at the annual Technology in Education (TechEd) Conference, Pasadena, CA.

Piña, A. A., & Heydenburg, P. R. (2008). *Learning management systems: Platform-independent course design*. Paper presented at the annual Supporting Learning and Technology in Education (SLATE) Conference, Chicago, IL.

Piña, A. A., Green, S., & Eggers, M. R. (2008). *Learning management systems: Lessons from the front lines*. Paper presented at the annual Technology in Education (TechEd) Conference, Ontario, CA.

Piña, A. A., Sadowski, K. P., Scheidenhelm, C. L., & Heydenburg, P. R. (2008). SLATE: A community of practice for supporting learning and technology in education. *International Journal of Instructional Technology and Distance Learning, 5*(7).

Pinto, D., Civera, J., Barrón-Cedeòo, A., Juan, A., & Rosso, P. (2009). A statistical approach to crosslingual natural language tasks. *J. Algorithms, 64*(1), 51-60. Retrieved from DOI= http://dx.doi.org/10.1016/j.jalgor.2009.02.005

Piotrowski, M. (2009). *Document-Oriented E-Learning Components*. PhD Thesis, Otto-von-Guericke-University Magdeburg, Germany.

*PLT Scheme*. (2009). Retrieved July 12, 2009 from http://www.plt-scheme.org/

Pogrow, S. (1994). Helping students who just don't understand. *Educational Leadership, 52*(3), 62–66.

Polycom, V. S. X. *8000*. (n.d). Retrieved April 30, 2008, from http://www.ivci.com/videoconferencing_polycom_vsx_8000.html

Portes, A. (1998). Social Capital: Its Origins and Applications in Modern Sociology. *Annual Review of Sociology, 22*, 1–24. doi:10.1146/annurev.soc.24.1.1

Postgre, S. Q. L. *The world's most advanced open source database*. (2009). Retrieved July 12, 2009, from http://www.postgresql.org/

Powazek, D. (2006). *Where Am I?* Retrieved June 21, 2009, from http://www.alistapart.com/articles/whereami

Prechelt, L., Malpohl, G., & Philippsen, M. (2002). Finding Plagiarisms among a set of programs with JPlag. *Journal of Universal Computer Science, 8*(11), 1016–1038.

Prewitt, T. (1998, July). The Development of Distance Learning Delivery Systems. [from Academic Search Premier database.]. *Higher Education in Europe, 23*(2), 187. Retrieved July 23, 2009. doi:10.1080/0379772980230205

PricewaterhouseCoopers. (2008). *Information security breaches survey 2008*. Retrieved from http://www.pwc.co.uk/eng/publications/berr_information_security_breaches_survey_2008.html

*Python*. (2009). Retrieved July 12, 2009, from http://www.python.org/

*QAED LINK*. (2005). [Online]. Retrieved April 25, 2009, from: http://gti.upf.edu/leteos/newnavs/qaed.html

*Qt Product Overview – single source C++ cross-platform application development for Windows, Linux, Mac*. (2009). Trolltech Corp. Retrieved July 12, 2009, from http://www.trolltech.com/products/qt/index.html

*Quanta Plus*. (2009). Retrieved July 12, 2009, from http://quanta.kdewebdev.org/

*QueryForm Database Tool*. (2009). Retrieved July 12, 2009, from http://sourceforge.net/projects/qform/

Rainie, L., & Madden, M. (2005). *Podcasting catches on*. Washington, DC: Pew Research Center.

Ranie, L., & Keeter, S. (2006). *How Americans use their cell phones*. Washington, DC: The Pew Research Center.

Raymond, E. (1999). Shut Up and Show Them the Code. *Linux Today* [Online]. Retrieved April 25, 2009, from http://linuxtoday.com/stories/7196.html

Redden, E. (2009). Blackboard 9.0. *Inside Higher Ed*. Retrieved from http://insidehighered.com/news/2009/01/27/blackboard

Reeves, T. C., Herrington, J., & Oliver, R. (2004). A development research agenda for online collaborative learning. *Educational Technology Research and Development, 52*(4), 53–65. doi:10.1007/BF02504718

*Registered moodle sites*. (2009). Retrieved June 12, 2009, from http://moodle.org/sites.

Reinmann, G. (2005). *Blended Learning in der Lehrerbildung. Grundlagen für die Konzeption innovativer Lernumgebungen. Lengerich: Pabst Science Publishers. Richardson, W. (2006). Blogs, Wikis, Podcasts, and Other Powerful Web Tools for Classrooms*. Thousand Oaks, CA: Corwin Press.

Resnick, M. (2002). Rethinking learning in the digital age. In Kirkman, G. (Ed.), *The global information technology report: Readiness for the networked world*. Oxford, UK: Oxford University Press.

Resnick, P. (2002). Beyond Bowling Together: Socio-technical Capital. In Carroll, J. M. (Ed.), *HCI in the New Millennium* (pp. 247–272).

Riabov, V. (2000a). *Challenge in the Aerospace University Classroom. AIAA Paper, No. 2000-0802*. Washington, DC: AIAA Inc.

Riabov, V. (2000b). *New Challenges for Aerospace Education Programs in American and Russian Universities. AIAA Paper, No. 2000-0528*. Washington, DC: AIAA Inc.

Riabov, V. (2002a). Simulation Techniques in Hypersonic Aerodynamics. In *Proceedings of the 23rd ICAS Congress*, Paper 181, Toronto, Canada, September 8-13, 2002.

Riabov, V. (2002b). Computational and Analytical Methods in the Aircraft Design Curriculum. In *Proceedings of the 23rd ICAS Congress*, Paper 255p, Toronto, Canada, September 8-13, 2002.

Riabov, V. (2003). Applications of Exponential Box Schemes for Viscous Flows with Combustion and Blowing. In K. J. Bathe (Ed.), *Proceedings of the Second MIT Conference on Computational Fluid and Solid Mechanics,* June 17-20, 2003 (Vol. 1, pp. 1102-1105). Boston, MA: Elsevier.

Riabov, V. (2004). Storage Area Networks. In Bidgoli, H. (Ed.), *The Internet Encyclopedia* (*Vol. 3*, pp. 329–339). Hoboken, NJ: John Wiley & Sons, Inc.

Riabov, V. (2005a). Networking Software Studies with the Structured Testing Methodology. In Petratos, P., & Michalopoulos, D. (Eds.), *Computer Science and Information Systems* (pp. 261–276). Athens, Greece: ATINER.

Riabov, V. (2005b). Three Fast Computational Approximation Methods in Hypersonic Aerothermodynamics. In K. J. Bathe (Ed.), *Proceedings of the Third MIT Conference on Computational Fluid and Solid Mechanics,* June 13-14, 2005 (Vol. 1, pp. 819-823). Boston, MA: Elsevier.

Riabov, V. (2005c). Mathematics and Computer Science Lecture Series (2002 – 2005). In *Rivier College Online Academic Journal, 1*(1). Retrieved July 12, 2009, from http://www.rivier.edu/journal/RCOAJ-Fall-2005_table.htm

Riabov, V. (2005d). Professional Seminar in Computer Science (Spring 2005). In *Rivier College Online Academic Journal, 1*(1). Retrieved July 12, 2009, from http://www.rivier.edu/journal/RCOAJ-Fall-2005_table.htm

Riabov, V. (2006a). Challenging Projects and Virtual Labs in Web-enhanced Networking Technology Classes. *Journal of Computing Sciences in Colleges, 21*(6), 88–99.

Riabov, V. (2006b). Simple Mail Transfer Protocol (SMTP). In Bidgoli, H. (Ed.), *The Handbook of Information Security: Key Concepts, Infrastructures, Standards and Protocols* (*Vol. 1*, pp. 878–900). Hoboken, NJ: John Wiley & Sons, Inc.

Riabov, V. (2009). *Instructor's Website*. Retrieved July 12, 2009, from http://www.rivier.edu/faculty/vriabov/

Richey, R., Fields, D., & Foxon, M. (2001). *Instructional Design Competencies: The Standards* (3rd ed.). Syracuse, NY: ERIC.

Ritter, S. (2008). *Social Networking in Higher Education*. Penn State World Campus. Retrieved from http://blog.worldcampus.psu.edu/index.php/2008/09/04/social-networking-in-higher-education/

Robson, R. (1999). WWW-based course-support systems: The first generation. *International Journal of Educational Telecommunications, 5*(4), 271–282.

Rogers, E. (1995). *Diffusion of Iinnovations* (4th ed.). New York: The Free Press.

Romer, D. (1993). Do students go to class? Should they? *The Journal of Economic Perspectives, 7*(3), 167–174.

Romero, C., & Ventura, S. (2007). Educational data mining: A survey from 1995 to 2005. *Expert Systems with Applications, 33*(1), 135–146. doi:10.1016/j.eswa.2006.04.005

Romero, C., & Ventura, S., & Garcia, E. (2007). Data mining in course management systems: Moodle case. *Computers & Education*. Retrieved from doi:10.1016/j.compedu.2007.05.016.

Rossett, A. (1991). Needs assessment. In Anglin, G. J.

(Ed.), *Instructional technology*. Englewood, CO: Libraries Unlimited.

Rosson, M. B., & Carroll, J. M. (2002). *Usability Engineering: Scenario-based Development of Human-Computer Interaction*. San Francisco: Morgan Kaufmann.

Rosson, M. B., Carroll, J. M., & Rodi, C. (2004). Case Studies for Teaching Usability Engineering. *Proceedings of 35th SIGCSE Technical Symposium on Computer Science Education,* Norfolk, VA, March 3-7 (pp. 36-40). New York: ACM Press.

Rovai, A. P. (2002). Building a Sense of Community at a Distance. *International Review of Research in Open and Distance Learning (On-Line), 3*(1). Retrieved April 19, 2008, from http://www.irrodl.org/content/v3.1/rovai.html

Ruf, U., & Gallin, P. (2003). *Dialogisches Lernen in Sprache und Mathematik, Volume 1: Austausch unter Ungleichen. Grundzüge einer interaktiven und facherubergreifenden Didaktik*. Seelze-Velber: Kallmeyer.

Ruf, U., & Weber, C. (2005). Dialogisches Lernen auf Lernplattformen: Den Dialog in Lehrveranstaltungen an der Hochschule fördern. *Das Hochschulwesen, 53*(6), 243–247.

Sabin, M., & Higgs, B. (2007). Teaching and Learning in Live Online Classrooms. In *Proceedings of the 8th ACM SIG - Information Technology Education (SIGITE 2007) Conference*, Sandestin, FL, October 18-20, 2007 (pp. 41-48).

Sabin, M., Higgs, B., Riabov, V., & Moreira, A. (2005). Designing and Running a Pre-College Computing Course. *Journal of Computing Sciences in Colleges, 20*(5), 176–187.

Safe Assign. (2009). *Safe assign anti-plagiarism service from Blackboard*. Retrieved May 14, 2009, from http://www.safeassign.com

Salmon, G. (2000). *E-Moderating: the key to teaching and learning online*. London: Kogan Page.

Salmon, G. (2002). *E-tivities. The Key to Active Online Learning*. London: Kogan Page.

Sanchez, J. (2009, February). A Social History of Virtual Worlds. [from Academic Search Premier database.]. *Library Technology Reports, 45*(2), 9–12. Retrieved July 23, 2009.

Santo, S. A. (2005, December). *Learning style and Personality Resources*. Retrieved March 1, 2006, from the USD Web Site: http://www.usd.edu/~ssanto/styles.html

Santos, O. C., Gonzalez-Boticario, J., & Barrera, C. (2005). aLFanet: An adaptive and standard-based learning environment built upon dotLRN and other open source developments. In Proc. Foro hispano de LRN. Madrid: Congreso de usuarios y desarrolladores de LRN.

Savoy, A., Proctor, R. W., & Salvendy, G. (2009). *Information retention from PowerPoint and traditional lectures*. Computers & Education.

Sayago, S. (2005). *User Manual v1.0 QAED: Assessments and Question editor (1.0)*. [Online]. Retrieved April 25, 2009, from http://www.tecn.upf.es/~ssayag/QAed10_User_Manual.pdf

ScanSoft Productivity Applications. (2004). Case study: Scan to PC desktop – The University of Michigan. Peabody, MA: ScanSoft division of Xerox Corporation.

*Scheme*. (2009). Retrieved July 12, 2009, from http://groups.csail.mit.edu/mac/projects/scheme/

Schleimer, S., Wilkerson, D., & Aiken, A. (2003 June). Winnowing: Local Algorithms for Document Fingerprinting. In *Proceedings of the ACM SIGMOD International Conference on Management of Data* (pp. 76-85).

Schmidt, K. (2002). The web-enhanced classroom. *Journal of Information Technology, 18*(2).

Schön, D. A. (1983). *The Reflective Practitioner: How Professionals Think in Action*. New York: Basic Books.

Schrire, S. (2002). Interaction and cognition in asynchronous computer conferencing. *Instructional Science, 32*, 475–502. doi:10.1007/s11251-004-2518-7

Schrire, S. (2006). Knowledge building in asynchronous discussion groups: going beyond quantitative analysis. *Computers & Education, 46*(1), 49–70. doi:10.1016/j.compedu.2005.04.006

Science, D. E. a. (2000). Learning for life: white paper on adult education.

*Screem HTML/XML Editor*. (2009). Retrieved July 12, 2009, from http://www.screem.org/

*SeaMonkey*. (2009). Retrieved July 12, 2009 from http://www.seamonkey-project.org/

*Second Life*. (n.d.). Retrieved from http://www.secondlife.com

Secretariat, I. E. T. F. (2009). Retrieved July 12, 2009, from http://www.ietf.org/rfc.html

Sengstag, C., & Miller, D. (2005). Von der klassischen Vorlesung zur Bologna-kompatiblen Lehrveranstaltung: Redesign einer Lehrveranstaltung. *Zeitschrift für Hochschuldidaktik, 2005*(4), 63–74.

*Servlet 2.3 and JavaServer Pages 1.2 Specifications*. (2009). Retrieved July 12, 2009, from http://www.jcp.org/aboutJava/communityprocess/first/jsr053/

Serway, R., Faughn, J., & Moses, C. (2003). *College Physics*. Melbourne, Australia: Thomson-Brooks/Cole.

Sharda, R., Romano, N., Lucca, J., Weiser, M., Scheets, G., Chung, J., & Sleezer, C. (2004). Foundation for the Study of Computer-Supported Collaborative Learning Requiring Immersive Presence. *Journal of Management Inquiry*, *20*(4), 31–63.

Shelton, K., & Saltsman, G. (2005). *An administrator's guide to online education*. Charlotte, NC: Information Age Publishing.

Sherwood, B. (1981). Speech synthesis applied to language teaching. *Studies in Language Learning*, *3*(1), 171–181.

Shieh, R. S., Gummer, E., & Niess, M. (2008, November/December). The quality of a web-based course: Perspectives of the instructor and the students. *TechTrends*, *52*(6), 61–69. doi:10.1007/s11528-008-0220-3

Shin, M., & Lee, Y. (2009, January). Changing the landscape of teacher education via online teaching and learning. *Techniques*, *84*(1), 32–34.

Shott, S. (1979). Emotion and social life: A symbolic interactionist analysis. *American Journal of Sociology*, *84*, 1317–1334. doi:10.1086/226936

Shulock, N. (2004). No Less Than a Cultural Shift. *The Chronicle of Higher Education, 51*. Retrieved January 6, 2006, from http://chronicle.com/weekly/v51/i02/02b00601.htm

Siegfried, D. (2001). *Biology for Dummies*. New York: Hungry Minds.

Siemens, G. (2004). *Learning management systems: The wrong place to start learning*. Retrieved May 4, 2009, from http://www.elearnspace.org/Articles/lms.htm

Simba Information. (2007). Pearson expands with eCollege. *Electronic Education Report, 14*(10).

Sloan Consortium. (2008). *Staying the course: Online education in the United States, 2008*. Retrieved August 2, 2009, from http://www.sloan-c.org/publications/survey/pdf/staying_the_course.pdf

Sluijsmans, D., Prins, F. J., & Martens, R. L. (2006). The Design of Competency-Based Performance Assessment in E-Learning. *Learning Environments Research*, *9*, 45–66.

Smart, K. A., & Meyer, K. A. (2003). Changing course management systems: Lessons learned. *EDUCAUSE Quarterly*, *28*(2).

Smart, K. A., & Meyer, K. A. (2005). Changing course management systems: Lesson learned. *EDUCAUSE Quarterly*, *2*, 68–70.

Smathers, D. (2007). *Phi Kappa Phi and the Distance Learner*. Retrieved from http://www.phikappaphi.org/web/Publications/DistanceLearner/index.html

Snogles, D. (2005). Personal Encrypted Talk - Securing Instant Messaging with a Java Application. *Rivier College Online Academic Journal, 1*(1). Retrieved July 12, 2009, from http://www.rivier.edu/journal/RCOAJ-Fall-2005_table.htm

Soler-Lahuerta, F., Cubero, S., López, A., Roig, D., & Roca, S. (2005). Desarrollo del módulo fichas en la estructura de la UVEG para la herramienta groupware LRN. In Proc. Foro hispano de.LRN. Madrid: Congreso de usuarios y desarrolladores de LRN.

Song, L., Singleton, E., Hill, J., & Koh, M. H. (2004). Improving online learning: Student perceptions of useful and challenging characteristics. *The Internet and Higher Education, 7*, 59–71. doi:10.1016/j.iheduc.2003.11.003

Sood, A. (2007a). Network Design by Using OPNET™ IT Guru Academic Edition. Software. *Rivier College Online Academic Journal, 3*(1). Retrieved July 12, 2009, from http://www.rivier.edu/journal/RCOAJ-Spring-2007_table.htm

Sood, A. (2007b). Network Access Control. *Rivier College Online Academic Journal, 3*(2). Retrieved July 12, 2009, from http://www.rivier.edu/journal/RCOAJ-Fall-2007_table.htm

*SourceEdit.* (2009). Retrieved July 12, 2009, from http://www.brixoft.net/prodinfo.asp?id=1

Specht, M. (2008). *Designing Contextualized Learning. Handbook on Information Technologies for Education and Training* (pp. 101–111). Berlin, Germany: Springer. doi:10.1007/978-3-540-74155-8_6

Spector, M. J. D. T. I. *(2001).* Competencies for on-line teaching. *Retrievedfrom*http://purl.access.gpo.gov/GPO/LPS26199

Spickard, I. A., Alrajeh, N., Cordray, D., & Gigante, J. (2002). Learning about screening using an online or live lecture: Does it matter? *Journal of General Internal Medicine, 17*(7), 540–545.

*SQLExplorer Plug-in for Eclipse.* (2009). Retrieved July 12, 2009, from http://sourceforge.net/projects/eclipsesql

Squirre, L. *SQL Client.* (2009). Retrieved July 12, 2009, from http://sourceforge.net/projects/squirrelsql

Stahl, G., Koschmann, T., & Suthers, D. (2006). Computer-supported collaborative learning: An historical perspective. In Sawyer, R. K. (Ed.), *Cambridge handbook of the learning sciences* (pp. 409–426). Cambridge, UK: Cambridge University Press.

Stanford Law School Case Studies Collection. (n.d.). Retrieved April 19, 2008 from http://www.law.stanford.edu/publications/casestudies/

Stansfield, M., & Connolly, T. (2009). *Institutional Transformation through Best Practices in Virtual Campus Development: Advancing E-Learning Policies.* Hershey, PA: IGI.

Stein, S. J., Isaacs, G., & Andrews, T. (2004). Incorporating authentic learning experiences within a university course. *Studies in Higher Education, 29*(2), 239–258. doi:10.1080/0307507042000190813

Steiner, G. (2001). Lernen und Wissenserwerb. In Krapp, A., & Weidenmann, B. (Eds.), *Pädagogische Psychologie* (4th ed., pp. 137–205). Weinheim: Beltz.

Stepien, W., & Gallagher, S. (1993). Problem-based learning: As authentic as it gets. *Educational Leadership, 50*, 25–28.

Sterling, M. (2001). *Algebra for Dummies.* New York: Hungry Minds.

Stern, C. (2002). *Information literacy unplugged: teaching information literacy without technology.* White paper prepared for UNESCO, the US NCLIS and National Forum for Information Literacy. Retrieved June 17, 2009, from http://www.nclis.gov/libinter/

Stevens, J. A. (1999). *A Modern Approach to Intermediate Statistic* (2nd ed.). Mahwah, NJ: LEA.

Stewart, B. (2007). Why Athabasca chose moodle. *Distance Education Report, 11*(3).Changing Course Management Systems: Lessons Learned Changing Course Management Systems: Lessons LearnedChanging

Stifle, J. E. (1978). *The PLATO V terminal.* CERL Report X-50, University of Illinois, Urbana, IL.

Stringer, E. (1999). *Action research: A handbook for practitioners* (2nd ed.). Thousand Oaks, CA: Sage.

Sturgeon, J. (2007). Talkin' Up a Storm. *T.H.E. Journal*. Retrieved April 6, 2008, from http://www.thejournal.com/articles/21531

Swan, K. (2001). Virtual Interactivity: design factors affecting student satisfaction and perceived learning in as ynchronous online courses. *Distance Education, 22*(2), 306–331. doi:10.1080/0158791010220208

Talbot, C. (2007). *Studying at a Distance- A guide for students*. Berkshire, UK: Open University Press.

Talor, M. (1997). Learning styles. *Inquiry, 1*(1), 45–48.

Taylor, J. C. (2002). Teaching and learning online: The workers, the lurkers and the shirkers. *Conference on Research in Distance & Adult Learning in Asia*. Retrieved May 14, 2009, from http://www.ouhk.edu.hk/CRIDAL/cridala2002/speeches/taylor.pdf

*TCL Manual of Work*. (2008). Retrieved April 25, 2009, from http://tmml.sourceforge.net/doc/tcl/index.html

Teferra, D. (2001). Academic dishonesty in African universities-trends, challenges, and repercussions an Ethiopian case study. *International Journal of Educational Development, 21*(2), 163–178. doi:10.1016/S0738-0593(00)00037-7

*TextPad, The Text Editor for Windows*. (2009). Retrieved July 12, 2009, from http://www.textpad.com/

The Campus Computing Project. (2008). *2008 national survey of information technology in higher education*. Retrieved from http://www.campuscomputing.net

*The Common Gateway Interface*. (2009). Retrieved July 12, 2009, from http://hoohoo.ncsa.uiuc.edu/cgi/overview.html

The Electronic Hallway. (n.d.). *Public affair case study collection*. Retrieved April 19, 2008, from https://hallway.org

The Request For Comments Editor. (2009). Retrieved July 12, 2009, from http://www.rfc-editor.org/

Thomas, J. W. (2000). *A Review of Research on Project-based Learning*. Retrieved June 17, 2009, from http://www.bobpearlman.org/BestPractices/PBL_Research.pdf

Thompson, M. D. (2004). Faculty self-study research project: Examining the online workload. *Journal of Asynchronous Learning Networks, 8*(3).

Thompson, P., & Randal, B. (2001). Can E-Learning Spur Creativity, Innovation and Entrepreneurship? *Educational Media International, 38*(4), 289–292. doi:10.1080/09523980110105105

*TinySQL*. (2009). Retrieved July 12, 2009, from http://sourceforge.net/projects/tinysql/

Tobin, K., & Fraser, B. J. (1998). A distance education learning environment survey. *Distance Education, 22*, 306–331.

Tomasello, M. (1999). *The Cultural Origins of Human Cognition*. Cambridge, MA: Harvard University Press.

Tomasello, M., Carpenter, M., Call, J., Behne, T., & Moll, H. (2005). Understanding and sharing intentions: The origins of cultural cognition. *The Behavioral and Brain Sciences, 28*(5), 675–691. doi:10.1017/S0140525X05000129

*TopStyle Lite*. (2009). Retrieved July 12, 2009, from http://www.newsgator.com/individuals/topstyle/default.aspx

Torrance, M., Thomas, G. V., & Robinson, E. J. (1994). The writing strategies of graduate research students in the social sciences. *Higher Education, 27*, 379–392.

*Trellian WebPage*. (2009). Retrieved July 12, 2009, from http://www.trellian.com/webpage/

Trull, B. R. (2006). An Overview of Broadband over Power Line. *Rivier College Online Academic Journal, 2*(1). Retrieved July 12, 2009, from http://www.rivier.edu/journal/RCOAJ-Spring-2006_table.htm

Turnitin. (2009). *Turnitin brochure*. Retrieved May 14, 2009, from http://turnitin.com/static/pdf/Turnitin_brochure.pdf

*Twitter*. (n.d.). Retrieved from http://twitter.com/

U. S. Department of Labor. (1999, March). *Issues in labor statistics.* Washington, DC: U.S. Government Printing Office.

U.S. Department of Education Institute of Education Sciences. (2009). *Web-related legal issues and policies, weaving a secure web around education: A guide.* Retrieved July 3, 2009, from http://nces.ed.gov/pubs2003/secureweb/ch_3.asp

Ullman, C., & Rabinowitz, M. (2004). Course management systems and the reinvention of instruction. *Technological Horizons in Education Journal.* Retrieved May 11, 2009, from http://www.thejournal.com/articles/17014

Underwood, J., & Szabo, A. (2003). Academic offences and e-learning: individual propensities in cheating. *British Journal of Educational Technology, 34*(4), 467–477. doi:10.1111/1467-8535.00343

United Kingdom Department for Education and Skills. (2005). *Learning platforms (Secondary).* Ref. 2102-2005DBW-E.

University of Illinois Springfield. (2009). *The office of technology-enhanced learning at UIS.* Retrieved July 10, 2009, from http://otel.uis.edu/Portal/index.asp

University of Illinois. (2006). *What Makes a Successful Online Student?* Retrieved May 17, 2009, from http://www.ion.uillinois.edu/resources/tutorials/pedagogy/StudentProfile.asp

Using McCabe, Q. A. (1999). Columbia, MD: McCabe & Associates.

*V – A Freeware Portable C++ GUI Framework.* (2009). Retrieved July 12, 2009, from http://programming.ccp14.ac.uk/objectcentral/main.htm

Valade, J. (2002). *PHP & MySQL for dummies.* New York: Wiley.

Varga, A. (2004). *OMNeT++ Version 3.0 User Manual.* Retrieved July 12, 2009, from http://www.omnetpp.org/

Verkasalo, H. (2006). Empirical observations on the emergence of mobile multimedia services and applications in the U.S., & Europe. In *Proceedings of the 5th international Conference on Mobile and Ubiquitous Multimedia,* Stanford, California, December 04 - 06, 2006.

Violet, U. M. L. (Ed.). (2009). Retrieved July 12, 2009, from http://www.horstmann.com/violet/

Visser, J., & Visser-Valfrey, M. (2008). *Learners in a changing learning landscape: Reflections from a dialogue on new roles and expectations.* Dordrecht, The Netherlands: Springer.

Visual Paradigm, Inc. (2009). *Visual Paradigm for UML Community Edition (VP-UML CE).* Retrieved July 12, 2009, from http://www.visual-paradigm.com/product/vpuml/editions/community.jsp

Vonderwell, S., & Zachariah, S. (2005). Factors that influence participation in online learning. *Journal of Research on Technology in Education, 38*(2), 213–230.

Vygotsky, L. S. (1978). *Mind in society.* Cambridge, MA: Harvard University Press.

Vygotsky, L. S. (1986). *Thought and Language.* Cambridge, MA: MIT Press.

Wains, S. I., & Mahmood, W. (2008). Integrating m-learning with e-learning. In *Proceedings of the 9th ACM SIGITE conference on Information technology education,* Cincinnati, OH, USA.

Walker, S., & Fraser, B. (2005). Development and Validation of an Instrument for Assessing Distance Education Learning Environments in Higher Education: The Distance Education Learning Environment Survey (DELES). *Learning Environments Research, 8*(3). doi:10.1007/s10984-005-1568-3

Wang, Q. Y. (2008). A generic model for guiding the integration of ICT into teaching and learning. *Innovations in Education and Teaching International, 45*(4), 411–419. doi:10.1080/14703290802377307

Watson, A. H., & McCabe, T. J. (1996). *Structured Testing: A Testing Methodology Using the Cyclomatic Complexity Metric. NIST Special Publication, No. 500-235.* Gaithersburg, MD: National Institute of Standards and Technology.

Watson, W. R., & Watson, S. L. (2007). An argument for clarity: What are learning management systems, what are they not, and what should they become? *TechTrends*, *51*(2).

Watson, W. R., & Watson, S. L. (2007). An argument for clarity: what are Learning Management Systems, what are they not, and what should they become? *TechTrends*, *51*(2), 28–34. doi:10.1007/s11528-007-0023-y

Weaver, D. (2006). The challenges facing staff development in promoting quality online teaching. *International Journal on E-Learning*, *5*(2), 275–286.

*Web Developer*. (2009). Retrieved July 12, 2009, from https://addons.mozilla.org/en-US/firefox/addon/60

Webb, N. M. (1995). Group collaboration in assessment: multiple objectives, processes, and outcomes. *Educational Evaluation and Policy Analysis*, *17*(2), 239–261.

Weekes, S. (2007). Get a SECOND LIFE. *Training & Coaching Today*, 29.

Weinert, F. E. (2001). Concept of competence: A conceptual clarification. In Rychen, D. S., & Salganik, L. H. (Eds.), *Defining and selecting key competencies* (pp. 45–65). Seattle, WA: Hogrefe & Huber Publishers.

Weiss, M. A. (1999). *Data Structures & Algorithm Analysis in Java*. Boston, MA: Addison Wesley.

Weka. (2007). Retrieved from http://www.cs.waikato.ac.nz/ml/weka/

Wekhande, V. (2006). Wi-Fi Technology: Security Issues. *Rivier College Online Academic Journal, 2*(2). Retrieved July 12, 2009, from http://www.rivier.edu/journal/RCOAJ-Fall-2006_table.htm

Wellman, J. V. (2006). *Costs, prices and affordability: A background paper for the secretary's commission on the future of higher education (Report)*. Washington, DC: U.S. Department of Education.

Wenger, E. (1998). *Communities of Practice: Learning, Meaning, and Identity*. Cambridge, UK: Cambridge University Press.

Wenger, E., & Snyder, W. (2000). Communities of Practice: The Organizational Frontier.

Wertsch, J. (1991). *Voices of the mind: A sociocultural approach to mediated action*. Cambridge, MA: Harvard University Press.

Wertsch, J. V. (1985). *Vygotsky and the social formation of mind*. Cambridge, MA: Harvard University Press.

*What is the MediaTone™ Network?* (n.d.). Retrieved April 22, 2008, from http://www.webex.com/smb/media-tone.html

Wheeler, S. (2006). Learner support needs in online problem-based learning. *The Quarterly Review of Distance Education*, *7*(2), 175–184.

White, B. Y., Shimoda, T., & Frederiksen, J. (1999). Enabling students to construct theories of collaborative inquiry and reflective learning: Computer support for metacognitive development. *International Journal of Artificial Intelligence in Education*, *10*, 151–182.

Whyte, W. F., Greewood, D. J., & Lazes, P. (1991). Participatory action research: Through practice to science in social research. In W. F. White (ed.), Participatory Action Research (pp. 19-55). New York: Sage.

Wiepcke, C. (2006). *Computergestützte Lernkonzepte und deren Evaluation in der Weiterbildung. Blended Learning zur Förderung von Gender Mainstreaming.* Hamburg, Germany: Kovac.

WiFi Alliance. (2009). Retrieved July 12, 2009, from http://www.wi-fi.org/OpenSection/index.asp

Wikipedia. (n.d.). *Project-based learning.* Retrieved June 17, 2009, from http://en.wikipedia.org/wiki/Project-based_Learning

Wild, M., & Quinn, C. (1997). Implications of educational theory for the design of instructional multimedia. *British Journal of Educational Technology*, *29*(1), 73–82. doi:10.1111/1467-8535.00047

Williams, S. M. (1992). Putting Case-Based Instruction Into Context: Examples From Legal and Medical Education. *Journal of the Learning Sciences*, *2*, 367–427. doi:10.1207/s15327809jls0204_2

Wilson, J. M. (1994). The CUPLE Physics Studio. *The Physics Teacher*, *32*, 518–523. doi:10.1119/1.2344100

Winn, W. (1990). Some implications of cognitive theory for instructional design. *Instructional Science*, *19*, 53–69. doi:10.1007/BF00377985

*WinSCP*. (2009). Retrieved July 12, 2009, from http://winscp.net/eng/index.php

*Wireshark: A Network Protocol Analyzer*. (2009). Retrieved July 12, 2009, from http://www.wireshark.org/

Wong, H. K. (2003). *How to Improve Student Achievement*. Indianapolis, IN: Harry K. Wong Publications.

Wooley, B. A. (2002). Utilizing a Computing Lab to Improve Retention and Recruiting of Computer Science and Computer Information Science Students. In *Proceedings of the 18th Annual CCSC Eastern Conference*, Bloomsburg, PA, October 2002.

Woolfolk, A. E. (1993). *Educational Psychology* (5th ed.). Boston: Allyn & Bacon.

*WordPress*. (2009). Retrieved July 12, 2009, from http://wordpress.org/

Wu, C. T. (2006). *An Introduction to Object-Oriented Programming with Java* (4th ed.). New York: McGraw-Hill Higher Education.

Xiao, L., Carroll, J. M., & Rosson, M. B. (2007). Support Case-Based Authentic Learning Activities: A Collaborative Case Commenting Tool and a Collaborative Case Builder. *Human-Computer Interaction* (. *LNCS*, *4553*, 371–380.

Xythos Software, Inc. (n.d.) *Document management 2.0 – Web-based collaboration and the road to compliance*. Retrieved April 24, 2009, from http://www.xythose.com/_media/pdf/xythose_wp_docmgmt20.pdf

Yeaman, A. R. (2004). Professional ethics: The misuse of technology. *TechTrends*, *48*(5), 16-18. Zaïane, O. R., Xin, M., & Han, J. (1998). Discovering web access patterns and trends by applying OLAP and data mining technology on web logs. In *Proceedings of the Advances in Digital Libraries Conference* (April 22 - 24). Washington, DC: IEEE Computer Society.

Yong, J. (2007). Security modelling for e-Learning. In *Proceedings of the 1st International Symposium on Information Technologies and Applications in Education (ISITAE 07)* (pp. 1-5). Washington, DC: IEEE.

Young, J. (2009, June 1). Colleges consider using blogs instead of blackboard. *Chronicle of Higher Education*.

Young, J. R. (2009). Why Blackboard's plan to buy a rival sparked a campus uproar. *Chronicle of Higher Education*. Retrieved from http://chronicle.com/free/v55/i3/37a00102.htm

Zeilenga, K. D. (2006). *Lightweight Directory Access Protocol (LDAP). Technical Specification Road Map. RFC 4510*. Internet Engineering Task Force.

Zhao, F. (2003). Quality assurance in education. *Bradford*, *11*(4), 218-222.

Zimmermann, T. (in press). Online-Diskussionen als elektronische Leistungsnachweise Mit Blended Learning vom Test- zum Assessment-Paradigma. In Mandel, S., & Ruedel, C. (Eds.), *E-Assessment*. Münster: Waxmann.

Zimmermann, T., et al. (2008). Dialog mit 200 Studierenden—geht das? Blended Learning in einer Vorlesung mit hoher Teilnehmerzahl. *Das Hochschulwesen*, *2008*(6), 179–185. Retrieved July 30, 2009, from https://www.zora.uzh.ch/11297/

Zimmermann, T., Haab, S., & Schneider-Lastin, W. (2008). *Leitfaden zur E-Moderation*. Zürich: E-Learning-Koordination der Philosophischen Fakultät der Universität Zürich. Retrieved June 20, 2009, from http://www.phil.uzh.ch/institute/elearning/angebot/leitfaeden/ELK-Leitfaden_E-Moderation.pdf

Zimmermann, T., Hurtado, D., Berther, M., & Winter, F. (2008). Dialog mit 200 Studierenden – geht das? Blended Learning in einer Vorlesung mit hoher Teilnehmerzahl. *Das Hochschulwesen 56*(6), 179–185. Retrieved June 20, 2009, from https://www.zora.uzh.ch/11297/2/Zimmermannua_DialogV.pdf

Zsohar, H., & Smith, J. A. (2008). Transition from classroom to the web: Successful strategies for teaching online. *Nursing Education Perspectives*, *29*(1), 23–28.

Zupko, R. J. (2007). Introduction to the IEEE 802.16 Standard: Wireless Broadband Access. *Rivier College Online Academic Journal, 3*(1). Retrieved July 12, 2009, from http://www.rivier.edu/journal/RCOAJ-Spring-2007_table.htm

Zurita, L., & Ryberg, T. (2005). Towards a Collaborative Approach of Introducing e-Learning in Higher Education Institutions. How Do University Teachers Conceive and React to Transitions to e-Learning? In *Proc. 8th IFIP World Conference on Computers in Education.*

# About the Contributors

**Yefim Kats** received his PhD from the Graduate Center of the City University of New York. His interests include Informatics, Data and Knowledge Management, e-Learning, and Social Implications of Technology. Dr. Kats has an extensive curriculum development and teaching experience with traditional, hybrid, and online courses. He authored a number of publications, several of them on curriculum development and instructional methodology. Dr. Kats is a recipient - as the Principle Investigator - of the federal research grants from the National Research Council and the National Institute of Health. He is currently an Assistant professor in Computer Science and Information Systems at Rivier College, New Hampshire, USA.

* * *

**Michael F. Beaudoin**, Ed.D. is Professor of Education at the University of New England, (USA), where he was previously founding dean of a new college. He has held senior administrative positions and faculty appointments at institutions in the US and abroad, and has provided leadership in designing and directing numerous innovative academic offerings, including distance education degree programs. He has been a visiting scholar at institutions in Germany, China, Mexico, and Japan, and a Fulbright Fellow in Ghana. With 100 publications and presentations, including two books, he has written extensively in the field of distance education and related areas, presents frequently at conferences, and serves as an evaluator and consultant for distance education programs and courses. He lives in Portland and Cliff Island, Maine, is a competitive masters athlete, and has traveled in over fifty counties. He can be reached at mbeaudoin@une.edu

**Danilo M. Baylen** is an Associate Professor of Instructional Technology in the Department of Media and Instructional at the University of West Georgia. He completed his graduate degrees in counseling (master's) and instructional technology (doctorate) from Northern Illinois University. He has a second master 's degree in elementary education from Florida Gulf Coast University and undergraduate work in economics from the University of the Philippines. Prior work experiences include instructional design, curriculum development, technology training, and project management. Also, he taught English as a second language to Indochinese refugees in Southeast Asia.

**Karen-Lynn Bucher** holds an M.A. in Philosophy, Media Studies and Political Science from the University of Konstanz, Germany. She has worked as an e-learning coordinator at the Faculty of Arts at the University of Zurich, Switzerland. She is currently a research associate at the University of Teacher

Copyright © 2010, IGI Global, distributing in print or electronic forms without written permission of IGI Global is prohibited.

Education Central Switzerland (PHZ) and teaches at a school for secondary education. In her research she focuses on blended learning in teacher education and continuing education including collaborative learning and the role and techniques of moderators in online learning environments.

**Sergey Butakov** is an Assistant Professor at the Solbridge International School of Business, Woosong University, Daejeon, South Korea. Dr. Butakov holds PhD in Computer Science from Altai State University, Russia. He has several awards for teaching and research from universities in Russia, Nigeria and South Korea. Dr. Butakov authored and co-authored more than 30 publications. In 2001 – 2004 he served as principal researcher in two projects funded by Russian Foundation for Basic Research and the last five years he participated in a number of industry and university funded research projects. He is interested in plagiarism detection, text search, software development, network infrastructure monitoring and development, semantic web, artificial intelligence and neural networks.

**John M. Carroll** was a founder of human-computer interaction, the youngest of the 9 core areas of Computer Science identified by the Association for Computing Machinery (ACM). He served on the program committee of the 1982 Bureau of Standards Conference on the Human Factors of Computing Systems that in effect inaugurated the field, and was the direct predecessor of the field's flagship conference series, the ACM CHI Conferences. Through the past two decades, Carroll has been a leader in the development of the field of Human-Computer Interaction. Through the past two decades, Carroll has been a leader in the development of the field of Human-Computer Interaction. In 1984 he founded the User Interface Institute at the IBM Thomas J. Watson Research Center, the most influential corporate research laboratory during the latter 1980s. Now, he is still actively involved in HCI research, leading the computer-supported collaboration and learning (CSCL) lab in Pennsylvania State University.

**Mary Angela Coleman** currently serves as Director of Educator Preparation for the University System of Georgia's Office of Educator Preparation, Innovation and Research. She is on leave from the University of West Georgia as an Assistant Professor of Educational Research. She received her master's degree in creative writing from Temple University and doctorate in educational research and evaluation from Virginia Commonwealth University.

**Vickie Cook** serves as Dean of the School of Education/Director of Online Learning at Greenville College. Previously she was an assistant professor of Educational Leadership at University of Illinois at Springfield. She holds a Bachelor's Degree from Western Illinois University and Master's degree in Education and Ph.D. in Higher Education Leadership from Capella University. Her research interests include the scholarship of teaching and the impact of distributed learning on the adult learner.

**Kara McElwrath** serves as a Training Specialist in the Information Technology Services department at the University of Illinois at Springfield, where she provides instructional support and training to faculty, staff, and students. She designs and delivers a variety of technology-based workshops, assists faculty in integrating technology into their curriculum, assists staff in leveraging technology to be more productive, and supports students with course-related projects. In addition, she is an adjunct instructor for the Teacher Education department. She has a Bachelor of Science in Elementary Education from Butler University and a Master of Science in Instructional Technology and Telecommunications from Western Illinois University. She is a Microsoft Certified Application Specialist and a Certified Technical Trainer.

**Ian Douglas** is Associate Professor in both the Learning Systems Institute and the College of Communications and Information at Florida State University. The Learning Systems Institute is a multidisciplinary research unit of Florida State that has been involved in educational research and reform for almost forty years. Dr Douglas has over twenty years experience in the design and development of information technology systems. He began his career working in Artificial Intelligence and was one of the first people to integrate intelligent tutoring into a commercial simulation-based training system. In addition to his experience in e-learning, he has expertise in user-centered design, software engineering, and knowledge management. Dr. Douglas has research publications in a number of different disciplines and has won two awards for the innovative use of technology in learning.

**Craig H. Ganoe** is an instructor in Information Sciences and Technology at The Pennsylvania State University. His research interests include multiple-device interactions, computer supported cooperative work and learning, collaboration in community network contexts, and ubiquitous computing. Ganoe has a BS in Computer Science from The Pennsylvania State University, and an MS in Computer Science from Virginia Tech. He is a member of the ACM.

**Christian M. Graham**, MSBE: Christian Graham is a Lecturer of Management Information Systems at the University of Maine Business School. He is currently working towards the completion of a PhD in Information Systems from Nova Southeastern University in Ft. Lauderdale, Florida. Christian's research interests include online learning environments (OLE), information security (InfoSec), computer privacy, and telecommunication technology.

**Mary Hancock** is an Assistant Professor of Educational Psychology in the Department of Counseling and Educational Psychology at the University of West Georgia. She received her master's and doctoral degrees in educational psychology from Indiana University and undergraduate degree from University of Indianapolis.

**Bryan Higgs** holds a B.Sc. in Physics from the University of Liverpool, UK, and M.Phil. and Ph.D. degrees in Physics from Yale University. In 1974, he moved into the computer industry and joined Digital Equipment Corporation where he worked for 20 years in a variety of positions. In 1994, Dr. Higgs transferred to Oracle, together with the rest of Digital's Database group, and worked in their Software Engineering group for 7 years. In 2002, Dr. Higgs joined the full-time Computer Science faculty at Rivier College, and retired in 2008. His major interests are database systems, programming languages (especially Java and C/C++), computer security, and Web development. Recently, he has become interested in scripting languages, including Perl, Python, JavaScript, PHP, and Ruby.

**Wolfgang Hommel** heads the identity management research team at the Leibniz Supercomputing Centre (LRZ) in Germany and teaches information security at Ludwig Maximilians University and Technische Universität München. His research focuses on ICT security and privacy management in large distributed systems, including national identity federations and international Grid infrastructures. Emphasis is put on a holistic perspective, i.e. the problems and solutions are analyzed from the design phase through software engineering, deployment in heterogeneous infrastructures, and during the operation and change phases according to IT service management process frameworks, such as ISO/IEC 20000. Being both a regional computing centre for higher education institutions with more than

100,000 users and a national supercomputing centre, the LRZ offers a plethora of real world scenarios and large projects to apply and refine the research results in practice, including the hosting of Learning Management Systems for various German universities.

**Daniel Hurtado** is a teaching assistant at the Institute of Teacher Education at the University of Zurich. His research concentrates on both didactical and technical aspects of online learning and teaching. Besides, he is also active in the field of migration history, concentrating on Philippine migration.

**Hao Jiang** is a Ph.D student from human-computer interaction center in Pennsylvania State University, USA. His research targets on design science, computer-supported collaborative work (CSCW), computer-supported collaborative learning (CSCL), and community informatics. He graduated from Beijing University in China, 2001, with a bachelor degree majored in Library Science. He joined the computer-supported collaboration and learning (CSCL) lab in 2005, participating in several research and system development projects. He has a strong interest in exploring interactions between social factors (social capital, social networks, social identity, etc) and information technology in various levels of social interactions, and developing knowledge to enhance designing in society.

**Nory B. Jones**, Ph.D.: Dr. Jones is an Associate Professor of Management Information Systems and Director of Graduate Programs at the University of Maine Business School. She received her Ph.D. in Information Systems from the University of Missouri-Columbia. Her research interests include knowledge management, collaborative technologies, and organizational learning.

**David Long** is a doctoral student in the Management program at the University of Florida. He holds a Bachelor's degree in Business Administration (BS) from Presbyterian College, and a Master's degree in Business Administration (MBA) from The University of Florida. David served for 8 years in the US Navy as a Naval Flight Officer, and worked for 4 years in management at The Home Depot. He is currently a Navy Reservist and he serves as a navigator on the P-3 airplane. His research interests include justice, leadership, work motivation, and overall job attitudes. David lives in Gainesville, Florida with his wife and two daughters.

**Adrian Lu** is a Learning Designer with the Centre for Educational Development and is part of the team providing learning support for faculty members and students at NTU. Prior to joining NTU, he has worked in the corporate sector as an Instructional Designer developing eLearning courseware and training materials for clients from diverse industries ranging from the telecommunications to the finance sector. Adrian holds a Masters degree in Instructional Design Technology from NTU and his areas of interest include web and graphic design, online courseware development, and the innovative use of technologies for teaching and learning.

**Cerstin Mahlow** holds an MA in Computational Linguistics, Spanish and Politics from Friedrich-Alexander University Erlangen-Nuremberg, Germany. For several years she worked as e-learning coordinator at the Faculty of Humanities and Arts at the University of Zurich, Switzerland, supporting and consulting e-learning projects at the faculty with respect to didactics, technology, assessment, collaboration, and communication. She designed and implemented database applications for the faculty. Since 2009 she is Senior Scientific Officer responsible for e-learning at the School of Social Work at the

University of Applied Sciences Northwestern Switzerland. There she is responsible for the design and implementation of an e-learning strategy and consults e-learning projects. Her research interests include linguistically supported text processing, computer-aided assessment and the integration of e-learning into university teaching. She serves as reviewer for several international e-learning conferences and is author of various research papers in the fields of Computational Linguistics and E-Learning.

**Juley McGourty** currently works at the International Equine Institute, University of Limerick in the role of desktop publisher of module learning materials and coordinator of learner support for the Equine Science distance learning programs. After graduating from the University of Limerick in 2001 with a BSc Degree in Equine Science, she traveled to the USA for a few years. On returning to Ireland she completed an MSc in Interactive Media from the University of Limerick in 2007. She is currently involved in integrating on-line technologies, which include Sulis and Adobe Connect Pro, into the Equine Science distance learning programs.

**Paloma Moreno-Clari** has a M.Sc. in Telecommunications Engineering from the Polytechnic University of Valencia and is now Assistant Professor in the area of Telematic Engineering, Department of Computer Science of the University of Valencia. She has been researcher of the Institute of Robotics of the University of Valencia (2001-2004). Her principal area of interest is the use of e-learning tools in high education. She got her Ph.D. from the University of Valencia and developed her Dissertation on the analysis of the use of platforms of learning management in higher education.

**Luvai F. Motiwalla** is currently a Professor and Chair of Operations and Information Systems Department in the College of Management at the University of Massachusetts, Lowell. He has a Ph.D. and M.S. in Management Information Systems from the University of Arizona (1989) and a B.B.A. from Penn State University (1983). His current research and teaching areas include information privacy protection, ERP, mobile applications, speech recognition, e-learning, and outcomes assessment. He has published over 35 articles in several national and international journals including Communications of ACM, Information & Management, Computers & Education, Jour of Electronic Commerce, Jour of MIS, and others. He recently authored a text book on Enterprise System for Management with Pearson Prentice Hall publications. Professor Motiwalla has consulted or worked on research projects funded by NSF, Davis Foundation, Department of Education, Connecticut Department of Health Services, IBM, NCR and U.S. Army, John Hancock Insurance, APC, Corp., Skillsoft, Inc. and others.

**Carol M. Mullen** is a master's degree student in community counseling, and a graduate research assistant in the Department of Media and Instructional Technology at the University of West Georgia. She received her bachelor's degree in psychology from Purdue University. Prior work experiences include community-based case management with adults dealing with mental and substance abuse issues, and families with developmental needs.

**Anthony Piña** is Dean of Online Studies for the Sullivan University System. He has both academic and industry experience and has been a professional consultant to Fortune 500 corporations, small businesses, government agencies, university consortia and the U.S. Military. Tony is a member of the board of directors of a charter school, the advisory board of two higher education institutions, the editorial board of two refereed journals. He has taught at the secondary, community college and university levels and has received eight national awards in the area of instructional technology. Tony is the author of the book "Distance Learning and the Institution" and has published several peer-reviewed journal articles and book chapters on instructional technology and distance learning. He has delivered over 100 invited presentations at professional conferences. Tony is a Past President of the Division of Distance Learning of the Association for Educational Communications & Technology (AECT).

**Michael Piotrowski** is a senior research associate at ZHAW Zurich University of Applied Sciences in Winterthur, Switzerland and a lecturer at the University of Zurich. He holds a doctorate in computer science from Otto von Guericke University, Magdeburg, Germany, and a master's degree in computational linguistics from Friedrich-Alexander-University Erlangen Nuremberg, Germany. He is one of the designers of the eduComponents, an open-source component-based e-learning platform based on the free content management system Plone. His research interests are in the areas of e-learning and computational linguistics, and he is the author of numerous peer-reviewed articles in both fields.

**Rademacher Mena** is a native Chilean who obtained his Physics PhD in 2002 and is the founder and CEO of Futur-E-Scape, LLC. Created in 2004, this company is dedicated to the creation and research of educational virtual worlds. In the last five years, Futur-E-Scape has received several grants and has been the subject of several published papers on the subject of educational virtual worlds. Related to his company's mission, Dr. Rademacher also teaches and develops courses for various pure-online universities. He also speaks at various educational and entertainment conferences and has written several publications on these subjects.

**Angélica Rísquez** BA (psych), MBS. Angélica has been a researcher at the Centre for Teaching and Learning at the University of Limerick since 2003. She is committed to the use of innovative teacher and student support mechanisms at third level, and is directly involved on the implementation and support of the learning management system at UL as a technology enhanced learning facilitator. She also supports the use of plagiarism prevention software, and is involved in initiatives involving using technology to promote transferable skills (intercultural education, research skills, employability, etc.). She is currently finalizing her PhD studies in the Spanish National Open University (UNED) in the area of peer electronic mentoring

**Vladimir V. Riabov** is a professor of Computer Science at Rivier College. He teaches Networking Technologies, Software Engineering, and System Simulation and Modeling. His research interests include Computational Algorithms, Complexity Code Studies, Networking Software Test Strategies, and Object-Oriented System Design. He received a Ph.D. in Applied Mathematics and Physics from Moscow Institute of Physics and Technology and M.S. in Computer Information Systems from Southern New Hampshire University. Vladimir worked as a senior researcher and the Deputy Director at the Central Aero-Hydrodynamic Institute (TsAGI, Moscow, Russia), taught at Worcester Polytechnic Institute, Daniel Webster College, and the University of New Hampshire, and was a senior software engineer at Newbridge Networks Inc. and Nortel Networks Inc. He is a co-founder of the Russian Branch of the Open University of Great Britain, and a founder and the Editor-in-Chief of the Rivier Academic Journal. Vladimir is a member of ACM, AIAA, IEEE, and MAA.

**Esteban Sanchis-Kilders** was born in Valencia, Spain, in 1967. He obtained his M.Sc. degree on Physics, specialisation Electronics, 1990 and his Ph.D. in July 1997 both from the University of Valencia (Spain). His employment experience includes two years in the European Space Agency (Noordwijk, The Netherlands) and five years as assistant professor at the University of Valencia. Now he is working as an associate professor at the University of Valencia (Spain) and belongs to the Laboratory of Industrial Electronics and Instrumentation (LEII) at the University of Valencia. His main interests are space power systems, industrial applications and new lecturing tools for higher education.

**Clark Shah-Nelson** has developed, designed, taught, and coordinated distance and online learning programs. He is currently Coordinator of Online Education at SUNY Delhi in the Catskills of New York, where he oversees strategic planning, faculty and policy development, as well as instructional design and support of online courses. Other current projects include consulting with the SUNY Center for Online International Learning (COIL) for the European Humanities University in Vilnius, Lithuania. Past awards include a 2006 Excellence in Online Teaching award and his German learning videos from MyGermanClass.com were one of the top 100 People's Choice Video Podcasts from Apple's iTunes. Clark was a featured presenter at the New Media Consortium's 2007 Online Conference on the Convergence of Web Culture and Video and has recently presented at SLOAN-C ALN, Educause's NERCOMP, and the SUNY CIT and COIL Center Conferences. Prior to his move to New York, he spent over seven years teaching online German for Colorado Online Learning, and seven years building the Denver Public Schools' Distance Learning Network. He has a Master's in Instructional Technology from the University of Colorado.

**Melanie Shaw** has over ten years of educational experience ranging from classroom and graduate level teaching to counseling and administration. She is currently an Associate Professor and Lead Mentor of Research and Dissertation Courses at Northcentral University. She mentors graduate and post graduate learners in the School of Education, and serves as a dissertation committee member and chair. In addition, Melanie is an Education Consultant for the national healthcare staffing organization, Temps Inc. Melanie's primary research interests include online teaching and learning, alternative teacher certification, curriculum, and traditional instructional practices. She is the author of the book, Alternative Teacher Certification: A Remedy for Teacher Shortages. She has also written numerous articles including, Instructional Strategies for Multiage Classrooms; Violence Prevention Group; Where in the World is Mr. Maraca?; Categorizing Music Using the Information Processing Model; and Baroque Composition Discovery.

**Vladislav Shcherbinin** Instructor, School of Information Technology and Communication, American University of Nigeria, Nigeria, American University of Nigeria, Nigeria. Vladislav Shcherbinin holds MS in Computer Science from Altai State Technical University, Russia. He is a professional software developer and technical translator. He has extensive experience in databases design and development, software development and software projects management. His research interests are in the area of data warehouses, data mining, OLAP, text mining and software engineering. Vladislav is a Microsoft Certified Professional, member of the Russian SQL Server Club.

**Daniel Tan** is the Director/Centre for Educational Development and Associate Professor/School of Electrical and Electronic Engineering, Nanyang Technological University. He obtained his BSc from University of Aston, Birmingham, England. He subsequently achieved a PhD from University of Manchester Institute of Science and Technology and a post-graduate Diploma in Teaching in Higher Education from Nanyang Technological University. Daniel Tan led a team to develop and implement edveNTUre – eLearning campus eco-system, edUtorium – staff development program, and eUreka – automated online project work management system. Awards received included Intelligent20 Award 2003, Honouree of CIO Asia100 Award 2004, EMC Best Practice Award 2004, ZDNet Asia Smart50 Award 2006, SiTF eLearning Organization of the Year 2006, APICTA 2007, ascilite Award 2008, Leadership Award for 'Best in Project Work Management' in the IMS Global Learning Impact Awards 2009. He was conferred The Public Administration Medal (Silver)/Singapore National Day Awards 2007.

**Steven F. Tello** is Assistant Professor in the College of Management at the University of Massachusetts Lowell and Associate Director at the Massachusetts Medical Device Development Center. These joint appointments allow him to combine his passion for teaching with research examining how to best promote the commercialization of emerging technologies. Steven co-founded the University's undergraduate concentration in Entrepreneurship in 2007 and is now developing an interdisciplinary graduate program in Technological Entrepreneurship. His research interests include assessment in online education, technology innovation and entrepreneurship, and technology management and planning. Steve co-founded Lowell's online education program in 1996 and oversaw faculty development, course development, technology management and program evaluation until his transition to the faculty in 2005. He received his Doctorate of Education from UMass Lowell in 2002 and his MBA from UMass Amherst. His course load at the University includes teaching technology innovation and entrepreneurship, leadership processes and business ethics.

**Kam Hou Vat** is currently an invited lecturer in the Department of Computer and Information Science, under the Faculty of Science and Technology, at the University of Macau, Macau SAR, China. His current research interests include learner-centered design with constructivism in Software Engineering education, architected applications developments for Internet software systems, information systems for learning organization, information technology for knowledge synthesis, and collaborative technologies for electronic organizations and virtual communities.

**Kelley Walters** has more than 20 years in the field of education. She has served as a teacher, elementary principal, and language arts consultant and now teaches online courses while operating a consulting company. She holds a Masters degree from Eastern Michigan University and earned her doctorate in Reading and Language Arts from Oakland University. She is currently an Associate Professor and Lead Mentor of Instructional Leadership at Northcentral University. In addition to serving as dissertation chair and dissertation committee member, she teaches courses in classroom management, school leadership, literacy, and research in the School of Education. Her research interests include early literacy, word study, online teaching and learning, and teacher leadership.

**Sheryl E. Wong** is part of the team of Instructional Designers in the Teaching and Learning with Technology (TLT) Section of CED, providing end user support for the eLearning Portal in NTU. She provides training and consultancy services to faculty in the design, development and delivery of online lessons. The scope includes content design for online delivery and recommendations in leveraging web technology to enhance eLearning. She is also involved in projects on the design and development of integrated systems to better cater to eLearning support for the NTU community, one of such projects is eUreka, for which she is the project lead. Sheryl holds a Bachelor Degree in Accountancy and Post-graduate Diploma in Education from National University of Singapore and Master Degree in Education (IT, Multimedia and Education) from University of Leeds, UK.

**Tobias Zimmermann** holds an M.A. in German Literature and Linguistics, Secondary Education and History from the University of Zurich, where he is a research associate at the Institute of Teacher Education and has worked as an e-learning coordinator for five years. Currently, he is also working at the University of Teacher Education Zurich (PHZH), where he is a research associate at the Centre for Higher and Adult Education. He is completing his doctoral thesis on Hybrid Dialogs and how they best can support learning.

# Index



Copyright © 2010, IGI Global, distributing in print or electronic forms without written permission of IGI Global is prohibited.

## D

## E



## F

## L



## M

## N



## O



## P

## T



## U



## V